William R. Aford

ORGANIZATIONAL SYSTEMS
General Systems Approaches
to Complex Organizations

Organizational Systems

General Systems Approaches to Complex Organizations

Edited by

FRANK BAKER, Ph.D.

Harvard Medical School

1973

 RICHARD D. IRWIN, INC. *Homewood, Illinois* 60430

IRWIN-DORSEY INTERNATIONAL
London, England WC2H 9NJ

IRWIN-DORSEY LIMITED
Georgetown, Ontario L7G 4B3

© RICHARD D. IRWIN, INC., 1973

All rights reserved. No part of this publication may be
reproduced, stored in a retrieval system, or transmitted,
in any form or by any means, electronic, mechanical,
photocopying, recording, or otherwise, without the prior
written permission of the publisher.

First Printing, July 1973

ISBN 0-256-00236-3
Library of Congress Catalog Card No. 72–98124

Printed in the United States of America

Preface

THE ASSUMPTION that organizations can most usefully be viewed as open systems having properties in common with other living systems has in recent years been gaining widespread acceptance. Although there is as yet no single generally accepted specific theory of organizational systems, I believe the time is ripe for pulling together the best of the work done so far and presenting it in a useful format for the wide audience concerned with applying general systems concepts to the study and management of complex organizations. That is the purpose of this book.

The editor of a book of readings has an obligation to present his criteria for selecting the papers included in his book. A book of readings should provide a more convenient source for the reader than is available elsewhere. Relative lack of general accessibility is therefore one of the criteria I have employed in choosing selections. Thus, in some cases, I chose not to include a paper which had been frequently cited and reproduced in other anthologies, but decided instead to include an equally important but less readily available paper. An anthology should also include the most recent significant papers available at the time the book is edited, and so I have continued to search the new literature for appropriate papers right up to the last possible opportunity for their inclusion. However, it is also important to include older papers which continue to be relevant because they are classics in the field, so I have not hesitated to include some key papers which have appeared earlier, but continue to retain their currency and importance.

Relevance to the main orientation of the book is another important criterion. Thus, I have excluded some otherwise good papers which make only passing faddish reference to general systems theory without being primarily oriented to an open systems approach to organization research and theory.

The clarity of writing has also been a consideration in choosing papers to include in this book. Since mathematical treatment seems premature in an area just evolving conceptually and empirically, I have tended to favor papers which do not involve extensive mathematics but do offer clear exposition. In the past, the typical social scientist and manager have been somewhat wary of "systems" approaches because of an identification of the term with operations research, systems engineering, and computerology. This book emphasizes articles which are more sociological and behavioral in orientation, although the potential usefulness of other approaches is recognized.

In addition, I believe the issue of redundancy among some articles bears discussion. Any editor of an anthology has to make a choice about abridging specific papers so as to exclude material which may be covered in other papers. Having eliminated papers which grossly overlap, I nevertheless have not tampered with the papers included which briefly review or cite concepts, models, and results presented more completely in other papers. The reason for this decision lies in my belief that it is important to show the ways in which a field grows and how later scientific work builds on and relates to earlier efforts.

One advantage of a set of readings is that it can convey the different ways in which related issues are formulated and addressed in a developing field. However, a collection of readings can be frustrating to those who seek an integrated view of a field. I have attempted to deal with this problem in the introductory chapter, which presents a broad overview of general systems theory and its application to complex organizations, relating the selections included in the book to each other and to other major writings in the field. The section introductions also are intended to help show the relations of articles included in the various major parts of the book and to present brief introductory summaries to the papers in each section.

Finally, it is my hope that this volume will be both stimulating and useful to practicing managers, theoreticians, researchers, and students alike. I would hope that the individual papers and introductory material would be particularly helpful to those who are not familiar with or have only limited experience with the concepts and perspectives presented here. As for those who are already more deeply involved, I hope they will be challenged to extend the work in this field so that the further development and testing of general systems approaches to complex organizations will receive the kind of attention I believe is called for.

ACKNOWLEDGMENTS

I would like to thank Dr. Gerald Caplan for his support and encouragement of my work in attempting to apply general systems theory to organization evaluation of human services over the past eight years.

While preparing this book I was supported by NIMH Research Grant MH 18382. I would like to thank Dr. Harold Halpert of NIMH for his encouragement in exploring the application of general systems theory to organization theory and research.

I am indebted to my students who have over the years stimulated my interest and thinking in this area.

In addition, I would like to thank the authors of the papers and the copyright holders thereof for their permissions to reprint. A special vote of thanks goes to those authors who wrote papers specifically for this volume.

I am greatly indebted to Joyce Olesen for her editorial assistance and to Ericka Stackhouse for her excellent secretarial assistance.

My wife, Adrienne, has provided me with aid and encouragement throughout the preparation of this volume, and I am happy to express my deep gratitude and appreciation for her continuing support.

Boston, Massachusetts　　　　　　　　　　　FRANK BAKER
June 1973

Contributors

Russell L. Ackoff, Ph.D., Professor of Systems Sciences, Wharton School of Finance and Commerce, University of Pennsylvania, Philadelphia, Pennsylvania.

Howard Aldrich, Ph.D., Assistant Professor of Organizational Behavior, School of Industrial and Labor Relations, Cornell University, Ithaca, New York.

Frank Baker, Ph.D., Assistant Professor of Psychology in Department of Psychiatry; and Head, Program Research Unit, Laboratory of Community Psychiatry, Harvard Medical School, Boston, Massachusetts.

Warren G. Bennis, Ph.D., President, University of Cincinnati, Cincinnati, Ohio.

Ann B. Blalock, M.A., Coeditor, *Methodology in Social Research;* Seattle, Washington.

Hubert M. Blalock, Jr., Ph.D., Professor of Sociology, Department of Sociology, University of Washington, Seattle, Washington.

Warren B. Brown, Ph.D., Head, Department of Personnel and Industrial Management, Graduate School of Management and Business, University of Oregon, Eugene, Oregon.

F. E. Emery, Ph.D., Fellow of the Research School of Social Sciences, Australian National University, and former Chairman, Human Resources Center, Tavistock Institute of Human Relations, London, England.

Amitai Etzioni, Ph.D., Professor, Department of Sociology, Columbia University, New York, New York.

Wendell French, Ed.D., Professor of Management and Organization, Graduate School of Business, University of Washington, Seattle, Washington.

Gabarro, John J., D.B.A., Assistant Professor of Organizational Behavior, Harvard Business School, Boston, Massachusetts.

Basil S. Georgopoulos, Ph.D., Program Director and Coordinator of Research, Organizational Behavior Program, Institute for Social Research; and Professor of Psychology, University of Michigan, Ann Arbor, Michigan.

Everett E. Hagen, Ph.D., Professor of Economics and Political Science, Massachusetts Institute of Technology, Cambridge, Massachusetts.

Richard A. Johnson, D.B.A., Associate Dean, Graduate Programs, and Professor of Operations and Systems Analysis, University of Washington, Seattle, Washington.

Fremont E. Kast, D.B.A., Professor of Management and Organization, University of Washington, Seattle, Washington.

Daniel Katz, Ph.D., Professor of Psychology, Department of Psychology, University of Michigan, Ann Arbor, Michigan.

Sol Levine, Ph.D., University Professor; and Professor, Department of Sociology, Boston University, Boston, Massachusetts.

Rolf P. Lynton, Ph.D., Associate Professor of Mental Health, School of Public Health; and Faculty Associate, Carolina Population Center, University of North Carolina, Chapel Hill, North Carolina.

Eric J. Miller, Ph.D., Consultant, Centre for Applied Social Research, Tavistock Institute of Human Relations, London; and Visiting Professor of Organizational Behavior, Manchester Business School, Manchester, England.

James G. Miller, M.D., Ph.D., Vice President and Director, Division of Instructional Technology, Academy for Educational Development; and Lecturer in Psychiatry, Department of Psychiatry and Behavioral Sciences, Johns Hopkins University, Baltimore, Maryland.

Eileen Morley, Ed.D., Associate in Administration for Careers; and Lecturer in Interpersonal Behavior, Harvard Business School, Boston, Massachusetts.

Gregory M. St. L. O'Brien, Ph.D., Director, Human Services Design Laboratory, School of Applied Social Sciences, Case Western Reserve University, Cleveland, Ohio.

A. K. Rice, Senior staff member, Centre for Applied Social Research, Tavistock Institute of Human Relations, London, England, until his death in 1969; a key figure in developing concepts of primary task and sociotechnical systems as tools for the analysis of different types of enterprise; author of *Productivity and Social Organisation, The Enterprise and Its Environment,* and *Systems of Organisation,* as well as numerous articles in the field of industrial social science.

Ned A. Rosen, Ph.D., Professor and Chairman, Department of Organizational Behavior, New York State School of Industrial and Labor Relations, Cornell University, Ithaca, New York.

James E. Rosenzweig, Ph.D., Professor of Management and Organization, University of Washington, Seattle, Washington.

Herbert C. Schulberg, Ph.D., Assistant Clinical Professor of Psychology in the Department of Psychiatry, Laboratory of Community Psychiatry, Harvard Medical School; and Associate Director, United Community Services, Boston, Massachusetts.

William G. Scott, D.B.A., Professor of Management and Organization, Graduate School of Business Administration, University of Washington, Seattle, Washington.

Stanley E. Seashore, Ph.D., Professor, Institute for Social Research, University of Michigan, Ann Arbor, Michigan.

Alan Sheldon, M.B., B.Chir., D.P.M., S.M. (Hyg.), Associate Professor of Business Administration, Harvard Business School, Boston, Massachusetts.

Ralph M. Stogdill, Ph.D., Professor of Management Sciences and Psychology; Director, Research in Leadership and Organizational Behavior, Ohio State University, Columbus, Ohio.

Robert L. Swinth, Ph.D., Professor, School of Business, University of Kansas, Lawrence, Kansas.

Shirley Terreberry, Ph.D., Associate Professor, School of Community Service and Public Affairs, University of Oregon, Eugene, Oregon.

James D. Thompson, Ph.D., Professor and Acting Chairman, Department of Sociology and Anthropology, Vanderbilt University, Nashville, Tennessee.

E. L. Trist, M.A., Professor of Organizational Behavior and Ecology, Department of Statistics and Operations Research; and Chairman, Management and Behavioral Science Center, Wharton School of Finance and Commerce, University of Pennsylvania, Philadelphia, Pennsylvania.

Paul E. White, Ph.D., Acting Chairman, Department of Behavioral Sciences, School of Hygiene and Public Health, Johns Hopkins University, Baltimore, Maryland.

Ephraim Yuchtman, Ph.D., Senior Lecturer, Department of Sociology, Tel-Aviv University, Tel-Aviv, Israel.

Contents

Section IV

Section V

Section VI

Section VII

Frank Baker

1. Introduction: Organizations as Open Systems

CLASSICAL and neoclassical organization theories have been found wanting largely because of their emphasis on organizations as fragmented and closed social systems acting independently of external forces. Although the development of general systems theory began in the field of biology, with its emphasis on systems interdependencies and the importance of openness to the environment, its development has resulted in the application of open systems approaches to the study of all living systems, including such large social systems as complex organizations. As Scott has observed (Chapter 6), the distinctive feature of modern organization theory lies in its conceptualization of an organization as an open system.

The major thrust of the view that living systems are essentially "open systems," as opposed to "closed systems," comes from an article in *Science* published by the theoretical biologist Ludwig von Bertalanffy in 1950. Bertalanffy had been a pioneer in the promotion of an organismic view in biology and first developed his "general system theory" in the thirties. However, he did not publish his general system ideas until the conclusion of World War II, later explaining that he waited until biology was more receptive to theory and model building (Bertalanffy, 1968, p. 90). Bertalanffy is responsible both for introducing the term "general systems theory," and for initiating the intellectual movement for a unified science, which goes by that name as well.

To many people the "systems movement" suggests only the military and industrial applications of systems concepts in such applied technologies as systems engineering, systems analysis, and systems design. However, there is a broader aspect of systems development in science which is related to the renewed quest for a "general" theory. In a time of ever-increasing specialization, the appeal of a "general" theory has drawn together individuals from a variety of widely differing disciplines who seek to bridge the gaps among disciplines and who feel the need

1

for a body of organized theoretical constructs which can be employed to discuss the general relationships of the empirical world, that is, a general systems theory.

By the late forties and early fifties, Bertalanffy not only found that his ideas were being well received, but also found that a number of scientists were independently evolving similar approaches. In 1954 he discussed the formation of a scientific society centering around general systems with a multidisciplinary group at the Center for Advanced Study in the Behavioral Sciences at Stanford. In 1955, Bertalanffy and his colleagues from Stanford organized the Society for the Advancement of General Systems Theory at the annual meeting of the American Association for the Advancement of Science in Berkeley, California. In addition to Bertalanffy, the multidisciplinary organizing committee included Kenneth E. Boulding, economist; Ralph W. Gerard, biologist; and Anatol Rapoport, mathematician. In 1956 the Society began publication of a yearbook, *General Systems*, and through the ensuing years it has been one of the key publication outlets for papers which develop, modify, and apply the general systems theory described by Bertalanffy and other key investigators, including Boulding (1956), Ashby (1958), and Rapoport and Horvath (1959). The parallel development of Wiener's cybernetics (1948), the information theory of Shannon and Weaver (1949), and the game theory of von Neumann and Morgenstern (1947) produced the important idea of cybernetic feedback and information concepts that enriched the general systems formulations arising from biology and the economic and social sciences.

Two other multidisciplinary groups should be mentioned in this brief review of the history and development of general systems theory. During the period from 1951 to 1956, a group of biologists, psychologists, psychiatrists, and social scientists met for twelve biannual conferences in Chicago. Roy R. Grinker edited a volume based on the transcript of discussions at the early meetings. This symposium volume, entitled *Toward a Unified Theory of Human Behavior*, was published in 1956 and contains a number of important early attempts to integrate different disciplines using systems concepts.

Another multidisciplinary group of importance in the development of general systems research and theory was formed by James G. Miller at the University of Chicago in 1949. Miller, a psychiatrist and psychologist, and his associates sought to develop what he called a "general systems behavior theory." Later, in 1955, Miller moved to the University of Michigan Medical School where he started the Mental Health Research Institute. In 1956 the journal *Behavioral Science*, the official publication of the Institute, was started, and over the years the structural and behavioral properties of many different systems have been analyzed in articles published in it. Miller has published a series of his own related articles in this journal summarizing the results of his work and that of his associates over the years. Miller's writings have done much to advance general systems research and to clarify and elaborate the assumptions, definitions, and propositions of general systems theory (Chapter 2).

The literature dealing with systems has continued to expand at an accelerated pace in the last few years. Recent books making valuable applications of general systems theory in the social sciences include the work of David Easton (1965) in political science, Walter Buckley (1967) in sociology, and F. K. Berrien (1968) in social psychology.

The shift in recent years in American organization theory and research from an emphasis on traditional concepts of individual psychology and interpersonal relations to open systems concepts is in large measure related to the impact of the major contribution of Katz and Kahn to the field of organization theory and research in their work *The Social Psychology of Organizations,* published in 1966. Katz and Kahn had been involved with major empirical research in testing the human relations approach, and after encountering the general systems theory of Bertalanffy and his followers and the sociotechnical systems approach of the Tavistock group in England, they adopted an open systems approach to organizations. Their book provides a convincing description of the advantages of an open system perspective for examining the important relations of an organization with its environment and the ways in which feedback processes enable an organization to survive in a changing environment.

The trend to adapt general systems concepts in developing modern organization theory may be seen in numerous recent attempts to conceptualize organizations as complex open systems (see, for example, Morley & Sheldon, Chapter 8; Stogdill, Chapter 9; and other chapters in this volume). Today, general systems theory and modern organization theory are closely related and many of the concepts being applied to the study of organizations are taken from general systems theory.

BASIC CONCEPTS OF GENERAL SYSTEMS THEORY

It is significant that the founders of the Society for General Systems Theory later changed the name to the less pretentious Society for General Systems Research. The term "theory" has perhaps fallen victim to overuse in recent years and scientists have become increasingly reluctant to apply the term (Caws, 1968). While general systems theory may not yet meet the purist's definition of a theory, it at least consists of a set of useful concepts and working hypotheses. It is also a vigorous approach to the basic similarities believed to exist between certain properties of all systems.

In his attempt to survey the work that had been done in the field of general systems theory, Young (1964) concluded that the basic key to general systems theory lies in the types of concepts and polarities of concepts that have been elaborated to describe, explain, and predict the behavior of a general system. Young observed that these concepts form the basis of all work done to date in the field and it is to these concepts that one must turn in thinking about applications of general systems theory to any specific discipline or problem. It seems wise to follow this advice in reviewing the application of the conceptual framework of general systems theory to the field of organizational research. Accordingly, some of these basic concepts will be reviewed before discussing the application of this framework to the field of complex organizations.

Definition of a "System"

A system has been defined as "the totality of elements in inter-action with each other" (Bertalanffy, 1956), "the totality of objects together with their mutual interactions" (Hall & Fagen, 1956), "unity consisting in mutually interacting parts" (Ackoff, Chapter 5), and a "recognizably delimited aggregate of dynamic elements that are in some way interconnected and interdependent and that continue to operate together according to certain laws and in such a way as to produce some characteristic total effect" (Boguslaw, 1965). Other similar definitions have been given and it seems that each writer in discussing systems must answer the question "What is a system?" before feeling he can assume such knowledge on the part of his reader. These defini-tions in essence agree that a system is a set of units or elements which are actively interrelated and which operate in some sense as a bounded unit.

General systems theory is, then, primarily concerned with problems of relationships, of structure, and of interdependence rather than with the constant attributes of objects. Older formulations of systems con-structs had dealt with the closed systems of the physical sciences, in which relatively self-contained structures could be treated successfully as if they were independent of external forces. As a biologist, Bertalanffy was more interested in living systems which are open to, and acutely dependent upon, an external environment.

Systems Levels

The concept of hierarchical levels of systems is basic in the writing of Bertalanffy (1968), James G. Miller (Chapter 2), and other general systems theorists. Boulding (1956) has offered a nine-level classifica-tion of systems. The first three levels in this hierarchy of systems are made up of static structures, simple dynamic systems, and cybernetic systems. These may be classified as physical and mechanical systems and are the domain of the physical sciences. The fourth, fifth, and sixth levels include open systems, genetic-societal systems, and animal sys-tems. They are the concern of biologists, botanists, and zoologists since these levels are in the realm of biological systems. The last three levels constitute human systems, social systems, and transcendental systems. They are the concern of the social scientist since they comprise human social systems, and they are also the domain of the arts, humanities, and religion. Scott (Chapter 6) and Johnson, Kast, and Rosenzweig (Chapter 23) the full list of Bouldings systems levels with some description of each level. The use of analogies between lower and higher levels and the attempt to find universals which cut across all levels help to develop the understanding of higher levels.

Bertalanffy, while attempting to point to the isomorphies which cut across systems levels, has emphasized that part of general systems theory which he calls "open systems." As he points out, living systems differ from nonliving systems in that the former are open to their environments, the latter are relatively closed. In presenting his thorough and well-developed exposition of a general systems behavior theory, J. G. Miller (Chapter 2) limits his concern to the subset of living systems: cells, organs, organisms, groups, organizations, societies, and supranational systems.

Closed and Open Systems

An ideal closed system would be one which receives no energy from an outside source and from which no energy is released to its surroundings. This, however, describes a special case, i.e., a system having impermeable boundaries through which no matter-energy or information transmissions of any sort can occur. No actual system found in nature is ever in fact completely closed; systems are therefore only relatively closed. According to the second law of thermodynamics, every closed system finally attains a state of equilibrium with maximum entropy and minimum free energy; the system tends to run down and the elements composing it become randomly arranged. Bertalanffy (1968) rejected the second law of thermodynamics and its assumption of a "closed system" in his description of open systems, which characterize living entities, including individuals, groups, and organizations. In contrast to closed systems, open systems are those in which a continuing flow of component materials from the environment and a continuous output of products of the system's action back to the environment occurs.

Entropy and Negentropy

In order to survive, open systems must move against the tendency toward entropy (maximum disorganization or disorder). They must therefore acquire a steady state of "negentropy," or negative entropy, even though entropic changes occur in them as they do everywhere else. This is accomplished by taking in inputs higher in complexity of organization, i.e., lower in entropy than their outputs. In this way open systems restore their own energy and repair breakdowns in their own organization. There is then a general trend in an open system, as long at it is alive, to maximize this ratio of imported to expended energy. Open systems typically seek to improve their survival position and to acquire in their reserves a comfortable margin in operation.

Input, Transformation, Output

The pattern of activities of the energy exchange of an open system has a cyclic character. Open systems take in input, i.e., they import some form of energy from the external environment, and then transform or reorganize it through the application of throughput processes. For example, the organization creates a new product, processes materials, trains people, or provides a service. Open systems export some products into the environment, i.e., they produce outputs. The outputs of one system become available for use as inputs for another system. This basic conception of an open system as a cycle of input-conversion-output facilitates the analysis of living systems at a variety of levels from the cell to the society.

Boundaries and Interface

As a distinct systemic entity, an organization must maintain some discontinuity with its external environment in order to continue to exist as a separate system. Boundaries are not only conceived of as physical lines, or a skin surrounding the parts of a system, although at some levels of living systems this may be so. E. J. Miller and A. K. Rice (1967) suggest that the boundary should be considered a region rather than a line. In the case of an organization, a boundary may be difficult to define in terms of physical factors and may be more appropriately considered in terms of the discontinuity in patterns and clusterings of human interactions. Katz and Kahn (1966) offer a succinct definition of boundaries as "the demarcation lines or regions for the definition of appropriate system activity, for admission of members into the system, and for other imports into the system" (pp. 60–61).

A related term which is useful in understanding boundary relationships is the concept of interface, which may be defined as the area of contact between one system and another. An organizational system engages in numerous transactional processes at the interface across the boundaries of systems, including those involving the transfer of matter, energy, information, and people.

Subsystems

In every system it is possible to identify an element or functional component of the larger system which fulfills the conditions of a system in itself, but which also plays a specialized role in the operation of a larger system. Taking a particular living system as the focus, one may define the totality of all the structures in the focal system which carry out a particular process as a subsystem. Thus, at least three types of subsystems can be identified for an open system—an input subsystem, a conversion or operating subsystem, and an output subsystem. There is not necessarily a one-to-one relationship between process and structure, and one or more processes may be carried on by two or more subsystems.

James Miller (1965a, 1965b, 1965c) makes a distinction between subsystems, which he calls process units, and components, which he calls structural units. Guest (1962) observed that the survival of any living organism is dependent upon the ongoing interaction of its parts. As Miller explains, relationships among subsystems or components involve either structures or processes; the structural relationships are spatial in character and the process relationships are either purely temporal or involve spatial changes in time.

Suprasystem and Environment

Just as a living system may be analyzed in terms of its components or subsystems, so it may also be viewed as part of a higher-level system, the suprasystem, in which it then plays a subsystem role (Miller, Chapter 2). As an example, the suprasystem of a cell or tissue is the organ containing it; the suprasystem of an individual is the group of which he is a member.

The suprasystem is not to be confused with the environment. The immediate external environment of a system is the suprasystem of which it is a part, minus the system itself. The entire environment includes this immediate environment plus the suprasupra-system and the systems at all the higher levels which contain it. In order to survive, a system must interact with and somehow adjust to the other parts of the suprasystem which comprise its environment. Blalock and Blalock (Chapter 3) call this the "first perspective" in systems analysis, in which whatever level of social phenomena one wants to consider as an entity is examined in relation to its environment.

Black Box Analysis

In focusing on the relationships between the system and its environment, it is possible to neglect the internal properties of the system, either because the internal structure is unknown or because it is inconvenient to consider what is going on within the system. The system is treated as a "black box" that for some reason is not to be opened. The term "black box" originated in electrical engineering where it was convenient to provide physical black boxes containing electronic components with just a set of input-output specifications, that is, instructions for connections and performance tests that did not require opening the box and taking the hardware apart. In this "behavioristic" approach it is assumed that useful prediction can be made of output from consideration of the input, within the limits of the need of the system analyst.

Closed System Analysis

Similarly, one may wish to focus on what is going on inside the system, ignoring the environment. In the traditional approach to organizations, which tended to focus on the internal properties of an organizational system to the point of excluding the consideration of the systems relations with the environment, the system was treated as if it were closed. Hagen (Chapter 4) notes that it may be necessary for purposes of analysis to "close" or fix a system in time to examine the operation of the system in terms of environmental conditions as they relate to interrelations among units of the system. Forrester (1969) has described a "closed-boundary" approach to systems analysis in which the analyst chooses a boundary, for purposes of computer or mathematical simulation study, inside of which only those interacting components necessary to generate the modes of behavior of interest are included.

The closed boundary approach does not mean that the system is assumed to be unaffected by outside events, but that it may be convenient to view environmental occurrences as essentially random happenings that impinge on the system and do not themselves give the system its intrinsic growth and stability characteristics. Although it may be convenient at times to treat a system as if it were a "black box," or to analyze it as if it were closed or isolated, the general systems or open systems approach to complex organizations emphasizes the consideration of the relationships between a system and its environment as well as what goes on within the system.

Differentiation and Integration

A movement occurs in open systems in the direction of differentiation and elaboration in which specialized functions replace diffuse global patterns. For example, organizations move in the direction of multiplication and elaboration of roles with greater specialization of functions. In this way, some parts of an open system develop to cope with different parts of the external environment. Other parts perform specialized tasks related to input, throughput, output, and other critical system processes. In order to maintain the unity of parts as components of the whole, a reciprocal process of integration must occur. Without a sufficient degree of integration, a system would break down into separate elements.

Feedback

The system continually receives information from the environment which helps it adjust and allows it to take corrective actions to rectify deviations from a prescribed course. Feedback describes a critical system process in which a portion of the output (e.g., behavior) is returned to the system as input so as to modify succeeding outputs of the system. A living system may be viewed as self-regulating because within it input not only affects output, but output often adjusts input. *Negative feedback* is informational input looping back to the system in such a way as to decrease the deviation of output from a steady state. *Positive* feedback occurs when signals are fed back over a feedback channel in such a manner that they increase the deviation of the output from a steady state. Miller (1965b) emphasizes the critical nature of negative feedback in his proposition which states that when a system's negative feedback discontinues, its steady state disappears, its boundaries vanish, and the system can no longer survive. If there is no corrective device to get the system back on its course, it will expend or ingest too much energy and no longer continue as a system.

1. Introduction: Organizations as Open Systems

Equifinality

Bertalanffy (1968) has suggested that open systems are characterized by the principle of equifinality. While in any closed system the final state is unequivocally determined by initial conditions, according to this principle, an open system may reach the same final state from different initial conditions and in different ways. The amount of equifinality in an open system may be reduced as the system moves toward regulatory mechanisms to control its operations.

Areas of Interdependency

In attempting to develop a comprehensive understanding of system behavior it is necessary to gain knowledge of three areas of interdependency (Emery & Trist, Chapter 10). These three areas include (1) the exchanges between the system and its environment; (2) the processes within the system; and (3) the processes through which parts of the environment become related to each other. Each of these three sets of interdependencies—transactional interdependencies, internal interdependencies, and interdependencies within the environment itself—will now be discussed under the topics Organizational Environments, Intraorganizational Systems, and Interorganizational Systems.

ORGANIZATIONAL ENVIRONMENTS

Living systems, as was noted above, differ from nonliving systems in terms of their openness to the external environment in which they exist. Like other living systems, organizations themselves are changed in the course of interacting with and adjusting to their environment and also change that environment.

Emery and Trist (Chapter 10) have performed an important service by describing and classifying the range and type of system environments. They introduced the concept of "causal texture" to refer to those processes and interdependencies that occur within the environment itself. They propose four ideal types of environment which vary in the degree to which environmental components are connected as a system. The first type is a "placid, randomized" environment in which there is little or no connection between environmental parts. For a particular organizational system existing within such an environment, the environmental systems which offer rewards or adverse effects are relatively unchanging in themselves and are randomly distributed. This type of environment corresponds to the economist's "classical market." The second ideal type is a "placid, clustered" environment in which "goods" and "bads" are relatively unchanging in themselves, but they are clustered. This type of environment corresponds to the economist's "imperfect competition." The third type of environment proposed by Emery and Trist is the "disturbed-reactive" environment and corresponds to the economist's "oligopolic market," which is similar to the environment of "imperfect competition," except that more than one organization of the same kind exists, and the existence of these similar systems in the environmental field constitutes a major qualitative difference. The fourth ideal type of environment is called by Emery and Trist a "turbulent field." In this type of environment dynamic processes arise from the field itself and not simply from the interaction of the organizations and groups which comprise the environment. Emery and Trist point out that the actions of component organizations and linked sets of organizations "are both persistent and strong enough to induce autochthonous processes in the environment," and they liken the effect produced by organizations in the environmental field to that of "a company of soldiers marching in step over a bridge."

Each of these four ideal types of environment differs in terms of the degree of behavioral predictability available to the focal organization regarding other organizations and groups in its environment. Emery and Trist point out that each type of environment calls for a different organizational strategy. In the first type of environment, organizational adaptation is easiest and an organization can proceed by trial and error and still survive. However, in the second type of environment, intelligence about the environment for the development of organizational strategy becomes crucial for survival, and location within the environmental field becomes very important. In the third type of environment, because of the existence of similar organizations, overlapping effects of the action of other organizations become important, and an organization must calculate the actions and reactions of the other systems. The fourth type is still more complicated and environmental effects are uncertain.

In reviewing the literature on organizations, Shirley Terreberry (Chapter 11) concludes that an increasing number of organizational systems find themselves in environments of the fourth type. She suggests that a turbulent field may be described as a situation in which the accelerating rate and complexity of interactive effects exceeds the capacities for prediction of the organizational systems which make up the environment and, hence, these systems tend to lose control of the compounding consequences of their actions. For Terreberry, the dominant characteristic of turbulent fields is uncertainty. Several other organizational theorists have emphasized the point that an essential issue in the viable coping of an organization with its environment is the problem of dealing with uncertainty (Crozier, 1964; Thompson, 1967).

The uncertainty dimension of environment is related to two other major dimensions which have been identified in studies of organizational adaptation to environmental change: the environment's characteristic rate of change and the diversity of the environment. These dimensions are related to uncertainty because rapid change and diversity are themselves sources of uncertainty.

Several empirical and theoretical studies have shown that in comparing organizations relating to different types of environments, different types of organizational structure and patterns of system behavior are required of the organization for it to be effective (Burns & Stalker,

1961; Harvey, 1968; Perrow, 1967; Rice, 1963; Lawrence & Lorsch, 1967, 1969; Thompson, 1967). Gabarro (Chapter 12) has also pointed out that if an environment changes, the patterns of organization necessary for its continuing effectiveness must also change, and that an organizational pattern which may initially have been well suited to its environment would not continue to be so if the environment underwent major changes. Lawrence and Lorsch (1969) have labeled studies taking this approach as subscribing to a "contingency theory" since they hold the view that differences in internal system states and processes of effective organizations are contingent on differences in their external environments. Lawrence and Lorsch (1967) also offer empirical evidence of a relationship among such external environmental variables as certainty and diversity of the environment and internal system states of differentiation and integration and the processes of conflict resolution. In general, the Lawrence and Lorsch findings emphasize the importance of organization-environment "fit," i.e., if an organization's internal system states and processes are consistent with external environmental demands, the system will be more effective in dealing with its environment.

In an environment increasingly characterized by diversity, differentiated subsystems assume primary importance. Subsystems which deal with boundary-spanning transactions are particularly needed to deal effectively with such an environment. Thompson (Chapter 13) has offered a typology of specialized output roles which function in terms of linking the organization to its varied subenvironments. Brown (Chapter 14) has described some of the boundary problems which an organizational system must deal with in terms of controlling the impact of environmental forces.

Duncan (1972) has raised questions about the clarity of definition of organizational environment and the elements comprising it. He notes, for example, that Lawrence and Lorsch, while studying how organizations relate to segments of their environment, have conceptualized the environment as a total entity but have actually looked only at the environment from the organization outward. They define environment as the information available to administrators pursuing organizational goals and admit that they make no attempt to distinguish between the real attributes of the environment and management's perceptions of these attributes (Lawrence & Lorsch, 1967, p. 4). Dill (1958) offered the concept of "task environment," which also focused only on those parts of the organization's external environment that were relevant or potentially relevant to goal setting and goal attainment. Dill describes a task environment as the information that is inputted from external sources—which is not the same as the "task" of the organization, by which he means a "cognitive formulation consisting of a goal and usually also of constraints on behaviors appropriate for reaching the goal" (1958, p. 411). He also distinguishes the activities of an organization, i.e., the things that it does, from its task, i.e., the things that the organization sets itself to do, as well as from the task environment, which he defines as the stimuli the organization might respond to.

Levine and White (Chapter 20) have offered a more inclusive, although perhaps less rigorous, definition of environment which they call the organization's "task domain." They describe the task domain as the "claims which an organization stakes out for itself" in terms of clients, tasks, services, and important groups, whether they are inside or outside the organization. Gabarro (Chapter 12) observes that the definition offered by Levine and White is more inclusive than that used by Lawrence and Lorsch, who conceive of the environment primarily as uncertainty of information, as well as broader than the definition employed by Dill (1958) and Thompson (1967) who include only external groups and organizations. Gabarro uses a definition of organizational environment similar to that of Levine and White, with the intent of providing a way of considering all of the factors that might affect the organization, as well as behavior which is required for adaptation. In the conception of task domain used by Gabarro, uncertainty of information as well as other aspects of information coming from the environment are included, together with the issues posed both by external and by internal groups relevant to the organization.

Duncan offers a definition of the environment as the "totality of physical and social factors that are taken directly into consideration in the decision making behavior of individuals in the organization" (1972, p. 314). Like Gabarro, Duncan includes factors within the boundaries of the organization or specific decision-making subsystems, as well as factors outside the boundaries of the organization. Rice (1963) offered an important distinction between the internal environment, in which he included the interpersonal relations of organizational members and their interactions with each other, and the external environment, in which he included other individuals, groups, and institutions. Duncan notes the importance of recognizing a distinction between internal and external environment, and further separates in his definition those relevant physical and social factors within the boundaries of the organization, or specific decision units, from those outside the organization which must be taken into consideration in terms of decision-making behavior.

Further work is clearly needed to refine definitions of organizational system environment. Most organization theorists and researchers have tended to define the nature of the external environment in terms of the information which was obtained from within the organization rather than on the basis of some external "objective" criteria. Although it is difficult to obtain operational measures of the "real" attributes of the environment which are distinct from the perceptions of this environment by organizational members, future theoretical and empirical efforts in this area should give more attention to separating out variables descriptive of the internal environment, the external environment, and those aspects of the external environment which the organization perceives and to which it reacts.

INTRAORGANIZATIONAL SYSTEMS

In an article in *Behavioral Science* of about 200 pages, J. G. Miller (1972) presents a comprehensive analysis of present knowledge concerning the organizational level of living systems. Miller observes that the accepted view of what the basic units of organizational structure are and how they are arranged does not as yet serve organization theorists to the degree that the concepts of the cell and the organism serve biologists (p. 7). He suggests that his version of general systems theory can identify the critical subsystems of all organizations and relate them to their analogues at other system levels.

Miller divides the critical subsystems necessary for organizational survival into three major groupings: (1) subsystems which process both matter-energy and information; (2) matter-energy processing subsystems; and (3) information-processing subsystems. (Chapter 2 in this volume presents a detailed discussion of the nineteen critical subsystems which Miller defines as necessary for the life of all living systems.) He recognizes, however, that organizations are not explicitly subdivided into departments and other units according to his set of subsystems. He does assert that all living systems require the life functions associated with his nineteen specified subsystems and have either a complement of the critical subsystems or an intimate association and effective interaction with systems which carry out the missing life functions for them.

In applying his critical subsystem model to the organization, Miller is able to list specific structures which make up all the critical subsystems, except for the associator and supporter subsystems. He concludes that the associator subsystem does not exist at the organizational level but is downwardly dispersed to individuals or occasionally to outside consultants. He further notes that the supporter subsystem is also unknown at this level, although he notes that organizations must make use of nonliving supporters (such as buildings, roads, ships, and parcels of land) in order to maintain proper spatial relationship among components so that these components can interact without hindering each other.

Katz and Kahn (1966) have listed five basic organizational subsystems: (1) production subsystems concerned with getting the work of the organization done; (2) supportive subsystems for procurement, disposal, and institutional relations; (3) maintenance subsystems such as personnel administration which tie people into their functional roles; (4) adaptive subsystems concerned with organizational change; and (5) managerial subsystems for the direction, adjudication, and control of the various subsystems and activities of the organization. This categorization differs significantly from that of Miller. The Katz and Kahn production subsystems would include a number of Miller's matter-energy and information-processing subsystems. However, the decider subsystem hypothesized by Miller would carry out several of the maintenance, adaptive, and managerial activities which Katz and Kahn describe as the functions of different subsystems. Rosen (Chapter 18), adopting the theoretical model proposed by Katz and Kahn (1966), reports a field experiment which tests selected aspects of their theory.

Perhaps the most extensive body of theoretical and empirical work applying the systems perspective at the level of internal organizational processes is the sociotechnical systems theory developed over the past two decades by behavioral scientists at the Tavistock Institute of Human Relations in London. The concept of a sociotechnical system was developed from consideration of production systems and recognition that such systems require both technological components (e.g., machinery, raw materials, and plant layout) as well as social components which structure the work relationship relating the human operators both to the technology and to each other (Emery and Trist, Chapter 15). In a systemic approach to work organizations, both the technological and the social components of a productive organization must be considered, since technology and social systems mutually influence one another. Rice (1958) further developed the concept of a production system as a sociotechnical system, introduced by Trist and Bamforth (1951).

In Rice's book, based on his study of calico mills in India, he first introduced the concept of "primary task" to discriminate between the varied goals of industrial enterprises. He defined the primary task as the task an organization had been created to perform. In a later book, Rice (1963) recognized the difficulty of treating the organization as if it had a single goal or task and he cited the teaching hospital, the prison, and mental health services in particular as examples of institutions which carry out many tasks at the same time. He redefines primary task as the task that an organization must perform to survive and observes that all enterprises perform many tasks at the same time, noting that some organizations have no one primary task but, rather, have many, each of which may be primary at a given time.

Essentially, Rice conceives of the enterprise as an open system defined by its dominant import-conversion-export processes. He describes the organization as differentiated into two major types of subsystems, the operating systems which perform the primary task of the enterprise and the managing subsystem which is external to the operating systems and is required to control and service them. Rice subdivides operating subsystems into three types: (*a*) the import subsystem which is concerned with the acquisition of raw materials; (*b*) the conversion subsystem which is concerned with the transformation of imports into exports; and (*c*) the export subsystem which is concerned with the disposal of the results of import and conversion. The second major type of subsystem according to Rice, the management subsystem, may be differentiated according to service and control functions.

Eric Miller (Chapter 16) elaborates the Rice model of a sociotechnical system and asserts that any production system may be defined along three dimensions—territory, technology, and time—which he views as intrinsic to the structure of the system task. He sees any large complex production system as divided into progressively smaller subsystems along these three dimensions and argues that task performance is impaired if the subsystems are differentiated along dimensions other than these. Lynton (Chapter 17) observes that organizations also require subsystems whose function is to develop innovations for the whole system and that this task differs from the task of other subsystems in terms of technology, territory, and time.

In summarizing his system model of the organization, Rice (Chapter 19) observes that an organization containing more than one operating subsystem must also have a differentiated managing subsystem in order to control, coordinate, and service the activities of the different operating subsystems. This managing subsystem will, in turn, have responsibilities for managing the total system, each operating system, and also those nonoperating subsystems which do not directly perform any part of the primary task of the total system but which provide controls over and services to the operating systems. Rice sees management as being responsible for regulating task system boundaries, sentient system boundaries, and the relations between task and sentient systems.

INTERORGANIZATIONAL SYSTEMS

Until about a decade ago, research and theory concerning organizations had been concerned principally with intraorganizational phenomena. In the early 1960s several organizational researchers pointed out the lack of investigation in the area of interaction of organizations in the literature (Etzioni, 1960; Litwak and Hylton, 1962). Efforts in recent years have produced a growing body of research and conceptualization in this field. Here again, an open system model of organizational behavior which emphasizes that organizations are embedded in an environment made up of other organizations and social systems is paramount.

The exchanges which occur between organizations have provided one of the foci for studying organization interaction. In their studies of relations among health organizations, Levine and White (Chapter 20) have conceptualized a suprasystem comprised of a number of organizational systems which vary in the types and frequency of their exchange relationships with one another. Levine and White define organizational exchange as "any voluntary activity between two organizations which has consequences, actual or anticipated, for the realization of their respective goals or objectives." Exchange relations among organizations within the health system are determined by three factors: (1) the objective of each organization and the particular functions it carries out, which in turn determine the goods or elements it needs; (2) the access which each organization has to necessary elements from sources outside the system of health agencies, which may determine an agency's dependence upon other parts of the health system; and (3) the degree to which domain consensus exists within the health system. They define organizational domain in the health field as the area staked out by individual organizations in terms of the population served, the problems of diseases covered, and the types of services given. To the degree that some consensus can be reached among organizations which comprise such a suprasystem regarding what roles are to be allocated to what organizations, cooperation may be expected to occur among the organizations in the system. Without a consensus concerning respective goals and functions of organizations within the suprasystem, however, competition would be an expected outcome.

Evan (1966) attempted to apply Merton's notion of role-set to interorganizational phenomena by introducing the concept of organization-set. Whereas Merton (1957) selected a particular role as the unit of analysis and then charted the complex of role relationships in which the role occupant was involved, Evan selects an organization or a class of organizations as the unit of analysis and traces interactions with various organizations in its environment, i.e., with elements of its organization-set.

Incorporating his concept of organization-set into a systems model, Evan (1972) divides the complement of organizations with which a focal organization interacts in its environment into its "input-organization-set" and its "output-organization-set." As the terms suggest, the input-organization-set refers to those organizations that provide resources to the focal organization, and the output-organization-set refers to all those organizations which receive the products of goods, services, and decisions produced by the focal organization. Evan points out that analyzing an organization as an open system should also include concern for feedback effects from the output-organization-set to the focal organization and from there to the input-organization-set, or directly from the output-organization-set to the input-organization-set. For Evan, an "interorganizational system" includes a focal organization, its input-organization-set, its output-organization-set, and feedback effects which may be positive or negative as well as anticipated or unanticipated.

Thus, Evan includes within his interorganizational model the focal organizational system itself, including its subsystems, and part of the suprasystem within which it is embedded, i.e., those organizations with which it engages in either input or output exchange. He also seems to include in his "interorganizational system" model those interactions between organizations which engage in some input or output exchanges with the focal organization even when these exchanges exclude direct interaction with the focal organization.

Evan is concerned with the problems of treating organizations as disembodied entities and therefore suggests that it is important to move to a different level of social system and examine the system linkages observable in the role-set relations of boundary personnel. He suggests a number of hypotheses dealing with the characteristics of boundary personnel and the input and output organization-sets with which they relate (1972). Like Thompson (Chapter 13), who offers a typology of "output transactions" between members of an organization and its nonmembers, Evan also provides a model of interorganizational relations which takes into account the behavior of the members of the organization.

In applying an open-systems or organization-environment perspective, Aldrich (Chapter 21) notes that the distinction between members and nonmembers is an important defining element of an organization and emphasizes the importance of considering organizations as boundary-maintaining systems. For Aldrich, the exercise of authority for setting conditions of entry into the system and exit from the system are of particular importance for the organizational system in dealing with other organizations in its environment. In discussing interorganizational conflict, he points out that two strategies are available for heightening the member participation necessary for providing the resources used in interorganizational competition. One strategy is to constrict the boundaries of the organization by strengthening the requirements for participation, and the other strategy is to expand the boundaries of the organization in order to take in persons from competing organizations.

Each of the models of interorganizational behavior reviewed thus far may be examined in terms of the degree to which the model assumes that the action of a focal organizational system must be considered in terms of its role in a richly joined supraorganizational system. Levine and White view individual organizations as system parts in an inclusive community health system. Their emphasis is on the interdependence of organizations as part of a local health agency system, although they recognize that a particular organization may have relative independence because of the availability of scarce resources from outside the system. Evan also recognizes that a focal organizational system may be viewed as a part of a larger suprasystem, but his system model emphasizes the interaction between a focal organizational system and limited sets of external systems which are interdependent with the focal system in terms of providing inputs or accepting outputs from that system. Aldrich, in his emphasis on competitive interorganizational relations and the importance of boundary maintenance, does not assume that an organizational system operates within a well-defined, highly interdependent suprasystem complex.

Like Levine and White, Baker and O'Brien (Chapter 22) are concerned with health and welfare organizations, as contrasted with models like Evan's, which purport to describe interrelations among formal organizations in general. Nevertheless, Baker and O'Brien question the assumption of cooperative independence which has been the focus of much of the theory and research on health organizations. They note that "there is an inherent danger in the assumption that there is one coherent system of human service caregivers whose boundaries can be well defined both conceptually and empirically." They propose, instead, an intersystem model which focuses on the existence of autonomy and variations in the degree of interdependence in the interactions between two related systems. From an intersystem perspective they observe that the existence of a suprasystem can be tested empirically in terms of the degree of component interdependence existing among various organizational systems which are hypothesized to make up a larger system. Further, they note the usefulness of the concept of the interorganizational field, which does not imply a denied membership in a larger suprasystem (Warren, 1967).

Baker and O'Brien propose that the relations among systems in an interorganizational field may differ in terms of the particular type of interaction which is being examined to assess the interdependence of organizations. Thus organizations may display a high degree of inter-dependent exchange in terms of information flow but may maintain autonomy with regard to financial exchange—a distinction which may have important implications for a planner attempting to develop a highly interdependent and well-integrated community suprasystem, or for the manager of a particular organizational system who must recognize the need for different organizational policies for different sets of organizations in the environment which vary according to the media of exchange relations.

SYSTEMS THEORY AND ORGANIZATIONAL MANAGEMENT

The implications of organizational systems theory for the manager become important as he conceptualizes his organization as an open system, as he thinks about his job, and as he organizes his activities. Seiler (1967) observes that the idea of "system" leads the manager to substitute the habit of conceiving of behavior as caused by multiple interdependent forces for the habit of thinking in terms of single causes. He notes that systemic thinking frees one from oversimplification by making the understanding of complexity manageable. Johnson, Kast, and Rosenzweig (Chapter 23) also note that systems theory may be useful in thinking about the job of management by providing a framework for visualizing internal and external environmental factors as an integrated whole and for recognizing the proper place and function of subsystems. In a book emphasizing the systems approach to organization and management, Kast and Rosenzweig (1970) present another advantage in considering organizations as open systems. They note that it "provides a means of concentrating on the synergistic aspects of the whole system" (p. 22). By synergy, they mean that the whole is greater than, or at least different from, the sum of its parts. They also emphasize the advantage of a conceptual scheme which allows the consideration of a number of levels of phenomena, including interaction of individuals, small group dynamics, and large group behavior, all within the constraints of an external environmental system.

In examining the manager's job from a systems point of view, Tilles (1963) views the management task as consisting of four basic parts, each necessarily related to all the others. Tilles points out that a common error committed by general managers, especially in smaller companies, lies in equating the notion of viewing things in their totality with doing everything oneself. When the manager examines the total system, it becomes evident that he must delegate in order to carry out his primary responsibility, that is, thinking about how various system parts fit together and where to define the boundaries of the company in terms of who and what are included as part of the organizational system. Tilles sees the second major task of the manager as establishing system objectives. One advantage of viewing a company as an open system is that this perspective broadens the area of relationships which a manager should properly consider. A system view encourages the manager to look outward to a wider environment and upward to the suprasystems of which the organization is a part, rather than to be overly preoccupied with events taking place within the organizational system. As a part of this task, the manager is charged with the responsibility of not only identifying wider systems of which the organization is a part, but also of setting up system criteria which may be used in determining how well the organization has performed in regard to each system with which it relates. As a third basic task of management, Tilles sees the creation of formal subsystems, by which he means the officially established groups and entities created and changed to carry on the organization's activities. Viewing the company as a set of formal subsystems offers the benefit of focusing attention on the essential relatedness of activities carried out by specific individuals, as well as emphasizing the importance of grouping employees to form meaningful subsystems related to the requirements of the total organization. A fourth major task is systems integration, by which Tilles refers to activities which bring together organizational subunits as well as activities that integrate different models of corporate life.

Other organizational systems analysts have emphasized the responsibilities of management for integration, i.e., the development of collaboration and mutual understanding among the subunits of an organizational system, as well as its twin process of differentiation, which involves the evolution of change to the extent to which organizational subunits differ from one another (Lawrence & Lorsch, 1969; Sayles & Chandler, 1971; Allen, 1971). Lawrence and Lorsch observed that it is not possible to have both a high degree of differentiation and extremely precise integration within an organizational system. In examining the problems of managing large systems, Sayles and Chandler observe that if integration is to proceed successfully, it must take place in conjunction with differentiation, since the two are actually inverse processes. They suggest that changes in subsystem components should be made in light of the interaction between subsystems and that the processes of differentiation and integration should constitute "an oscillating system in which each element continuously takes the other into account" (p. 253).

In a comparative analysis of corporate divisional relationships in multiunit organizations, Allen (1971) defined integrative effort as the amount of working time that managers devoted to achieving integration. Through a comparison of high- and low-performing firms faced with high levels of environmental diversity, Allen found that the high-performing firms had high levels of integration and relatively low levels of integrative effort, while the low performers were characterized by lower levels of differentiation than their environmental diversity suggested was required. Yet the latter were found to devote more effort to integration while achieving a lower level of integration. In dealing with these counterintuitive findings, Allen suggests that there are upper and lower limits to the benefits obtainable from managers investing their time in integrative efforts. He suggests that those organizations which seemed to be performing on a lower level had passed this upper limit and were overloading communication channels rather than directing energies needed to deal with their own particular environments. Allen's research implies that management may spend too much time and effort attempting to develop collaboration among various specialized subsystems and cites the need to find a balance between the requirements for interdependence among internal subunits while maintaining the degree of internal diversity necessary for dealing with the external requirements of a highly diverse environment.

French (Chapter 24) also attempts to develop a model of the manager's job by considering system processes. He suggests that the manager needs to anticipate the extent to which enterprise subprocesses are systematized, as this relates to changes in the flexibility of the total system in meeting internal and external demands for change. Presenting the view that the less structured a system, the more that unsystematized processes can be inferred, French emphasizes the necessity for an open system to remain relatively loosely joined in order to flexibly change to meet a diverse and changing environment. This point of view is compatible with Allen's observation and with prior discussion of interorganizational relations.

Swinth discusses the problems of setting goals in interdependent systems, a major task of management, and presents the results of an experiment which supports the contention that organizations can be designed, at least in part, with a focus on the task to be performed rather than in terms of the question of power and who controls whom. He suggests a useful principle for the manager of an organizational system: By using a goal-setting technique of allocating goals which bridge subsystem interdependencies while not imposing an authority hierarchy, a situation may develop in which both parties working together on the decision can work closely without the necessity of an overseer, and motivation can be maintained to contribute to the success of the total system.

Another major management task is the ability to cope effectively with organizational change. Schein (1965) described a six-stage "adaptive-coping cycle" for dealing with change in the internal or external environment and thus maintaining organization effectiveness. The adaptive-coping cycle includes the following stages or processes: (1) sensing environmental change; (2) inputting relevant information about the change to subsystems which can act upon it; (3) changing internal conversion processes according to the information; (4) stabilizing internal change and managing undesired side effects; (5) delivering outputs which are more consonant with changes sensed in the environment; and (6) obtaining feedback on the effectiveness of the change. All of the stages of the organizational adaptive-coping cycle are important, and there are problems and pitfalls connected with each stage, many of which have been mentioned at various points in this chapter. However, the last stage, in which the outputs and the processes of change are evaluated, has not yet been specifically discussed. Examination of this topic will conclude this introductory chapter.

EVALUATION

Conceptualizing an organization as an open system also has implications for the model employed in evaluating organizational effectiveness. Perhaps the most popular model for organizational evaluation has been the goal-attainment model. Yuchtman and Seashore (Chapter 28) observe that most researchers employing this traditional approach to organizational effectiveness make two implicit or explicit assumptions: (1) that organizations have an "ultimate" goal, mission, or function which they are attempting to realize; and (2) that the ultimate goal can be empirically identified and further, that progress toward it can be measured.

Etzioni (Chapter 26) has pointed out two major shortcomings of the goal-attainment model related to the stereotyped findings of studies which arise out of the assumptions implicit in this model. Many goal-attainment evaluations conclude that the focal organization does not effectively realize its goals and/or the organization has "displaced" its goals, or is assumed to have goals which are different from those it claims to have.

One of the major assets of the goal attainment-model has been to free the researcher from imposing the bias of his own goals as criteria of the organization's ineffectiveness or effectiveness by turning to the goal statements of organizational spokesmen. However, this asset may be mythical in nature. Etzioni points out that the "public goals," i.e., the goals which an organization claims to pursue, fail to be realized because they are not meant to be realized.

The evaluator may be misled even if he attempts to avoid naive acceptance of the public fictions of official pronouncements, the ideal statements of a charter and other official documents, or the official reports of organizational leaders, and instead attempts to use "private goals" (i.e., the goals the organization actually seems to follow) as criteria for performance evaluation. This relates to the multifunctional character of organizations, i.e., an organization has multiple goals which are not necessarily compatible and, additionally, an organization invests its resources in nongoal as well as goal functions.

Etzioni suggests that a system model approach to evaluation does not start with the goal itself. Instead, the system model of evaluation is concerned with establishing a working model of a social unit which is capable of achieving a goal. Unlike the study of a single goal, or even a set of goal activities, the system model described by Etzioni is that of a multifunctional unit. It recognizes that an organization must fulfill at least four important functions for survival. In addition to the achievement of goals and subgoals, the system model is concerned with: the effective coordination of organizational subunits; the acquisition and maintenance of necessary resources; and the adaptation of the organization to the environment and to its own internal demands. The system model assumes that some of the organization's means must be devoted to such nonobvious functions as custodial activities, including means employed for maintenance of the organization itself. From the viewpoint of the system model, such activities are functional and actually increase organizational effectiveness.

In contrast to the goal-attainment model of evaluation, which is concerned with degree of success in reaching a specific objective, the system model establishes the degree to which an organization realizes its goals under a given set of conditions. Etzioni indicates that the key question is: Under the given conditions, how close does the organizational allocation of resources approach an optimum distribution? Optimum is the key word, and what counts is a balanced distribution of resources among all organizational objectives, not maximal satisfaction of any one goal. From this perspective, just as a lack of resources for any one goal may be dysfunctional, an excess of resources for the goal may be equally dysfunctional. In the latter instance, superfluous attention to one goal leads to depressed concern for the others, and problems of coordination and competition will arise.

In applying an open system model of evaluation, the evaluator moves from the more or less exclusive emphasis on output in the goal-attainment model to a concern for throughput as well as input processes. Bennis (Chapter 29) points out that the traditional goal-attainment models provide static indicators of output without adequate attention to crucial dynamic processes of problem solving. Bennis takes the view that the processes through which an organization searches for, adapts to, and solves its changing goals are crucial to making inferences about the organization's effectiveness in meeting its major challenge—coping with environmental stress and change. He emphasizes the inherent inadequacy of any single, time-slice measures of organizational performance as a basis for providing valid indicators of organizational health.

Yuchtman and Seashore (Chapter 28), while recognizing the other major dimensions of throughput and output in an open system model, choose to emphasize the input phase of this model of organizational behavior as a basis for defining the effectiveness of an organization. They define the effectiveness of an organization "in terms of its bargaining position, as reflected in the ability of the organization, in either absolute or relative terms, to exploit its environment in the acquisition of scarce and valued resources."

Baker and Schulberg (Chapter 27), building on Etzioni, emphasize all three aspects of the major cyclic phases in the open systems model of organizational evaluation, in addition to recognizing the importance of feedback and the role of the organizational system as a component of subsystems in a larger system. Baker and Schulberg also discuss some of the specific data collection methods required in a systems approach to evaluation, and they observe that this model of organizational evaluation is a more demanding and expensive one for the researcher.

. . . .

It is hoped that this general introduction to general systems theory as applied to the study of complex organizations will aid the reader in understanding the common threads which tie together the separate papers presented here. The chapters which follow are focused in the following manner: Section One is concerned with general theory of living systems and perspectives on social systems analysis; in Section Two, the emphasis shifts to specific concern with organizational systems theory; Section Three explores the environments of organizations as they influence and are influenced by organizational systems; internal organizational systems provide the focus of Section Four; and Section Five is concerned with interorganizational systems relations; systems theory as it relates to organizational management is examined in Section Six; finally, the evaluation of organizations from a systems point of view is discussed in Section Seven.

REFERENCES

Allen, S. A. A Comparative Analysis of Corporate-Divisional Relationships in Multi-Unit Organizations. Paper prepared for meeting of the Institute of Management Sciences, College on Organization, March 22–24, 1971, Washington, D.C.

Ashby, W. R. "General Systems Theory as a New Discipline." *General Systems*, 1958, *3*, 1–17.

Berrien, F. K. *General and Social Systems*. New Brunswick, N.J.: Rutgers University Press, 1968.

Bertalanffy, L. von. "General Systems Theory." *General Systems*, 1956, *1*, 1–10.

Bertalanffy, L. von. *General Systems Theory*. New York: George Braziller, 1968.

Boguslaw, W. *The New Utopians*. Englewood Cliffs, N.J.: Prentice-Hall, 1965.

Boulding, K. E. "General Systems Theory: The Skeleton of Science." *General Systems*, 1956, *1*, 11–17.

Buckley, W. *Sociology and Modern Systems Theory*. Englewood Cliffs, N.J.: Prentice-Hall, 1967.

Burns, T. & Stalker, G. M. *The Management of Innovation*. London: Tavistock, 1961.

Caws, P. "Science and System: On the Unity and Diversity of Scientific Theory." *General Systems*, 1968, *13*, 3–12.

Crozier, M. *The Bureaucratic Phenomenon*. Chicago: University of Chicago Press, 1964.

Dill, W. "Environment as an Influence on Managerial Autonomy." *Administrative Quarterly*, 1958, *2*, 409–43.

Duncan, R. B. "Characteristics of Organizational Environments and Perceived Environmental Uncertainty." *Administrative Science Quarterly*, 1972, *17*, 313–27.

Easton, D. *A Systems Analysis of Political Life*. New York: Wiley, 1965.

Etzioni, A. "New Directions in the Study of Organization and Society. *Social Research,* 1960, *27,* 223–28.

Evan, W. M. "The Organization Set: Toward a Theory of Interpersonal Relations." In J. D. Thompson (Ed.), *Approaches to Organizational Design.* Pittsburgh: University of Pittsburgh Press, 1966.

Evan, W. M. "An Organization-Set Model of Interorganizational Relations." In M. Tuite, R. Chisholm & M. Radnor (Eds.), *Interorganizational Decision Making.* Chicago: Aldine, 1972.

Forrester, J. W. *Urban Dynamics.* Cambridge: M.I.T. Press, 1969.

Grinker, R. R. (Ed.) *Toward a Unified Theory of Human Behavior.* New York: Basic Books, 1956.

Guest, R. M. *Organizational Change: The Study of Effective Leadership.* Homewood, Ill.: Irwin-Dorsey, 1962.

Hall, A. D. & Fagen, R. E. "Definition of System." *General Systems,* 1956, *1,* 18–29.

Harvey, E. "Technology and the Structure of Organization." *American Sociological Review,* 1968, *33,* 247–59.

Kast, F. E. & Rosenzweig, J. E. *Organization and Management.* New York: McGraw-Hill, 1970.

Katz, D. & Kahn, R. L. *The Social Psychology of Organizations.* New York: Wiley, 1966.

Lawrence, P. R. & Lorsch, J. W. "Differentiation and Integration in Complex Organizations." *Administrative Science Quarterly,* 1967, *12,* 1–47.

Lawrence, P. R. & Lorsch, J. W. *Organization and Environment.* Homewood, Ill.: Irwin, 1969.

Litwak, E. & Hylton, L. F. "Interorganizational Analysis: A Hypothesis on Coordinating Agencies." *Administrative Science Quarterly,* 1962, *6,* 395–420.

Merton, R. K. *Social Theory and Social Structure.* Glencoe, Ill.: Free Press, 1957.

Miller, E. J. & Rice, A. K. *Systems of Organization.* London: Tavistock, 1967.

Miller, J. G. "Living Systems: Basic Concepts." *Behavioral Science,* 1965a, *10,* 193–237.

Miller, J. G. "Living Systems: Structure and Process." *Behavioral Science,* 1965b, *10,* 337–79.

Miller, J. G. "Living Systems: Cross-level Hypotheses." *Behavioral Science,* 1965c, *10,* 380–411.

Miller, J. G. "Living Systems: The Organization." *Behavioral Science,* 1972, *17,* 1–182.

Perrow, C. "A Framework for the Comparative Analysis of Organizations." *American Sociological Review,* 1967, *32,* 194–208.

Rapoport, A. & Horvath, W. J. "Thoughts on Organization Theory and a Review of Two Conferences." *General Systems,* 1959, *4,* 87–93.

Rice, A. K. *Productivity and Social Organization: The Ahmedabad Experiment.* London: Tavistock, 1958.

Rice, A. K. *The Enterprise and Its Environment.* London: Tavistock, 1963.

Sayles, L. R. & Chandler, M. K. *Managing Large Systems.* New York: Harper & Row, 1971.

Schein, E. H. *Organizational Psychology.* Englewood Cliffs, N.J.: Prentice-Hall, 1965.

Seiler, J. A. *Systems Analysis in Organizational Behavior.* Homewood, Ill.: Irwin, 1967.

Shannon, C. & Weaver, W. *The Mathematical Theory of Communication.* Urbana, Ill.: University of Illinois Press, 1949.

Thompson, J. D. *Organizations in Action.* New York: McGraw-Hill, 1967.

Tilles, S. "The Manager's Job: A Systems Approach." *Harvard Business Review,* 1963, *41,* 73–81.

Trist, E. L. & Bamforth, K. W. "Some Social and Psychological Consequences of the Long-Wall Method of Coal-Getting." *Human Relations,* 1951, *4,* 3–38.

von Neumann, J. & Morgenstern, O. *Theory of Games and Economic Behavior.* Princeton, N.J.: Princeton University Press, 1947.

Warren, R. "The Interorganizational Field as a Focus for Investigation." *Administrative Science Quarterly,* 1967b, *12,* 396–419.

Wiener, N. *Cybernetics.* New York: Wiley, 1948.

Young, O. R. "A Survey of General Systems Theory." *General Systems,* 1964, *9,* 61–80.

Section I

GENERAL SYSTEMS THEORY AND SYSTEMS ANALYSIS

GENERAL SYSTEMS THEORY is concerned with the development of a set of interrelated theoretical constructs which apply across every level of phenomena in the empirical world. Particularly when applied to living systems, general systems theory offers some unique advantages in viewing that class of living systems which we call social systems. In the first paper in this section, James G. Miller, one of the earliest and most enlightening exponents of general systems theory, outlines what he calls general systems behavior theory—that part of general systems theory which specifically relates to living systems.

In the next paper, the Blalocks explain some of the basic concepts and ideas used in the analysis of systems. They offer the view that in analyzing social systems, three perspectives may be employed: (1) that which involves the relationship between system and environment; (2) that which involves interaction between several systems; and (3) that which involves one type of system composed of other types of systems. These writers also discuss the concept "structure" and "equilibrium" as they apply to systems analysis, pointing out that social scientists have used these concepts often in a rather vague and inconsistent fashion.

Hagen points out some of the logical requirements of general systems analysis. According to Hagen, an analytical model requires definition of mutually interacting or functionally related variables, each with a single dimension. Although he recognizes that all natural systems interact with the environment and are therefore open, this author argues that in analysis, a system must be "closed." He also recognizes, however, that in the process of analysis, a closed system will of necessity be opened so that the theorist can observe the effects of change in the environment. Although it is useful to construct analytical models which are in a state of stable equilibrium, systems need not be in either stable or unstable, static or moving equilibrium. Analysis must include systems not in equilibrium in order to gain empirical relevancy and analytical power. Hagen argues that previous models of societal systems have reflected a lack of understanding of these logical requirements.

James G. Miller

2. The Nature of Living Systems

GENERAL SYSTEMS behavior theory is a set of related definitions, assumptions, and propositions which deal with reality as an integrated hierarchy of organizations of matter and energy. General systems behavior theory is concerned with a special subset of all systems, the living ones.

Even more basic to this presentation than the concept of "system" are the concepts of "space," "time," "matter," "energy," and "information," because the living systems discussed here exist in space and are made of matter and energy organized by information.

1. SPACE AND TIME

In the most general mathematical sense, a *space* is a set of elements which conform to certain postulates. The *conceptual spaces* of mathematics may have any number of dimensions.

Physical space is the extension surrounding a point. Classically the three dimensional geometry of Euclid was considered to describe accurately all regions in physical space. The modern general theory of relativity has shown that physical space-time is more accurately described by a geometry of four dimensions, three of space and one of time.

This presentation of a general theory of living systems will employ two sorts of spaces in which they may exist, *physical* or *geographical space* and *conceptual* or *abstracted spaces*.

1.1 Physical or Geographical Space

This will be considered as Euclidean space, which is adequate for the study of all aspects of living systems as we now know them. Among the characteristics and constraints of physical space are the following: (*a*) From point *A* to point *B* is the same distance as from point *B* to

From *Behavioral Science*, July 1971, vol. 16, pp. 278–301. Reprinted with permission of the author and the publisher.
References and notes begin on page 58.

point *A*. (*b*) Matter or energy moving on a straight or curved path from point *A* to point *B* must pass through every intervening point on the path. This is true also of markers bearing information. (*c*) In such space there is a maximum speed of movement for matter, energy, and markers bearing information. (*d*) Objects in such space exert gravitational pull on each other. (*e*) Solid objects moving in such space cannot pass through one another. (*f*) Solid objects moving in such space are subject to friction when they contact another object.

The characteristics and constraints of physical space affect the action of all concrete systems, living and nonliving. The following are some examples: (*a*) On the average, people interact more with persons who live near to them in a housing project than with persons who live far away in the project. (*b*) The diameter of the fuel supply lines laid down behind General Patton's advancing American Third Army in World War II determined the amount of friction the lines exerted upon the fuel pumped through them, and therefore the rate at which fuel could flow through them to supply Patton's tanks. This was one physical constraint which limited the rate at which the army could advance, because they had to halt when they ran out of fuel. (*c*) Today information can flow worldwide almost instantly by telegraph, radio, and television. In the seventeenth century it took weeks for messages to cross an ocean. A government could not send messages so quickly to its ambassadors then as it can now because of the constraints on the rate of movement of the marker bearing the information. Consequently ambassadors of that century had much more freedom of decision than they do now.

Physical space is a common space, for the reason that it is the only space in which all concrete systems, living and nonliving, exist (though some may exist in other spaces simultaneously). Physical space is shared by all scientific observers, and all scientific data must be collected in it. This is equally true for natural science and behavioral science. Most people learn that physical space exists, which is not true of many spaces I shall mention in the next section. They can give the location of objects in it.

1.2 Conceptual or Abstracted Spaces

Scientific observers often view living systems as existing in spaces which they conceptualize or abstract from the phenomena with which they deal. Examples of such spaces are: (*a*) Peck order in birds or other animals. (*b*) Social class space (lower lower, upper lower, lower middle, upper middle, lower upper, and upper upper classes). (*c*) Social distance among ethnic or racial groups. (*d*) Political distance among political parties of the right and the left. (*e*) Sociometric space, e.g., the rating on a scale of leadership ability of each member of a group by every other member. (*f*) A space of time costs of various modes of transportation, e.g., travel taking longer on foot than by air, longer upstream than down.

These conceptual and abstracted spaces do not have the same characteristics and are not subject to the same constraints as physical space. Each has characteristics and constraints of its own. These spaces may be either conceived of by a human being or learned about from others. Interpreting the meaning of such spaces, observing relations, and measuring distances in them ordinarily require human observers. Consequently the biases of individual human beings color these observations.

Social and some biological scientists find conceptual or abstracted spaces useful because they recognize that physical space is not a major determinant of certain processes in the living systems they study. E.g., no matter where they enter the body, most of the iodine atoms in the body accumulate in the thyroid gland. The most frequent interpersonal relations occur among persons of like interest or like attitudes rather than among geographical neighbors. Families frequently come together for holidays no matter how far apart their members are. Allies like England and Australia are often more distant from each other in physical space than they are from their enemies.

It is desirable that scientists who make observations and measurements in any space other than physical space should attempt to indicate precisely what are the transformations from their space to physical space. Other spaces are definitely useful to science, but physical space is the only common space in which all concrete systems exist.

1.3 Time

This is the fundamental "fourth dimension" of the physical space-time continuum. *Time* is the particular instant at which a structure exists or a process occurs, or the measured or measurable period over which a structure endures or a process continues. For the study of all aspects of living systems as we know them, for the measurement of durations, speeds, rates, and accelerations, the usual absolute scales of time—seconds, minutes, days, years—are adequate. A concrete system can move in any direction on the spatial dimensions, but only forward —never backward—on the temporal dimension.

2. MATTER AND ENERGY

Matter is anything which has mass (m) and occupies physical space. *Energy* (E) is defined in physics as the ability to do work. The principle of the conservation of energy states that energy can be neither created nor destroyed in the universe, but it may be converted from one form to another, including the energy equivalent of rest-mass. Matter may have (*a*) *kinetic* energy, when it is moving and exerts a force on other matter; (*b*) *potential* energy, because of its position in a gravitational field; or (*c*) *rest-mass* energy, which is the energy that would be released if mass were converted into energy. Mass and energy are equivalent. One can be converted into the other in accordance with the relation that rest-mass energy is equal to the mass times the square of the velocity of light. Because of the known relationship between matter and energy, throughout this article the joint term *matter-energy* is used except where one or the other is specifically intended. Living systems require matter-energy, needing specific types of it, in adequate amounts. Heat, light, water, minerals, vitamins, foods, fuels, and raw materials of various kinds, for instance, may be required. Energy for the processes of living systems is derived from the breakdown of molecules (and, in a few recent cases, of atoms as well). Any change of state of matter-energy or its movement over space, from one point to another, is *action*. It is one form of process.

3. INFORMATION

Throughout this presentation *information* (H) will be used in the technical sense first suggested by Hartley in 1928.[1] Later it was developed by Shannon in his mathematical theory of communication.[2] It is not the same thing as meaning or quite the same as information as we usually understand it. *Meaning* is the significance of information to a system which processes it: It constitutes a change in that system's processes elicited by the information, often resulting from associations made to it on previous experience with it. *Information* is a simpler concept: the degrees of freedom that exist in a given situation to choose among signals, symbols, messages, or patterns to be transmitted. The total of all these possible categories (the alphabet) is called the *ensemble*. The amount of information is measured by the binary digit, or *bit* of information. It is the amount of information which relieves the uncertainty when the outcome of a situation with two equally likely alternatives is known. Legend says the American Revolution was begun by a signal to Paul Revere from Old North Church steeple. It could have been either one or two lights "one if by land or two if by sea." If the alternatives were equally probable, the signal conveyed only one bit of information, resolving the uncertainty in a binary choice. But it carried a vast amount of meaning, meaning which must be measured by other sorts of units than bits.

The term *marker* refers to those observable bundles, units, or changes of matter-energy whose patterning bears or conveys the informational symbols from the ensemble or repertoire.[3] These might be the stones of Hammurabi's day which bore cuneiform writing, parchments, writing paper, Indians' smoke signals, a door key with notches, punched cards, paper or magnetic tape, a computer's magnetized ferrite core memory, an arrangement of nucleotides in a DNA molecule, the molecular structure of a hormone, pulses on a telegraph wire, or waves emanating from a radio station. The marker may be static, as in a book or in a computer's memory. Communication of any sort, however, requires that the marker move in space, from the transmitting system to the receiving system, and this movement follows the same physical laws as the movement of any other sort of matter-energy. The advance of communication technology over the years has been in the direction of decreasing the matter-energy costs of storing and transmitting the markers which bear information. The efficiency of information processing can be increased by lessening the mass of the markers, making them smaller so they can be stored more compactly and transmitted more rapidly and cheaply. Over the centuries engineering progress has altered the mode in markers from stones bearing cuneiform to magnetic tape bearing electrons, and clearly some limit is being approached.

In recent years systems theorists have been fascinated by the new ways to study and measure information flows, but matter-energy flows are equally important. Systems theory deals both with information theory and with energetics—such matters as the muscular movements of people, the flow of raw materials through societies, or the utilization of energy by brain cells.

It was noted above that the movement of matter-energy over space, *action,* is one form of process. Another form of process is information processing or *communication,* which is the change of information from one state to another or its movement from one point to another over space. Communications, while being processed, are often shifted from one matter-energy state to another, from one sort of marker to another. If the form or pattern of the signal remains relatively constant during these changes, the information is not lost. For instance, it is now possible to take a chest X-ray, storing the information on photographic film; then a photoscanner can pass over the film line by line, from top to bottom, converting the signals to pulses in an electrical current which represent bits; then those bits can be stored in the core memory of a computer; then those bits can be processed by the computer so that contrasts in the picture pattern can be systematically increased; then the resultant altered patterns can be printed out on a cathode ray tube and photographed. The pattern of the chest structures, the information, modified for easier interpretation, has remained largely invariant throughout all this processing from one sort of marker to another. Similar transformations go on in living systems.

One basic reason why communication is of fundamental importance is that informational patterns can be processed over space and the local matter-energy at the receiving point can be organized to conform to, or comply with, this information. As already stated, if the information is conveyed on a relatively small, light, and compact marker, little energy is required for this process. Thus it is a much more efficient way to accomplish the result than to move the entire amount of matter-energy, organized as desired, from the location of the transmitter to that of the receiver. This is the secret of success of the delivery of "flowers by telegraph." It takes much less time and human effort to send a telegram from London to Paris requesting a florist in the latter place to deliver flowers locally, than it would to drive or fly with the flowers from the former city to the latter.

Shannon was concerned with mathematical statements describing the transmission of information in the form of signals or messages from a sender to a receiver over a channel such as a telephone wire or a radio band.[4] These channels always contain a certain amount of noise. In order to convey a message, signals in channels must be patterned and must stand out recognizably above the background noise.

Matter-energy and information always flow together. Information is always borne on a marker. Conversely there is no regular movement in a system unless there is a difference in potential between two points, which is negative entropy or information. Which aspect of the transmission is most important depends upon how it is handled by the receiver. If the receiver responds primarily to the material or energic aspect, it is a matter-energy transmission; if the response is primarily to the information, it is an information transmission. For example, the banana eaten by a monkey is a nonrandom arrangement of specific molecules, and thus has its informational aspect, but its use to the monkey is chiefly to increase the energy available to him. So it is an energy transmission. The energetic character of the signal light that tells him to depress the lever which will give him a banana is less important than the fact that the light is part of a nonrandom, patterned organization which conveys information to him. So it is an information transmission. Moreover, just as living systems must have specific forms of matter-energy, so they must have specific patterns of information. For example, some species of animals do not develop normally unless they have appropriate information inputs in infancy. As Harlow showed, for instance, monkeys cannot make proper social adjustment unless they interact with other monkeys during a period between the third and sixth months of their lives.[5]

4. SYSTEM

The term *system* has a number of meanings. There are systems of numbers and of equations, systems of value and of thought, systems of law, solar systems, organic systems, management systems, command and control systems, electronic systems, even the Union Pacific Railroad system. The meanings of "system" are often confused. The most general, however, is: A *system* is a set of interacting units with relationships among them.[6] The word "set" implies that the units have some common properties, which is essential if they are to interact or have relationships. The state of each unit is constrained by, conditioned by, or dependent on the state of other units.[7] The units are coupled.

4.1 Conceptual Systems

4.1.1. Units. *Units* of a *conceptual system* are terms, such as words (commonly nouns, pronouns, and their modifiers), numbers, or other symbols, including those in computer simulations and programs.
4.1.2 Relationships. A relationship of a conceptual system is a set of pairs of units, each pair being ordered in a similar way. E.g., the set of all pairs consisting of a number and its cube is the cubing relationship. Relationships are expressed by words (commonly verbs and their modifiers), or by logical or mathematical symbols, including those in computer simulations and programs, which represent operations, e.g., inclusion, exclusion, identity, implication, equivalence, addition, subtraction, multiplication, or division. The language, symbols, or com-

puter programs are all concepts and always exist in one or more concrete systems, living or nonliving, like a scientist, a textbook, or a computer.

4.2 Concrete System

A *concrete system* is a nonrandom accumulation of matter-energy, in a region in physical space-time, which is organized into interacting, interrelated subsystems or components.

4.2.1 Units. The units (subsystems, components, parts, or members) of these systems are also concrete systems.[8]

4.2.2 Relationships. Relationships in concrete systems are of various sorts, including spatial, temporal, spatiotemporal, and causal.

Both units and relationships in concrete systems are empirically determinable by some operation carried out by an observer. In theoretical verbal statements about concrete systems, nouns, pronouns, and their modifiers typically refer to concrete systems, subsystems, or components; verbs and their modifiers usually refer to the relationships among them. There are numerous examples, however, in which this usage is reversed and nouns refer to patterns of relationships or processes, such as "nerve impulse," "reflex," "action," "vote," or "annexation."

4.2.3 Open System. Most concrete systems have boundaries which are at least partially permeable, permitting sizeable magnitudes of at least certain sorts of matter-energy or information transmissions to cross them. Such a system is an *open system*. Such inputs can repair system components that break down and replace energy that is used up.

4.2.4 Closed System. A concrete system with impermeable boundaries through which no matter-energy or information transmissions of any sort can occur is a *closed system*. No actual concrete system is completely closed, so concrete systems are either relatively open or relatively closed. Whatever matter-energy happens to be within the system is all there is going to be. The energy gradually is used up and the matter gradually becomes disorganized. A body in a hermetically sealed casket, for instance, slowly crumbles and its component molecules become intermingled. Separate layers of liquid or gas in a container move toward random distribution. Gravity may prevent entirely random arrangement.

4.2.5 Nonliving System. Every concrete system which does not have the characteristics of a living system is a *nonliving system*.

4.2.6 Living Systems. The *living systems* are a special subset of the set of all possible concrete systems, composed of the plants and the animals. They all have the following characteristics:

a) They are open systems.
b) They use inputs of foods or fuels to restore their own energy and repair breakdowns in their own organized structure.
c) They have more than a certain minimum degree of complexity.
d) They contain genetic material composed of deoxyribonucleic acid

(DNA), presumably descended from some primordial DNA common to all life, or have a charter, or both. One or both of these is the template—the original "blueprint" or "program"—of their structure and process from the moment of their origin.

e) They are largely composed of protoplasm including proteins and other characteristic organic compounds.

f) They have a decider, the essential critical subsystem which controls the entire system, causing its subsystems and components to interact.

g) They also have certain other specific critical subsystems or they have symbiotic or parasitic relationships with other living or nonliving systems which carry out the processes of any such subsystem they lack.

h) Their subsystems are integrated together to form actively self-regulating, developing, reproducing unitary systems, with purposes and goals.

i) They can exist only in a certain environment. Any change in their environment of such variables as temperature, air pressure, hydration, oxygen content of the atmosphere, or intensity of radiation, outside a relatively narrow range which occurs on the surface of the earth, produces stresses to which they cannot adjust. Under such stresses they cannot survive.

4.3 Abstracted System

4.3.1 Units. The units of *abstracted systems* are relationships abstracted or selected by an observer in the light of his interests, theoretical viewpoint, or philosophical bias. Some relationships may be empirically determinable by some operation carried out by the observer, but others are not, being only his concepts.

4.3.2 Relationships. The relationships mentioned above are observed to inhere and interact in concrete, usually living, systems. In a sense, then, these concrete systems are the relationships of abstracted systems. The verbal usages of theoretical statements concerning abstracted systems are often the reverse of those concerning concrete systems: the nouns and their modifiers typically refer to relationships and the verbs and their modifiers (including predicates) to the concrete systems in which these relationships inhere and interact. These concrete systems are empirically determinable by some operation carried out by the observer. A theoretical statement oriented to concrete systems typically would say "Lincoln was President," but one oriented to abstracted systems, concentrating on relationships or roles, would very likely be phrased "The Presidency was occupied by Lincoln."[9]

An abstracted system differs from an *abstraction*, which is a concept (like those that make up conceptual systems) representing a class of phenomena all of which are considered to have some similar "class characteristic." The members of such a class are not thought to interact or be interrelated, as are the relationships in an abstracted system.

Abstracted systems are much more common in social science theory than in natural science.

Parsons has attempted to develop general behavior theory using abstracted systems. To some a social system is something concrete in space-time, observable and presumably measurable by techniques like those of natural science. To Parsons the system is abstracted from this, being the set of relationships which are the form of organization. To him the important units are classes of input-output relationships of subsystems rather than the subsystems themselves.

4.4 Abstracted versus Concrete Systems

One fundamental distinction between abstracted and concrete systems is that the boundaries of abstracted systems may at times be conceptually established at regions which cut through the units and relationships in the physical space occupied by concrete systems, but the boundaries of these latter systems are always set at regions which include within them all the units and internal relationships of each system.

A science of abstracted systems certainly is possible and under some conditions may be useful. When Euclid was developing geometry, with its practical applications to the arrangement of Egyptian real estate, it is probable that the solid lines in his figures were originally conceived to represent the borders of land areas or objects. Sometimes, as in Figure 1, he would use dotted "construction lines" to help conceptualize

FIGURE 1

A Euclidean Figure

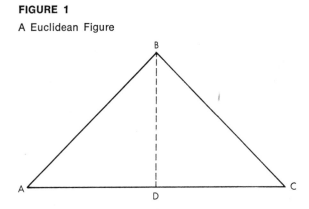

a geometric proof. The dotted line did not correspond to any actual border in space. Triangle *ABD* could be shown to be congruent to Triangle *CBD*, and therefore the angle BAD was equal to the angle BCD. After the proof was completed, the dotted line might well be erased, since it did not correspond to anything real and was useful only for the proof. Such construction lines, representing relationships among real lines, were used in the creation of early forms of abstracted systems.

If the diverse fields of science are to be unified, it would help if all disciplines were oriented either to concrete or to abstracted systems. It is of paramount importance for scientists to distinguish clearly between them. To use both kinds of systems in theory leads to unnecessary problems. It would be best if one type of system or the other were generally used in all disciplines.

All three meanings of "system" are useful in science, but confusion results when they are not differentiated. A scientific endeavor may appropriately begin with a conceptual system and evaluate it by collecting data on a concrete or on an abstracted system, or it may equally well first collect the data and then determine what conceptual system it fits. Throughout this article and the next the single word "system," for brevity, will always mean "concrete system." The other sorts of systems will always be explicitly distinguished as either "conceptual system" or "abstracted system."

5. STRUCTURE

The *structure* of a system is the arrangement of its subsystems and components in three-dimensional space at a given moment of time. This always changes over time.[10] It may remain relatively fixed for a long period or may change from moment to moment, depending upon the characteristics of the process in the system. This process halted at any given moment, as when motion is frozen by a high-speed photograph, reveals the three-dimensional spatial arrangement of the system's components as of that instant.

6. PROCESS

All change over time of matter-energy or information in a system is *process*. If the equation describing a process is the same no matter whether the temporal variable is positive or negative, it is a *reversible* process; otherwise it is *irreversible*. Process includes the ongoing *function* of a system, reversible actions succeeding each other from moment to moment. Process also includes *history*, less readily reversed changes like mutations, birth, growth, development, aging, and death; changes which commonly follow trauma or disease; and the changes resulting from learning which are not later forgotten. Historical processes alter both the structure and the function of the system. The statement "less readily reversed" has been used instead of "irreversible" (although many such changes are in fact irreversible) because structural changes sometimes can be reversed. E.g., a component which has developed and functioned may atrophy and finally disappear with disuse; a functioning part may be chopped off as a hydra and regrow. History, then, is more than the passage of time. It involves also accumulation in the system of residues or effects of past events (structural changes, memories, and learned habits). A living system carries its history with it in the form of altered structure, and consequently of altered function also. So there is a circular relation among the three primary aspects of systems—structure changes momentarily with functioning, but when such change is so great that it is essentially irreversible, a historical process has occurred, giving rise to a new structure.

7. TYPE

If a number of individual living systems are observed to have similar characteristics, they often are classed together as a *type*. Types are abstractions. Nature presents an apparently endless variety of living things which man, from his earliest days, has observed and classified—first, probably, on the basis of their threat to him, their susceptibility to capture, or their edibility, but eventually according to categories which are scientifically more useful. Classification by species is applied to organisms, plants or animals, or to free-living cells, because of their obvious relationships by reproduction. These systems are classified together by taxonomists on the basis of likeness of structure and process, genetic similarity and ability to interbreed, and local interaction, often including, in animals, ability to respond appropriately to each other's signs.

There are various types of systems at other levels of the hierarchy of living systems besides the cell and organism levels, each classed according to different structural and process taxonomic differentia. There are, for instance, primitive societies, agricultural societies, and industrial societies. There are epithelial cells, fibroblasts, red blood cells, and white blood cells, as well as free-living cells.

8. LEVEL

The universe contains a hierarchy of systems, each higher *level* of system being composed of systems of lower levels.[11] *Atoms* are composed of *particles; molecules,* of atoms; *crystals* and *organelles,* of molecules. About at the level of crystallizing *viruses,* like the tobacco mosaic virus, the subset of living systems begins. Viruses are necessarily parasitic on cells, so cells are the lowest level of living systems. *Cells* are composed of atoms, molecules, and multimolecular organelles; *organs* are composed of cells aggregated into *tissues; organisms,* of organs; *groups* (e.g., herds, flocks, families, teams, tribes), of organisms; *organizations,* of groups (and sometimes single individual organisms); *societies,* of organizations, groups, and individuals; and *supranational systems,* of societies and organizations. Higher levels of systems may be of mixed composition, living and nonliving. They include *planets, solar systems, galaxies,* and so forth. It is beyond the scope of this paper to deal with the characteristics—whatever they may be—of systems below and above those levels which include the various forms of life, although others have done so.[12] The subset of living systems includes cells, organs, organisms, groups, organizations, societies, and supranational systems.

It would be convenient for theorists if the hierarchical levels of living systems fitted neatly into each other like Chinese boxes. The facts are more complicated. No one can argue that there are exactly these seven levels, no more and no less. For example, one might conceivably separate tissue and organ into two separate levels. Or one might maintain that the organ is not a level, since no organ exists that can exist independent of other organs.

What are the criteria for distinguishing any one level from the others? They are derived from a long scientific tradition of empirical observation of the entire gamut of living systems. This extensive experience of the community of scientific observers has led to a consensus that there are certain fundamental forms of organization of living matter-energy. Indeed the classical division of subject-matter among the various disciplines of the life or behavioral sciences is implicitly or explicitly based upon this consensus.

It is important to follow one procedural rule in systems theory, in order to avoid confusion. Every discussion should begin with an identification of the level of reference, and the discourse should not change to another level without a specific statement that this is occurring.[13] Systems at the indicated level are called systems. Those at the level above are *suprasystems*, and at the next higher level, *suprasuprasystems*. Below the level of reference are *subsystems*, and below them *subsubsystems*. For example, if one is studying a cell, its organelles are the subsystems, and the tissue or organ is its suprasystem, unless it is a free-living cell whose suprasystem includes other living systems with which it interacts.[14]

8.1 Intersystem Generalization

A fundamental procedure in science is to make generalizations from one system to another on the basis of some similarity between the systems, which the observer sees and which permits him to class them together. For example, since the nineteenth century, the field of "individual differences" has been expanded, following the tradition of scientists like Galton in anthropometry and Binet in psychometrics. In Figure 2, states of separate specific individual systems on a specific

FIGURE 2

Individual, Type, Level

$$I_1 \cdots I_n$$
$$T_1 \cdots T_n$$
$$L_1 \cdots L_n$$

structural or process variable are represented by I_1 to I_n. For differences among such individuals to be observed and measured, of course, a variable common to the type, along which there are individual variations, must be recognized (T_1). Physiology depends heavily, for instance, upon the fact that individuals of the type (or species) of living organisms called cats are fundamentally alike, even though minor variations from one individual to the next are well recognized.

Scientists may also generalize from one type to another (T_1 to T_n). An example is cross-species generalization, which has been commonly accepted only since Darwin. It is the justification for the labors of the white rat in the cause of man's understanding of himself. Rats and cats, cats and chimpanzees, chimpanzees and human beings are similar in structure, as comparative anatomists know, and in function, as comparative physiologists and psychologists demonstrate.

The amount of variance among species is greater than among individuals within a species. If the learning behavior of cat Felix is compared with that of mouse Mickey, we would expect not only the sort of individual differences which are found between Mickey and Minnie Mouse, but also greater species differences. Cross-species generalizations are common, and many have good scientific acceptability, but in making them inter-individual and interspecies differences must be kept in mind. The learning rate of men is not identical to that of white rats, and no man learns at exactly the same rate as another.

The third type of scientific generalization indicated in Figure 2 is from one level to another. The basis for such generalization is the assumption that each of the levels of life, from cell to society, is composed of systems of the previous lower level. These cross-level generalizations will, ordinarily, have greater variance than the other sorts of generalizations, since they include variance among types and among individuals. But they can be made, and they can have great conceptual significance.

That there are important uniformities, which can be generalized about, across all levels of living systems is not surprising. All are composed of comparable carbon-hydrogen-nitrogen constituents, most importantly a score of amino acids organized into similar proteins, which are produced in nature only in living systems. All are equipped to live in a water-oxygen world rather than, for example, on the methane and ammonia planets so dear to science fiction. Also they are all adapted only to environments in which the physical variables, like temperature, hydration, pressure, and radiation, remain within relatively narrow ranges.[15] Moreover, they all presumably have arisen from the same primordial genes or template, diversified by evolutionary change. Perhaps the most convincing argument for the plausibility of cross-level generalization derives from analysis of this evolutionary development of living systems. Although increasingly complex types of living systems have evolved at a given level, followed by higher levels with even greater complexity, certain basic necessities did not change. All these systems, if they were to survive in their environment, had, by some means or other, to carry out the same vital subsystem processes. While free-living cells, like protozoans, carry these out with relative simplicity, the corresponding processes are more complex in multicellular organisms like mammals and even more complex at higher levels. The same processes are *"shredded out"* to multiple components in a more complex system, by the sort of *division of labor* which Parkinson has made famous as a law.[16] This results in formal identities across levels of systems, more complex subsystems at higher levels carrying out the same fundamental processes as simpler subsystems at lower levels.

A formal identity among concrete systems is demonstrated by a procedure composed of three logically independent steps: (*a*) recognizing an aspect of two or more systems which has comparable status in those systems; (*b*) hypothesizing a quantitative identity between them; and (*c*) demonstrating that identity within a certain range of error by collecting data on a similar aspect of each of the two or more systems being compared. It may be possible to formulate some useful generalizations which apply to all living systems at all levels. A comparison of systems is complete only when statements of their formal identities are associated with specific statements of their interlevel, intertype, and interindividual disidentities. The confirmation of formal identities and disidentities is done by research.

What makes interindividual, intertype, or interlevel formal identities among systems important and of absorbing interest, is that—if they can be conclusively demonstrated—very different structures, which carry out similar processes, may well turn out to carry out acts so much alike that they can be quite precisely described by the same formal model. Conversely, it may perhaps be shown as a general principle that subsystems with comparable structures but quite different processes may have quantitative similarities as well.

8.2 Emergents

The more complex systems at higher levels manifest characteristics, more than the sum of the characteristics of the units, not observed at lower levels. These characteristics have been called *emergents*. Significant aspects of living systems at higher levels will be neglected if they are described only in terms and dimensions used for their lower-level subsystems and components.

A clear-cut illustration of emergents can be found in a comparison of three electronic systems. One of these—a wire connecting the poles of a battery—can only conduct electricity, which heats the wire. Add several tubes, condensers, resistors, and controls, and the new system can become a radio, capable of receiving sound messages. Add dozens of other components, including a picture tube and several more controls, and the system becomes a television set which can receive sound and a picture. And this is not just more of the same. The third system has emergent capabilities the second system did not have, emergent from its special design of much greater complexity, just as the second has capabilities the first lacked. But there is nothing mystical about the colored merry-go-round and racing children on the TV screen—it is the output of a system which can be completely explained by a complicated set of differential equations such as electrical engineers write, including terms representing the characteristics of each of the set's components.

9. ECHELON

This concept may seem superficially similar to the concept of level, but is distinctly different. Many complex living systems, at various levels, are organized into two or more *echelons* (in the military sense of a step in the "chain of command," not in the other military sense of arrangement of troops in rows in physical space). In living systems with echelons the components of the decider, the decision-making subsystem, are hierarchically arranged so that usually certain types of decisions are made by one component of that subsystem and others by another. Each is an echelon. All echelons are within the boundary of the decider subsystem. Ordinarily each echelon is made up of components of the same level as those which make up every other echelon in that system. Characteristically the decider component at one echelon gets information from a source or sources which process information primarily or exclusively to and from that echelon. It may be that at some levels of living systems—e.g., cells—there are no cases in which the decider is organized in echelon structure.

After a decision is made at one echelon on the basis of the information received, it is transmitted, often through a single subcomponent which may or may not be the same as the decider, but possibly through more than one subcomponent, upward to the next higher echelon, which goes through a similar process, and so on to the top echelon. Here a final decision is made and then command information is transmitted downward to lower echelons. Characteristically information is abstracted or made more general as it proceeds upward from echelon to echelon and it is made more specific or detailed as it proceeds downward. If a given component does not decide but only passes on information, it is not functioning as an echelon. In some cases of decentralized decision-making, certain types of decisions are made at lower echelons and not transmitted to higher echelons in any form, while information relevant to other types of decisions is transmitted upward. If there are multiple parallel deciders, without a hierarchy that has subordinate and super-ordinate deciders, there is not one system but multiple ones.

10. SUPRASYSTEM

10.1 Suprasystem and Environment

The *suprasystem* of any living system is the next higher system in which it is a component or subsystem. For example, the suprasystem of a cell or tissue is the organ it is in; the suprasystem of an organism is the group it is in at the time. Presumably every system has a suprasystem except the "universe." The suprasystem is differentiated from the *environment*. The immediate environment is the suprasystem minus the system itself. The entire environment includes this plus the suprasuprasystem and the systems at all higher levels which contain it. In order to survive the system must interact with and adjust to its environment, the other parts of the suprasystem. These processes alter both the system and its environment. Living systems adapt to their environment, and in return mold it. The result is that, after some period of inter-action, each in some sense becomes a mirror of the other.

10.2 Territory

The region of physical space occupied by a living system, and fre-quently protected by it from an invader, is its territory.[17] Examples are a bowerbird's stage, a dog's yard, a family's property, a nation's land.

11. SUBSYSTEM AND COMPONENT

In every system it is possible to identify one sort of unit, each of which carries out a distinct and separate process, and another sort of unit, each of which is a discrete, separate structure. The totality of all the structures in a system which carry out a particular process is a *subsystem*. A subsystem, thus, is identified by the process it carries out. It exists in one or more identifiable structural units of the system. These specific, local, distinguishable structural units are called *components* or *members* or *parts*. Reference has been made to these com-ponents in the definition of a concrete system as "a nonrandom ac-cumulation of matter-energy, in a region in physical space-time, which is organized into interacting, interrelated subsystems or components." There is no one-to-one relationship between process and structure. One or more processes may be carried out by two or more components. Every system is a component, but not necessarily a subsystem of its suprasystem. Every component that has its own decider is a system at the next lower level, but many subsystems are not systems at the next lower level, being dispersed to several components.

The concept of subsystem process is related to the concept of *role* used in social science.[18] Organization theory usually emphasizes the functional requirements of the system which the subsystem fulfills, rather than the specific characteristics of the component or components that make up the subsystem. The typical view is that an organization specifies clearly defined roles (or component processes) and human beings "fill them."[19] But it is a mistake not to recognize that character-istics of the component—in this case the person carrying out the role —also influence what occurs. A role is more than simple "social posi-tion," a position in some social space which is "occupied." It involves interaction, adjustments between the component and the system. It is a multiple concept, referring to the demands upon the component by the system, to the internal adjustment processes of the component, and to how the component functions in meeting the system's requirements. The adjustments it makes are frequently compromises between the re-quirements of the component and the requirements of the system.

The way living systems develop does not always result in a neat dis-tribution of exactly one subsystem to each component. The natural arrangement would appear to be for a system to depend on one structure for one process. But there is not always such a one-to-one relationship.

Sometimes the boundaries of a subsystem and a component exactly overlap, are congruent. Sometimes they are not congruent. There can be (a) a single subsystem in a single component; (b) multiple subsystems in a single component; (c) a single subsystem in multiple comnents; or (d) multiple subsystems in multiple components.

Systems differ markedly from level to level, type to type, and perhaps somewhat even from individual to individual, in their *patterns of allocation* of various subsystem processes to different structures. Such process may be (a) localized in a single component; (b) combined with others in a single component; (c) dispersed laterally to other components in the system; (d) dispersed upwardly to the suprasystem or above; (e) dispersed downwardly to subsubsystems or below; or (f) dispersed outwardly to other systems outside the hierarchy it is in. Which allocation pattern is employed is a fundamental aspect of any given system. For a specific subsystem function in a specific system one strategy results in more efficient process than another. One can be better than another in maximizing effectiveness and minimizing costs. Valuable studies can be made at each level on optimal patterns of allocation of processes to structures. In all probability there are general systems principles which are relevant to such matters. Possible examples are: (a) Structures which minimize the distance over which matter-energy must be transported or information transmitted are the most efficient. (b) If multiple components carry out a process, the process is more difficult to control and less efficient than if a single component does it. (c) If one or more components which carry out a process are outside the system, the process is more difficult to integrate than if they are all in the system. (d) Or if there are duplicate components capable of performing the same process, the system is less vulnerable to stress and therefore is more likely to survive longer, because if one component is inactivated, the other can carry out the process alone.

11.1 Critical Subsystem

Certain processes are necessary for life and must be carried out by all living systems that survive or be performed for them by some other system. They are carried out by the following *critical subsystems* listed in Table 1.

The definitions of the critical subsystems are as follows:

11.1.1 Subsystems Which Process Both Matter-Energy and Information. *Reproducer*, the subsystem which is capable of giving rise to other systems similar to the one it is in. *Boundary*, the subsystem at the perimeter of a system that holds together the components which make up the system, protects them from environmental stresses, and excludes or permits entry to various sorts of matter-energy and information.

11.1.2 Matter-Energy Processing Subsystems. *Ingestor*, the subsystem which brings matter-energy across the system boundary from the environment. *Distributor*, the subsystem which carries inputs from

TABLE 1

The Critical Subsystems

Matter-Energy Processing Subsystems	Subsystems Which Process Both Matter-Energy and Information	Information Processing Subsystems
	Reproducer	
	Boundary	
Ingestor		Input transducer
		Internal Transducer
Distributor		Channel and net
Converter		Decoder
Producer		Associator
Matter-energy storage		Memory
		Decider
		Encoder
Extruder⎫ Motor ⎬ Supporter⎭		Output transducer

outside the system or outputs from its subsystems around the system to each component. *Converter,* the subsystem which changes certain inputs to the system into forms more useful for the special processes of that particular system. *Producer,* the subsystem which forms stable associations that endure for significant periods among matter-energy inputs to the system or outputs from its converter, the materials synthesized being for growth, damage repair, or replacement of components of the system, or for providing energy for moving or constituting the system's outputs of products or information markers to its suprasystem. *Matter-energy storage,* the subsystem which retains in the system, for different periods of time, deposits of various sorts of matter-energy. *Extruder,* the subsystem which transmits matter-energy out of the system in the forms of products and wastes. *Motor,* the subsystem which moves the system or parts of it in relation to part or all of its environment or moves components of its environment in relation to each other. *Supporter,* the subsystem which maintains the proper spatial relationships among components of the system, so that they can interact without weighting each other down or crowding each other.

11.1.3 Information Processing Subsystems. *Input transducer,* the sensory subsystem which brings markers bearing information into the system, changing them to other matter-energy forms suitable for transmission within it. *Internal transducer,* the sensory subsystem which receives, from all subsystems or components within the system, markers bearing information about significant alterations in those subsystems or components, changing them to other matter-energy forms of a sort which can be transmitted within it. *Channel and net,* the subsystem composed of a single route in physical space, or multiple interconnected routes, by which markers bearing information are transmitted to all parts of the system. *Decoder,* the subsystem which alters the code of information input to it through the input transducer or the internal transducer into a "private" code that can be used internally by the system. *Associator,* the subsystem which carries out the first stage of the learning process, forming enduring associations among items of information in the system. *Memory,* the subsystem which carries out the second stage of the learning process, storing various sorts of information in the system for different periods of time. *Decider,* the executive subsystem which receives information inputs from all other subsystems and transmits to them information outputs that control the entire system. *Encoder,* the subsystem which alters the code of information inputs to it from other information processing subsystems, from a "private" code used internally by the system into a "public" code which can be interpreted by other systems in its environment. *Output transducer,* the subsystem which puts out markers bearing information from the system, changing markers within the system into other matter-energy forms which can be transmitted over channels in the system's environment.

Of these critical subsystems only the decider is essential, in the sense that a system cannot be dependent on another system for its deciding. A living system does not exist if the decider is dispersed upwardly, downwardly, or outwardly.

Since all living systems are genetically related, have similar constituents, live in closely comparable environments, and process matter-energy and information, it is not surprising that they should have comparable subsystems and relationships among them. All systems do not have all possible kinds of subsystems. They differ individually, among type, and across levels, as to which subsystems they have and the structures of those subsystems. But all living systems either have a complement of the critical subsystems carrying out the functions essential to life or are intimately associated with and effectively interacting with systems which carry out the missing life functions for them.

11.2 Inclusion

Sometimes a part of the environment is surrounded by a system and totally included within its boundary. Any such thing not a part of the system's own living structure is an *inclusion.* Any living system at any level may include living or nonliving components. The amoeba, for example, ingests both inorganic and organic matter and may retain particles of iron or dye in its cytoplasm for many hours. A surgeon may replace an arteriosclerotic aorta with a plastic one and that patient may live comfortably with it for years. To the two-member group of one dog and one cat an important plant component is often added—one tree. An airline firm may have as an integral component a computerized mechanical system for making reservations which extends into all its offices. A nation includes many sorts of vegetables, minerals, buildings, and machines, as well as its land.

The inclusion is a component or subsystem of the system if it carries out or helps in carrying out a critical process of the system; otherwise it is part of the environment. Either way the system, to survive, must adjust to its characteristics. If it is harmless or inert it can often be left undisturbed. But if it is potentially harmful—like a pathogenic bacterium in a dog or a Greek in the giant gift horse within the gates of Troy—it must be rendered harmless or walled off or extruded from the system or killed. Because it moves with the system in a way the rest of the environment does not, it constitutes a special problem. Being inside the system it may be a more serious or more immediate stress than it would be outside the system's protective boundary. But also, the system that surrounds it can control its physical actions and all routes of access to it. For this reason international law has developed the concept of extraterritoriality to provide freedom of action to ambassadors and embassies, nations' inclusions within foreign countries.

11.3 Artifact

An *artifact* is an inclusion in some system, made by animals or man. Spider webs, bird nests, beaver dams, houses, books, machines, music, paintings, and language are artifacts. They may or may not be *prostheses*, inventions which carry out some critical process essential to a living system. An artificial pacemaker for a human heart is an example of an artifact which can replace a pathological process with a healthy one. Insulin and thyroxine are replacement drugs which are human artifacts. Chemical, mechanical, or electronic artifacts have been constructed which carry out some functions of all levels of living systems.

Living systems create and live among their artifacts. Beginning presumably with the hut and the arrowhead, the pot and the vase, the plow and the wheel, mankind has constructed tools and devised machines. The Industrial Revolution of the nineteenth century, capped by the recent harnessing of atomic energy, represents the extension of man's matter-energy processing ability, his muscles. A new Industrial Revolution, of even greater potential, is just beginning in the twentieth century, with the development of information and logic-processing machines, adjuncts to man's brain. These artifacts are increasingly becoming prostheses, relied on to carry out critical subsystem processes. A chimpanzee may extend his reach with a stick; a man may extend his cognitive skills with a computer. Today's prostheses include input transducers which sense the type of blood cells that pass before them and identify missiles that approach a nation's shores; photographic, mechanical, and electronic memories which can store masses of information over time; computers which can solve problems, carry out logical and mathematical calculations, make decisions, and control other machines; electric typewriters, high speed printers, cathode ray tubes, and photographic equipment which can output information. An analysis of many modern systems must take into account the novel problems which arise at man-machine interfaces.

Music is a special sort of human artifact, an information-processing artifact.[20] So are the other arts and cognitive systems which people share. So is language. Whether it be a natural language or the machine language of some computer system, it is essential to information processing. Often stored only in human brains and expressed only by human lips, it can also be recorded on nonliving artifacts like stones, books, and magnetic tapes. It is not of itself a concrete system. It changes only when man changes it. As long as it is used it is in flux, because it must remain compatible with the ever-changing living systems that use it. But the change emanates from the users, and without their impact the language is inert. The artifactual language used in any information transmission in a system determines many essential aspects of that system's structure and process.[21]

12. TRANSMISSIONS IN CONCRETE SYSTEMS

All process involves some sort of transmission among subsystems within a system, or among systems. There are *inputs* across the boundary into a system, *internal processes* within it, and *outputs* from it. Each of these sorts of transmissions may consist of either (*a*) some particular form of matter; (*b*) energy, in the form of light, radiant energy, heat, or chemical energy; or (*c*) some particular pattern of information.

13. STEADY STATE

When opposing variables in a system are in balance, that system is in equilibrium with regard to them. The equilibrium may be static and unchanging or it may be maintained in the midst of dynamic change. Since living systems are open systems, with continually altering fluxes of matter-energy and information, many of their equilibria are dynamic and are often referred to as *flux equilibria* or *steady states*. These may be *unstable*, in which a slight disturbance elicits progressive change from the equilibrium state—like a ball standing on an inverted bowl; or *stable*, in which a slight disturbance is counteracted so as to restore the previous state—like a ball in a cup; or *neutral*, in which a slight disturbance makes a change, but without cumulative effects of any sort—like a ball on a flat surface with friction.

All living systems tend to maintain steady states (or homeostasis) of many variables, keeping an orderly balance among subsystems which process matter-energy or information. Not only are subsystems usually kept in equilibrium, but systems also ordinarily maintain steady states with their environments and suprasystems, which have outputs to the systems and inputs from them. This prevents variations in the environment from destroying systems. The variables of living systems are constantly fluctuating, however. A moderate change in one variable may produce greater or lesser alterations in other related ones. These alterations may or may not be reversible.

13.1 Stress, Strain, and Threat

There is a *range of stability* for each of numerous variables in all living systems. It is that range within which the rate of correction of deviations is minimal or zero, and beyond which correction occurs. An input or output of either matter-energy or information, which by lack or excess of some characteristic, forces the variables beyond the range of stability, constitutes *stress* and produces a *strain* (or strains) within the system. Input lack and output excess both produce the same strain —diminished amounts in the system. Input excess and output lack both produce the opposite strain—increased amounts. Strains may or may not be capable of being reduced, depending upon their intensity and the resources of the system. The totality of the strains within a system resulting from its template program and from variations in the inputs from its environment can be referred to as its *values*. The relative urgency of reducing each of these specific strains represents its *hierarchy of values*.

Stress may be anticipated. Information that a stress is imminent constitutes a *threat* to the system. A threat can create a strain. Recognition of the meaning of the information of such a threat must be based on previously stored (usually learned) information about such situations. A pattern of input information is a threat when—like the odor of the hunter on the wind; a change in the acidity of fluids around a cell; a whirling cloud approaching the city—it is capable of eliciting processes which can counteract the stress it presages. Processes— actions or communications—occur in systems only when a stress or a threat has created a strain which pushes a variable beyond its range of stability. A system is a constantly changing cameo and its environment is a similarly changing intaglio, and the two at all times fit each other. That is, outside stresses or threats are mirrored by inside strains. Matter-energy storage and memory also mirror the past environment, but with certain alterations.

13.1.1 Matter-energy stress.   There are various ways for systems to be stressed. One class of stresses is the *matter-energy stresses*, including: (*a*) matter-energy input lack or underload—starvation or inadequate fuel input; (*b*) input of an excess or overload of matter-energy; and (*c*) restraint of the system, binding it physically. [This may be the equivalent of (*a*) or (*b*).]

13.1.2 Information stress.   Also there are *information stresses*, including: (*a*) information input lack or underload, resulting from a dearth of information in the environment or from improper function of the external sense organs or input transducers; (*b*) injection of noise into the system, which has an effect of information cutoff, much like the previous stress; and (*c*) information input excess or overload. Informational stresses may involve changes in the rate of information input or in its meaning.

13.2 Adjustment Processes

Those processes of subsystems which maintain steady states in systems, keeping variables within their ranges of stability despite stresses, are *adjustment processes*. In some systems a single variable may be influenced by multiple adjustment processes. As Ashby has pointed out, a living system's adjustment processes are so coupled that the system is ultrastable.[22] This characteristic can be illustrated by the example of an army cot. It is made of wires, each of which would break under a 300-pound weight, yet it can easily support a sleeper of that weight. The weight is applied to certain wires, and as it becomes greater, first nearby links and then those farther and farther away, take up part of the load. Thus a heavy weight which would break any of the component wires alone can be sustained. In a living system, if one component cannot handle a stress, more and more others are recruited to help. Eventually the entire capacity of the system may be involved in coping with the situation.

13.2.1 Feedback. The term *feedback* means that there exist two channels carrying information, such that Channel B loops back from the output to the input of Channel A and transmits some portion of the signals emitted by Channel A (see Figure 3).[23] These are tell-tales or

FIGURE 3

Negative Feedback

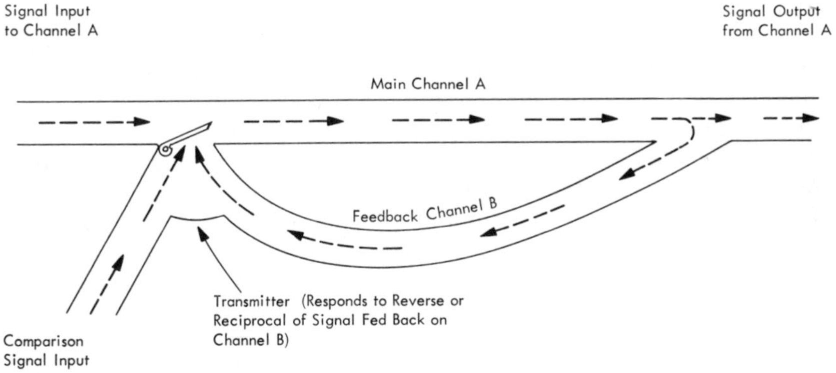

Signal Input
to Channel A

Signal Output
from Channel A

Main Channel A

Feedback Channel B

Transmitter (Responds to Reverse or
Reciprocal of Signal Fed Back on
Channel B)

Comparison
Signal Input

monitors of the outputs of Channel A. The transmitter on Channel A is a device with two inputs, formally represented by a function with two independent variables, one the signal to be transmitted on Channel A and the other a previously transmitted signal fed back on Channel B. The new signal transmitted on Channel A is selected to decrease the strain resulting from any error or deviation in the feedback signal from a criterion or comparison reference signal indicating the state of the output of Channel A which the system seeks to maintain steady. This provides control of the output of Channel A on the basis of actual rather than expected performance.

When the signals are fed back over the feedback channel in such a manner that they increase the deviation of the output from a steady state, *positive feedback* exists. When the signals are reversed, so that they decrease the deviation of the output from a steady state, it is *negative feedback*. Positive feedback alters variables and destroys their steady states. Thus it can initiate system changes. Unless limited, it can alter variables enough to destroy systems. At every level of living systems numerous variables are kept in a steady state, within a range of stability, by negative feedback controls. When these fail, the structure and process of the system alter markedly—perhaps to the extent that the system does not survive. Feedback control always exhibits some oscillation and always has some lag. When the organism maintains its balance in space, this lag is caused by the slowness of transmissions in the nervous system, but is only of the order of hundredths of seconds. An organization, like a corporation, may take hours to correct a breakdown in an assembly line, days or weeks to correct a bad management decision. In a society the lag can sometimes be so great that, in effect, it comes too late. General staffs often plan for the last war rather than the next. Governments receive rather slow official feedbacks from the society at periodic elections. They can, however, get faster feedbacks from the press, other mass media, picketers, or demonstrators. Public opinion surveys can accelerate the social feedback process. The speed and accuracy of feedback have much to do with the effectiveness of the adjustment processes they mobilize.

13.2.2 Power. In relation to energy processing, *power* is the rate at which work is performed, work being calculated as the product of a force and the distance through which it acts. The term also has another quite different meaning. In relation to information processing, *power* is control, the ability of one system to elicit compliance from another, at the same or a different level. A system transmits a command signal or message to a given address with a signature identifying the transmitter as a legitimate source of command information. The message is often in the imperative mode, specifying an action the receiver is expected to carry out. It elicits compliance at the lower levels because the electrical or chemical form of the signal sets off a specific reaction. At higher levels the receiving system is likely to comply because it has learned that the transmitter is capable of evoking rewards or punishments from the suprasystem, depending on how the receiver responds.

13.2.3 Purpose and Goal. By the information input of its charter or genetic input, or by changes in behavior brought about by rewards and punishments from its suprasystem, a system develops a preferential hierarchy of values that gives rise to decision rules which determine its preference for one internal steady state value rather than another. This is its *purpose*. It is the comparison value which it matches to information received by negative feedback in order to determine whether the variable is being maintained at the appropriate steady state value. In this sense it is normative. The system then takes one alternative action rather than another because it appears most likely to maintain the steady state. When disturbed, this state is restored by the system by successive approximations, in order to relieve the strain of the disparity recognized internally between the feedback signal and the comparison signal. Any system may have multiple purposes simultaneously.

A system may also have an external *goal,* such as reaching a target in space, or developing a relationship with any other system in the environment. Or it may have several goals at the same time. Just as there is no question that a guided missile is zeroing in on a target, so there is no question that a rat in a maze is searching for the goal of food at its end or that the Greek people under Alexander the Great were seeking the goal of world conquest. As Ashby notes, natural selection permits only those systems to continue which have goals that enable them to survive in their particular environments.[24] The external goal may change constantly, as when a hunter chases a moving fox or a man searches for a wife by dating one girl after another, while the internal purpose remains the same.

A system's hierarchy of values determines its purposes as well as its goals. It is not difficult to distinguish purposes from goals, as the terms have been used: An amoeba has the purpose of maintaining adequate energy levels and therefore it has the goal of ingesting a bacterium; a boy has the purpose of keeping his body temperature in the proper range and so he has the goal of finding and putting on his sweater; Poland had the purpose in March 1939 of remaining uninvaded and autonomous and so she sought the goal of a political alliance with Britain and France in order to have assistance in keeping both Germany and Russia from crossing her borders.

13.2.4 Costs and Efficiency. All adjustment processes have their *costs,* in energy of nonliving or living systems, in material resources, in information (including in social systems a special form of information often conveyed on a marker of metal or paper money), or in time required for an action. Any of these may be scarce. (Time is a scarcity for mortal living systems.) Any of these is valued if it is essential for reducing strains. The costs of adjustment processes differ from one to another and from time to time. They may be immediate or delayed, short-term or long-term.

How successfully systems accomplish their purposes can be determined if those purposes are known. A system's *efficiency,* then, can be determined as the ratio of the success of its performance to the costs involved. A system constantly makes economic decisions directed toward increasing its efficiency by improving performance and decreasing costs. How efficiently a system adjusts to its environment is determined by what strategies it employs in selecting adjustment processes and whether they satisfactorily reduce strains without being too costly. This decision process can be analyzed by game theory, a mathematical approach to economic decisions. This is a general theory concerning the best strategies for weighing "plays" against "pay-offs," for selecting actions which will increase profits while decreasing losses, increase rewards while decreasing punishments, improve adjustments of variables to appropriate steady state values, or attain goals while diminishing costs. Relevant information available to the decider can improve such decisions. Consequently such information is valuable. But there are costs to obtaining such information. A mathematical theory on how to calculate the value of relevant information in such decisions was developed by Hurley.[25] This depends on such considerations as whether it is tactical (about a specific act) or strategic (about a policy for action); whether it is reliable or unreliable, overtly or secretly obtained, accurate, distorted, or erroneous.

14. HYPOTHESES

A large number of general hypotheses which apply to two or more levels can be stated about structure and process of living systems. These are propositions concerning cross-level formal identities which can be demonstrated empirically. A few out of many such hypotheses that could be stated are listed below. These are selected because they are referred to in the next article.* Many of the hypotheses were suggested by the work of others, though usually the others thought of them as related to one level only and not as general systems hypotheses. It must be remembered that when they are applied to a specific system, allowance must be made for the disanalogies among systems. The variables involved show regular changes with level and type of system, and from one individual system of the same level and type to another. The hypotheses are numbered by the same procedure, distinguishing the various subsystems processing matter-energy and information and separating structure and process, which I employ in numbering the sections of the next article.

Hypothesis 2–3. The more isolated a system is the more totipotential it must be.

Hypothesis 3.1.2.2–1. When the boundary (except those portions containing the openings for the ingestor or the extruder) of one living system, *A*, is crossed by another, smaller living or nonliving system, *B*, of significant size, i.e., no smaller than the subsystems or subcomponents of *A*, more work must be expended than when *B* is transmitted over the same distance in space immediately inside or outside the boundary of *A*.

Hypothesis 3.1.2.2–2. More work must be expended in moving the marker bearing an information transmission over the boundary of a system at the input transducer than in making such a transmission

* Editor's Note: This article was originally published in *Behavioral Science* as the first in a series of articles in which a common numbering system for sections was employed throughout the series, each successive paper dealing with various levels of living systems, e.g., group, organization, etc. For the first general article, which appears in this volume, Miller explained in an editorial preface: "At each level there are scientists who apply systems theory in their investigations. They are systems theorists but not necessarily general systems theorists. They are general systems theorists only if they accept the more daring and controversial position that—though every living system and every level is obviously unique—there are important formal identities of large generality across levels. These can potentially be evaluated quantitatively, applying the same model to data collected at two or more levels. This possibility is the chief reason why the author has used the same outline with identically numbered sections to analyze the present knowledge about each of the seven levels of living systems."

over the same distance in the suprasystem immediately outside the boundary or in the system immediately inside it.

Hypothesis 3.1.2.2–3. The amount of information transmitted between points within a system is significantly larger than the amount transmitted across its boundary.

Hypothesis 3.3.3.1–1. The structures of the communication networks of living systems at various levels are so comparable that they can be described by similar mathematical models of nonrandom nets.

Hypothesis 3.3.3.2–2. There is always a constant systematic distortion between input and output of information in a channel.

Hypothesis 3.3.3.2–4. A system never completely compensates for the distortion in information flow in its channels.

Hypothesis 3.3.3.2–9. Use of multiple parallel channels to carry identical information, which farther along in the net can be compared for accuracy, is commoner in more essential components of a system than in less essential ones.

Hypothesis 3.3.3.2–17. When a channel has conveyed one signal or message, its use to convey others is more probable.

Hypothesis 3.3.4.2–1. As a system matures it uses increasingly efficient codes, e.g., codes which require fewer binary digits or equivalent signals per input signal. These codes approach but never actually reach the theoretical minimum number of symbols required to transmit the information. Efficient codes also have the following characteristics:

a) Simple symbols are used for the most frequent messages and more complex ones for the less frequent ones.

b) The symbols are selected to minimize confusion among them.

c) The symbols are chunked in long rather than short blocks.

d) Limitations on the transmitter of the signal are taken into account. E.g., if it transmits highly redundant signals, each one is not coded, but some of the redundancy is removed.

e) Limitations on the receiver are taken into account. E.g., distinctions to which the receiver cannot react are neglected.

Hypothesis 3.3.4.2–2. If a transmitter of information is putting out information coded to have H bits per symbol and a channel has a capacity (in bits per second) of C, then the channel cannot transmit at a rate faster than C/H symbols per second, though it is possible to encode the message so as to transmit at a rate of $C/H—\epsilon$ symbols per second, where ϵ is a positive fraction, less than one and usually small, of C.

Hypothesis 3.3.4.2–3. The quantity ϵ (see Hypothesis 3.3.4.2–2) decreases as a system matures and associates, gaining practice in coding information.

Hypothesis 3.3.4.2–4. If a transmitter with an information transmisson rate (in bits per second) of R is transmitting over a noisy channel—and all living channels are noisy—with a capacity (in bits per second) of C, and if R is less than C, there is a code which can make the transmission almost free of errors, and as the system matures and associates, gaining practice, it gradually approaches such transmission.

Hypothesis 3.3.4.2–6. As the noise in a channel increases, a system encodes with increasing redundancy in order to reduce error in the transmission.

Hypothesis 3.3.4.2–7. If messages are so coded that they are transmitted twice, errors can be detected by comparing every part of the first message with every part of the second, but which of the two alternative transmissions is correct cannot be determined. If they are transmitted three times, they can be both detected and corrected by accepting the alternative on which two of the three transmissions agree.

Hypothesis 3.3.4.2–8. Over time a system tends to decrease the amount of recoding necessary within it, by developing more and more common system-wide codes.

Hypothesis 3.3.4.2–9. As the amount of information in an input decreases (i.e., as it becomes more ambiguous), the input will more and more tend to be interpreted (or decoded) as required to reduce strains within the system.

Hypothesis 3.3.4.2–10. As the strength of a strain increases, information inputs will more and more be interpreted (or decoded) as required to reduce the strain.

Hypothesis 3.3.5.2–3. A system does not form associations without (*a*) feedback as to whether the new output relieves strains or solves problems, and (*b*) reinforcement, i.e., strain reduction by the output.

Hypothesis 3.3.5.2–6. In general, association is slower the higher the level of the system.

Hypothesis 3.3.6.2–1. The longer information is stored in memory, the harder it is to recall and the less likely it is to be correct but the rate of loss is not regular over time.

Hypothesis 3.3.7.2–1. Every adaptive decision is made in four stages: (*a*) establishing the purpose or goal whose achievement is to be advanced by the decision; (*b*) analyzing the information relevant to the decision; (*c*) synthesizing a solution selecting the alternative action or actions most likely to lead to the purpose or goal; and (*d*) issuing a command signal to carry out the action or actions.

Hypothesis 3.3.7.2–16. The deciders of a system's subsystems and components satisfice (i.e., make a sufficiently good approximation to accomplishment in order to survive in its particular environment) shorter-term goals than does the decider of the total system.

Hypothesis 3.3.7.2–21. The higher the level of a system the more correct or adaptive its decisions are.

Hypothesis 5.1–42. A minimum rate of information input to a system must be maintained for it to function normally.

Hypothesis 5.2–2. The greater a threat or stress upon a system, the more components of it are involved in adjusting to it. When no further components with new adjustment processes are available, the system function collapses.

Hypothesis 5.2–3. When variables in a system return to a steady state after stress, the rate of return and the strength of the restorative forces are functions, with increasing first derivatives greater than 1, of the amount of displacement from the range of stability.

Hypothesis 5.2–6. Positive feedback produces continuous increments of outputs which give rise to "spiral effects" destroying one or more equilibria of a system.

Hypothesis 5.2–7. When a barrier stands between a system under strain and a goal which can relieve that strain, the system ordinarily uses the adjustment processes of removing the barrier, circumventing it, or otherwise mastering it. If these efforts fail, less adaptive adjustments may be tried, including: (*a*) attacking the barrier by energic or informational transmissions; (*b*) displacing aggression to another innocent but more vulnerable nearby system; (*c*) reverting to primitive, nonadaptive behavior; (*d*) adopting rigid, nonadaptive behavior; and (*e*) escaping from the situation.

Hypothesis 5.2–13. Under threat or stress, a system that survives, in the common good of total system survival, temporarily subordinates conflicts among subsystems or components until the threat or stress is relieved, when internal conflicts recur.

Hypothesis 5.2–26. If a system has multiple purposes and goals and they are not placed in clear priority and commonly known by all components or subsystems, conflict among them will ensue.

Hypothesis 5.4.1–2. Growing systems develop in the direction of: (*a*) more differentiation of subsystems; (*b*) more decentralization of decision making; (*c*) more interdependence of subsystems; (*d*) more elaborate adjustment processes; (*e*) sharper subsystem boundaries; (*f*) increased differential sensitivity to inputs; and (*g*) more elaborate and patterned outputs.

Hypothesis 5.4.3–1. For the same level of system output, more transmission of information is necessary to coordinate segregated systems than integrated systems.

Hypothesis 5.4.3–7. Up to a certain level of stress, systems do more centralized deciding when under stress than when not under stress. Beyond that level deciding becomes increasingly decentralized until the system terminates or the stress abates.

Hypothesis 5.4.3–8. A component will comply with a system's purposes and goals to the extent that those functions of the component directed toward the goal are rewarded and those directed away from it are punished.

Hypothesis 5.6–1. If a system's negative feedback discontinues and is not restored by that system or by another on which it becomes parasitic or symbiotic, it decomposes into multiple components and its suprasystem assumes control of them.

15. CONCLUSIONS

This analysis of living systems uses concepts of thermodynamics, information theory, cybernetics, and systems engineering, as well as the classical concepts appropriate to each level. The purpose is to produce a description of living structure and process in terms of input and output, flows through systems, steady states, and feedbacks, which will clarify and unify the facts of life. The approach generates hypotheses relevant to single individuals, types, and levels of living systems, or relevant across individuals, types, and levels. These hypotheses can be confirmed, disconfirmed, or evaluated by experiments and other empirical evidence.

REFERENCES AND NOTES

1. Hartley, R. V. L.   Transmission of Information. *Bell Sys. Tech. J.*, 1928, 7, 535.

2. Shannon, C. E.   A Mathematical Theory of Communication. *Bell Sys. tech. J.*, 1948, 27, 379–423 and 623–56.

3. von Neumann, J.   *The Computer and the Brain.* New Haven, Conn.: Yale Univ. Press, 1958, 6–7.

    NOTE: Christie, Luce, and Macy (Christie, L. S., Luce, R. D., & Macy, J., Jr.   *Communication and Learning in Task-Oriented Groups.* Cambridge, Mass.: Research Lab. of Electronics, MIT, Tech. Rep. No. 231, May 13, 1952) call the physical form which the communication takes the "symbol design," and the information itself the "symbol contents."

4. Shannon, C. E.   Op. cit., 380–82.

5. Harlow, H. F. & Harlow, M. K.   Social Deprivation in Monkeys. *Sci. Amer.*, 1962, 207(5), 137–46.

6. Bertalanffy, L. v.   General Systems Theory. *Yearb. Soc. Gen. Sys. Res.* 1956, 1, 3.

    NOTE: Bertalanffy suggests that systems can be defined much as I define them, as "sets of elements standing in interaction." And he says that this definition is not so vague and general as to be valueless. He believes these systems can be specified by families of differential equations.

7. NOTE: Rothstein, J.   *Communication, Organization, and Science.* Indian Hills, Colo.: Falcon's Wing Press, 1958, 34–36, deals with the constraints among units of organized systems in terms of entropy and communication as information processing:

    "What do we mean by an organization? First of all an organization presupposes the existence of parts, which, considered in their totality, constitute the organization. The parts must interact. Were there no communication between them there would be no organization, for we would merely have a collection of individual elements isolated from each other. Each element must be associated with its own set of alternatives. Were there no freedom to choose from a set of alternatives, the corresponding element would be a static, passive cog rather than an active unit. We suggest the following general characterization of organization. Consider a set of elements, each associated with its own set of alternatives. We now define a complexion as a particular set of alternatives. There are, of course, as many complexions as there are ways of selecting a representative from each set of alternatives. The set of complexions then has an entropy which is merely the sum of the entropies of the individual sets of alternatives so long as the elements do not interact. Complexion entropy is a maximum for independent elements. Maximal entropy, i.e., zero coupling, will be said to constitute the condition of zero organization."

Ashby [Ashby, W. R. Principles of the Self-Organizing System. In H. von Foerster & G. W. Zopf (Eds.). *Principles of Self-Organization*. New York: Pergamon Press, 1962, 255–57] also deals with this. He says, speaking of what "organization" means as applied to systems, "The hard core of the concept is, in my opinion, that of 'conditionality.' As soon as the relation between two entities A and B become conditional on C's value or state then a necessary component of 'organization' is present. Thus *the theory of organization is partly co-extensive with the theory of functions of more than one variable.*"

He goes on to ask when a system is not a system or is not organized:

"The converse of 'conditional on,' is 'not conditional on,' so the converse of 'organization' must therefore be, as the mathematical theory shows as clearly, the concept of 'reducibility.' (It is also called 'separability.') This occurs, in mathematical forms, when what looks like a function of several variables (perhaps very many) proves on closer examination to have parts whose actions are *not* conditional on the values of the other parts. It occurs in mechanical forms, in hardware, when what looks like one machine proves to be composed of two (or more) sub-machines, each of which is acting independently of the others. . . .

"The treatment of 'conditionality' (whether by functions of many variables, by correlation analysis, by uncertainty analysis, or by other ways) makes us realize that the essential idea is that there is first a product space—that of the possibilities—within which some sub-set of points indicates the actualities. This way of looking at 'conditionality' makes us realize that it is related to that of 'communication,' and it is, of course, quite plausible that we should define parts as being 'organized' when 'communication' (in some generalized sense) occurs between them. (Again the natural converse is that of independence, which represents non-communication.)"

"Now 'communication' from A to B necessarily implies some constraint, some correlation between what happens at A and what at B. If, for a given event at A, all possible events may occur at B, then there is no communication from A to B and no constraint over the possible (A, B) couples that can occur. Thus the presence of 'organization' between variables is equivalent to the existence of a constraint in product-space of the possibilities."

8. See Hall, A. D. & Fagen, R. E. Definition of System. *Yearb. Soc. Gen. Sys. Res.*, 1956, 1, 18.

9. NOTE: In Cervinka, V., A Dimensional Theory of Groups, *Sociometry*, 1948, 11, 100–107, the author very precisely distinguishes, at the group level, between a concrete system, which he calls a "socius," that is a single person in a group together with all his relationships, and a "groupoid," an abstracted system, which is a pattern of attachments of a single kind of relation selected by an observer, which interrelates a set of people.

10. NOTE: This definition is consistent with the usage of Weiss [Weiss, P. A. In R. W. Gerard (Ed.). Concepts of Biology. *Behav. Sci.*, 1958, 3, 140]. Murray [Murray, H. A. Preparations for a scaffold of comprehensive system. In S. Koch (Ed.). *Psychology: A Study of a*

Science, Vol. 3. New York: McGraw-Hill, 1959, 24] prefers the word "configuration" for an instantaneous spatial arrangement of subsystems or components of a system (or "entity," in his terms) and "structure" for an enduring arrangement. He distinguishes these clearly from an "integration" of recurrent temporal relations of component processes, a patterning of temporal variables.

11. NOTE: This concept is not a product of our times. It developed long ago. For instance, in the middle of the Nineteenth Century, Virchow (Virchow, R. Atome und Individuen, *Vier Reden Über Leben und Krankstein*. Berlin, 1862. Trans. by L. J. Rather as: Atoms and individuals. In *Disease, Life, and Man, Selected Essays by Rudolph Virchow*. Stanford, Calif.: Stanford Univ. Press, 1958, 120–141) wrote that the scope of the life sciences must include the cellular, tissue, organism, and social levels of living organization. In modern times the concept of hierarchical levels of systems is, of course, basic to the thought of Bertalanffy and other general systems theorists (see Bertalanffy, L. v. Op. cit., 7). Even some scientists not explicitly of such persuasion, who have perhaps been skeptical in the past (see Simon, H. A. The architecture of complexity. *Proc. Amer. Phil. Soc.*, 1962, 106, 467), recognize value in such an approach. For example, Simon (Ibid., 467–68) writes:

"A number of proposals have been advanced in recent years for the development of 'general systems theory' which, abstracting from properties peculiar to physical, biological, or social systems, would be applicable to all of them. We might well feel that, while the goal is laudable, systems of such diverse kinds could hardly be expected to have any nontrivial properties in common. Metaphor and analogy can be helpful, or they can be misleading. All depends on whether the similarities the metaphor captures are significant or superficial.

"It may not be entirely vain, however, to search for common properties among diverse kinds of complex systems. The ideas that go by the name of cybernetics constitute, if not a theory, at least a point of view that has been proving fruitful over a wide range of applications. It has been useful to look at the behavior of adaptive systems in terms of the concepts of feedback and homeostasis, and to analyze adaptiveness in terms of the theory of selective information. The ideas of feedback and information provide a frame of reference for viewing a wide range of situations, just as do the ideas of evolution, of relativism, of axiomatic method, and of operationalism."

He goes on to assert that "hierarchic systems have some common properties that are independent of their specific content. . . .

"By a hierarchic system, or hierarchy, I mean a system that is composed of interrelated subsystems, each of the latter being, in turn, hierarchic in structure until we reach some lowest level of elementary subsystem. In most systems in nature, it is somewhat arbitrary as to where we leave off the partitioning, and what subsystems we take as elementary. Physics makes much use of the concept of 'elementary particle' although particles have a disconcerting tendency not to remain elementary very long. Only a couple of generations ago, the atoms themselves were elementary particles; today, to the nuclear physicist they are complex systems. For certain purposes of astronomy, whole stars, or even galaxies, can be regarded as elementary subsystems. In one kind of biological research, a cell may be treated as an elementary subsystem; in another, a protein molecule; in still another, an amino acid residue.

"Just why a scientist has a right to treat as elementary a subsystem that is in fact exceedingly complex is one of the questions we shall take up. For the moment, we shall accept the fact that scientists do this all the time, and that if they are careful scientists they usually get away with it."

Leake sees value in the concept of levels for contemporary theory about biological organization (see Leake, C. D. The Scientific Status of Pharmacology. *Science,* 1961, 134, 2076). He writes:

"Life begins with complex macromolecules such as genes and viruses, and here the principles of physics and chemistry directly apply. Macromolecules may be organized and integrated with many other chemical materials to form cells, which at Virchow's time were thought to be the basic units of life. Cells, however, may be organized into tissues or organs, with specific integrations serving their specific functions. These tissues and organs may further be integrated into organisms, constituting individuals such as human beings. Human beings, and indeed many other organisms, are capable of further integration and organization into societies. These societies in turn may be integrated with a more or less limited ecological environment."

The view is also well stated by de Chardin (de Chardin, P. T. *The Phenomenon of Man.* New York: Harper, 1959, 43–44):

"The existence of 'system' in the world is at once obvious to every observer of nature, no matter whom.

"The arrangement of the parts of the universe has always been a source of amazement to men. But this disposition proves itself more and more astonishing as, every day, our science is able to make a more precise and penetrating study of the facts. The farther and more deeply we penetrate into matter, by means of increasingly powerful methods, the more we are confounded by the interdependence of its parts. Each element of the cosmos is positively woven from all the others; from beneath itself by the mysterious phenomenon of 'composition,' which makes it subsistent throughout the apex of an organized whole; and from above through the influence of unities of a higher order which incorporate and dominate it from their own ends.

"It is impossible to cut into this network, to isolate a portion without it becoming frayed and unravelled at all its edges.

"All around us, as far as the eye can see, the universe holds together, and only one way of considering it is really possible, that is, to take it as a whole, in one piece."

Kaplan (Kaplan, M. A. *System and Process in International Politics.* New York: Wiley, 1957, 12) has applied the concept of a hierarchy of systems to international relations:

"The same variables will be used at different system levels. The international system is the most inclusive system treated by this book. National and supranational systems are subsystems of the international system. They may, however, be treated separately as systems, in which case inputs from the international system would function as parameters. This holds also for subsystems of nation states and even for personality systems."

The Panel on Basic Research and Graduate Education of the President's Science Advisory Committee of the United States in 1960 appeared also to recognize value in a general systems approach [see Seaborg, G. T. (Chairman) Panel on Basic Research and Graduate Education of the President's Science Advisory Committee. Scientific progress and the federal government. *Science,* 1960, 132, 1810]. They wrote: ". . . we suggest that there is great promise in such an emerging subject as a general study of complex systems in action, within which such very large questions as the communication sciences, cognition, and large parts of biology itself might conceivably be treated as special cases."

A textbook of psychology has been written which embodies a conceptualization of a hierarchy of living systems like that I advance in the present work (see Coleman, J. C. *Personality Dynamics and Effective Behavior.* Chicago: Scott-Foresman, 1960).

A presidential address of the Association of American Medical Colleges included a passage emphasizing the desirability of synthesizing the medical curriculum around the concept of the relations among levels of living systems. Hubbard (Hubbard, W. N., Jr. Janus revisited. *J. Med. Educ.,* 1967, 42, 1079) wrote: "For the medical student . . . the significance of descriptions at the molecular and submolecular level must be presented in the context of their relationship to the more complex organizations of these same living systems at the level of the organ, the individual, and the family group."

And there is widespread scientific and popular implicit recognition of hierarchical levels of living systems. As one instance out of many, six banners in one of the halls of the United Nations Palais des Nations in Geneva depict six levels of social organization. They say: Family, Village, Clan, Medieval State, Nation, and Federation.

12. Newman, J. & Scott, E. L. On a Mathematical Theory of Populations Conceived as Conglomerations of Clusters. *Yearb. Soc. Gen. Sys. Res.,* 1958, 3, 180–92.
      Also J. Newman, E. L. Scott & C. D. Shane. Statistics of Images of Galaxies with Particular Reference to Clustering. *Yearb. Soc. Gen. Sys. Res.,* 1958, 3, 193–219.

13. Cf. Herbert, P. G. Situation dynamics and the theory of behavior systems. *Behav. Sci.,* 1957, 2, 28. Herbert makes it clear that one should make the level of reference explicit. He says that often, in writing on group research, for instance, an author will change his level of reference from the leader (organism) to the group and back to a group member (organism) again without explicitly referring to the change. This produces confusing conceptual ambiguity.

14. NOTE: Illustrative of the similarities between the approach outlined here and current thinking about electronic system design is the following statement by Goode (Goode, H. H. Intracompany Systems Management. *IRE Trans. Engng. Mgmt.,* EM-7, 1960, 15) concerning the need to identify the level of reference:
      "Confusion . . . arises from consideration of the level of design. System design may be done:

1) At the *set* level: that is, a radar, an ignition system, a navigation set. Any of these may be designed on a system engineering basis, given a need and the necessary analysis of requirements.

2) At the *set of sets* level: thus an airplane, a telephone exchange, a missile system, each is itself a set of sets and is subject to system design.

3) At the *set of sets of sets* level: thus an over-all weapon system, a telephone system, an air traffic system, represent such sets of sets of sets."

In a similar analysis Malcolm [Malcolm, D. G. Reliability maturity index (RMI)—an extension of PERT into reliability management. *J. industr. Engng.*, 1963, 14, 4–5] distinguishes eight hierarchical levels in a large weapon system: system, subsystem, component, assembly, subassembly, unit, unit component, and part.

15. Henderson, L. J. *The Fitness of the Environment: an Inquiry into the Biological Significance of the Properties of Matter.* Boston: Beacon, 1958.

16. Parkinson, C. N. *Parkinson's Law.* Boston: Houghton Mifflin, 1957, 2–13.

17. See Ardrey, R. *The Territorial Imperative.* New York: Atheneum, 1966.

18. See Levinson, D. J. Role, Personality, and Social Structure in the Organizational Setting. *J. Abnorm. Soc. Psychol.*, 1959, 58, 170–80.

19. Weber, M. *The Theory of Social and Economic Organization.* (Trans. by A. M. Henderson and T. Parsons.) New York: Oxford Univ. Press, 1947.

20. See Meyer, L. B. Meaning in Music and Information Theory. *J. Aesthet. Art. Crit.*, 1957, 15, 412–24.
Also J. E. Cohen. Information Theory and Music. *Behav. Sci.*, 1962, 7, 137–63.

21. See Whorf, B. L. *Language, Thought, and Reality.* Cambridge, Mass.: Technology Press, 1956.

22. Ashby, W. R. *Design for a Brain.* (2nd ed. rev.) New York: Wiley, 1960, 153–58, 210–11.

23. Rosenblueth, A., Wiener, N., & Bigelow, J. Behavior, Purpose and Teleology. *Philos. Sci.*, 1943, 10, 19.

24. Ashby, W. R. *Cybernetics Today and Its Future Contribution to the Engineering-Sciences.* New York: Foundation for Instrumentation, Engineering and Research, 1961, 6–7.

25. Hurley, W. V. *A Mathematical Theory of the Value of Information.* Report 63–3. New York: Port of New York Authority, Engineering Department, Research and Development Division, May, 1963.

H. M. Blalock, Jr.
Ann B. Blalock

3. Toward a Clarification of System Analysis in the Social Sciences

THE POINT of view taken in this paper is that on the most general level what can be termed "system analysis" involves a way of thinking which is common to all sciences, whether explicitly recognized or not. It is our contention also that although sociologists and other social scientists have employed system analysis, there has frequently been a lack of understanding of the basic ideas which are commonly involved in its use in the natural sciences. This has often led to much confusion and controversy. It is hoped that efforts to clarify some of these basic ideas and concepts will help to reduce this confusion and to increase the utility of system analysis in the study of social as well as natural phenomena.

To indicate the generality of system analysis in scientific thinking and to outline some of the important ideas it involves we can point to three perspectives or ways of studying reality which are in common use in all sciences and which when taken together provide insight into the nature of system analysis. This discussion will serve to introduce the concept "structure," an important tool in system analysis. Finally, the writers will discuss briefly how the concept "equilibrium" is used in the analysis of systems.

1. First Perspective: Forcings, Systems, and Responses. The first perspective can be made explicit by means of a simple diagram adapted from diagrams introduced by J. D. Trimmer ([13], pp. 1–3). Briefly, Trimmer's thesis is that in a large number of sciences it is our habitual mode of thought to conceive of something *a* being done to something *b,* and in turn this something *b* doing something *c.* Each discipline has its own vocabulary for describing this process, and Trimmer arbitrarily selects the terms "forcing," "system," and "response" to stand for *a, b,* and *c* respectively. He uses the term "law" to describe the interrelationships between forcings, the properties of systems, and responses. The process can be diagrammed as in Figure 1.

From *Philosophy of Science,* 1959, vol. 26, pp. 84–92. Reprinted with permission of the authors and the publisher.

FIGURE 1

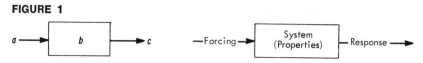

Depending on the purposes of analysis, we arbitrarily make a distinction between a system and its environment. The system is anything one wishes to study as an entity; in the social sciences it may be a person or a group such as a family or a large business organization. The environment is everything outside the system. Attention can be focused on the relationships *between* the system and the environment, with certain parts of the environment providing the forcings, and the system in turn responding to the environment in a kind of feedback process. If one wishes to focus on what is going on *within* the system by interrelating its properties, the environment (and therefore forcings and responses) can be neglected and the system taken to be closed or isolated.

2. Second Perspective: Interaction Between Systems. It is also possible to study the *interaction* between two or more systems. In Figure 2,

FIGURE 2

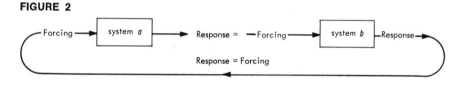

Trimmer's diagram has been extended to include the concept of interaction. Here the responses of one system can be thought of as forcings for the other and vice versa. The perspective has changed from a focus on one system to a focus on several systems and the interaction between them. For example, if we are treating the person as a system we are in this second perspective focusing not on one person and his relationships with everything else (the environment) but on several persons simultaneously. On the operational level one may of necessity focus on one system at a time, describing the interaction as a sequence of forcings and responses.[1] However, the change in focus requires an entirely new set of concepts for the description and understanding of the interaction process.

3. Third Perspective: A System Composed of Other Systems. A third general perspective involves thinking in terms of one kind of "something" composed of another kind of "something." In sociology one often takes the group as the larger system and thinks of it as being composed

---

[1] Perhaps the most systematic methodological attempts to describe interaction in small groups have been developed by Bales (1). Bales' method essentially involves recording in detail the sequence of forcings and responses for each of the group members.

of interacting persons. Trimmer's diagram can again be modified (Figure 3). Two different kinds of systems are represented here, the one (System A) being composed of other kinds of systems (systems *a* and *b*) as its elements.[2] In practice, of course, it is confusing for the analyst to think of both A and *a, b* as systems at the same time. Ac-

FIGURE 3

System A

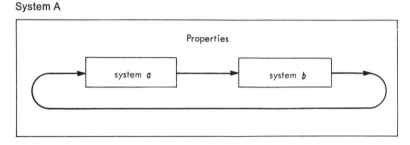

cording to one's focus at the moment, a and b can be thought of as elements of System A, or they can be conceived not only as mere elements of A but as systems which are themselves made up of elements. The properties of System A may consist of:

1. summarizing measures of the *individual properties* of systems *a* and *b* (*e.g.*, sex ratios or frequency distributions of physical characteristics or attitudes);
2. summarizations of *interaction* between systems *a* and *b* (*e.g.*, the processes of cooperation or competition); or
3. summarizing measures of the individual *behavior* (responses) of *a* and *b* (*e.g.*, the amount of work accomplished).

It is also possible within this perspective to develop more complex elaborations of Trimmer's original diagrams. Two very important but distinctly different points of view can be illustrated by Figures 4 and 5. In Figure 4 we are studying a System₁ composed of Systems A and B as two of its elements, the latter systems being in turn composed of systems *a, b, c* and *d, e* in interaction. Let us suppose that systems *a, b, c* and *d, e* are persons, and Systems A and B are groups or collectivities of persons. The "forcing" and "response" lines indicate that we are conceiving of groups A and B as actors, as responding to the forcings of one another. Interaction here refers to what is going on between these systems as elements of System₁ and *not* between the personnel of the two groups—although one still may think of interaction as occurring between persons *within* either A or B. The Talcott Parsons school of sociological theorists, when referring to collectivities as actors, ap-

---

[2] Ideally all elements of System A should be specified so that the system's boundaries are well defined.

FIGURE 4

System₁

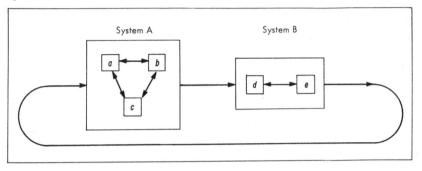

parently subscribes to this point of view in the analysis of social systems.[3]

The second point of view is represented in Figure 5. Advocates of this perspective are opposed to thinking of collectivities as actors and would prefer to limit the term "actor" to persons. In this diagram persons (*a, b, c, d, e*) are conceived of as being in direct interaction, and we can delineate *A* and *B* as subsystems without necessitating our thinking in terms of interaction between these subsystems per se. The elements of System₂ are not *A* and *B* but *a,b,c* and *d,e*. *A* and *B* are

FIGURE 5

System₂

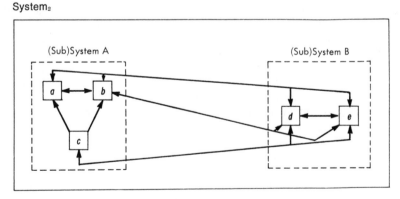

included in but are not *elements of* system₂. This second point of view expresses the present writers' conception of the "social system." This would seem to be the simpler point of view conceptually and operationally. Regardless of what one conceives as the social system, the failure to make the distinction between the two points of view has often led to

---

[3] For example, see Parsons and Shils (11), pp. 4, 39, 66, 101.

considerable confusion. The nominalist-realist controversy over the reality of human groups seems to have involved, at least in part, certain semantic difficulties stemming from a failure to recognize this distinction. In sociology the same concepts have been used to refer to both groups and individuals, a state of affairs which can hardly be expected to contribute to conceptual clarification. It should be clearly recognized that operations appropriate to the group as the unit need to be distinguished from those operations appropriate to the person. Operations used in determining the "status" of a group, for instance, will differ from those used in obtaining an individual's "status." Nor does a group "behave" in the same sense that a person behaves.

4. Relationships between Elements.   Implied though not explicitly stated in the discussion of the three perspectives is the study of *relationships* between the elements of systems, whether these elements be persons, groups, or some other kind of unit. The analyst of systems is interested not only in studying the interaction between elements of a system but is also concerned with relating these elements with respect to certain variables. For example, if we are studying a social system we are interested in relating persons with respect to such variables as status and power as well as describing their interaction. If we have knowledge about how these persons are related with respect to status and power we are in a better position to understand this interaction. Thus knowledge about the properties of several systems enables us to relate the systems and to predict the nature of their interaction.

5. Structure.   It is in discussing relationships between elements that we begin to talk about the structure of a system. In the biological and physical sciences the concept structure is frequently used to refer to an arrangement or configuration of a system's elements. Most generally it is a tool used to describe relationships between elements in a system with respect to certain variables, the variables chosen being dependent upon the purposes of the analysis. (Eddington [5], pp. 137–44).

In the social sciences as well, it may be said that the structure of a system has been specified whenever certain determinate relationships between the elements of that system have been described. In the biological and physical sciences it is often commonly understood exactly which elements and variables are involved in the description of the structure of a given system. That this is seldom the case in sociology and other social sciences seems to have led to certain difficulties in the use of this concept.

In order to specify the structure of a system it must first be decided (1) what the elements are that are being used, and (2) what the variables are which are being considered, values of which need to be determined. It therefore seems meaningless to think in terms of *the* structure of a system or *the* elements (or "parts") of a system. (Eddington [5], pp. 118–20). It cannot be assumed that the elements and variables involved are uniform from one type of system to another or from one analysis to the next. It would be more useful to speak of the structure *with respect to* elements of a given type and variables X, Y, and Z.

Another difficulty in the use of the concept is the inclusion of *time* as one of the variables defining the structure of a system. Implicit in the definitions and discussions of the concept structure often given by social scientists is the notion that a system cannot be said to have a structure unless relationships between elements are relatively fixed and unchanging over a considerable period of time.[4] Using the concept in this restricted sense means that the identification of the structure of a system is dependent upon the demonstration of relatively stable relationships between elements.[5] This use of the concept supports the contention that an analysis of the structure of a system is a static analysis and that the concept structure cannot be used to study the dynamics of change. The notion of stability is not necessarily involved in the use made of the concept in other sciences, and it is thought by the present writers that the most useful referent for the concept is those relationships among the elements of a system found to exist *at any given point in time.* The only requirement is that the relationships remain constant long enough to be measured. The particular set of relationships existing at the time of the study is that system's structure at that time. On this basis we can legitimately say that the study of structure is not necessarily limited to the study of statics; the study of change in structure over time is made possible.

6. Equilibrium. The concept equilibrium is often quite helpful in the analysis of systems and seems to be fairly popular with present-day sociological theorists. Often the term has been used rather loosely by social scientists in the sense that *what* it is that is in equilibrium is not precisely specified. Also there is a tendency to assume a special type of equilibrium, "stable equilibrium," holding under conditions which often remain unspecified.

Certain general conclusions can be drawn about how the concept equilibrium is used in physics, the most advanced of the natural sciences.[6] *First,* the term equilibrium is used in a number of different senses; there are several types of equilibria discussed by writers in different branches of physics. *Second,* the notion of equilibrium always seems to imply that *something* is remaining unchanged or is assumed

---

[4] Many writers do not explicitly define the concept structure. For examples of definitions which imply relative stability see Bennett and Tumin (2), p. 69; Cartwright and Zander (4), p. 416; and Williams (14), p. 20.

[5] This use of the concept raises the question, "When can a set of relationships among elements in a system be said to be relatively fixed and therefore a structure?" If one wished to study the family as a system it is apparent that this use of the concept would entail studying the family several times over a period of time in order to determine whether or not certain relationships (for instance with respect to power or solidarity) changed over time or remained relatively fixed. If they remained constant over this period the family could be said to have a structure, but one could say only that it exhibited a structure at *that* period. If, however, relationships were characterized by change, than supposedly no structure could be said to exist for that period, and there would be no general concept to describe these changing relationships. This illustrates some of the limitations of this use of the concept.

[6] For example, see Planck (12), p. 65; Miller (9), p. 19; Zemansky (15), pp. 23–27; and Fermi (6), pp. 1, 55, 57, 80, and 83. These conclusions are not mentioned explicitly in any of the works cited above and seem to be often taken for granted by the physicist.

to remain unchanged over some period of time (however short). *Third,* these "somethings" (remaining unchanged) are always clearly specified by the physicist. It is seldom if ever implied that *all* variables under consideration are necessarily fixed. It is misleading to say that a given system is "in equilibrium." As with the concept structure, it would be preferable to say that a system is in equilibrium *with respect to* specific variables. The implication is clear: social scientists who use the concept equilibrium need to make efforts to specify precisely what it is about the system they are assuming to remain unchanged.

A *fourth* conclusion is that once assumptions have been made about equilibrium, or what variables are remaining constant, the scientist is able to state more exactly relatively simple relationships between two or more variables under these equilibrium conditions. Stating that these relationships hold under specific equilibrium conditions seems to be a much more precise way of saying that they hold, "other things being equal." When equilibrium conditions are specified one knows precisely what the "other things" are. A quotation from a well-known text in thermodynamics may be helpful at this point.

Imagine for the sake of simplicity a constant mass of gas in a vessel so equipped that the pressure, volume, and temperature may easily be measured. If we fix the volume at some arbitrary value and cause the temperature to assume an arbitrarily chosen value, then we shall not be able to vary the pressure at will. Once V and θ are chosen by us, the value of P at equilibrium is determined by nature. Similarly if P and θ are chosen arbitrarily, then the value of V at equilibrium is fixed. That is, of the three thermodynamic coordinates P, V, and θ, only two are independent variables. This implies that there exists an equation of equilibrium which connects the thermodynamic coordinates and which robs one of them of its independence. (Zemansky [15], pp. 23–24).

Notice that Boyle's and Charles' Laws hold only when certain equilibrium states can be assumed. Zemansky, in other passages, is careful to point out precisely what is assumed to remain constant under thermodynamic equilibrium. The essential point is that one does not simply state a relationship between the three variables temperature, volume, and pressure. It is possible to state a relationship between these three variables only when certain assumptions can be made about *other* variables. Equilibrium conditions, when used in this manner, amount essentially as a set of qualifications to a generalization about these three variables. Stated in less precise form, propositions relating temperature, volume, and pressure often make no mention of equilibrium or of other factors which are assumed to remain fixed. It is when precision is required and when experiments show that the hypothesized relationships do not always hold that the notion of equilibrium is necessary. This may suggest how the concept equilibrium may prove helpful to the social scientist.

The *fifth* general conclusion is that the use of the concept equilibrium does not necessarily imply that a state of equilibrium is in any sense "natural" or desirable, or that a system automatically tends toward equilibrium. The physicist makes use of the notion as an idealization or model. He does not necessarily expect this model to correspond exactly to reality. He assumes certain variables are remaining temporarily constant, even though he may realize this is not actually the case.[7] By making simplifying assumptions of this sort he is able to solve for unknowns more easily.

Some social scientists, especially those associated with the "structure-function" school, have been accused of a conservative bias in assuming equilibrium states as natural.[8] This tendency may have resulted, in part at least, from the borrowing of the concept "function" from the biological sciences where the idea of survival of the system has been associated with "homeostasis," *i.e.*, a stable equilibrium model in which the system has a tendency to maintain its original state. (W. B. Cannon [3], pp. 24, 299–300). That this is not the only type of equilibrium model has been pointed out above. A closer examination of the assumptions underlying the use of functional analyses may be of value to the social scientist who wishes to use the concept equilibrium in the study of social systems.

7. Concluding Remarks. Some of the basic ideas in system analysis have been discussed. It has been suggested that this analytic approach and the use of certain related conceptual tools are common to scientific thinking generally, and that properly used and understood, system analysis can contribute to the clarification of theoretical thinking in the social sciences. Certain implications of the use of system analysis have emerged from this discussion and can be stated briefly.

1) It is necessary to keep clearly in mind the nature of the system studied. This means that its elements (persons, groups, etc.) must be clearly specified. To reduce confusion, the same concepts should not be used to describe the properties of different kinds of systems.

2) In using the terms "structure" and "equilibrium" it is always necessary to specify the variables and elements concerned; it is therefore meaningless to speak of *the* structure of a system.

3) The study of structure can just as legitimately be a study of changing relationships among the elements of a system as it can be a study of relatively fixed relationships if the identification of a system's structure is not made dependent on the demonstration of stability over time.

4) It is helpful to be aware of assumptions made in the use of any particular equilibrium model. "Stable equilibrium" is only one type of equilibrium. The concept "function" has apparently been borrowed from a science in which this particular type of equilibrium model is most appropriate.

---

[7] He may compromise with reality by thinking in terms of a "moving equilibrium" and by replacing a smooth curve by a step function, assuming a variable to remain constant for a short period of time and then taking on a new value.

[8] See Homans (7), pp. 301–5; Merton (8), pp. 38–40, 53–54; and Myrdal (10), pp. 1065–67.

REFERENCES

1. Bales, R. F.: *Interaction Process Analysis.* Cambridge: Addison-Wesley Press, 1950.
2. Bennett, J. W. and Tumin, M. M.: *Social Life.* New York: Knopf, 1948.
3. Cannon, W. B.: *The Wisdom of the Body.* New York: Norton & Co., 1939.
4. Cartwright, D. and Zander, A. (eds.): *Group Dynamics.* Evanston: Row, Peterson, 1953.
5. Eddington, A. S.: *The Philosophy of Physical Science.* New York: Macmillan, 1939.
6. Fermi, E.: *Thermodynamics.* New York: Prentice-Hall, 1937.
7. Homans, G. C.: *The Human Group.* New York: Harcourt, Brace, 1950.
8. Merton, R. K.: *Social Theory and Social Structure.* Glencoe: The Free Press, 1949.
9. Miller, C. W.: *An Introduction to Physical Science.* New York: Wiley, 1935.
10. Myrdal, G.: *An American Dilemma.* New York: Harper, 1944.
11. Parsons, T. and Shils, E. (eds.): *Toward a General Theory of Action.* Cambridge: Harvard University Press, 1952.
12. Planck, M.: *General Mechanics: Introduction to Theoretical Physics.* Vol. 1, London: Macmillan, 1933.
13. Trimmer, J. D.: *Response of Physical Systems.* New York: Wiley, 1950.
14. Williams, R. M.: *American Society.* New York: Knopf, 1952.
15. Zemansky, M. W.: *Heat and Thermodynamics.* New York: McGraw-Hill, 1943.

Everett E. Hagen

4. Analytical Models in the Study of Social Systems[1]

As JUDGED by the history of the physical, biological, and social sciences, study in any field is apt to begin with a none-too-ordered description of phenomena in the field, followed by a cataloguing of them on bases that seem to make sense. As understanding grows, the systems of classification become more closely related to the functioning of interacting elements. Gradually, generalizations about functioning are reached which are useful in predicting future events. As the generalizations gain rigor, they take the form of analytical models of the behavior of the elements being studied. An analytical model is a mental construct consisting of a set of elements in interrelation, the elements and their interrelations being precisely defined.

The first stage in the analysis of functioning is usually study of processes at narrowly defined points within the general area of the science. Attention is focused on how the elements at the point being examined would function in the absence of change elsewhere. Then a mental model of the processes at this point is formed, which is a simplification of reality that retains only the features essential for predicting similar processes elsewhere. Such analysis of a narrowly defined point in a system may be termed "partial analysis."

Later comes the development of more comprehensive analytical models, which in some sense encompass a complete system rather than simply one point in relationships. Such a model is termed a "general system" or "analytical model"; its construction and use are "general analysis" or "system analysis." There is no sharp distinction between partial and general models, for analysis of a general system also holds in abeyance change beyond certain boundaries. As a science is able to move to more and more comprehensive systemic analysis, its power increases greatly.

From *The American Journal of Sociology*, 1961–1962, vol. 67, pp. 144–51. Reprinted with permission of the author and the publisher. © 1962 by The University of Chicago. All rights reserved.

[1] I am grateful to Robert Solow for comments on the first draft of this paper.

Since the work of Willard Gibbs in the physical sciences and mathematics some three-quarters of a century ago, thinking in the physical sciences has been self-consciously in terms of models. Since Walras, theoretical analysis in economics also has rapidly come to be stated exclusively in terms of models. The analytical model has only recently become prominent in the other social sciences, however, probably mainly because of their greater complexity but also because of their slighter contact with the general science of relationships among magnitudes, that is, mathematics.

In the evolution of theory, concepts found useful at various stages are later discarded as analysis grows in rigor. In the study of social systems, many early concepts, for example, those which reified society, have been sloughed off. But certain concepts and methodologies remain which, it seems, are incompatible with rigorous analysis of causal relationships.

LOGICAL REQUIREMENTS OF GENERAL SYSTEM ANALYSIS

The following requirements of general system analysis are of most interest here:

1. An analytical model is defined by defining the elements and their interrelations.[2] The relationships among the elements of a system are statements of the alternative values (magnitudes) or states of one of the elements associated with alternative values or states of one or more of the other elements. Because the elements are assumed to vary in magnitude or state, they are termed variables—which, broadly, includes constants—that is, the variation in some may be zero. If two variables are related in this way, each is said to be a function of the other without regard to the direction of causation between them. While the flow of causation between any two elements may be in one direction and not the other, among all of the elements taken as a group, apart from the impact of forces from outside the system, all depend on all. Let it be noted clearly that this concept of mutual interdependence or interaction does not involve circular reasoning or indeterminacy.[3]

2. The variables of a system must exist either in conceptually measurable amounts, or in one or another set of definable states. It is impossible to conceive of variation in one element associated with

---

[2] Physical scientists refer to a set of elements in interaction as an "analytical system" or simply a "system." They include as systems entities of the real world. I use the term analytical model to emphasize that the concept relevant in theoretical analysis is one of an intellectual construct.

For definitions and discussions of their properties, see the following articles in *General Systems*, Vol. I (1956): Ludwig von Bertalanffy, "General System Theory," pp. 1–10, reprinted with footnotes added from *Main Currents of Modern Thought*, LXXI, 17 ff.; Kenneth Boulding, "General Systems Theory—the Skeleton of Science," pp. 11–17, reprinted from *Management Science*, II (April 1956), 197–208; and A. D. Hall and R. E. Fagen, "Definition of System," pp. 18–28. See also W. Ross Ashby, "General Systems Theory as a New Discipline," *General Systems*, III (1958), 1–6; and *Decision Processes*, ed. R. M. Thrall, C. H. Coombs, and R. L. Davis (New York: John Wiley & Sons, 1954).

[3] Henderson's mechanical example provides a beautifully simple visual illustration of mutual dependence or interaction: Lawrence J. Henderson, *Pareto's General Sociology* (Cambridge, Mass.: Harvard University Press, 1937), p. 14.

variation in another if the two cannot be conceived of as varying by measurable amounts, or from one state or structural form to another. If a variable (such as "community spirit" or "love of family") is not defined so as to be conceptually measurable or as existing in one or another set of definable states, it cannot have a precise reasoning, in an analytical model or otherwise.

A variable is a single dimension of an entity, not the entity itself. Thus a variable is not a physical body but one of its qualities, for example, length; in a model of society, it is not an individual, but, say, each value and each need (motive) attributed to him and each component of his perception of the nature of the world. The individual as a group of interacting elements may be a subsystem within the model.

3. For use in analysis, a system must be "closed." A system which is interacting with its environment is an "open" system: All systems of "real life" are therefore open systems. For analysis, however, it is necessary in the intellectual construct to assume that contact with the environment is cut off[4] so that the operation of the system is affected only by given conditions previously established by the environment and not changing at the time of analysis, plus the relationships among the elements of the system.

Elements of the system whose magnitudes are wholly determined by the environment, and which are therefore constant rather than variable so long as the system is insulated from change in the environment, are termed *parameters*. For example, in some analyses in economics the size of the population and per capita income are parameters, that is, it is assumed that they remain constant.

In the process of analysis a closed system is not assumed to remain closed. Only extremely limited analysis is possible except as the theorist opens the system to a change in the environment, and observes its effect. Thus in a model of a society, the sequence of effects when some one type of relationship with other societies is changed may be analyzed.

4. It is often useful to construct a model which is in equilibrium, and in stable rather than unstable equilibrium.

Equilibrium in its simplest sense refers to a condition in which the variables in the system are in such a relationship to each other that all remain constant in value, not by assumption, but by their interaction.

If the magnitude of a variable has been changed by a temporary change in the magnitude of some external force that affects the system, its change will necessarily cause at least temporary changes in the magnitudes of other variables, because of the functional relationships among them. (If change in one variable affects no other variable, then that one variable is not in any significant sense a part of the system.) These changes will in turn react on the magnitudes of the variable which first changed, and of each other. The equilibrium of the system is

---

[4] Technically, that there is no exchange of energy in any form, in the broadest definition of the term energy, to include, for example, information. For this reason, Ashby (loc. cit.) suggests that instead of being termed "closed," such a system should be termed "energy-tight," "information-tight," or "noise-tight," the last term of course coming from the terminology of the modern study of communication.

stable if the final result of this interaction, after an initial temporary disturbance, is a return to the initial values. The equilibrium is unstable if a temporary disturbance causes the values of some or all variables to move cumulatively farther from the initial equilibrium.

The equilibrium of a system may, of course, be stable with respect to one type of disturbance and not with respect to another. Further, the equilibrium of a system may be stable with respect to a small disturbance ("stability in the small") but not with respect to a large disturbance ("stability in the large"). Stability of equilibrium, moreover, implies only that the equilibrium values of the variables will remain unchanged as long as the system remains closed except for temporary "disturbances." If permanent changes in the environment are communicated to the system, there will be corresponding permanent changes in the equilibrium values of variables in the system (that is, the values they will have when the system has settled down into the new equilibrium).

To illustrate equilibrium and related concepts, suppose that deposits in a certain commercial bank are at a "normal" level, and a rumor that the bank is unsafe (a "disturbance") arises. The rumor causes a few depositors to withdraw their money (a movement away from equilibrium). If the total network of circumstances is such that the withdrawals do not lead to a spread of anxiety, but instead the outflow stops, and the funds are redeposited, then by definition the equilibrium was stable; if the initial withdrawals cause a run on the bank so that it fails, the equilibrium was unstable. The bank's equilibrium might be stable with respect to a rumor that a nearby bank was about to close, but not with respect to the unexplained disappearance of the cashier.

The stability of equilibrium is caused, not by (*a*) the degree of confidence depositors had in the bank, nor by (*b*) the magnitude of the net demand for withdrawals by depositors, nor by (*c*) the ability and willingness of the bank to supply funds taken alone, but by the interrelationships among the three—the magnitude of rise in net demand for withdrawals caused by a given decrease in confidence; the ability of the bank to supply an increase of this magnitude in the demand for funds, and the seeming lack of concern with which it does so; the effect of withdrawals of this magnitude in causing further decline in confidence; and the effect of the bank's readiness to supply funds (and the attitude of its officers) in increasing confidence.

Suppose that total income in the community increased because of opening of a new factory. With this change in one of the parameters, deposits in the bank may be expected to rise to a new higher level, at which their value will be in a new stable equilibrium.

Comparison of the equilibrium positions of the variables of a model under two differing values of one or more of the parameters—in the example above, comparison of the level of bank deposits at the two different levels of income—is termed *comparative statics* in economics. There is no comparable term in the other social sciences.

If one or more of the parameters of a system goes through a process of continuing change—for example, if total income in the community steadily increases—the values of the variables at which they are in equilibrium may be expected to change continuously. We may then refer to a "moving equilibrium."

It is especially important, in the application of models to the study of societies, to note that the presence or absence of equilibrium in a system and the stability or instability of equilibrium are results of the inter-relationships among the variables. Equilibrium or its absence, and its stability or instability, cannot be caused by the nature of one variable considered without relation to the others. If we knew the interrelation-ships accurately we could tell in advance whether equilibrium would be restored after a given temporary disturbance. Stability of equilibrium is not merely an *ex post facto* fact.

5. It may also be fruitful to study a system which is not in equilib-rium.

Often we are concerned only with the conditions for equilibrium. We may solve a set of equations to determine the value each variable will have in equilibrium. But we may also be interested in a sequence of change, in time, in the values of the variables. A change in the position of one variable has an effect on one or more other variables only after a time interval.[5] For example, a change in birth rates will affect the age composition of the population throughout many generations. A change in the environment in which the children of a group are brought up will affect their personalities as adults only after the lapse of years nec-essary for them to become adults, and through their impact on their children, will continue to cause alteration in adult personality for gen-erations thereafter.[6]

A new equilibrium will be reached only after a time interval.

In contrast to comparative statics, analysis may be made of the path of change of the several variables of the system (presumably from one equilibrium to another) when a change in a parameter occurs. Such analysis is termed "dynamics," and a model whose process of change is being analyzed is said to be a dynamic model. The term "diachronic analysis" in anthropology is apparently identical in meaning with dy-namic analysis.

It is unfortunate that the terms *dynamic* and *dynamics* are used in this sense with reference to analytical models and in quite a different sense in contemporary psychology. Both usages are so well entrenched that they must be lived with. Freudian psychology introduced, or gave increased emphasis to, two elements in psychological theory. One was the study of the formation of personality, that is, change in personality. This is a study of "personality dynamics" in a sense precisely analogous to that in which the term "dynamics" is used above; hence, the terms

[5] It is sometimes said that, when we consider only the conditions of equilibrium, our analysis is as though the causal effect of change in each variable on each other were instantaneous. This statement may give some "feel" of the nature of analysis of equilibriums, but it is not literally correct. Nothing happens instan-taneously, and analysis does not really assume so.

[6] We select a length of time in which we assume one step of change occurs and treat this as the unit time period. Where the value of a variable is determined by the values of other variables in past periods, and not by the values of other vari-ables in the same time period, the value of the variable is said to be "predeter-mined," and the variable is said to be a "predetermined variable."

"personality dynamics" and "dynamic psychology" came to be applied. The other new element was emphasis on unconscious motivation. These two new elements appeared at once, and by terminological inaccuracy, "dynamic" became a synonym for "motivational." Thus the terms "personality structure" and "personality dynamics" are sometimes used interchangeably, and the term "personality dynamics" is also used to refer to the study of the influences which cause a person to behave as he does.

Sociology has taken over, somewhat out of context, this extension of psychological terminology. Parsons, for example, frequently refers to dynamic factors or processes in any social system, including one which is in stationary equilibrium. And, on the other hand, sociology has no technical terms for the distinction between a social system in equilibrium and one in movement not in equilibrium. Parsons discusses such movement; he entitles the relevant chapter of *The Social System* simply "The Processes of Change in Social Systems." In this essay, to prevent confusion, I shall avoid the psychological-sociological use of the term *dynamics*.

6. When the system moves to a new position of equilibrium, not all the variables necessarily change in value. The interrelationships of the system may be such that, in spite of permanent change in one or more parameters, some of the variables, after being temporarily disturbed, will return to their initial magnitudes. This is the condition termed homeostasis: It is usually illustrated by organic or mechanical examples. If the temperature in the environment of an organism falls, the fall will cause heat to drain more rapidly from the organism, which, in turn, will activate a mechanism which will increase the body's generation of heat, so that unless the fall in external temperature is too great, the temperature of the organism, after a temporary fall, returns to normal.

Homeostasis (or an analogue, if it is preferred to reserve the term for reference to biological or mechanical cases) may also be illustrated by an example from economics. Suppose that in a certain city, the price charged for putting new rubber heels on a pair of shoes is 75 cents. Suppose that the city now grows rapidly; because shoe-repair shops find themselves flooded with business, they can and do obtain 90 cents for putting new rubber heels on shoes, and obtain similarly increased prices for other shoe repairing. The increased profit margin, however, draws more artisans into the shoe-repair business, so that after a time the supply of these services increases so much that it is no longer possible to obtain more than 75 cents for putting on heels. The new equilibrium of the price of equipping a pair of shoes with rubber heels is the same as the old: A "negative feedback mechanism" has restored the former price.

Note, however, that one variable (the body temperature or the price for supplying a pair of rubber heels) could return to its old value only if another one (the bodily consumption of energy and generation of heat, or the quantity of shoe-repair services available) changed permanently in magnitude. This is an aspect of homeostasis sometimes overlooked. Homeostasis with respect to one variable necessarily implies an altered position of another—"heterostasis"—for as long a period as the changed external condition that brought the homeostatic mechanism into play prevails.[7]

CONCEPTS AND METHODOLOGY

It will be obvious that these requirements of analytical models are necessary characteristics of the interrelationships within any set of variables in any field. Hence concepts concerning society which contradict them are either logically mistaken or, at best, not useful. Concepts which either contradict the logical requirements or, at best, are ambiguous are found, however, in the writings not merely of lesser students but of some of the most creative and influential of recent theorists.

Some of these errors or instances of imprecise formulation of concepts may have arisen originally from a state of mind characteristic of the early stages of anthropology and sociology, social constancy being regarded as good, and social change (perhaps because it created tensions, or because it was imposed, willy-nilly, from without), as bad. Some may have arisen from concentration on social structure rather than on social processes. Perhaps the explanation of their persistence is that the study of societies has not yet fully reached the stage of precise definition of variables. In any event, the social sciences are now moving toward greater precision in the definition of variables, analysis of functional relationships, and creation of models, and in this transitional phase it may be useful to call attention to some concepts that seem obsolescent.

1. In much sociological writing, the concept of society is viewed as necessarily involving stable equilibrium (either static or moving). Thus in *Toward a General Theory of Action*, Parsons writes, with Shils:

The most general and fundamental property of a system is the interdependence of parts or variables. Interdependence consists in the existence of determinate relationships among the parts or variables as contrasted with randomness of variability. In other words, interdependence is *order* in the relationship among the components which enter into a system. This order must have a tendency to self-maintenance, which is very generally expressed in the concept of equilibrium. That is, if the system is to be permanent enough to be worth study, there must be a tendency to maintenance of order except under exceptional circumstances. It need not, however, be a static self-maintenance or a stable equilibrium. It may be an ordered process of change—a process following a determinate pattern rather than random variability relative to the starting point. This is called a moving equilibrium and is well exemplified by growth. Furthermore, equilibrium, even when stable, by no means implies that process is not going on; process is continual even in stable systems, the stabilities residing in the interrelations involved in the process.[8]

[7] The term *heterostasis* is from R. C. Davis, "The Domain of Homeostasis," *Psychological Review*, LXV (January 1958), 8–13.

[8] Talcott Parsons and Edward A. Shils (eds.), *Toward a General Theory of Action* (Cambridge, Mass.: Harvard University Press, 1954), p. 107.

Note that the word *stable* here is used to refer to static as distinguished from moving equilibrium, and not to stable equilibrium in the sense in which the term is defined above.[9]

In this statement, it is not entirely clear whether the reference to moving equilibrium is to that concept as defined above (continuing shift in equilibrium caused by continuing change in an exogenous force) or to a dynamic process (change in time in the values of variables in the system caused by the relationships within the system). However, in the latter case the statement is tautologous, and means merely, "Do not study a system unless it is a system," for the only possible states of a system are equilibrium and dynamic. Hence, and because his own analytical models are purely static,[10] I conclude that Parsons is warning against formulation of models not in equilibrium.

Related is the implication, in the terms "dysfunction," "eufunctional," and "dysfunctional," by Merton and by Levy[11] that tendency toward equilibrium is somehow good and toward disequilibrium somehow bad. While these authors may disavow the implication, the use of the prefixes *eu* and *dys* conveys an inescapable suggestion.

Parsons does not assert, it should be noted, that a society necessarily possesses stable equilibrium, but only that it is not worth studying unless it does. The restriction is thus logically permissible. It also has empirical relevance. Some societies certainly have had a tendency toward stability of equilibrium; their internal dynamics have at best brought rather slow change. But adherence to this model as the general case unnecessarily limits the domain of sociological theory, and excludes from sociological theory important problems that ought to be treated within it. First, it tends to exclude from the theoretical system consideration of what kind of force is necessary (and sufficient) to push the society away from the equilibrium and lead to the disruption of the social system. Second, it excludes study of the relationships within the society which will determine the nature of the sequence of change in time, once the equilibrium has been disturbed—or the nature of a sequence of change in time in a society conceived of as never having been in equilibrium, but rather under continuing change from its own dynamics. Virtually all societies in the world at present are in a process of change which, however it began, is best analyzed as continuing partly by virtue of the dynamics within the system itself. A model of stable equilibrium is not a satisfactory theoretical analogue for use in studying their behavior.

[9] Parsons presents a similar formulation, though one which may be interpreted as presenting stable equilibrium as a basis for Parsons' work rather than as a general theoretical requirement, in *The Social System* (Glencoe, Ill.: Free Press, 1951), p. 481.

[10] Parsons discusses social change, not only in specific "empirical" discussions, but more generally in chap. xi of *The Social System*, and in chap. vii of Parsons and Bales, *Family, Socialization, and Interaction Process* (Glencoe, Ill.: Free Press, 1954), but in each case this discussion is an addendum, not a part of the analytical system which he presents in *The Social System* and in *Toward a General Theory of Action*.

[11] Robert K. Merton, *Social Theory and Social Structure* (rev. and enlarged; Glencoe, Ill.: Free Press, 1957), p. 51; Marion J. Levy, Jr., *The Structure of Society* (Princeton, N.J.: Princeton University Press, 1952), passim.

If sociologists are to analyze change in a society as a whole, rather than merely to describe it loosely, they must go beyond models in equilibrium and construct models involving dynamic processes.

Further, even with regard to a society—or a model—in stable equilibrium, or changing only very slowly, study of the relationships that bring about the stability or quasistability may be extremely fruitful. It is illuminating to isolate the network of relationships which, if a temporary disturbance brings change within a society, determines whether the change will be cumulative or whether the system will return to the initial equilibrium. Out of studying precisely this question great advances in the understanding of societies, even of societies in stable equilibrium, may come.

2. Perhaps underlying these difficulties are undesirably vague definitions of *function*. The question of the meaning of the concept has been much speculated upon since Radcliffe-Brown's essay of 1935,[12] yet in 1954 Parsons had not yet arrived at a precise meaning. Referring to testing the significance of processes, he states "That test of significance takes the form of the 'functional' relevance of the process. The test is to ask the question, what would be the differential consequences for the system of two or more alternative outcomes of a dynamic process."[13]

And in 1957, Merton defined function as "the observed consequences" of "a sociological item" for "the social or cultural system in which it is implicated,"[14] a statement which in addition to being vague suggests that the nature of a single "item" can lead to stability or instability.

These statements are not incorrect, only vague. In them precisely what does the word *consequences* refer to?

These writers have made great contributions to sociological theory, as Radcliffe-Brown did to anthropology. It implies no lack of appreciation of their work to suggest that, in the most fruitful usage, only a quality can be a variable; only a variable has a functional relationship; a functional relationship consists of the change in the magnitude (or state) of one variable (not in a system) associated with change in the magnitude (or state) of another; and no single variable, but only the func-

---

[12] A. R. Radcliffe-Brown, "On the Concept of Function in Social Science," *American Anthropologist*, Vol. XXXVII (1935), reprinted in his *Structure and Function in Primitive Society: Essays and Addresses* (Glencoe, Ill.: Free Press, 1952).

[13] *The Social System*, pp. 21–22. The term *dynamic* here does not refer to change in the equilibrium of the system, but to a process of action by an individual or a group in a role. In neither *The Social System* nor *Toward a General Theory of Action* does Parsons define *function*.

[14] This definition is arrived at by joining phrases from two sentences, on the assumption that *function* and *dysfunction* as contrasting concepts are subcategories of the general concept *function:*
"We have observed two prevailing types of confusion enveloping several current conceptions of 'function':
"(1) The tendency to confine sociological observations to the *positive* contributions of a sociological item to the social or cultural system in which it is implicated; . . .
"Functions are those observed consequences which make for the adaptation or adjustment of a given system; and dysfunctions, those observed consequences which lessen the adaptation or adjustment of the system." *Social Theory and Social Structure*, p. 51.

tional relationships among variables can lead to stability (adaptation, adjustment) or instability in a system.

Though system analysis is used more in psychological theory than in that relating to societies or communities as wholes[15] failure to state theory relating to social systems in terms of variables, functions, and a general system has not been for lack of sophistication. Rather, the difficulty has been one of substantive complexity. The applicability of functional analysis, in the mathematical sense, in any science does not become apparent to students of that science until they have been able to arrive at a certain precision and breadth of understanding of causal relationships. This is why each discipline slowly and stumblingly rediscovers concepts concerning method already discovered long ago in other disciplines—why, for example, economics clumsily and painfully groped its way to the concept of marginal productivity and only subsequently realized that it was merely applying elementary calculus to its problems; and why anthropologists groped toward the concepts of synchronic and diachronic analysis, and not all anthropologists realize fully even today that they are referring to static and dynamic analysis of an analytical system. Scholars are not apt to realize the applicability of the concepts of variable, function, and general system until they understand functional relationships in their field of study well enough so that their images of the phenomena in their field of analysis begin to resemble variables in interaction.

Freudian and post-Freudian analysis of personality formation has made it possible to formulate plausible and useful models of individual personality and its formation. If we incorporate these subsystems in models of society, the time when we may formulate useful though heroically simplified models of society in equilibrium and of societal change does not seem far off. Even a model which, with respect to many functional relationships, merely indicated that between two certain variables a relationship must exist, while admitting ignorance of its nature, could be a highly useful vehicle for the furthering of substantive analysis.

---

[15] For a recent discussion by a psychologist which takes aim beyond the boundaries of his discipline, see James G. Miller, "Toward a General Theory for the Behavior Sciences," *American Psychologist*, X (September 1955), 513–31, reprinted in *The State of the Social Sciences*, ed. Leonard D. White (Chicago: University of Chicago Press, 1956).

Section II

ORGANIZATIONAL SYSTEMS THEORY

IN THE FIRST paper in this section, Ackoff offers a definition of an organization as a partially self-controlled system, characterized by its content, structure, communications, and decision-making procedures. He points out the major disciplinary approaches to each of these four central characteristics of organizations and observes that in order to construct truly interdisciplinary models of systems, it is necessary to first conceptually relate the variables dealt with in each of the disciplines involved in systems research. Complex systems research requires development of a conceptual system which not only can pull together concepts applied to the systems by various disciplines, but also can reduce them to quantities which are measurable on compatible scales. Ackoff further recommends that a methodology better adapted to the unique aspects of systems research be developed and points out the need for an innovative educational program to train researchers in conducting systems research.

Reviewing the development of organizational theory up to the beginning of the past decade, Scott joins with Ackoff in noting the need for new tools of analysis in conceptual frameworks and agrees that these will come from applications of general systems theory to the analysis of social systems such as complex organizations.

In *The Social Psychology of Organizations* (1966) Katz and Kahn attempted to apply and extend open systems theory to human organizations in a work of far-reaching influence. In the paper included in this section, Katz and Georgopoulos examine the structure and functions of organizations as open social systems in a world characterized by rapid social change and dependence on organizational social structures. They show how an examination of organizations as open systems can be useful in understanding the interaction of particular organizations in a psychosociocultural environment which rapidly challenges traditional organizational forms.

Organizations constitute a major environmental context for human work behavior. As Katz and Georgopoulos point out, work organizations today must deal with the value conflicts that trouble the larger society. Morley and Sheldon, in a paper especially prepared for this volume, discuss human work behavior as a crucial system variable affected by values introduced from outside the work organization which must be supplemented by internal values. They use systems theory in considering the ways in which work organizations deal with employee behavior, and they develop a perspective which treats both individual and organization as systems in their own right, although at different levels. A single set of terms is presented for considering phenomena at both the individual and organizational level which is consistent with the goal of developing general systems theory.

Ralph Stogdill attempts to reconcile classical and behavioral theories of organizations by regarding the organization as an input-processing-output system. He notes that all organizations may be viewed as utilizing human performances, expectations, and interactions as inputs, if material and financial resources are disregarded. These three variables act upon each other in newly formed groups and organizations to generate structure, operations, and interpersonnel variables, which in turn become differentiated and are utilized in task performance. These variables also act as processes to further transform the inputs into outputs of product, drive, and cohesiveness. Stogdill attempts to show that the hypothesized relationships between output variables are in accord with research findings and that these relationships reconcile long-standing discrepancies.

Russell L. Ackoff

5. Systems, Organizations, and Interdisciplinary Research

"Speak English!" said the Eaglet, "I don't know the meaning of half those long words, and, what's more, I don't believe you do either!" And the Eaglet bent down its head to hide a smile: some of the other birds tittered audibly.

Alice's Adventures in Wonderland
Lewis Carroll

INTRODUCTION

WHEN THE ANNOUNCEMENT was made of the establishment of a Systems Research Center at Case a number of my associates in Operations Research asked me how the activity of the Center was to differ from that of the Operations Research Group at Case. My colleagues in the Control (or Systems) Engineering Group at Case were asked similar questions concerning the relationship of their group to the Center. The question could be answered by saying that the Center is designed to faciltate cooperative research and educational activities among the Operations Research Group, the Control Engineering Group, and other systems-oriented activities at Case, particularly the Computing Center.

This answer may satisfy the curiosity of some and discourage probing by others. It is not enough, however, to satisfy or discourage probing by those of us who have some responsibility for the development of this Center. Much more than good will among men is required to make this Center play a significant role in research and education. Part of what is required is a philosophy and a program. A philosophy and program for the Center cannot be expected to spring into existence in a mature state; it must evolve out of proposals, discussion, reformulations, and experience. I should like here to formulate an initial philosophy and program which I hope will lead to constructive discussion, not only at this conference, but afterwards in other cells in which the systems movement is taking shape. There is no doubt in my mind that centers such as we are forming here at Case will develop in profusion in other academic institutions and in industrial and governmental organizations.

I will use my own interdiscipline, Operations Research, as my springboard. But before I take the leap I would like to make some general observations about the systems movement.

First, I believe the systems movement will reach its fruition in an interdiscipline of wider scope and greater significance than has yet been attained. I should like to emphasize that my concern is not with what Systems Research is, but rather with what we can make of it. I consider Operations Research as an intermediate step toward this fruition, a step away from traditional science. Correspondingly, I take Systems Engineering to be an intermediate step toward the same objective, a step away from traditional engineering. I believe Systems Engineering and Operations Research are rapidly converging. What more fitting title for the convergence than "Systems Research."

Operations Research is concerned with increasing the effectiveness of operations of organized man-machine systems. A complete understanding of the significance of this too brief characterization requires at least definitions of system, operations, and organization. I shall deal with the first two very lightly,[1] only enough for my immediate purposes. I shall, however, deal with the concept of organization in more detail because I shall use it as the key to the philosophy and program which I hope to develop. It is in the context of organized man-machine systems, I believe, that we find the most comprehensive demands for departure from the existing context and structure of science and technology.

Now, to the task.

SYSTEMS AND OPERATIONS

The term "system" is used to cover a wide range of phenomena. We speak, for example, of philosophical systems, number systems, communication systems, control systems, educational systems, and weapon systems. Some of these are conceptual constructs and others are physical entities. Initially we can define a system broadly and crudely as *any entity, conceptual or physical, which consists of interdependent parts.* Even without further refinement of this definition it is clear that in systems research we are interested only in those systems which can display activity; that is *behavioral* systems.

It is also apparent that systems research is only concerned with behavioral systems which are subject to control by human beings. Consequently, the solar system—although it may be on the verge of becoming so—is not yet a part of the subject matter of systems research. The relevant domain of such research, then, is controllable behavioral systems.

The essential characteristic of a behavioral system is that it consists of parts each of which displays behavior. Whether or not an entity with parts is considered as a system depends on whether or not we are concerned with the behavior of the parts and their interactions.

---

[1] From *General Systems*, 1960, vol. 5, pp. 1–8. Reprinted with permission of the author and the publisher.

A behavioral system, then, is a conceptual construct as well as a physical entity since such a system may or may not be treated as a system, depending on the way it is conceptualized by the person treating it. For example, we would not normally think of a man who starts the car as a system because we do not distinguish the parts of the man involved in the component acts. We may, however, consider man as a biological system when studying the metabolic process. A physical entity is considered as a system if the outcome of its behavior is conceptualized as the product of the interactions of its parts. Therefore, many entities may be studied either as elements or as systems; it is a matter of the researcher's choice.

The behavior displayed by a system consists of a set of interdependent acts which constitute an operation. "*Operation*" is a complex concept which I do not want to deal with in detail here. Loosely put, a set of acts can be said to constitute an operation if each act is necessary for the occurrence of a desired outcome and if these acts are interdependent. The nature of this interdependence can be precisely defined. The relevant outcome and acts involved in an operation can each be defined by a set of properties which can be treated as variables. The acts are interdependent relative to the outcome if the rate of change of any outcome variable affected by change in any variable describing one of the acts depends on (i.e., is a function of) all the other relevant act-variables. Therefore, if the variables can all be represented by continuous quantities, the derivative of an outcome-variable with respect to any act-variable (if it exists) is a function of all other act-variables. In ordinary language, then, an outcome is the product of a set of interdependent acts if it is more than the sum of (or difference between) these acts.

ORGANIZATION

An organization can be defined as an at least partially self-controlled system which has four esssential characteristics:

1. *Some of its components are animals.* Of particular interest to us, however, are those systems in which the animals are human beings. Wires, poles, switchboards, and telephones may constitute a communication system, but they do not constitute an organization. The employees of a telephone company make up the organization that operates the communication system. Men and equipment together constitute a more inclusive (man-machine) system that we can refer to as organized. Since most organizations utilize machines in a significant way in order to achieve their objectives, the discussion here will be directed to organized man-machine systems.

2. *Responsibility for choices from the sets of possible acts in any specific situation is divided among two or more individuals or groups of individuals.* Each subgroup (consisting of one or more choices of action and the set of choices is divided among two or more subgroups. The classes of action and (hence) the subgroups may be individuated by a

variety of types of characteristics; for example: (*a*) *by function* (e.g., the department of production, marketing, research, finance, and personnel of an industrial organization), (*b*) by geography (e.g., areas of responsibility of the Army), and (*c*) *by time* (e.g., waves of an invading force). The classes of action may, of course, also be defined by combinations of these and other characteristics.

It should be noted that the individuals or groups need not carry out the actions they select; the actions may be performed by machines or other human beings which are programmed or controlled by the individuals so that they act as desired. It should also be noted that the equipment involved and the subgroups may also be considered as systems; that is, as subsystems.

3. *The functionally distinct subgroups are aware of each other's behavior either through communication or observation.* In many laboratory experiments, for example, subjects are given interrelated tasks to perform and are rewarded on the basis of an outcome which is determined by their collective choices. The subjects, however, are not permitted to observe or communicate with each other. In such cases the subjects are unorganized. Allow them to observe each other or communicate and they may become an organization. Put another way, in an organization the human subgroups must be capable of responding to each other either directly or indirectly.

4. *The system has some freedom of choice of both means (courses of action) and ends (desired outcomes).* This implies that at least some parts have alternative courses of action under at least some possible sets of conditions. The simplest type of system, the binary type, has only two possible states: "off" and "on" (e.g., a heating system in a home). More complex *adaptive* systems can behave differently under different conditions, but only in one way under any particular set of conditions (e.g., a ship operated by automatic pilot). Still others are free to choose their means to an end but have no choice of this end (e.g., a computer programmed to play chess). Finally, there are those which are free to choose *how* they will act in any situation (means-free) and why (ends-free). To be sure, such systems are usually constrained in their choices by larger systems which contain them (e.g., government restrictions on a company's behavior). Their efficiency is also affected by either the behavior of other systems (e.g., competition in industry) or natural conditions (e.g., weather).

The four essential characteristics of an organization, then, can be briefly identified as its (*a*) content, (*b*) structure, (*c*) communications, and (*d*) decision-making (choice) procedures.

DESIGN AND OPERATION ORGANIZED (MAN-MACHINE SYSTEMS)

Now we want to consider the significance of these characteristics to one who wants either to create an effective organized system or to improve the operation of an existing one. He has four basic types of approach to organizational effectiveness and combinations thereof. The basic types of approach correspond to the four essential characteristics of organizations.

Content

The content (men and machines) of an organization can be changed. The study of organizational personnel—their selection, training, and utilization—has come to be the domain of *industrial psychology.*[2]

Three fundamentally different approaches to personnel problems have developed within industrial psychology. The first, *personnel* psychology, is primarily concerned with selecting the right man for a specified job. Its principal activity, therefore, is directed toward specifying the relevant characteristics of a job, determining which individual properties are related to its performance, and selecting those individuals who are best equipped for the job. The personnel psychologist, therefore, takes the task-to-be-done as fixed and varies the men.

The personnel psychologist is also interested in modifying man so that he is better capable of performing the task. He attempts such modification through education and training. Here he partially overlaps with the *industrial engineer* who tries to modify the behavior of man more directly. On the basis of time and motion studies the industrial engineer attempts to find those movements which optimize the individual's operations. Industrial engineers, therefore, are preoccupied with manual operations whereas the personnel psychologist tends to concentrate on communication and decision making.

The second psychological approach is that of the human engineer. The human engineer tries to modify the job-to-be-done so that it can be done better by the people available to do it. Here the men are taken to be fixed and the task is taken to be variable. Hence, human engineers, like industrial engineers, are concerned with the acts to be performed, but they try to modify them through the design of the equipment involved in these tasks. It is only natural, therefore, that there has been an increasing convergence of these two approaches.

A third psychological approach takes both the man and the job to be fixed, but the psychological and social environment to be variable. This type of approach yields studies of motivation, incentive systems, interpersonal relationships, group identification or alienation, and the like, and the effect of such variables on human productivity, job satisfaction, and morale. These studies are essentially social-psychological in nature and are epitomized by the early work of Mayo and Roethlisberger and Dickson. Studies of the social environment frequently consider the effect of the noncontent aspects of organization (structure, communication, and control) on human performance. For example, the effect of various types of communication networks on the performance of an individual in the network has been extensively explored. Clearly, such

---

[2] For a very penetrating review of this field see Mason Haire's "Psychology and the Study of Business: Joint Behavioral Sciences," in *Social Science Research on Business: Product and Potential* by Robert A. Dahl, Mason Haire, and Paul F. Lazarsfeld, Columbia University Press, New York, 1959.

studies are related to ones directed at structure, communication, and control; but the emphasis of most of them is on the *individual's* performance rather than on the performance of the organization as a whole.

The other part of the content of man-machine systems is equipment. We have already observed that human engineers are concerned with modifying equipment so that it can be better operated by available personnel. They seldom, however, completely design this equipment. Normally they collaborate with representatives of the traditional branches of engineering in design activity so that the latter can take the capabilities of the operators into account more effectively. Human engineers, therefore, do not replace, they supplement the traditional engineer in his design function.

The individual piece of equipment can frequently be studied as a system. Engineers have increasingly tended to so regard the machine and the weapon. In equipment incorporating automatic controls the systems approach is almost inescapable. In addition, engineers have become increasingly concerned with the interactions of equipment in machine and weapon complexes and so they have become concerned with larger and larger equipment systems. Out of this concern the interdiscipline of *systems engineering* has emerged. The engineer, of course, can no more ignore the human operator than the personnel psychologist can ignore the machine to be operated. The variables which they manipulate, however, remain distinct.

Structure

The second major approach to organizational effectiveness is through its structure, that is, to the way that the necessary physical and mental labor is divided. Although political scientists, economists, and sociologists have concerned themselves with organizational structure, there is as yet no organized body of theory or doctrine of practice on which a unified disciplinary or interdisciplinary applied-research activity can be based. As a consequence most studies of organizational structure, such as those leading to reorganization of a system, are generally done by managers or management consultants whose approach involves more art and common sense than science.

Within the last few decades there has been increasing experimental study of organizational structure. More recently there has begun to appear a body of mathematical theory of organizational structure. Haire has pointed out, however, that as yet:

We do not have much in the way of systematic behavioral data collected for the purpose of testing hypotheses or quantifying variables used in models. For example, we have models dealing with the cost of decentralized decision-making in abstract terms, but we know nothing about the information and decision load that can be supported, or how individuals vary along this dimension . . . We know little about the effect of various communication structures and practices on alternative forms of organizations and their cohesiveness . . . We should be just on the brink of a period of exciting systematic data collection (1959, p. 72).

We may have reached that brink in a provocative new development: operational gaming.[3] In operational gaming organized groups are given problems analogous to real ones, usually with a collapse of the time dimension, and are observed under controlled conditions. We appear to be developing a way of experimenting quantitatively with at least small organizations under conditions which appear to be relevant to actual operations.[4] Difficult problems remain concerning inferences from the game to the real situation, but there is little doubt that within the next few years significant reduction of these difficulties will occur.

Communication

The effectiveness of an organization depends in part on its having "the right information at the right place at the right time." The study of organizational communication is in much the same stage of development as the study of organizational structure. It has no organized body of theory, but it has been developing a doctrine of practice. *Systems and procedures analysts,* stimulated by the numerous installations of automatic data processing systems, have been perfecting techniques of qualitative analysis of information and its flow. It may seem peculiar that this work is predominantly quantitative in light of the highly developed mathematical theory of communication (based to a large extent on the work of Claude Shannon[5]) and its pervasive application to the design of physical communication and information-processing systems.

This theory, however, concerns itself exclusively with the physical aspects of communication and has no relevance to problems involving the meaning of the communication. In Shannon's theory, for example, the measure of information contained in a message is a function of the number of distinct physical messages that could have been sent and the probability associated with the selection of each. The measure makes no reference to the content or significance of the message.

The same thing has been said very well by Haire in his discussion of an article by Rapoport:

He [Rapoport] points out that in dealing with communication among linked individuals we have tended to use the information theory on "bits" developed for the communication engineering. Such a formulation is useful for determining channel capacity . . . but it is not maximally useful for studying decision-making in groups. Here one needs a model of the cognitive

---

[3] For a detailed discussion of the product and potential of this technique see Clayton J. Thomas and Walter L. Deemer, Jr., "The Role of Operational Gaming in Operations Research," Operations Research, 5, 1–27 (February 1957). For illustrative applications see Harold Guetzkow, "The Development of Organizations in a Laboratory," *Management Science*, 3, 380–402 (July 1957).

[4] For some work along these lines performed at Case see D. F. Clark and R. L. Ackoff, "A Report on Some Organizational Experiments," *Operations Research*, 7 (May–June 1959), pp. 279–93.

[5] Claude E. Shannon and Warren Weaver, *The Mathematical Theory of Communication*, The Unviersity of Illinois Press, Urbana, 1949.

aspects of communication theory—a way to indicate the potential of bits for reducing uncertainty about a real state of affairs. Such an approach contrasts with the definition of information in terms of the probabilities of selecting a certain class of messages from a source with given statistical characteristics (1959, p. 7).

There is a growing body of experimental work on the effect of different types of communication networks on organizational (rather than individual) performance, particularly on small groups. (Such experimentation has been stimulated to a large extent by the pioneering work of Alex Bavelas.[6] In addition, the body of special theories is rapidly expanding so that we may well be on the verge of a major break-through in this area. (This work is very effectively summarized in the recent work of Colin Cherry.)[7]

Beginnings toward the construction of a behavioral theory of communication have been made here as Case.[8] This theory has two essential characteristics. First, it does not equate the transmission of information with communication but recognizes three types of message content: information, instruction, and motivation. Information is defined and measured in terms of the effect on the receiver's possibilities and probabilities of choice. Instruction is defined and measured in terms of the effect on the efficiency of the receiver's action, and motivation in terms of the effect of the message on the values which the receiver places on possible outcomes of his choices. A single message may combine all three types of content.

The second essential aspect of this theory is that it provides separate measures of the amount and value of information, instruction, and motivation contained in a message. It therefore distinguishes between information and misinformation, effective and ineffective instruction and motivation.

A theory with these characteristics, whether the one developed at Case or another, increases the possibility of useful quantitative treatment of organizational communication problems.

Decision Making

The last type of approach to organizational problems involves its decision-making procedures. An organization with good personnel and equipment, and an effective structure and communication system may still be inefficient because it does not make effective use of its resources. That is, the operations of the organization may not be efficiently controlled. Control is a matter of setting objectives and directing the or-

---

[6] See, for example, A. Bavelas, "Communication Patterns in Task-oriented Groups," *Journal of Acoustical Society of America*, 22 (1950), pp. 725–30.

[7] Colin Cherry, *On Human Communication*, Technology Press and Wiley, New York, 1957.

[8] See R. L. Ackoff, "Toward a Behavioral Theory of Communication," *Management Science* 4, 218–34 (April 1958).

ganization toward them. It is obtained by efficient decision making by those who manage the operation.

Study of the effective utilization of economic resources in industrial and public organizations is a well-established domain of interest to that splinter group in economics which concerns itself with *micro-economics* and *econometrics*. In the last decade it has produced a rapidly expanding body of theory and research techniques. Concurrent with this development there has been another which deals with a broader class of resources than do the economists alone and, consequently, with a wider variety of organizational decision-problems. This broader interdisciplinary approach to organizational control has come to be known as *operations research*.

The essential characteristics of this interdisciplinary activity lie in its methodology. Out of an analysis of the desired outcomes, objectives of the organization, it develops a measure of performance (P) of the system. It then seeks to model the organization's behavior in the form of an equation in which the measure of performance is equated to some function of those aspects of the system which are subject to management's control (C_i) and which affect the desired outcome, and to those controlled aspects of the system (U_j) which also affect the outcome.

Thus the model takes the form:

$$P = f(C_i, U_j).$$

From the model, values of the control variables are found which maximize (or minimize) the measure of the systems performance:

$$C_i = g(U_j).$$

The solution, therefore consists of a set of rules, one for each control variable, which establishes the value at which that variable should be set for any possible set of values of the uncontrolled variables. In order to employ these rules it is necessary to set up procedures for determining or forecasting the values of the uncontrolled variables.

It will be recognized that this procedure is one by which equipment systems should be ideally designed. In design one should also develop a consolidated measure of system performance, and identify the variables which the designer can control as well as those uncontrolled aspects of the system or its environment which will affect its performance. Unfortunately, in many cases such a model of a desired equipment system cannot be constructed because of our ignorance. For example, I have not yet seen a good single consolidated measure proposed for the performance of an aircraft. Nor is there sufficient knowledge to relate any of the less perfect available measures of performance to the large number of design variables of such craft. As a consequence, design is currently accomplished by a combination of scientific analysis, intuition, and esthetic considerations. It should be recognized, however, that current design procedures are only an evolutionary stage which will be replaced as rapidly as possible by effective modelling and the extraction of solutions from the resulting models.

INTEGRATED RESEARCH INTO ORGANIZED
MAN-MACHINE SYSTEMS

As we have seen there is a large group of disciplines and interdisciplines dedicated to studying various aspects of organized man-machine systems. The fact that the subject is so dissected leads to several residual problems. Suppose that an organizational problem is completely solvable by one of the disciplines we have considered. How is the manager who controls the system to know which one? Or, for that matter, how is a practitioner of any one discipline to know in a particular case if another discipline is better equipped to handle the problem that is his? It would be rare indeed if a representative of any one of these disciplines did not feel that his approach to a particular organizational problem would be very fruitful, if not the most fruitful. The danger that results can perhaps best be illustrated by a report that may be apocryphal but which makes the point very well.

The manager of a large office building received an increasing number of complaints about the elevator service in the building. He engaged a group of engineers to study the situation and to make recommendations for improvements if they were necessary. The engineers found that the tenants were indeed receiving poor service and considered three possible ways of decreasing the average waiting time. They considered adding elevators, replacing the existing ones by faster ones, and assigning elevators to serve specific floors. The latter turned out to be inadequate and the first two were prohibitively expensive to the manager. He called together his staff to consider the report by the engineers. Among those present was his personnel director, a psychologist.

This young man was struck by the fact that people became impatient with a wait which seemed so short to him. On reflection he became convinced that their annoyance was due to the fact that they had to stand inactive in a crowded lobby for this period. This suggested a solution to him which he offered to the manager, and because it was so inexpensive the manager decided to try it. Complaints stopped immediately. The psychologist had suggested installing large mirrors on the walls of the lobbies where people waited for the elevators.

Those who have worked with systems can recall many such incidents, that is, many except those in which they played a role similar to that of the engineers in the one just recounted. There is undoubtedly a considerable waste of research effort and a considerable failure to obtain successful solutions to system problems just because the wrong discipline was involved. How can this be avoided? We will return to this question in a moment, but now let's look at the second residual problem.

In most problems involving organized man-machine systems each of the disciplines we have mentioned might make a significant improvement in the operations. But as systems analysts know, few of the problems that arise can adequately be handled within any one discipline. Such systems are not fundamentally mechanical, chemical, biological, psychological, social, economic, political, or ethical. These are merely different ways of looking at such systems. Complete understanding of such systems requires an integration of these perspectives. By integration I do not mean a synthesis of results obtained by independently conducted undisciplinary studies, but rather results obtained from studies in the process of which disciplinary perspectives have been synthesized. The integration must come during, not after, the performance of the research.

We must stop acting as though nature were organized into disciplines in the same way that universities are. The division of labor along disciplinary lines is no longer an efficient one. In fact, it has become so inefficient that even some academic institutions have begun to acknowledge the fact. What can be done about it?

If the various disciplines involved in studying systems are brought together organizationally, this would help solve the first type of problem because it would then be possible to have each discipline examine each problem that arises. Presumably, by discussion the interdisciplinary group could determine which of the disciplines is best suited to handle a particular problem if it could be handled exclusively by one discipline.

This type of proximity between the disciplines is not enough, however, to effect a truly interdisciplinary approach to systems. The various disciplines must be able to work together effectively on the problem, not merely before and after the problem is studied. To accomplish this some specific steps should be taken.

First, it will be necessary to construct mathematical models of systems in which content, structure, communication, and decision variables all appear. For example, several cost variables are usually included in a typical operations research model. These are either taken as uncontrollable or as controllable only by manipulating such other variables as quantity purchased or produced, time of purchase or production, number and type of facilities, and so on. These costs, however, are always dependent on human performance, but the relevant variables dealing with personnel, structure, and communication seldom appear in such models. To a large extent this is due to the lack of operational definitions of many of these variables, and, consequently, to the absence of suitable measures in terms of which they can be characterized.

In order to be able to construct truly interdisciplinary models of systems, then, it will be necessary to relate conceptually the variables dealt with in each of the disciplines which should be involved in systems research. This is a formidable task, but a beginning has been made. At the Institute of Experimental Method (which operated at the University of Pennsylvania in 1946 and 1947) a monograph was produced which attempted an interdisciplinary conceptual system.[9] More recently Rudner and Wolfson have been extending this work, particularly as it applies to organizations.[10]

---

[9] C. West Churchman and Russell L. Ackoff, *Psychologistics* (mimeographed) University of Pennsylvania Research Fund, 1947.

[10] Richard S. Rudner and Robert J. Wolfson, *Notes on a Constructional Framework for a Theory of Organizational Decision Making*, Working Paper No. 3, Management Science Nucleus, Institute of Industrial Relations, University of California, Berkeley (1958).

The second requirement is for the self-conscious development of a sound methodology for systems research. This can be accomplished by turning systems research in on itself since systems research is itself an operation performed by man-machine systems. The methods and techniques of the traditional sciences and technologies are not good enough for the job which must be done. Let me illustrate this point briefly by reference to only one of many methodological problems that might be discussed.

The performance objectives of most systems can be stated in terms of a number of variables. For example, in a truck we seek such characteristics as speed, rapid rate of acceleration, long range, large pay-load, low cost of operation, and so on. We cannot really optimize the design of an aircraft unless we can in some way amalgamate these performance considerations into a single measure of performance. In a production system, for example, we may have to amalgamate a measure of cost, a measure of the length of time required to fill orders, and a measure of the frequency and duration of shortages. In order to accomplish such an amalgamation of measures we must be able to transform all the scales involved into some common (standard) scale. We have much to learn about how to find the appropriate transformations or "trade-off" functions.

The criterion of "best performance" can be shown to depend further on our ability to find the relative value (or utility) of increments along the scales used to measure performance. For example, to a man who is destitute, the value of twenty dollars is clearly not twice that of ten dollars. If it were he would prefer a 51% chance of getting twenty dollars to the certainty of getting ten dollars. We have experimental evidence that this is not the case. It is important, therefore, to increase our ability to measure the value of increments of performance along whatever scale(s) they are measured.

The third requirement for effective systems research is effective education and organization of representatives from practically all the scientific and technological disciplines. Systems research will not be the only beneficiary of such education and organization. The contributing disciplines will be significantly benefited. It is not accidental that so much of the important work currently going on in many disciplines is being done by persons trained in other disciplines. For example, the most important work being done in the behavioral sciences, in my opinion, is that being reported in the two new journals, *Behavioral Science* and *Conflict Resolution*. Most of the contributors to these journals were not trained in the behavioral sciences. Major work in learning theory, for example, is being done by Merrill M. Flood,[11] a mathematician at the University of Michigan and Frederick Mosteller,[12] a statistician at Harvard. On the other hand, measurement theory,

---

[11] See Merrill M. Flood, "Game-Learning Theory and Some Decision-Making Experiments," in *Decision Processes*, edited by R. M. Thrall, C. H. Coombs, and R. L. Davis, Wiley, New York, 1954.

[12] See, for example, R. R. Bush, Frederick Mosteller, and G. L. Thompson, "A Formal Structure for Multiple-Choice Situations," Ibid.

which has been thought of as the domain of physics since the work of Norman Campbell[13] has been significantly extended by psychologists such as S. S. Stevens[14] and Clyde Coombs[15] and philosophers of science such as C. West Churchman.[16]

The implications of these observations to the educational process are important.

No single individual can be educated so as to be expert in all the disciplinary approaches to systems. It is difficult enough to make him expert in one. We can, however, educate him to an awareness of what others know and can do in systems work and motivate him to desire to work collaboratively with them. Scientific snobbery must go. Systems research cannot thrive where it prevails.

It is my feeling that the two most important steps that can be taken to break down barriers to effective interdisciplinary collaboration are:

1. to elevate those trained in each discipline to a uniformly high level of competence in mathematics and statistics, and
2. to educate all students in science and technology to a thorough understanding of scientific method in its most general sense.

Mathematics is the language of science and like all languages it molds the concepts and thinking processes of those who are familiar with it. In my opinion, the behavioral sciences are less mathematically oriented than are the physical sciences not so much because of the difference between the types of phenomena they study as because of the difference of language in which their practitioners think about these types of phenomena. On the other hand, existing mathematics is not adequate to provide a complete basis for quantification in the behavioral sciences because it was developed as a handmaiden of the physical sciences. The greatest challenges to mathematics, it seems to me, are increasingly to be found in the behavioral rather than the physical sciences.

Through an exposure to the accomplishments and problems of scientific method, the student can best come to understand the underlying unity of science and hence of its disciplines. Only by a thorough analysis of research procedures in each of the sciences can one come to an appreciation of the interdependence of the sciences. For example, in this way a student can come to realize that progress in physical science involves (among other things) continuous reduction of observer errors and that perceptual psychology and human engineering have a great deal to contribute to such error reduction. He can also come to understand that the social environment of a physical laboratory affects the reliability of measurements of even simple physical quantities. Through

---

[13] Norman Robert Campbell, Foundations of Science (Formerly titled: *Physics the Elements*), Dover Publications, Inc., New York, 1957.

[14] See "Mathematics, Measurements, and Psychophysics," in *Handbook of Experimental Psychology*, edited by S. S. Stevens, Wiley, New York, 1951.

[15] See Coomb, C. H., Raiffa, H., and Thrall, R. M., "Some Views on Mathematical Models and Measurement Theory," in *Decision Processes* (see Note 9).

[16] See *Measurement: Definitions and Theories*, edited by C. West Churchman and Philburn Ratoosh, Wiley, New York, 1959.

the study of scientific method, then, he can begin to see the scientific crusade for the reduction of error as one that is necessarily interdisciplinary in character.

For the systems researcher methodological self-consciousness has an added importance because, as already observed, research itself is frequently an operation performed by an organized system. As such it is susceptible to the same kind of analysis as are other systems. The possibility of so studying research holds great promise for future increases in research effectiveness in all areas of science and technology.

In summary, then, if systems research is to develop the capacity to conduct effective research on complex as well as simple types of systems we must do the following:

1. Develop a conceptual system which relates the concepts applied to systems by various disciplines and reduce them to quantities which are measurable along compatible scales.
2. Develop methodology better adapted to unique aspects of systems research.
3. Design and put in operation an educational program which produces the kind of researcher who can conduct systems research in an interdisciplinary context.

The era of systems research—and I think this is an era—can become one in which not only science is effectively reorganized, but one in which the educational process is similarly reorganized. The exciting and challenging character of systems research, then, is not to be found so much in what it is, but in what can be made of it and the research and educational institutions that house it.

William G. Scott

6. Organization Theory: An Overview and an Appraisal

MAN IS INTENT on drawing himself into a web of collectivized patterns. "Modern man has learned to accommodate himself to a world increasingly organized. The trend toward ever more explicit and consciously drawn relationships is profound and sweeping; it is marked by depth no less than by extension."[1] This comment by Seidenberg nicely summarizes the pervasive influence of organization in many forms of human activity.

Some of the reasons for intense organizational activity are found in the fundamental transitions which revolutionized our society, changing it from a rural culture, to a culture based on technology, industry, and the city. From these changes, a way of life emerged characterized by the *proximity* and *dependency* of people on each other. Proximity and dependency, as conditions of social life, harbor the threats of human conflict, capricious antisocial behavior, instability of human relationships, and uncertainty about the nature of the social structure with its concomitant roles.

Of course, these threats to social integrity are present to some degree in all societies, ranging from the primitive to the modern. But, these threats become dangerous when the harmonious functioning of a society rests on the maintenance of a highly intricate, delicately balanced form of human collaboration. The civilization we have created depends on the preservation of a precarious balance. Hence, disrupting forces impinging on this shaky form of collaboration must be eliminated or minimized.

Traditionally, organization is viewed as a vehicle for accomplishing goals and objectives. While this approach is useful, it tends to obscure the inner workings and internal purposes of organization itself. Another fruitful way of treating organization is as a mechanism having the ultimate purpose of offsetting those forces which undermine human

---

From *Academy of Management Journal*, 1961, vol. 4, pp. 7–26. Reprinted with permission of the author and the publisher.

[1] Roderick Seidenburg, *Post Historic Man* (Boston: Beacon Press, 1951), p. 1.

collaboration. In this sense, organization tends to minimize conflict, and to lessen the significance of individual behavior which deviates from values that the organization has established as worthwhile. Further, organization increases stability in human relationships by reducing uncertainty regarding the nature of the system's structure and the human roles which are inherent to it. Corollary to this point, organization enhances the predictability of human action, because it limits the number of behavioral alternatives available to an individual. As Presthus points out:

Organization is defined as a system of structural interpersonal relations . . . individuals are differentiated in terms of authority, status, and role with the result that personal interaction is prescribed. . . . Anticipated reactions tend to occur, while ambiguity and spontaneity are decreased.[2]

In addition to all of this, organization has built-in safeguards. Besides prescribing acceptable forms of behavior for those who elect to submit to it, organization is also able to counterbalance the influence of human action which transcends its established patterns.[3]

Few segments of society have engaged in organizing more intensively than business.[4] The reason is clear. Business depends on what organization offers. Business needs a system of relationships among functions; it needs stability, continuity, and predictability in its internal activities and external contacts. Business also appears to need harmonious relationships among the people and processes which make it up. Put another way, a business organization has to be free, relatively, from destructive tendencies which may be caused by divergent interests.

As a foundation for meeting these needs rests administrative science. A major element of this science is organization theory, which provides the grounds for management activities in a number of significant areas of business endeavor. Organization theory, however, is not a homogeneous science based on generally accepted principles. Various theories of organization have been, and are being evolved. For example, something called "modern organization theory" has recently emerged, raising the wrath of some traditionalists, but also capturing the imagination of a rather elite *avant-garde*.

The thesis of this paper is that modern organization theory, when stripped of its irrelevancies, redundancies, and "speech defects," is a logical and vital evolution in management thought. In order for this

---

[2] Robert V. Presthus, "Toward a Theory of Organizational Behavior," *Administrative Science Quarterly*, June 1958, p. 50.

[3] Regulation and predictability of human behavior are matters of degree varying with different organizations on something of a continuum. At one extreme are bureaucratic type organizations with tight bonds of regulation. At the other extreme are voluntary associations, and informal organizations with relatively loose bonds of regulation.

This point has an interesting sidelight. A bureaucracy with tight controls and a high degree of predictability of human action appears to be unable to distinguish between destructive and creative deviations from established values. Thus the only thing which is safeguarded is the *status quo*.

[4] The monolithic institutions of the military and government are other cases of organizational preoccupation.

thesis to be supported, the reader must endure a review and appraisal of more traditional forms of organization theory which may seem elementary to him.

In any event, three theories of organization are having considerable influence on management thought and practice. They are arbitrarily labeled in this paper as the classical, the neoclassical, and the modern. Each of these is fairly distinct; but they are not unrelated. Also, these theories are on-going, being actively supported by several schools of management thought.

THE CLASSICAL DOCTRINE

For lack of a better method of identification, it will be said that the classical doctrine deals almost exclusively with the *anatomy of formal organization*. This doctrine can be traced back to Frederick W. Taylor's interest in functional foremanship and planning staffs. But most students of management thought would agree that in the United States, the first systematic approach to organization, and the first comprehensive attempt to find organizational universals, is dated 1931 when Mooney and Reiley published *Onward Industry*.[5] Subsequently numerous books, following the classical vein, have appeared. Two of the more recent are Brech's, *Organization*[6] and Allen's, *Management and Organization*.[7]

Classical organization theory is built around four key pillars. They are the division of labor, the scalar and functional processes, structure, and span of control. Given these major elements just about all of classical organization theory can be derived.

1. *The division of labor* is without doubt the cornerstone among the four elements.[8] From it the other elements flow as corollaries. For example, *scalar* and *functional* growth requires specialization and departmentalization of functions. Organization *structure* is naturally dependent upon the direction which specialization of activities travels in company development. Finally, *span of control* problems result from the number of specialized functions under the jurisdiction of a manager.

2. *The scalar and functional processes* deal with the vertical and horizontal growth of the organization, respectively.[9] The scalar process refers to the growth of the chain of command, the delegation of authority and responsibility, unity of command, and the obligation to report.

The division of the organization into specialized parts and the re-

---

[5] James D. Mooney and Alan C. Reiley, *Onward Industry* (New York: Harper and Brothers, 1931). Later published by James D. Mooney under the title *Principles of Organization.*

[6] E. F. L. Brech, *Organization* (London: Longmans, Green and Company, 1957).

[7] Louis A. Allen, *Management and Organization* (New York: McGraw-Hill Book Company, 1958).

[8] Usually the division of labor is treated under a topical heading of departmentation, see for example: Harold Koontz and Cyril O'Donnel, *Principles of Management* (New York: McGraw-Hill Book Company, 1959), chapter 7.

[9] These processes are discussed at length in Ralph Currier Davis, *The Fundamentals of Top Management* (New York: Harper and Brothers, 1951), chapter 7.

grouping of the parts into compatible units are matters pertaining to the functional process. This process focuses on the horizontal evolution of the line and staff in a formal organization.

3. *Structure* is the logical relationships of functions in an organization, arranged to accomplish the objectives of the company efficiently. Structure implies system and pattern. Classical organization theory usually works with two basic structures, the line and the staff. However, such activities as committee and liaison functions fall quite readily into the purview of structural considerations. Again, structure is the vehicle for introducing logical and consistent relationships among the diverse functions which comprise the organization.[10]

4. *The span of control* concept relates to the number of subordinates a manager can effectively supervise. Graicunas has been credited with first elaborating the point that there are numerical limitations to the subordinates one man can control.[11] In a recent statement on the subject, Brech points out, "span" refers to ". . . the number of persons, themselves carrying managerial and supervisory responsibilities, for whom the senior manager retains his over-embracing responsibility of direction and planning, coordination, motivation, and control."[12] Regardless of interpretation, span of control has significance, in part, for the shape of the organization which evolves through growth. Wide span yields a flat structure; short span results in a tall structure. Further, the span concept directs attention to the complexity of human and functional interrelationships in an organization.

It would not be fair to say that the classical school is unaware of the day-to-day administrative problems of the organization. Paramount among these problems are those stemming from human interactions. But the interplay of individual personality, informal groups, intra-organizational conflict, and the decision-making processes in the formal structure appears largely to be neglected by classical organization theory. Additionally, the classical theory overlooks the contributions of the behavioral sciences by failing to incorporate them in its doctrine in any systematic way. In summary, classical organization theory has relevant insights into the nature of organization, but the value of this theory is limited by its narrow concentration on the formal anatomy of organization.

NEOCLASSICAL THEORY OF ORGANIZATION

The neoclassical theory of organization embarked on the task of compensating for some of the deficiencies in classical doctrine. The neoclassical school is commonly identified with the human relations movement. Generally, the neoclassical approach takes the postulates of the classical school, regarding the pillars of organization as givens. But

---

[10] For a discussion of structure see: William H. Newman, *Administrative Action* (Englewood Cliffs: Prentice-Hall, Incorporated, 1951), chapter 16.

[11] V. A. Graicunas, "Relationships in Organization," *Papers on the Science of Administration* (New York: Columbia University, 1937).

[12] Brech, op. cit., p. 78.

these postulates are regarded as modified by people, acting independently or within the context of the informal organization.

One of the main contributions of the neoclassical school is the introduction of behavioral sciences in an integrated fashion into the theory of organization. Through the use of these sciences, the human relationists demonstrate how the pillars of the classical doctrine are affected by the impact of human actions. Further, the neoclassical approach includes a systematic treatment of the informal organization, showing its influence on the formal structure.

Thus, the neoclassical approach to organization theory gives evidence of accepting classical doctrine, but superimposing on it modifications resulting from individual behavior, and the influence of the informal group. The inspiration of the neoclassical school were the Hawthorne studies.[13] Current examples of the neoclassical approach are found in human relations books like Gardner and Moore, *Human Relations in Industry*,[14] and Davis, *Human Relations in Business*.[15] To a more limited extent, work in industrial sociology also reflects a neoclassical point of view.[16]

It would be useful to look briefly at some of the contributions made to organization theory by the neoclassicists. First to be considered are modifications of the pillars of classical doctrine; second is the informal organization.

Examples of the Neoclassical Approach to the Pillars of Formal Organization Theory

1. The *division of labor* has been a long standing subject of comment in the field of human relations. Very early in the history of industrial psychology, study was made of industrial fatigue and monotony caused by the specialization of the work.[17] Later, attention shifted to the isolation of the worker, and his feeling of anonymity resulting from insignificant jobs which contributed negligibly to the final product.[18]

Also, specialization influences the work of management. As an organization expands, the need concomitantly arises for managerial motivation and coordination of the activities of others. Both motivation and coordination in turn relate to executive leadership. Thus, in part, stemming from the growth of industrial specialization, the neoclassical school has developed a large body of theory relating to motivation, co-

---

[13] See: F. J. Roethlisberger and William J. Dickson, *Management and the Worker* (Cambridge: Harvard University Press, 1939).

[14] Burleigh B. Gardner and David G. Moore, *Human Relations in Industry* (Homewood: Richard D. Irwin, 1955).

[15] Keith Davis, *Human Relations in Business* (New York: McGraw-Hill Book Company, 1957).

[16] For example see: Delbert C. Miller and William H. Form, *Industrial Sociology* (New York: Harper and Brothers, 1951).

[17] See: Hugo Munsterberg, *Psychology and Industrial Efficiency* (Boston: Houghton Mifflin Company, 1913).

[18] Probably the classic work is: Elton Mayo, *The Human Problems of an Industrial Civilization* (Cambridge: Harvard University, 1946, first printed 1933).

ordination, and leadership. Much of this theory is derived from the social sciences.

2. Two aspects of the *scalar and functional* processes which have been treated with some degree of intensity by the neoclassical school are the delegation of authority and responsibility, and gaps in or overlapping of functional jurisdictions. The classical theory assumes something of perfection in the delegation and functionalization processes. The neoclassical school points out that human problems are caused by imperfections in the way these processes are handled.

For example, too much or insufficient delegation may render an executive incapable of action. The failure to delegate authority and responsibility equally may result in frustration for the delegatee. Overlapping of authorities often causes clashes in personality. Gaps in authority cause failures in getting jobs done, with one party blaming the other for shortcomings in performance.[19]

The neoclassical school says that the scalar and functional processes are theoretically valid, but tend to deteriorate in practice. The ways in which they break down are described, and some of the human causes are pointed out. In addition the neoclassicists make recommendations, suggesting various "human tools" which will facilitate the operation of these processes.

3. *Structure* provides endless avenues of analysis for the neoclassical theory of organization. The theme is that human behavior disrupts the best laid organizational plans, and thwarts the cleanness of the logical relationships founded in the structure. The neoclassical critique of structure centers on frictions which appear internally among people performing different functions.

Line and staff relations is a problem area, much discussed, in this respect. Many companies seem to have difficulty keeping the line and staff working together harmoniously. Both Dalton[20] and Juran[21] have engaged in research to discover the causes of friction, and to suggest remedies.

Of course, line-staff relations represent only one of the many problems of structural frictions described by the neoclassicists. As often as not, the neoclassicists will offer prescriptions for the elimination of conflict in structure. Among the more important harmony-rendering formulae are participation, junior boards, bottom-up management, joint committees, recognition of human dignity, and "better" communication.

4. An executive's *span of control* is a function of human determinants, and the reduction of span to a precise, universally applicable ratio is silly, according to the neoclassicists. Some of the determinants of span are individual differences in managerial abilities, the type of

---

[19] For further discussion of the human relations implications of the scalar and functional processes see: Keith Davis, op. cit., pp. 60–66.

[20] Melville Dalton, "Conflicts between Staff and Line Managerial Officers," *American Sociological Review,* June 1950, pp. 342–51.

[21] J. M. Juran, "Improving the Relationship between Staff and Line," *Personnel,* May 1956, pp. 515–24.

people and functions supervised, and the extent of communication effectiveness.

Coupled with the span of control question are the human implications of the type of structure which emerges. That is, is a tall structure with a short span or a flat structure with a wide span more conducive to good human relations and high morale? The answer is situational. Short span results in tight supervision; wide span requires a good deal of delegation with looser controls. Because of individual and organizational differences, sometimes one is better than the other. There is a tendency to favor the looser form of organization, however, for the reason that tall structures breed autocratic leadership, which is often pointed out as a cause of low morale.[22]

The Neoclassical View of the Informal Organization

Nothing more than the barest mention of the informal organization is given even in the most recent classical treatises on organization theory.[23] Systematic discussion of this form of organization has been left to the neoclassicists. The informal organization refers to people in group associations at work, but these associations are not specified in the "blueprint" of the formal organization. The informal organization means natural groupings of people in the work situation.

In a general way, the informal organization appears in response to the social need—the need of people to associate with others. However, for analytical purposes, this explanation is not particularly satisfying. Research has produced the following, more specific determinants underlying the appearance of informal organizations.

1. The *location* determinant simply states that in order to form into groups of any lasting nature, people have to have frequent face-to-face contact. Thus, the geography of physical location in a plant or office is an important factor in predicting who will be in what group.[24]

2. *Occupation* is key factor determining the rise and composition of informal groups. There is a tendency for people performing similar jobs to group together.[25]

3. *Interests* are another determinant for informal group formation. Even though people might be in the same location, performing similar jobs, differences of interest among them explain why several small, instead of one large, informal organizations emerge.

4. *Special issues* often result in the formation of informal groups, but this determinant is set apart from the three previously mentioned. In this case, people who do not necessarily have similar interests, occupations, or locations may join together for a common cause. Once the

[22] Gardner and Moore, op. cit., pp. 237–43.

[23] For example: Brech, op. cit., pp. 27–29; and Allen, op. cit., pp. 61–62.

[24] See: Leon Festinger, Stanley Schachter, and Kurt Back, *Social Pressures in Informal Groups* (New York: Harper and Brothers, 1950), pp. 153–63.

[25] For example see: W. Fred Cottrell, *The Railroader* (Palo Alto: The Stanford University Press, 1940), chapter 3.

issue is resolved, then the tendency is to revert to the more "natural" group forms.[26] Thus, special issues give rise to a rather impermanent informal association; groups based on the other three determinants tend to be more lasting.

When informal organizations come into being they assume certain characteristics. Since understanding these characteristics is important for management practice, they are noted below:

1. Informal organizations act as agencies of *social control*. They generate a culture based on certain norms of conduct which, in turn, demands conformity from group members. These standards may be at odds with the values set by the formal organization. So an individual may very well find himself in a situation of conflicting demands.

2. The form of human interrelationships in the informal organization requires *techniques of analysis* different from those used to plot the relationships of people in a formal organization. The method used for determining the structure of the informal group is called sociometric analysis. Sociometry reveals the complex structure of interpersonal relations which is based on premises fundamentally unlike the logic of the formal organization.

3. Informal organizations have *status and communication* systems peculiar to themselves, not necessarily derived from the formal systems. For example, the grapevine is the subject of much neoclassical study.

4. Survival of the informal organization requires stable continuing relationships among the people in them. Thus, it has been observed that the informal organization *resists change*.[27] Considerable attention is given by the neoclassicists to overcoming informal resistance to change.

5. The last aspect of analysis which appears to be central to the neoclassical view of the informal organization is the study of the *informal leader*. Discussion revolves around who the informal leader is, how he assumes this role, what characteristics are peculiar to him, and how he can help the manager accomplish his objectives in the formal organization.[28]

This brief sketch of some of the major facets of informal organization theory has neglected, so far, one important topic treated by the neoclassical school. It is the way in which the formal and informal organizations interact.

A conventional way of looking at the interaction of the two is the "live and let live" point of view. Management should recognize that the informal organization exists, nothing can destroy it, and so the executive might just as well work with it. Working with the informal organization

---

[26] Except in cases where the existence of an organization is necessary for the continued maintenance of employee interest. Under these conditions the previously informal association may emerge as a formal group, such as a union.

[27] Probably the classic study of resistance to change is: Lester Coch and John R. P. French, Jr., "Overcoming Resistance to Change," in Schuyler Dean Hoslett (editor) *Human Factors in Management* (New York: Harper and Brothers, 1951) pp. 242–68.

[28] For example see: Robert Saltonstall, *Human Relations in Administration* (New York: McGraw-Hill Book Company, 1959), pp. 330–31; and Keith Davis, op. cit., pp. 99–101.

involves not threatening its existence unnecessarily, listening to opinions expressed for the group by the leader, allowing group participation in decision-making situations, and controlling the grapevine by prompt release of accurate information.[29]

While this approach is management centered, it is not unreasonable to expect that informal group standards and norms could make themselves felt on formal organizational policy. An honestly conceived effort by managers to establish a working relationship with the informal organization could result in an association where both formal and informal views would be reciprocally modified. The danger which at all costs should be avoided is that "working with the informal organization" does not degenerate into a shallow disguise for human manipulation.

Some neoclassical writing in organization theory, especially that coming from the management-oriented segment of this school, gives the impression that the formal and informal organizations are distinct, and at times, quite irreconcilable factors in a company. The interaction which takes place between the two is something akin to the interaction between the company and a labor union, or a government agency, or another company.

The concept of the social system is another approach to the interactional climate. While this concept can be properly classified as neoclassical, it borders on the modern theories of organization. The phrase "social system" means that an organization is a complex of mutually interdependent, but variable, factors.

These factors include individuals and their attitudes and motives, jobs, the physical work setting, the formal organization, and the informal organizations. These factors, and many others, are woven into an overall pattern of interdependency. From this point of view, the formal and informal organizations lose their distinctiveness, but find real meaning, in terms of human behavior, in the operation of the system as a whole. Thus, the study of organization turns away from descriptions of its component parts, and is refocused on the system of interrelationships among the parts.

One of the major contributions of the Hawthorne studies was the integration of Pareto's idea of the social system into a meaningful method of analysis for the study of behavior in human organizations.[30] This concept is still vitally important. But unfortunately some work in the field of human relations undertaken by the neoclassicists has overlooked, or perhaps discounted, the significance of this consideration.[31]

The fundamental insight regarding the social system, developed and applied to the industrial scene by the Hawthorne researchers, did not find much extension in subsequent work in the neoclassical vein. Indeed,

---

[29] For an example of this approach see: John T. Doutt, "Management Must Manage the Informal Group, Too," *Advanced Management*, May 1959, pp. 26–28.

[30] See: Roethlisberger and Dickson, op. cit., chapter 24.

[31] A check of management human relations texts, the organization and human relations chapters of principles of management texts, and texts on conventional organization theory for management courses reveals little or no treatment of the concept of the social system.

the neoclassical school after the Hawthorne studies generally seemed content to engage in descriptive generalizations, or particularized empirical research studies which did not have much meaning outside their own context.

The neoclassical school of organization theory has been called bankrupt. Criticisms range from, "human relations is a tool for cynical puppeteering of people," to "human relations is nothing more than a trifling body of empirical and descriptive information." There is a good deal of truth in both criticisms, but another appraisal of the neoclassical school of organization theory is offered here. The neoclassical approach has provided valuable contributions to lore of organization. But, like the classical theory, the neoclassical doctrine suffers from incompleteness, a shortsighted perspective, and lack of integration among the many facets of human behavior studied by it. Modern organization theory has made a move to cover the shortcomings of the current body of theoretical knowledge.

MODERN ORGANIZATIONAL THEORY

The distinctive qualities of modern organization theory are its conceptual-analytical base, its reliance on empirical research data and, above all, its integrating nature. These qualities are framed in a philosophy which accepts the premise that the only meaningful way to study organization is to study it as a system. As Henderson put it, the study of a system must rely on a method of analysis, ". . . involving the simultaneous variations of mutually dependent variables."[32] Human systems, of course, contain a huge number of dependent variables which defy the most complex simultaneous equations to solve.

Nevertheless, system analysis has its own peculiar point of view which aims to study organization in the way Henderson suggests. It treats organization as a system of mutually dependent variables. As a result, modern organization theory, which accepts system analysis, shifts the conceptual level of organization study above the classical and neoclassical theories. Modern organization theory asks a range of interrelated questions which are not seriously considered by the two other theories.

Key among these questions are: (1) What are the strategic parts of the system? (2) What is the nature of their mutual dependency? (3) What are the main processes in the system which link the parts together, and facilitate their adjustment to each other? (4) What are the goals sought by systems?[33]

Modern organization theory is in no way a unified body of thought. Each writer and researcher has his special emphasis when he considers the system. Perhaps the most evident unifying thread in the study of systems is the effort to look at the organization in its totality. Repre-

---

[32] Lawrence J. Henderson, *Pareto's General Sociology* (Cambridge: Harvard University Press, 1935), p. 13.

[33] There is another question which cannot be treated in the scope of this paper. It asks, what research tools should be used for the study of the system?

sentative books in this field are March and Simon, *Organizations*,[34] and Haire's anthology, *Modern Organization Theory*.[35]

Instead of attempting a review of different writers' contributions to modern organization theory, it will be more useful to discuss the various ingredients involved in system analysis. They are the parts, the interactions, the processes, and the goals of systems.

The Parts of the System and Their Interdependency

The first basic part of the system is the *individual*, and the personality structure he brings to the organization. Elementary to an individual's personality are motives and attitudes which condition the range of expectancies he hopes to satisfy by participating in the system.

The second part of the system is the formal arrangement of functions, usually called the *formal organization*. The formal organization is the interrelated pattern of jobs which make up the structure of a system. Certain writers, like Argyris, see a fundamental conflict resulting from the demands made by the system, and the structure of the mature, normal personality. In any event, the individual has expectancies regarding the job he is to perform; and, conversely, the job makes demands on, or has expectancies relating to, the performance of the individual. Considerable attention has been given by writers in modern organization theory to incongruencies resulting from the interaction of organizational and individual demands.[36]

The third part in the organization system is the *informal organization*. Enough has been said already about the nature of this organization. But it must be noted that an interactional pattern exists between the individual and the informal group. This interactional arrangement can be conveniently discussed as the mutual modification of expectancies. The informal organization has demands which it makes on members in terms of anticipated forms of behavior, and the individual has expectancies of satisfaction he hopes to derive from association with people on the job. Both these sets of expectancies interact, resulting in the individual modifying his behavior to accord with the demands of the group, and the group, perhaps, modifying what it expects from an individual because of the impact of his personality on group norms.[37]

Much of what has been said about the various expectancy systems in an organization can also be treated using status and role concepts. Part of modern organization theory rests on research findings in social-psychology relative to reciprocal patterns of behavior stemming from role demands generated by both the formal and informal organizations, and

---

[34] James G. March and Herbert A. Simon, *Organizations* (New York: John Wiley and Sons, 1958).

[35] Mason Haire, (editor) *Modern Organization Theory* (New York: John Wiley and Sons, 1959).

[36] See Chris Argyris, *Personality and Organization* (New York: Harper and Brothers, 1957), esp. chapters 2, 3, 7.

[37] For a larger treatment of this subject see: George C. Homans, *The Human Group* (New York: Harcourt, Brace and Company, 1950), chapter 5.

role perceptions peculiar to the individual. Bakke's *fusion process* is largely concerned with the modification of role expectancies. The fusion process is a force, according to Bakke, which acts to weld divergent elements together for the preservation of organizational integrity.[38]

The fifth part of system analysis is the *physical setting* in which the job is performed. Although this element of the system may be implicit in what has been said already about the formal organization and its functions, it is well to separate it. In the physical surroundings of work, interactions are present in complex man-machine systems. The human "engineer" cannot approach the problems posed by such interrelationships in a purely technical, engineering fashion. As Haire says, these problems lie in the domain of the social theorist.[39] Attention must be centered on responses demanded from a logically ordered production function, often with the view of minimizing the error in the system. From this standpoint, work cannot be effectively organized unless the psychological, social, and physiological characteristics of people participating in the work environment are considered. Machines and processes should be designed to fit certain generally observed psychological and physiological properties of men, rather than hiring men to fit machines.

In summary, the parts of the system which appear to be of strategic importance are the individual, the formal structure, the informal organization, status and role patterns, and the physical environment of work. Again, these parts are woven into a configuration called the organizational system. The processes which link the parts are taken up next.

The Linking Processes

One can say, with a good deal of glibness, that all the parts mentioned above are interrelated. Although this observation is quite correct, it does not mean too much in terms of system theory unless some attempt is made to analyze the processes by which the interaction is achieved. Role theory is devoted to certain types of interactional processes. In addition, modern organization theorists point to three other linking activities which appear to be universal to human systems or organized behavior. These processes are communication, balance, and decision making.

1. Communication is mentioned often in neoclassical theory, but the emphasis is on description of forms of communication activity, i.e., formal-informal, vertical-horizontal, line-staff. Communication, as a mechanism which links the segments of the system together, is overlooked by way of much considered analysis.

One aspect of modern organization theory is study of the communica-

---

[38] E. Wight Bakke, "Concept of the Social Organization," in *Modern Organization Theory*, Mason Haire, ed. (New York: John Wiley and Sons, 1959), pp. 60–61.

[39] Mason Haire, "Psychology and the Study of Business: Joint Behavioral Sciences," in *Social Science Research on Business: Product and Potential* (New York: Columbia University Press, 1959), pp. 53–59.

tion network in the system. Communication is viewed as the method by which action is evoked from the parts of the system. Communication acts not only as stimuli resulting in action, but also as a control and co-ordination mechanism linking the decision centers in the system into a synchronized pattern. Deutsch points out that organizations are composed of parts which communicate with each other, receive messages from the outside world, and store information. Taken together, these communication functions of the parts comprise a configuration representing the total system.[40] More is to be said about communication later in the discussion of the cybernetic model.

2. The concept of *balance* as a linking process involves a series of some rather complex ideas. Balance refers to an equilibrating mechanism whereby the various parts of the system are maintained in a harmoniously structured relationship to each other.

The necessity for the balance concept logically flows from the nature of systems themselves. It is impossible to conceive of an ordered relationship among the parts of a system without also introducing the idea of a stabilizing or an adapting mechanism.

Balance appears in two varieties—quasi-automatic and innovative. Both forms of balance act to insure system integrity in face of changing conditions, either internal or external to the system. The first form of balance, quasi-automatic, refers to what some think are "homeostatic" properties of systems. That is, systems seem to exhibit built-in propensities to maintain steady states.

If human organizations are open, self-maintaining systems, then control and regulatory processes are necessary. The issue hinges on the degree to which stabilizing processes in systems, when adapting to change, are automatic. March and Simon have an interesting answer to this problem, which in part is based on the type of change and the adjustment necessary to adapt to the change. Systems have programs of action which are put into effect when a change is perceived. If the change is relatively minor, and if the change comes within the purview of established programs of action, then it might be fairly confidently predicted that the adaptation made by the system will be quasi-automatic.[41]

The role of innovative, creative balancing efforts now needs to be examined. The need for innovation arises when adaptation to a change is outside the scope of existing programs for the purpose of keeping the system in balance. New programs have to be evolved in order for the system to maintain internal harmony.

New programs are created by trial and error search for feasible action alternatives to cope with a given change. But innovation is subject to the limitations and possibilities inherent in the quantity and variety of information present in a system at a particular time. New combinations of alternatives for innovative purposes depend on:

---

[40] Karl W. Deutsch "On Communication Models in the Social Sciences," *Public Opinion Quarterly*, 16 (1952), pp. 356–80.

[41] March and Simon, op. cit., pp. 139–40.

a)   The possible range of output of the system, or the capacity of the system to supply information.

b)   The range of available information in the memory of the system.

c)   The operating rules (program) governing the analysis and flow of information within the system.

d)   The ability of the system to "forget" previously learned solutions to change problems.[42] A system with too good a memory might narrow its behavioral choices to such an extent as to stifle innovation. In simpler language, old learned programs might be used to adapt to change when newly innovated programs are necessary.[43]

Much of what has been said about communication and balance brings to mind a cybernetic model in which both these processes have vital roles. Cybernetics has to do with feedback and control in all kinds of systems. Its purpose is to maintain system stability in the face of change. Cybernetics cannot be studied without considering communication networks, information flow, and some kind of balancing process aimed at preserving the integrity of the system.

Cybernetics directs attention to key questions regarding the system. These questions are: How are communication centers connected, and how are they maintained? Corollary to this question: What is the structure of the feedback system? Next, what information is stored in the organization, and at what points? And as a corollary: How accessible is this information to decision-making centers? Third, how conscious is the organization of the operation of its own parts? That is, to what extent do the policy centers receive control information with sufficient frequency and relevancy to create a real awareness of the operation of the segments of the system? Finally, what are the learning (innovating) capabilities of the system?[44]

Answers to the questions posed by cybernetics are crucial to understanding both the balancing and communication processes in systems.[45] Although cybernetics has been applied largely to technical-engineering problems of automation, the model of feedback, control, and regulation in all systems has a good deal of generality. Cybernetics is a fruitful area which can be used to synthesize the processes of communication and balance.

3. A wide spectrum of topics dealing with types of decisions in human systems makes up the core of analysis of another important process in organizations. Decision analysis is one of the major contributions of March and Simon in their book *Organizations*. The two major classes of decisions they discuss are decisions to produce and decisions to participate in the system.[46]

---

[42] Mervyn L. Cadwallader "The Cybernetic Analysis of Change in Complex Social Organization," *The American Journal of Sociology*, September 1959, p. 156.

[43] It is conceivable for innovative behavior to be programmed into the system.

[44] These are questions adapted from Deutsch, op. cit., 368–70.

[45] Answers to these questions would require a comprehensive volume. One of the best approaches currently available is Stafford Beer, *Cybernetics and Management* (New York: John Wiley and Sons, 1959).

[46] March and Simon, op. cit., chapters 3 and 4.

Decisions to produce are largely a result of an interaction between individual attitudes and the demands of organization. Motivation analysis becomes central to studying the nature and results of the interaction. Individual decisions to participate in the organization reflect on such issues as the relationship between organizational rewards versus the demands made by the organization. Participation decisions also focus attention on the reasons why individuals remain in or leave organizations.

March and Simon treat decisions as internal variables in an organization which depend on jobs, individual expectations and motivations, and organizational structure. Marschak[47] looks on the decision process as an independent variable upon which the survival of the organization is based. In this case, the organization is viewed as having, inherent to its structure, the ability to maximize survival requisites through its established decision processes.

The Goals of Organization

Organization has three goals which may be either intermeshed or independent ends in themselves. They are growth, stability, and interaction. The last goal refers to organizations which exist primarily to provide a medium for association of its members with others. Interestingly enough these goals seem to apply to different forms of organization at varying levels of complexity, ranging from simple clockwork mechanisms to social systems.

These similarities in organizational purposes have been observed by a number of people, and a field of thought and research called general system theory has developed, dedicated to the task of discovering organizational universals. The dream of general system theory is to create a science of organizational universals, or if you will, a universal science using common organizational elements found in all systems as a starting point.

Modern organization theory is on the periphery of general system theory. Both general system theory and modern organization theory studies:

1. The parts (individuals) in aggregates, and the movement of individuals into and out of the system.
2. The interaction of individuals with the environment found in the system.
3. The interactions among individuals in the system.
4. General growth and stability problems of systems.[48]

Modern organization theory and general system theory are similar in that they look at organization as an integrated whole. They differ, however, in terms of their generality. General system theory is concerned with every level of system, whereas modern organizational theory focuses primarily on human organization.

---

[47] Jacob Marschak, "Efficient and Viable Organizational Forms," in *Modern Organization Theory*, Mason Haire, editor (New York: John Wiley and Sons, 1959), pp. 307–20.

[48] Kenneth E. Boulding, "General System Theory—The Skeleton of a Science," *Management Science*, April 1956, pp. 200–202.

The question might be asked, what can the science of administration gain by the study of system levels other than human? Before attempting an answer, note should be made of what these other levels are. Boulding presents a convenient method of classification:

1. The static structure—a level of framework, the anatomy of a system; for example, the structure of the universe.
2. The simple dynamic system—the level of clockworks, predetermined necessary motions.
3. The cybernetic system—the level of the thermostat, the system moves to maintain a given equilibrium through a process of self-regulation.
4. The open system—level of self-maintaining systems, moves toward and includes living organisms.
5. The genetic-societal system—level of cell society, characterized by a division of labor among cells.
6. Animal systems—level of mobility, evidence of goal-directed behavior.
7. Human systems—level of symbol interpretation and idea communication.
8. Social system—level of human organization.
9. Transcendental systems—level of ultimates and absolutes which exhibit systematic structure but are unknowable in essence.[49]

This approach to the study of systems by finding universals common at all levels of organization offers intriguing possibilities for administrative organization theory. A good deal of light could be thrown on social systems if structurally analogous elements could be found in the simpler types of systems. For example, cybernetic systems have characteristics which seem to be similar to feedback, regulation, and control phenomena in human organizations. Thus, certain facets of cybernetic models could be generalized to human organization. Considerable danger, however, lies in poorly founded analogies. Superficial similarities between simpler system forms and social systems are apparent everywhere. Instinctually based ant societies, for example, do not yield particularly instructive lessons for understanding rationally conceived human organizations. Thus, care should be taken that analogies used to bridge system levels are not mere devices for literary enrichment. For analogies to have usefulness and validity, they must exhibit inherent structural similarities or implicitly identical operational principles.[50]

Modern organization theory leads, as it has been shown, almost in-

---

[49] Ibid., pp. 202–5.

[50] Seidenberg, op. cit., p. 136. The fruitful use of the type of analogies spoken of by Seidenberg is evident in the application of thermodynamic principles, particularly the entropy concept, to communication theory. See: Claude E. Shannon and Warren Weaver, *The Mathematical Theory of Communication,* (Urbana: The University of Illinois Press, 1949). Further, the existence of a complete analogy between the operational behavior of thermodynamic systems, electrical communication systems, and biological systems has been noted by: Y. S. Touloukian, *The Concept of Entropy in Communication, Living Organisms, and Thermodynamics,* Research Bulletin 130, Purdue Engineering Experiment Station.

evitably into a discussion of general system theory. A science of organization universals has some strong advocates, particularly among biologists.[51] Organization theorists in administrative science cannot afford to overlook the contributions of general system theory. Indeed, modern organization concepts could offer a great deal to those working with general system theory. But the ideas dealt with in the general theory are exceedingly elusive.

Speaking of the concept of equilibrium as a unifying element in all systems, Easton says, "It (equilibrium) leaves the impression that we have a useful general theory when in fact, lacking measurability, it is a mere pretence for knowledge."[52] The inability to quantify and measure universal organization elements undermines the success of pragmatic tests to which general system theory might be put.

Organization Theory: Quo Vadis?

Most sciences have a vision of the universe to which they are applied, and administrative science is not an exception. This universe is composed of parts. One purpose of science is to synthesize the parts into an organized conception of its field of study. As a science matures, its theorems about the configuration of its universe change. The direction of change in three sciences, physics, economics, and sociology, are noted briefly for comparison with the development of an administrative view of human organization.

The first comprehensive and empirically verifiable outlook of the physical universe was presented by Newton in his *Principia*. Classical physics, founded on Newton's work, constitutes a grand scheme in which a wide range of physical phenomena could be organized and predicted. Newtonian physics may rightfully be regarded as "macro" in nature, because its system of organization was concerned largely with gross events of which the movement of celestial bodies, waves, energy forms, and strain are examples. For years classical physics was supreme, being applied continuously to smaller and smaller classes of phenomena in the physical universe. Physicists at one time adopted the view that everything in their realm could be discovered by simply subdividing problems. Physics thus moved into the "micro" order.

But in the nineteenth century a revolution took place motivated largely because events were being noted which could not be explained adequately by the conceptual framework supplied by the classical school. The consequences of this revolution are brilliantly described by Eddington:

From the point of view of philosophy of science the conception associated with entropy must I think be ranked as the great contribution of the nineteenth century to scientific thought. It marked a reaction from the view that

---

[51] For example see: Ludwig von Bertalanffy, *Problem of Life* (London: Watts and Company, 1952).

[52] David Easton, "Limits of the Equilibrium Model in Social Research," in *Profits and Problems of Homeostatic Models in the Behavioral Sciences*, Publication 1, Chicago Behavioral Sciences, 1953, p. 39.

everything to which science need pay attention is discovered by microscopic dissection of objects. It provided an alternative standpoint in which the centre of interest is shifted from the entities reached by the customary analysis (atoms, electric potentials, etc.) to qualities possessed by the system as a whole, which cannot be split up and located—a little bit here, and a little bit there. . . .

We often think that when we have completed our study of *one* we know all about *two*, because "two" is "one and one." We forget that we have still to make a study of "and." Secondary physics is the study of "and"—that is to say, of organization.[53]

Although modern physics often deals in minute quantities and oscillations, the conception of the physicist is on the "macro" scale. He is concerned with the "and," or the organization of the world in which the events occur. These developments did not invalidate classical physics as to its usefulness for explaining a certain range of phenomena. But classical physics is no longer the undisputed law of the universe. It is a special case.

Early economic theory, and Adam Smith's *Wealth of Nations* comes to mind, examined economic problems in the macro order. The *Wealth of Nations* is mainly concerned with matters of national income and welfare. Later, the economics of the firm, micro-economics, dominated the theoretical scene in this science. And, finally, with Keynes' *The General Theory of Employment Interest and Money*, a systematic approach to the economic universe was reintroduced on the macro level.

The first era of the developing science of sociology was occupied by the great social "system builders." Comte, the so-called father of sociology, had a macro view of society in that his chief works are devoted to social reorganization. Comte was concerned with the interrelationships among social, political, religious, and educational institutions. As sociology progressed, the science of society compressed. Emphasis shifted from the macro approach of the pioneers to detailed, empirical study of small social units. The compression of sociological analysis was accompanied by study of social pathology or disorganization.

In general, physics, economics, and sociology appear to have two things in common. First, they offered a macro point of view as their initial systematic comprehension of their area of study. Second, as the science developed, attention fragmented into analysis of the parts of the organization, rather than attending to the system as a whole. This is the micro phase.

In physics and economics, discontent was evidenced by some scientists at the continual atomization of the universe. The reaction to the micro approach was a new theory or theories dealing with the total system, on the macro level again. This third phase of scientific development seems to be more evident in physics and economics than in sociology.

The reason for the "macro-micro-macro" order of scientific progress lies, perhaps, in the hypothesis that usually the things which strike man

[53] Sir Arthur Eddington, *The Nature of the Physical World* (Ann Arbor: The University of Michigan Press, 1958), pp. 103–4.

first are of great magnitude. The scientist attempts to discover order in the vastness. But after macro laws or models of systems are postulated, variations appear which demand analysis, not so much in terms of the entire system, but more in terms of the specific parts which make it up. Then, intense study of microcosm may result in new general laws, replacing the old models of organization. Or, the old and the new models may stand together, each explaining a different class of phenomenon. Or, the old and the new concepts or organization may be welded to produce a single creative synthesis.

Now, what does all this have to do with the problem of organization in administrative science? Organization concepts seem to have gone through the same order of development in this field as in the three just mentioned. It is evident that the classical theory of organization, particularly as in the work of Mooney and Reiley, is concerned with principles common to all organizations. It is a macro-organizational view. The classical approach to organization, however, dealt with the gross anatomical parts and processes of the formal organization. Like classical physics, the classical theory of organization is a special case. Neither are especially well equipped to account for variation from their established framework.

Many variations in the classical administrative model result from human behavior. The only way these variations could be understood was by a microscopic examination of particularized, situational aspects of human behavior. The mission of the neoclassical school thus is "micro-analysis."

It was observed earlier, that somewhere along the line the concept of the social system, which is the key to understanding the Hawthorne studies, faded into the background. Maybe the idea is so obvious that it was lost to the view of researchers and writers in human relations. In any event, the press of research in the micro-cosmic universes of the informal organization, morale and productivity, leadership, participation, and the like forced the notion of the social system into limbo. Now, with the advent of modern organization theory, the social system has been resurrected.

Modern organization theory appears to be concerned with Eddington's "and." This school claims that its operational hypothesis is based on a macro point of view; that is, the study of organization as a whole. This nobility of purpose should not obscure, however, certain difficulties faced by this field as it is presently constituted. Modern organization theory raises two questions which should be explored further. First, would it not be more accurate to speak of modern organization theories? Second, just how much of modern organization theory is modern?

The first question can be answered with a quick affirmative. Aside from the notion of the system, there are few, if any, other ideas of a unifying nature. Except for several important exceptions,[54] modern organization theorists tend to pursue their pet points of view,[55] suggest-

---

[54] For example: E. Wight Bakke, op. cit., pp. 18–75.

[55] There is a large selection including decision theory, individual-organization interaction, motivation, vitality, stability, growth, and graph theory, to mention a few.

ing they are part of system theory, but not troubling to show by what mystical means they arrive at this conclusion.

The irony of it all is that a field dealing with systems has, indeed, little system. Modern organization theory needs a framework, and it needs an integration of issues into a common conception of organization. Admittedly, this is a large order. But it is curious not to find serious analytical treatment of subjects like cybernetics or general system theory in Haire's, *Modern Organizational Theory* which claims to be a representative example of work in this field. Beer has ample evidence in his book *Cybernetics and Management* that cybernetics, if imaginatively approached, provides a valuable conceptual base for the study of systems.

The second question suggests an ambiguous answer. Modern organization theory is in part a product of the past; system analysis is not a new idea. Further, modern organization theory relies for supporting data on microcosmic research studies, generally drawn from the journals of the last ten years. The newness of modern organization theory, perhaps, is its effort to synthesize recent research contributions of many fields into a system theory characterized by a reoriented conception of organization.

One might ask, but what is the modern theorist reorienting? A clue is found in the almost snobbish disdain assumed by some authors of the neoclassical human relations school, and particularly, the classical school. Reevaluation of the classical school of organization is overdue. However, this does not mean that its contributions to organization theory are irrelevant and should be overlooked in the rush to get on the "behavioral science bandwagon."

Haire announces that the papers appearing in *Modern Organization Theory* constitute, "the ragged leading edge of a wave of theoretical development."[56] Ragged, yes; but leading no! The papers appearing in this book do not represent a theoretical breakthrough in the concept of organization. Haire's collection is an interesting potpourri with several contributions of considerable significance. But readers should beware that they will not find vastly new insights into organizational behavior in this book, if they have kept up with the literature of the social sciences, and have dabbled to some extent in the esoterica of biological theories of growth, information theory, and mathematical model building. For those who have not maintained the pace, *Modern Organization Theory* serves the admirable purpose of bringing them up-to-date on a rather diversified number of subjects.

Some work in modern organization theory is pioneering, making its appraisal difficult and future uncertain. While the direction of this endeavor is unclear, one thing is patently true. Human behavior in organizations, and indeed, organization itself, cannot be adequately understood within the ground rules of classical and neo-classical doctrines. Appreciation of human organization requires a *creative* synthesis of massive amounts of empirical data, a high order of deductive reason-

---

[56] Mason Haire, "General Issues," in Mason Haire, ed. *Modern Organization Theory* (New York: John Wiley and Sons, 1959), p. 2.

ing, imaginative research studies, and a taste for individual and social values. Accomplishment of all these objectives, and the inclusion of them into a framework of the concept of the system, appears to be the goal of modern organization theory. The vitality of administrative science rests on the advances modern theorists make along this line.

Modern organization theory, 1960 style, is an amorphous aggregation of synthesizers and restaters, with a few extending leadership on the frontier. For the sake of these few, it is well to admonish that pouring old wine into new bottles may make the spirits cloudy. Unfortunately, modern organization theory has almost succeeded in achieving the status of a fad. Popularization and exploitation contributed to the disrepute into which human relations has fallen. It would be a great waste if modern organization theory yields to the same fate, particularly since both modern organization theory and human relations draw from the same promising source of inspiration—system analysis.

Modern organization theory needs tools of analysis and a conceptual framework uniquely its own, but it must also allow for the incorporation of relevant contributions of many fields. It may be that the framework will come from general system theory. New areas of research such as decision theory, information theory, and cybernetics also offer reasonable expectations of analytical and conceptual tools. Modern organization theory represents a frontier of research which has great significance for management. The potential is great, because it offers the opportunity for uniting what is valuable in classical theory with the social and natural sciences into a systematic and integrated conception of human organization.

Daniel Katz
Basil S. Georgopoulos

7. Organizations in a Changing World

OUR NATION has been aptly characterized as an *organizational society* (Presthus, 1962). Most of our working hours are spent in one organizational context or another. If the dominant institution of the feudal period was the church, of the early period of nationalism the political state, the dominant structure of our time is the organization. Even among those in revolt the old union line still works: "Organize the guys."

But organization forms today are under challenge and without creative modification may face difficulties in survival. On the one hand they are growing in size and complexity, with criss-crossing relationships with other systems and with increasing problems of coordination, integration, and adaptation. The traditional answer in organization structure is on the technical side, *more* computerized programs for feedback and coordination, *more* specialization of function, *more* centralization of control. Yet the social and psychological changes in the culture are increasingly at odds with the technological solution calling for more and more of the same. Technological efficiency far surpasses social efficiency in most cases, even though neither is a substitute for the other from the standpoint of organizational effectiveness; and the gap is widening so as to generate serious conflict within the system.

THE MAINTENANCE AND INTEGRATION OF THE SYSTEM

Before exploring this conflict and its implications in greater detail, let us examine briefly the structure and functioning of organizations as open social systems. We can distinguish among the subsystems which comprise the larger structure (Katz & Kahn, 1966). The production, or instrumental, subsystem is concerned with the basic type of work that

This paper was presented at the 77th Annual Convention of the American Psychological Association, held September 1969 in Washington, D.C.

Reproduced by special permission from *The Journal of Applied Behavioral Science,* "Organizations in a Changing World," vol. 7, pp. 342–70, 1971 © NTL Institute for Applied Behavioral Science.

gets done, with the "throughput"—the modification of inputs which result in products or services. Attached to the productive subsystem are the supportive services of procurement of supplies, material and resources, and the disposal of the outputs.

The maintenance, or social, subsystem is concerned not with the physical plant but with the social structure, so that the identity of the system in relation both to its basic objectives and the environment is preserved. People not only have to be attracted to the system and remain in it for some period of time but they have to function in roles which are essential to the mission of the organization. The maintenance subsystem is concerned with rewards and sanctions and with system norms and values that ensure the continuity of the role structure. In short, its function has to do with the psychological cement that holds the organizational structure together, with the integration of the individual into the system. The channeling of collective effort in reliable and predictable pathways is the basis of organizational structure.

The managerial subsystem cuts across all other subsystems as a mechanism of control, coordination, and decision making and as a mechanism for integrating the instrumental and maintenance functions of the organization. To meet environmental changes both with respect to inputs and receptivity for outputs and to handle system strains, the managerial subsystem develops adaptive structures of staff members as in the case of research, development, marketing, and planning operations.

Let us look at the maintenance function more closely, however, before considering the adaptation problem in the light of major societal changes. Over time it requires more than sheer police power and coercive sanctions. It depends upon some degree of integration or involvement of people in terms of their own needs. Values, norms, and roles tie people into the system at different psychological levels and in different ways (Parsons, 1960).

Values provide the deepest basis of commitment by their rational and moral statement of the goals of a group or system. To the extent that these values are accepted by individuals as their own beliefs, we speak of the internalization of group goals. The degree of internalization will vary among the members of any group, but it is important for all organizations to have some hard core of people dedicated to their mission both for accomplishment of many types of tasks and as models for others. Such value commitment can come about through self-selection into the system of those possessing beliefs congruent with its goals, through socialization in the general society or in the organization, and through participation in the rewards and decision making of the group. The internalization of group goals is facilitated by the perception of progress toward these objectives. Such progress is interpreted as an empirical validation of values.

Normative involvement refers to the acceptance of system requirements about specific forms of behavior. These requirements are seen as legitimate because rules are perceived as necessary and because in general the rules are equitable. A particular demand by a particular officer may be seen as unjust, but in general there is acceptance of the need for directives from those in positions of authority, provided that they have attained their positions properly and that they stay within their areas of jurisdiction in the exercise of their authority. In complex organizations, such as universities, hospitals, and industrial firms, the rules of the game can be improved, but they are universalistic and do not permit particularistic favoritism or discrimination.

At the level of *role behavior* people make the system function because of their interdependence with others, the rewards for performing their roles, and the socio-emotional satisfactions from being part of a role-interdependent group. In carrying out their roles, organizational members at all times but in varying degrees are interdependent both functionally and psychologically. Not all role performance provides expressive gratifications, however, and hence other rewards such as monetary incentives and opportunities for individual achievement, as well as group accomplishment and socio-emotional satisfactions, can be linked to adequate behavior in the given role. In most large organizations extrinsic rewards of pay, good working conditions, and so on are relied on heavily.

It is apparent that these three levels of member involvement are not necessarily intrinsically related. The values of a particular organization may have little to do with many of the roles in the system, and the norms of legitimacy are not necessarily specific to organizational values. A research organization may furnish a nice fit among values, norms, and roles for its research workers but a poor fit for its supportive personnel. Few organizations, however, can rely on value commitment alone to hold their members, and hence they maximize other conditions and rewards to compete with other systems. The development of universalistic norms under the impact of bureaucratic organization forms has provided great mobility for people in an expanding economic society, and thus has contributed to its growth—at the expense of value commitment.

In a well-integrated system, however, there is some relationship among these levels so that they are mutually reinforcing. In an ideal hospital, even the attendant can be affected by the values of saving lives and improving health, can perceive the normative requirements as necessary and fair, and derive satisfaction from his role, particularly if he is made part of a therapy team. Values can contribute to the strength of the normative system in providing a broader framework of justifiable beliefs about the rightness of given norms. Thus norms can be seen not only as equitable rules but as embodiments of justice and of equality. The strength of maintenance forces lies in the many mechanisms for supporting the role structure and for some degree of mutual reinforcement.

THE IMPACT OF SOCIETAL CHANGES ON ORGANIZATIONS

The problem of how to organize human effort most effectively in complex, specialized organizations within a rapidly changing sociocultural environment, while maintaining the integrity of the system, is of utmost significance and concern everywhere in our time. Its solution generally demands greater social-psychological sophistication, however, rather than a more sophisticated work technology (Georgopoulos, 1970). It requires social organization innovations and the testing of new forms and patterns of organization, or at least significantly modified organizational structures than those now in operation (Georgopoulos, 1969; Likert, 1967). It will not be achieved at acceptable levels simply by an even more perfect technology at our disposal. As profound changes continue to occur in society at rapid but uneven rates, organizational viability and effectiveness are in jeopardy unless the social efficiency of the system can more clearly match its technological efficiency in the vast majority of organizations, and unless the adaptability of organizations can be improved well beyond current levels.

Four major changes have occurred in our society which challenge both the production and social subsystems of organizations: (*a*) a break, at first gradual and now pronounced, with traditional authority and the growth of democratic ideology; (*b*) economic growth and affluence; (*c*) the resultant changes in needs and motive patterns; and (*d*) the accelerated rate of change. These changes are significant for organizations, since as open systems organizations are in continuing interaction with their environment both with respect to production inputs of material resources and social inputs from the culture and from the larger social structure.

Weakened Traditional Authority Forces

The break with the older pattern of authority has eroded some of the formerly dependable maintenance processes of organizations. Bureaucratic systems had long profited from the socialization practices of traditional society, in which values and legitimacy had a moral basis of an absolutist character. It was morally wrong to reject in word or deed the traditional teachings about American institutions. It was wrong to seek change other than through established channels. Not everyone, of course, lived up to the precepts, but deviance was easy to define and highly visible, and those who deviated generally felt guilty about their misconduct. If they did not, they were considered to be psychopaths. Organizations had the advantage of a degree of built-in conformity to their norms and in some cases to their values because of the general socialization in the society about agreed-on standards. This consensus, moreover, made nonconformity a matter of conscience. There was an all-or-none quality about virtue, honesty, and justice, and these values were not seen as relativistic or empirical generalizations. Member compliance with organizational norms and values no longer can be sustained on the basis of authority (Etzioni, 1964; Georgopoulos, 1966; Georgopoulos & Matejko, 1967).

The very growth of bureaucratic systems helped to demolish absolutist values of a moral character. As conscious attempts to organize collective enterprises, organizations were guided by rational objectives and empirical feedback. Pragmatism replaced tradition. Results and accomplishment were the criteria rather than internal moral principle. Furthermore, the normative system shifted, as Weber (1947) noted, from traditional authority to rational authority. Rules and laws were the instruments of men to achieve their purposes and lacked any transcendental quality. They could be changed at will as situations and needs changed or they become ineffective. Having undercut the traditional basis of authority, the bureaucratic system can no longer rely upon the older moral commitment to its directives.

The growth of organizations affected the larger society and its socialization practices and in turn was affected by it. The training of children in a rational and democratic framework further increased a nontraditional orientation to values and norms.

The decline of traditional authority has been accompanied by the growth of the Democratic Ethic and democratic practices. The source of power has been shifting from the heads of hierarchies and from oligarchies to the larger electorate. This process can be observed in the political system where restrictions have been removed on suffrage. Nonproperty owners, women, and now blacks are eligible to vote. Indirect mechanisms of control from above are changing as in the political conventions of major parties. Democratic ideas of governance have extended into other institutions as well.

Economic Growth and Affluence

The tremendous technological advances which have increased the productivity of the nation need no documentation. We are already using the phrase "postindustrial society" to characterize our era. This development raises questions about the basic functions of colleges and universities. Havighurst (1967) has pointed out that in the past, two functions have been dominant: the *opportunity function* and the *production function*. Education was a means of social mobility, the opportunity function. On the production side, education provided the training for professional, technical, and industrial roles in the society. Today, however, when we are over the economic hump, these two functions are less important and a third function comes to the fore: the *consumption function*. "Education as a consumption good is something people want to enjoy, rather than to use as a means of greater economic production" (Havighurst, 1967, p. 516). This means not only greater attention to the arts but also greater concern with education as it relates to living here and now.

One reason why the demands of black students are often easier to deal with in spite of the rhetoric is that they are directed in good part to the opportunity and production functions. These are understandable issues in our established ways of operating. Demands on the consumption side present new problems. For the blacks, however, there is sometimes the complexity of attempting to achieve all three objectives at the same time.

The case of educational institutions, moreover, is not unique. The interests and expectations of consumers of goods and services no longer remain disregarded either by industry or government. Outside pressures and demands are increasingly responded to with greater attention by most organizations, however inadequately or belatedly. In the health care field, for example, hospitals slowly are becoming more responsive to the health care expectations of an increasingly better educated and more demanding clientèle who now see comprehensive health care as a right. At the same time, as the costs of care continue to rise at staggering rates, the quest for quality care is accompanied by demands for public controls, higher organizational efficiency, and even reorganization of the entire health system (Bugbee, 1969; McNerney, 1969; Sibery, 1969; TIME, 1969; U.S. News and World Report, 1969). Hospitals are being pressured from all directions to innovate and experiment with new patterns of internal social organization and more effective forms of operation in the areas of administration, staffing, organizational rewards for members, community relations, and the utilization of both new health knowledge and new social-psychological knowledge (Georgopoulos, 1964, 1969). As a consequence, they are being forced to be not only more community-oriented but also more sensitive to the interests and contributions of their various groups of members at all levels (Georgopoulos & Matejko, 1967). More generally, partly as a result of affluence and economic growth, organizations in all areas are becoming more open systems and less immune to social forces in their environment.

Resultant Changes in Motive Patterns

Economic affluence and the decline in traditional authority are related to a shift in motive patterns in our society. Maslow (1943) developed the notion years ago of a hierarchy of motives ranging from biological needs, through security, love, and belongingness, to ego needs of self-esteem, self-development, and self-actualization. His thesis was that the motives at the bottom of the hierarchy were imperative in their demands and made the higher level motives relatively ineffectual. Once these lower level needs are assured satisfaction, however, the higher level needs take over and become all-important.

Maslow's thesis has abundant support among the young people in our educational system. They are less concerned with traditional economic careers than was once the case. A recent study reported only 14 percent of the graduates of a leading university planning business careers, compared with 39 percent five years earlier and 70 percent in 1928 (Marrow, Bowers, & Seashore, 1967). Engineering schools similarly are experiencing falling enrollments.

Our society has been called with considerable justification, the *achieving society* (McClelland, 1961). The content analysis of children's readers by de Charms and Moeller (1962) shows that a great rise in achievement themes occurred in the last part of the 19th century, but a great decline in this emphasis has occurred in recent decades.

The decline in the older motive patterns has on direct consequence for all organizations. Extrinsic rewards such as pay, job security, fringe benefits, and conditions of work are no longer so attractive. Younger people are demanding intrinsic job satisfactions as well. They are less likely to accept the notion of deferring gratifications in the interests of some distant career.

In most organizations today the dominant motives of members are the higher-order ego and social motives—particularly those for personal gratification, independence, self-expression, power, and self-actualization (Argyris, 1964; Black & Mouton, 1968; Georgopoulos, 1970; Georgopoulos & Matejko, 1967; Herzberg, 1968; Likert, 1967; Marrow, Bowers, & Seashore, 1967; McGregor, 1960; Schein, 1965). Increasingly, expressive needs and the pursuit of immediate and intrinsic rewards are outstripping economic achievement motives in importance both in the work situation and outside. Correspondingly, the dominant incentives and rewards required for member compliance, role performance, and organizational effectiveness are social and psychological rather than economic (Black & Mouton, 1968; Etzioni, 1964; Georgopoulos, 1970; Georgopoulos & Matejko, 1967; Herzberg, 1968; Likert, 1967; Marrow, Bowers, & Seashore, 1967; McGregor, 1960). Even at the rank-and-file level, where economic motives are especially strong, there is now more concern on the part of unions for other than bread-and-butter issues, and contract negotiations often stall on matters of policy, control, and work rules rather than money. As a result of these shifts, there is pressure for a place in the decision-making structure of the system from all groups and members in organizations, and there is a growing need for meaningful participation in the affairs of the organization by all concerned at all levels.

The forms which newly aroused ego motives take can vary, but at present there are a number of patterns familiar to all of us. First there is the emphasis upon self-determination or self-expression, or "doing one's thing." Second is the demand for self-development and self-actualization, making the most of one's own talents and abilities. Third is the unleashing of power drives. The hippies represent the first emphasis of self-expression, some of the leftist leaders the emphasis upon power. Fourth is the outcome of the other three, a blanket rejection of established values—a revolutionary attack upon the existing system as exploitative and repressive of the needs of individuals.

With the need for self-expression goes the ideology of the importance of spontaneity; of the wholeness of human experience; the reliance upon emotions; and the attack upon the fragmentation, the depersonalization, and the restrictions of the present social forms. It contributes to the anti-intellectualism of the student movement and is reminiscent of the romanticism of an older period in which Wordsworth spoke of the intellect as that false secondary power which multiplies delusions. Rationality is regarded as rationalization.

Accelerated Rate of Change

Not only are we witnessing significant shifts in the economic and value patterns of society but they are happening at a very fast rate. Probably there has always been some conflict between the older and younger generations, but in the past there has been more time to socialize children into older patterns and the patterns were of longer duration, thus preventing serious lags and social dislocation. History is becoming less relevant for predicting change. It is difficult to know what the generation now entering high school will be like when they enter college.

All organizations face a period of trouble and turmoil because of these changes, which affect all three levels of integration in social systems. Some of the basic values of the social systems are under fire, such as representative democracy of the traditional, complex type, the belief in private property, conventional morality, the importance of work and of economic achievement, the good life as the conventional enjoyment of the products of mass culture. The Protestant Ethic (Weber, 1904) is no longer pervasive and paramount in our society.

The norms legitimized by societal values of orderly procedures and of conformity to existing rules until they are changed by socially sanctioned procedures are also brought into question. The rebels emphasize not law and order but justice, and justice as they happen to see it. It is interesting that President Nixon, the spokesman for the Establishment, modified his plea for law and order by stressing law, order, and justice. The challenge to the norms of any system is especially serious, since it is genuinely revolutionary or anarchistic in implication, whether voiced by official revolutionaries or reformers. If the legitimate channels for change are abandoned and the resort is to direct action, then people are going outside the system. If enough do, the system collapses.

At the level of role integration there is also real difficulty. As has already been noted, extrinsic rewards have lost some of their importance in our affluent society. Moreover, the usual set of roles in an organization segmentalize individuals, and our ever-advancing technology adds to the problem. A role is only partially inclusive of personality at most levels of the organization. This fractionation runs counter to the needs for wholeness and for self-expression. Increasing specialization everywhere exacerbates the problem (Etzioni, 1964; Georgopoulos, 1966; Georgopoulos, 1970; Georgopoulos & Mann, 1962; Likert, 1967; J. D. Thompson, 1967; V. Thompson, 1961). It engenders coordination and integration difficulties for organizations and their members because it results in greater organizational complexity and more intensive interdependence among unlike participants who must relate to one another and to the system and whose efforts must be collectively regulated (Georgopoulos, 1966; Georgopoulos & Mann, 1962; Morton, 1964; V. Thompson, 1961; Wieland, 1965). Role specialization is the main social invention available with which man can cope with the problems of the explosion of knowledge in our times, for specialization makes possible both the utilization of available knowledge and the development of new knowledge. But, at the same time, specialization leads to fractionation and diversity that make the integration of members into the system all the more difficult to attain.

In linking the changed patterns of many of the younger generation to societal changes, we want to emphasize that it is an error to simplify the problem as a younger-older generation conflict. It is broader and deeper than that, and many of the developing trends predate the present student generation. In fact, the revolt started with people now in their sixties, if not earlier. *We* were the ones to attack the inequities of bureaucratic society, the ones to raise children in democratic practices and to think for themselves. The older generations furnished the ideology of the present student movements. Try to find any ideology in these movements which is not a bastardized version of old revolutionary and romantic doctrines. *We* started the rebellion and now we are astonished to find that we are the "establishment."

This is one reason why organizations have been so vulnerable to attack. Older citizens do not rally to their support because they feel that the rebels are in good part right. Or else why should we so often hear it stated, "We agree with your objectives but we don't like your tactics"? Nathan Glazer (1969) has shown, however, that this vague sentiment is based upon a failure to come to grips with the significant issues.

THE PROBLEM OF ADAPTATION

The dynamic nature of our society makes imperative greater attention to processes of adaptation. In the past, industrial organizations, because of their dependence upon a market, have developed adaptive subsystems of planning, research, and development. The major emphasis has been, however, upon production inputs, upon product development, upon finding new markets and exploiting old ones, upon technology in improving their productive system. Only minor attention has been placed upon social inputs or upon restructuring the organization to meet the psychological needs of members. Technological innovation without social innovation has been the rule, and exclusive concern with technical and economic efficiency has undermined the social and psychological efficiency of the system to the detriment of organizational effectiveness and adaptability.

Traditionally, organizations have shown much more concern for the technology of work than for their social inputs and human assets. They have been more concerned with providing a safe and attractive physical work environment than with creating and maintaining an equally attractive social and psychological work climate for the members. With the emphasis for technical and economic efficiency, they have paid much more attention to recruiting and selecting members with the "proper" training and aptitude for filling inflexibly defined jobs than to problems of member attitudes, needs, and values. The approach has been to fit the man to the job rather than the other way around, and organizational role redefinition has been largely disregarded as a problem-solving mechanism. Organizational restructuring to improve the adaptability of the system has been abhorred and resisted. Most organizations have avoided social innovation and renovation and have sought technological innovation as the answer to all of their problems. Correspondingly, in relating to their members, they have been concerned with authority-based, superior-subordinate relations more than with social relations, relying more on economic incentives and rewards and less on social-psychological motivation and compensation.

Because of the changes in society just discussed, however, the situation is now changing within organizations as well. The conditions for effective role performance, job satisfaction, member integration into the system, and organizational effectiveness and adaptability demand different organization-member relations than in the past. For organizations to survive and perform their functions effectively in the future, some sizable proportion of their resources will have to be committed to enlarging their adaptive subsystems to deal more adequately with external relations and new social inputs. Social effectiveness will have to be added to productive efficiency as an important objective.

Better adaptive subsystems are now needed not only in industrial organizations but in all complex organizations, including educational and health institutions. The case of hospitals is instructive. Continuous progress in medicine, nursing, and allied health professions and occupations, advances in medical technology, the professionalization of hospital administration, and the explosion of knowledge witnessed inside and outside the health field have made a strong impact upon the traditional social organization of the hospital system. The result is a gradual redefinition of the institutional role of the hospital as the health center in the total framework of health-related institutions. Such redefinition, however, is being forced by public expectations and demands from without and pressures from nonmedical members on the prevailing power structure, rather than from planned social innovation within the system (Georgopoulos, 1969; Georgopoulos, 1970; Georgopoulos & Matejko, 1967). Redefinition is taking place and must be accomplished along with proper internal organizational restructuring, however, in the context of current health trends and societal health conceptions—for example, the Medicare program, the recent development of regional medical programs, the growing emphasis on comprehensive health planning, the promulgated national goal of adequate health care for all, and the widespread concern for improvements in care coverage, quality, and cost.

These recent changes in the health field, along with the major changes in society discussed earlier, have strong and concrete implications for the kind of organizational restructuring that is feasible and appropriate for the hospital as a complex, sociotechnical, problem-solving system. Today's hospital is still ruled by three dominant decision-making centers —trustees, physicians, and administrators (Georgopoulos & Mann, 1962; Georgopoulos & Matejko, 1967). The above trends argue, however, for better recognition of the contributions of nurses and nonmedical groups and for a broader base of decision making. They argue for an interaction-influence structure which transcends the conventional tripartite arrangement and which can truly encompass all participants regardless of their professional affiliation or hierarchical position in the system. The traditional maintenance mechanisms and adaptive structures of the organization (formal authority, rule enforcement, medical dominance, identification with the system primarily on the basis of moral values and service motives, influence and rewards according to professional status and hierarchical position) which have been success-ful in the past are clearly becoming less and less effective (Georgo-poulos, 1969; Georgopoulos & Matejko, 1967). Inside and outside the system the premises of the traditional structure no longer remain un-challenged, and new bases for organizational adaptation are therefore required.

Similar problems, evident in all large-scale organizations, await solution. Without an adequate adaptive subsystem to modify and filter new inputs leading to planned change, two things can happen: The new potential inputs can be summarily rejected; the organizational structure becomes rigid and the problems are postponed and often intensified. Or the inputs slip into the system and are incorporated in undigested fashion; there is erosion of basic values and the system loses its identity. It does not acquire a formal death certificate but for practical purposes it has been replaced. If a university were to accept research inputs uncritically from the Defense Department, for example, it could end up as a branch of the military and not as an institution for advancing sci-ence. Sometimes it happens that in organizations the first response of blanket rejection and rigidity cannot be maintained over time and the opposite reaction of wholesale acceptance of any and all demands fol-lows. For example, a university may show rigidity to suggested reforms at first, and then, as pressure mounts, capitulate completely without critical evaluation of the suggested changes. To complicate matters, both rigidity and uncritical incorporation can occur in different parts of the same organization.

Most organizations suffer from a lack of adequate adaptive and in-tegrative structures concerned with their maintenance subsystems. We should like to indicate some of the lines of inquiry which adaptive sub-systems should follow and some directions in restructuring organiza-tions which seem consistent with the present state of knowledge in the field. The views of our critics from the left are at times helpful, though not original, in pointing up vulnerable aspects of bureaucratic struc-tures, but they are singularly lacking in constructive suggestions for re-form. Some of them, of course, are not interested in reform, but are committed to destruction of the system. In short, they make no attempt to come to grips with the problems of the one and the many or with the fact that the individual doing *his* "thing" may interfere with other in-dividuals doing *their* "things." Anarchy may have its philosophic appeal, but it cannot be practiced in crowded settings where millions of people live in constant interdependence.

The young dissenters have focused upon some fundamental weak-nesses in organizational structure which have been recognized before but which become more critical as the majority of the younger genera-tion no longer finds them acceptable. In the first place, there is a grow-ing dissatisfaction with the fragmentation of life in an organization, with the difficulties of being a whole personality and of finding personal satisfactions in relating to others in impersonal role relationships. The major point is that many organization arrangements and procedures in our society cut the individual into segments in his various role responsibilities. This is especially true once we leave the more satisfying roles for the elite groups at the top of the structure. In this process we move toward a disintegrated personality. Once his wholeness and unity are violated, the individual may become alienated or seek personality expression outside his role responsibilities (Argyris, 1964; McGregor, 1960). Then we attempt artificial devices such as mass leisure pursuits of sports, movies, and television, of company programs of recreation, or even human relations training for supervisors to remedy the weaknesses in the system. But this is the organizational fallacy *par excellence.* Once we have destroyed the integrated individual we no longer have the unified pieces to provide a truly integrated system. It is not like making an automobile out of pieces of steel, rubber, and other materials. One cannot have a truly integrated social system if the human pieces are not themselves integrated. It is not possible to have a moral society made up of immoral men.

The second major criticism concerns the exploitative character of bureaucratic structures: Namely, that the rewards of the system, both intrinsic and extrinsic, go disproportionately to the upper hierarchical levels and that the objectives of organizations are distorted toward the immediate interests of the elite and away from desirable social goals of the many.

ADAPTATION THROUGH STRUCTURAL REFORM

One important line of structural reform which can be significant with respect to the above weakness is the fuller extension of democratic principles to the operation of organizations. Many writers on organizations fail to address themselves to this problem in structural terms, but talk about improvement of interpersonal relations, sensitivity training, and consultative practices. The extreme left is also not concerned with democratic reform. Nonetheless, we believe that much can be done in organizational settings through democratic restructuring to improve their social effectiveness—i.e., their psychological returns to their members.

Two issues must be faced in the extension of democratic principles to organizational functioning. One is direct or representative democracy. The second is the appropriate area of decision making for various subgroups in different types of complex systems.

Representative democracy has been under fire because it can be elaborated through complex mechanisms to distort the wishes of the electorate and to give top decision makers great power. Such abuses do not negate its potential virtues. What is critical here is the number of hierarchical levels in the form of a pyramid similar to the administrative structure: for example, what has been the older practice in some states of the elected state legislators in turn electing senators. The general rule of never allowing more than two levels, that of the electorate and that of their duly chosen representatives, is gaining recognition in political organizations and can be applied to other organizations as well since it ensures more responsible and more responsive decision making.

Direct democracy of the town meeting sort is a cumbersome and ineffective mechanism for many pusposes, once the electorate is numbered in hundreds and thousands. In many of our universities, however, we still persist in town meetings of a faculty of over a thousand. It is small wonder that such meetings get stalemated on details and often fail to come to grips with the central issue. Representative assemblies of a smaller number of democratically elected delegates would be a more effective mechanism for decision making.

Direct democracy, however, is still a necessary part of the picture. In the first place, the full electorate should have the opportunity to veto major policy changes suggested by their elected leaders. In the second place, direct democracy can be more adequately utilized in smaller units of the system. Within a university, for example, there is a greater role for direct democracy at the departmental level than at the university level—although we still have a long way to go to achieve this goal. The great advantage of direct democracy within the smaller unit is that it ties the individual into his own group on matters of direct concern to him and thus permits the possibilities of tying him into the larger system of representative democracy. If he is not integrated through participation in his own group, then he is more likely to be apathetic toward or alienated from the larger structure. Democracy, like charity, begins at home, and in most organizations, home is the functional group where the individual spends most of his time.

An esoteric example of this model of combining direct and indirect democracy is embodied in the kibbutzim of Israel (Golomb, 1968). These utopian communities are remarkable in that they have survived, even prospered, for 60 years under the most difficult circumstances. Each kibbutz is a community with a great deal of autonomy, run by direct democracy, with town meetings, with direct election of all decision-making officers including the farm manager, and with rotation of such officers. In addition, the individual kibbutz belongs to a larger movement which can include 40 or 50 similar communities. The larger movement operates training centers, banking facilities, and other services which the individual community could not afford. The management of the movement is handled through representative democracy. The sense of community achieved in the small kibbutz contributes to the integration of the individual into the larger system. Remote as this example is from the size and complexity of the American scene, it is of some interest in that the 230 communities of the kibbutz federation involving 90,000 persons have no problems of violence, delinquency, crime, or unemployment. Psychotic breakdown is an unusual event and crime a rare occurence. Farm productivity is higher than the productivity of private farms in Israel.

A model of this sort combining direct and representative democracy has obvious limitations in its application to large-scale organizations which restrict areas of decision making in many ways. In the first place, a complex system tends to reduce the decision-making powers of any component group, including the top echelon. Even the president of a university will feel that his margin for decision making is within a fairly narrow band of possibilities; the same is true, only more so, of the hospital administrator. And the individual member will feel even less room for meaningful participation for determining policy. In the second place, the administrative agency of a public institution has to operate within the legislation of a representative democracy in which it has had little say. In the third place, activities that have to be thoroughly coordinated on a rigorous time schedule, as in the military or space program, or in the operating room of a hospital, heavily restrict the areas of decision making for component groups.

Nevertheless, we do not take full advantage of the opportunities available within these limitations, nor do we examine the nature of these restrictions to see to what extent they can be made less rigid. Though coordination by experts may be necessary after a policy decision, there is still room for the involvement of people in making that decision and reviewing its outcomes. Though legislation determines objectives, there is often considerable leeway in how these directives can be implemented. Though size and complexity limit the amount of decision making possible for any one group, matters that are not of significance to the overall system may loom large for individuals in their own work setting. For example, we are constantly expanding the plant in many organizations with new buildings and new construction. Yet the people who have to use these facilities are frequently not consulted about them with respect to their own needs and the uses to which the buildings will be put. There are instances of windowless structures dictated by considerations of economy and standardization from above which turn out to be frightfully inefficient. We need to analyze the assumptions about coordination and centralization as demanding decisions only from the top echelons.

The facts are that centralization *de jure* often leads to decentralization *de facto*. We need to be more critical of the whole centralization concept to see where tight controls are really necessary and whether they will be genuinely effective. Even with centralized controls in large organizations it is sometimes true that the right hand does not know what the left hand is doing. Hence the criterion should not be some abstract concept of centralization but objective data about how it operates in practice.

Moreover, we need to distinguish among the types of organizations and their objectives, as between mass-producing and other organizations, and particularly among organizations whose primary output is a physical product, some service, or information. In some organizations there is a necessary coordination of all effort so that there is a convergence of activities upon one outcome, as in the space program's objective of getting men on the moon or in a heart transplant operation. But many organizations do not have single products which require such convergence of the energies of all members of the system. In a university, for example, where we are concerned with training people and extending knowledge, there is a great variety of outcomes and hence much more freedom and degrees of autonomy within the total system. The traditional model of organizational structure deriving from the military and industry is not necessarily the appropriate model for all organizations, nor even for all aspects of industry or the military.

For example, the concept of job enlargement, or that of job enrichment (Herzberg, 1968), as it is more recently described, has been limited primarily to single roles. There are, however, serious limits beyond which we cannot go with a single job. What *is* possible is group responsibility for a meaningful cycle of work involving a number of related work roles simultaneously. It has been demonstrated by the Tavistock researchers that a cohesive group can be created about a task objective. With some reduction in specialized roles within the group, some rotation of roles, removal of status differences, and responsibility given to the group to get the job done, the results in such widely separated industries as a calico mill in India (Rice, 1958) and a coal mine in Great Britain (Trist, Higgin, Murray, & Pollock, 1963) have been spectacular in the improvement of productivity and morale. The findings are particularly important because they give us new leverage on an old problem. Many jobs in themselves are routine and without challenge. Overall they add up to something significant in the way of performance. If they are not rigorously delimited and assigned to particular people, they can be given to a group with the group itself assuming responsibility for the outcome. Thus the advantages of collective accomplishment become not the ideal of top management but the objective and psychological reality for group members.

Another structural reform to make roles more meaningful has to do with organizational divisions based upon process specialization. Supportive activities today are often separated off from production activities, as when we set apart persons performing a service such as personnel recruitment or typing services in a bureau or section of their own. We organize on the basis of process rather than purpose, to use a distinction made by Aristotle. Then we proceed to institutionalize the separation by removing the given service unit from its production counterpart physically and psychologically. The service people may in fact never have direct personal contact with production people. This separation, which results in overspecialization and unnatural interdependence at work, may be particularly damaging to the morale of various service units since the major production functions enjoy the greater prestige and often the more rewarding types of work. It is difficult for the girls in a typing pool to identify with their task or with organizational objectives when they have no meaningful relationship to them. Moreover, the service unit split off from major functions develops a compensatory defensive posture which often interferes with the effective functioning of the larger system. Staff-line conflicts in complex organizations are another familiar problem (Dalton, 1950). More generally, as Morton (1964) points out, where organizational members whose work is related are separated with physical barriers, social bonds are essential to effective organizational functioning, and physical bonds are important when members are socially separated.

The type of restructuring that needs to be considered here is the creation of teams and groups for accomplishing an objective in which service people have primary membership in the production unit and secondary membership in a service unit, and sometimes the secondary membership can be dispensed with. Inclusive, common membership at the workgroup level may be similarly desirable for staff and line personnel in most organizations.

We may need more subgroups than we now have because people can identify more readily with one another and with the group task in small settings than in large. What is important, however, is that the subgroups are broadly enough designed with respect to an objective so that people with different skills can cooperate with one another and identify with the group goal. Groups should be small but their task responsibility large. This runs counter to the traditional way of organizing for turning out automobiles, but it may be more appropriate for many types of organizations not mass producing a physical product.

For example, general hospitals in the future may profit greatly from a self-governing structure that would give every professional and occupational group and member the opportunity to participate meaningfully in the decision-making processes of the organization, taking fully into account both the functional and the social-psychological interdependence of the participants (Georgopoulos, 1969; Georgopoulos, 1970; Georgopoulos & Matejko, 1967). It would seem advantageous to develop decision-making mechanisms on an organization-wide basis, built on the principles of representative democracy and multilevel federalism of semiautonomous small groups that are highly attractive to their members and can contribute effectively to the solution of system problems at the same time. Such an organizational structure would provide for effective representation of all individual members and their respective groups in the system. It could be established so as to take account of group size; the specialized competencies, functions, and interests of members; and of their location in the system in relation to major problems but not solely on the basis of traditional power or formal position in the authority hierarchy. Every new unit probably should consist of a relatively small number of members who would choose their own representative to the next level. The highest unit in the organization would be a Board of Participants (Georgopoulos, 1969).

This Board of Participants, representing the interests of all hospital personnel, and the conventional Board of Trustees, representing the community's interests, would work cooperatively and on a continuing basis on matters of organizational policy, external relations, priorities and objectives for the institution, and major decisions affecting the hospital, its members, and the public (Georgopoulos, 1969). Hospital administration would function as an executive body, charged with the principal managerial, planning, financial, personnel, and coordinative functions, following and implementing the general policies and decisions made jointly by the Board of Trustees and the Board of Participants (on which the administration would be properly represented, as would doctors, nurses, and other groups) of the institution. Other structural models built on the same organizing principles also would be appropriate. The same applies to organizations other than hospitals, of course.

In general, the changes we have been discussing all point toward a looser role system with broader role definitions, more flexibility and openness of subsystem boundaries, with group responsibility for task objectives. Admittedly, such debureaucratization may add to the noise in the system, but some of what appears to be noise from the point of view of the formal chart maker may be meaningful activity directed toward important goals. It is doubtful whether the true volume of uncertainty in the system would rise very significantly. It should also be remembered that in supposedly tight structures the empirical system may be at variance with the formal organization, with real noise that goes undetected because the formal channels do not code it. What matters is that the facts of functional as well as social-psychological interdependence among the participants in the organization both be taken fully into account.

In a change process the techniques of sensitivity training and of organizational development can be utilized for the full implementation of social reform. These techniques are directed primarily at the improvement of interpersonal relationships and do not in themselves provide for permanent changes in social structure. What is critical is the legitimation of change in institutional arrangements and a restructuring of formal patterns to permit more democratic processes to function. After this has been achieved, methods for improving self-understanding and the understanding of others can help to ensure the success of the structural reforms. To a limited extent they can even prepare the way for these more basic changes.

REFORMULATION OF VALUES FOR SOCIAL REINTEGRATION

We have left to the last the most difficult problem of all: the system values which can bind the individual to the organization and furnish the ideology to justify organizational norms and requirements. Values seldom are static, and our dynamic period with its accelerated changes has seen fundamental challenges to the older belief structure. The task for the adaptive processes of an organization is one of the creative adaptation of central values to changing inputs. Such adaptation means the preservation of the basic nature of the system with modifications which clarify issues but do not destroy the system. Adaptation through genuine participation and active involvement based upon democratic principles and processes can still be successful.

The great need of our time is a reformulation of social values that would make possible a higher level of integration for all social systems. Organizational leaders should play a much more vigorous role as the responsible agents of the adaptive mechanisms in their social structures. In many instances they have given much time and energy to the adaptive function but generally in terms of mediation, negotiation, and compromise in crisis situations. Negotiation and compromise are important, but they are far from a complete answer. Compromise without consideration of principle can merely lead to a new round of demands. Various factions within and without an organization interpret concessions and compromises as an invitation to mobilize their forces for a new offensive. Politics has been called the art of compromise; yet if we rely wholly upon this process we leave everything to power and power-driven individuals. The conceptualization of values and the enunciation of basic principles should not be left to the extremists on either the left or the right. Organizational leadership has a challenge in meeting the rhetoric of the dissidents with a compelling statement of principles and an implementation of them in practice. Many people today are eager for such a formulation of values. When Senator Eugene McCarthy took a clear position on foreign policy, the popular response was of such magnitude as to confound political analysts.

In the past we have been mainly a pragmatic nation, and leaders as well as followers have tended to shy away from ideological discussion. And some of this pragmatic emphasis is to be found in the present demands for relevance and in the anti-intellectualism of the New Left. Nonetheless, values and principles which transcend the single case— the single individual, the single organization, the single faction—are critical to the maintenance of social order. The facts are that there are many assumptions, as well as practices of a moral character in contemporary society, which need reemphasis and reformulation. Without attempting to catalog them, may we cite a few examples?

In the first place, research and observation show that the norm of reciprocity (Gouldner, 1960), of cooperation, of mutual helpfulness, runs wide and deep. Organizations could not exist without many uncounted acts of cooperation which we take for granted. If people operated merely on a basis of role prescription, organizations would run poorly. And role prescriptions also take account of mutual interdependence. Berkowitz and Daniels (1963) have shown experimentally that people will respond to others who need their help. Hospital studies (Georgopoulos & Mann, 1962; Georgopoulos & Matejko, 1967; Georgopoulos & Wieland, 1964; Wieland, 1965) have shown the importance of this norm for group performance, organizational coordination, and patient care. Studies of citizenship orientation have shown that individuals see themselves as good citizens not if they are flag wavers but if they are cooperative and helpful toward their fellows. We cite this basic value of mutual helpfulness because we lose sight of it in the self-oriented push of some protesters.

In the second place, justice and fairness are not outmoded values. Justice as a value is evident in many forms of social exchange (Blau, 1964), and the underdog elicits sympathy partly because there is an assumption that he has not been fairly treated. In fact, justice and fairness are the ideological weapons of the dissenters, but they have no monopoly on them. It is essential to emphasize the importance of justice and fairness in the operation of an organization and to introduce reforms where inequity is the practice.

In the third place, social responsibility or involvement in matters of more than local concern has a potential that remains to be developed. It is no longer acceptable to brag about one's nonparticipation in political affairs. It is apparently less difficult to recruit candidates for public office than was once the case. There seems to be more concern about national decisions which the "little" people were once content to leave to the authorities. Again, the New Left has taken advantage of this broadened conception of political and social responsibility to urge direct action by students on all types of issues. But again the doctrine of social concern can be formulated to make people more aware of the social consequences of their actions. This social concern, which transcends the individual's own self-oriented needs, is reflected in the positive esteem achieved by leaders who show humanitarian values. Part of the late Robert Kennedy's popular appeal was the conviction that he was concerned—that he cared about the fate of others.

All of these values are related to, if not an integral part of, the Democratic Ethic which is still our basic creed. We have already noted the development of democratic practices in our political system and their extension to other social institutions. The democratic doctrine is invoked by the left in its attack upon the establishment but not in its own operations and program. With this group the use of democratic phrases seems to be more of a tactic than an ideological commitment. As the right mobilizes, there is some revival of reactionary beliefs. With the increasing polarization, the middle of the spectrum could profit greatly from a reformulation of the democratic creed by those who believe that democracy is not an outmoded concept—in spite of inadequacies in its application. Organization reform needs such a value base both as a set of social principles and as guidelines for action.

REFERENCES

Argyris, C. *Integrating the Individual and the Organization.* New York: Wiley, 1964.

Berkowitz, L., & Daniels, Louise R. "Responsibility and Dependency." *J. Abnorm. Soc. Psychol.,* 1963, 66, 429–437.

Blake, R. R., & Mouton, Jane S. *Corporate Excellence through Grid Organization Development.* Houston, Tex.: Gulf, 1968.

Blau, P. M. Justice in Social Exchange. *Sociolog. Inquiry,* 1964, 24, 199–200.

Bugbee, G. "Delivery of Health Care Services: Long Range Outlook." The Univer. of Michigan *Medical Center J.,* 1969, 35, 75–76.

Dalton, M. "Conflicts between Staff and Line Managerial Officers." *Amer. Sociolog. Rev.,* 1950, 15, 342–51.

de Charms, R., & Moeller, G. H. "Values Expressed in Children's Readers." *J. Abnorm. Soc. Psychol.,* 1962, 64, 136–42.

Etzioni, A. *Modern Organizations.* Englewood Cliffs, N.J.: Prentice-Hall, 1964.

Georgopoulos, B. S. "Hospital Organization and Administration." *Hospital Admin.,* 1964, 9, 23–35.

Georgopoulos, B. S. "The Hospital System and Nursing: Some Basic Problems and Issues." *Nursing Forum,* 1966, 5, 8–35.

Georgopoulos, B. S. "The General Hospital as an Organization: A Social-Psychological Viewpoint." The Univer. of Michigan *Medical Center J.,* 1969, 35, 94–97.

Georgopoulos, B. S. "An Open-System Theory Model for Organizational Research: The Case of the Contemporary General Hospital." In A. R. Negandhi and J. P. Schwitter (Eds.), *Organizational Behavior Models.* Kent, Ohio: Kent State Univer., 1970. Pp. 33–70.

Georgopoulos, B. S., & Mann, F. C. *The Community General Hospital.* New York: Macmillan, 1962.

Georgopoulos, B. S., & Matejko, A. "The American General Hospital as a Complex Social System." *Health Services Res.,* 1967, 2, 76–112.

Georgopoulos, B. S., & Wieland, G. F. *Nationwide Study of Coordination and Patient Care in Voluntary Hospitals,* No. 2178. Ann Arbor, Mich.: Institute for Social Research, The Univer. of Michigan, 1964.

Glazer, N. "The Campus Crucible: 1. Student Politics and the University." *Atlantic Monthly,* July 1969, 244, 43–53.

Golomb, N. "Managing Without Sanctions or Rewards." *Mgmt. of Personnel Q.,* 1968, 7, 22–28.

Gouldner, A. W. "The Norm of Reciprocity: A Preliminary Statement." *Amer. Sociolog. Rev.,* 1960, 25, 161–78.

Havighurst, R. J. "The Social and Educational Implications of Inter-institutional Cooperation in Higher Education." In L. C. Howard (Ed.), *Interinstitutional Cooperation in Higher Education.* Milwaukee: Univer. of Wisconsin, 1967. Pp. 508–23.

Herzberg, F. "One More Time: How Do You Motivate Employees?" *Harvard Bus. Rev.,* 1968, 46, 53–62.

Katz, D., & Kahn, R. L. *The Social Psychology of Organizations.* New York: Wiley, 1966.

Likert, R. *The Human Organization: Its Management and Value.* New York: McGraw-Hill, 1967.

Marrow, A. J., Bowers, D. G., & Seashore, S. E. *Management by Participation.* New York: Harper and Row, 1967.

Maslow, A. H. "A Theory of Human Motivation." *Psycholog. Rev.,* 1943, 50, 370–96.

McClelland, D. *The Achieving Society.* New York: D. Van Nostrand, 1961.

McGregor, D. M. *The Human Side of Enterprise.* New York: McGraw-Hill, 1960.

McNerney, W. J. "Does America Need a New Health System?" The Univer. of Michigan *Medical Center J.,* 1969, 35, 82–87.

Morton, J. A. "From Research to Technology." *Int. Sci. & Technol.,* May 1964, Issue No. 29, 82–92.

Parsons, T. *Structure and Process in Modern Society.* Glencoe, Ill.: Free Press, 1960.

Presthus, R. *The Organizational Society.* New York: Knopf, 1962.

Rice, A. K. *Productivity and Social Organization.* London: Tavistock Publications, 1958.

Schein, E. H. *Organizational Psychology.* Englewood Cliffs, N.J.: Prentice-Hall, 1965.

Sibery, D. E. "Our Social Responsibilities as Health Professionals and University Center Hospitals." The Univer. of Michigan *Medical Center J.,* 1969, *35,* 88–93.

Thompson, J. D. *Organizations in Action.* New York: McGraw-Hill, 1967.

Thompson, V. *Modern Organization.* New York: Knopf, 1961.

TIME Magazine, Medicine—The Plight of the U.S. Patient. February 21, 1969, *93,* 53–58.

Trist, E. L., Higgin, G. W., Murray, H., & Pollock, A. B. *Organizational Choice.* London: Tavistock Publications, 1963.

U.S. News & World Report. "How to Improve Medical Care."   March 24, 1969, *66,* 41–46.

Weber, M. *The Theory of Social and Economic Organization* (A. M. Henderson & T. Parsons transl.). New York: Oxford Univer. Press, 1947.

Weber, M. *The Protestant Ethic and the Rise of Capitalism* (T. Parsons transl.). New York: Scribner, 1958; original publication in German, 1904; first English edition, 1930.

Wieland, G. F. "Complexity and Coordination in Organizations."   Unpublished Doctoral Dissertation, The Univer. of Michigan, 1965.

Eileen Morley
Alan Sheldon

8. Work Systems and Human Behavior: Toward a Systems-Theoretic Model

IN TWENTIETH-CENTURY western society the predominant style of work is that of employment in a complex organization. Today's monolithic work systems have evolved as the result of extensive changes in the location, occupational patterning and content of work roles over the last hundred years, and of equivalent changes in the productive and managerial technologies necessary to command such a scale of enterprise. The fact that in the year 1969 eighty percent of all employment in U.S. manufacturing and mining was provided by a group of one thousand companies[1] illustrates the extent to which this concentration of employment has proceeded. Because the work organization is so influential a life environment, it becomes crucial to understand the ways in which it elicits, alters, fosters, and constrains individual human behavior. This paper describes a systems approach to the understanding of this intricate process.

In all societies explicit and implicit values exist, according to which behavior is defined as "good" or "bad," as acceptable or unacceptable. These values or norms are enforced by sanctions which reward desirable and penalize undesirable behavior. In our society such sanctions range from the pressure of public opinion to the weight of the legal system. Societal values exist at a series of levels. At the most global, certain kinds of deviant behavior are defined in terms of crime or delinquency and are proscribed, while other actions are merely tolerated as eccentricities. In subsectors of society, such as the educational or medical worlds, local values develop. In educational institutions, desired academic behavior is codified in formal standards for graduation and degree awards which remain relatively constant, even though

---

This material was prepared especially for this volume.
[1] "The Fortune Directory of the Second 500 Largest Industrial Corporations," *Fortune Magazine* June 1970, p. 98.

values governing interpersonal relationships in the same sector are presently in flux. In the health care sector, the clinical difference between health and illness is captured in diagnostic terminology. Here current values for what constitutes sickness and health are influenced by the current state of medical knowledge. At still lower social levels even small units develop particularized values, especially when their membership is constant over long periods as in the case of the family.

Obviously the work system imports these societal values for its own internal use, if for no other reason than that they are held by the people who constitute the system's membership. But because human work behavior constitutes a crucial system variable, values introduced from non-work settings must be supplemented by internal values, to ensure that the members of the system will behave in ways which provide the inputs of human knowledge and energy needed to bring about appropriate transformations in the state of matter and information.

The steady state value of behavior variables which evolve within a work system relate to task activities; to interpersonal relationships between individuals and groups; and to obligations of membership in system and subsystems. Those values which are uniform throughout a system tend to be the ones which have been imported from outside. Thus the kind of noisy drunken behavior which is unacceptable in a public place is similarly unacceptable on the job, while illness sufficiently acute to prevent a person from attending a social function will usually also legitimately excuse his absence from work.

The values which evolve within the system tend to vary, both vertically and horizontally. Values enforced at the level of the workgroup may differ from those established by the supervisor, or from those laid down at the level of organizational procedure. It is vertical variation of this kind which demarcates formal and informal organization (Homans, 1950). Values vary horizontally from subsystem to subsystem in relation to the task to be accomplished, the occupational characteristics of subsystem members, and the values intrinsic to a particular subsystem process. Thus behavior which is valued in one place may be barely tolerated in another, a fact which stands in the way of standardizing policy and practice across any large organization. Lawrence and Lorsch (1967) have shown that horizontal discrepancies of this kind are in part related to differences between the environments with which subsystems interact.

A SYSTEMS VIEW OF INDIVIDUAL BEHAVIOR

If we are to use systems theory as a context in which to consider the ways in which work organizations deal with human behavior, we need to be able to consider the individual person within that same framework: namely as a system in his own right. We define human behavior as the outcome of (a)motivation, and (b)possession of the states and competences necessary to perform a particular behavior. We define motivation as the strength of a person's predisposition to engage in a given behavior, according to both the strength of his preference for the predicted outcome, and the strength of his expectation that the outcome will actually occur (Vroom, 1964). This definition allows us to adopt an input-output model of individual behavior.

Inputs to the individual consist of information both about the behavior needed and about its likely consequences. Output from the individual consists of activities at work, in exchange for which the person receives a pleasantly rewarding or distasteful "outcome input," such as more or less pay, more or less satisfaction, esteem, responsibility, status, and so on.

Inputs of information about the behavior required and about likely consequences come to the individual from a number of sources including the workgroup, the supervisor, the organization at large, and the trade union where one is present. If an individual's behavior falls outside the range of stability established by prevailing values, it is by definition deviant. Some values may be explicit, precise, and even written—as is usually the case with personnel policies or inspection procedures. Others may be implicit, relating more vaguely to "the way we do things around here." They are "felt" as group norms.

We can think of the pattern of information flowing to each individual as a series of nesting feedback loops, each of which may run in both directions.

The Individual-Workgroup Loop. This loop runs from the workgroup to the individual. Workgroup values for individual behavior, often called informal norms, tend to relate to social as well as technical task activities. Sometimes feedback will be positive,[2] as in the case when the individual is conforming more precisely than the group judges necessary to, say, lunch-break time allowance, or formal work standards. At other times feedback may be negative, perhaps when the person's work is not as consistent and reliable as the group deems necessary.

If the deviant behavior continues uncorrected, feedback is likely to intensify. The message gets louder. If the individual still fails to change his ways, the problem which his deviation represents is likely to be solved by his rejection from group membership, though he may be tolerated as an isolate. But if he in turn feeds back information communicating acceptable grounds for his behavior, this may modify the signal in his favor.

The Individual-Supervisor Loop. Another loop runs between individual and supervisor. The supervisor constitutes a feedback channel from the formal organization to the individual. Along this channel "official" information is passed about the extent to which a man or woman's behavior is consistent with system and subsystem norms. This feedback will also inevitably reflect the supervisor's personal values. Important as it is for the individual to satisfy the norms of the group, it is even more important that he satisfy those of the supervisor. He can con-

---

[2] "Feedback" is here used in the strictly cybernetic sense. Positive feedback increases deviation, negative diminishes it.

tinue to be a member of the system as an isolate or deviate if the work-group rejects him, even though his life may be uncomfortable. But he cannot do so if the supervisor rejects him, unless he is protected by unusual circumstances such as tenure of appointment, or the sponsor-ship of an alternative source of power and authority more influential than the supervisor.

The Individual-System Loop.   A third and more diffuse information channel runs from the organization directly to the individual. It tends to be less task-related in content, and to vary a great deal from system to system in the amount of information carried. Examples of this chan-nel are formal announcements about company affairs; house news-papers and magazines; and so on. The upward loop is often formalized in the concrete shape of a suggestion box.

The Individual-Trade-Union Loop.   Where a trade union is present, a separate feedback link from the delegated official to the individual also exists. The existence of a trade union system as a counterpart to the work system provides another major source of approval, identification, status, and esteem for the individual.

The Formal Authoritarian Control System.   The second of these feed-back loops—that running from the supervisor in his role as formal representative of the organization to the individual—is one link in a chain of similar feedback loops running from the most senior to the most junior member of the system. This essentially authoritarian vertical form of feedback control has been the principal means by which industrial organizations have traditionally controlled human behavior. The extent to which such a system develops and is used varies greatly from system to system. Burns and Stalker (1961) have indicated that the emergence and use of this loop is a function of the stability of environmental demands and of the corresponding formalization of internal system structure.

If formal feedback channels run to the individual from sources other than the immediate superior, both the supervisor and the individual's role are likely to become conflict laden. The worker will be in the dif-ficult position of attempting to satisfy two masters. (Kahn, et al. 1964) The supervisor will experience the feedback which he is sending to the individual as potentially in conflict with that sent from another source. This occurs, for instance, when a supervisor is responsible for evaluat-ing performance, while responsibility for deciding about a pay increase lies with another role such as a personnel manager or wage and salary administrator. Other kinds of conflict may occur when upward and downward loops converging on a single role carry opposing informa-tion; as when the provision of support and protection for subordinates which is implicit in a foreman's role conflicts acutely with instructions received from superiors to lay off a proportion of his workgroup.

There is likely to be considerable variability between inputs of in-formation about required behavior which are received by the individual from each of the sources mentioned above. Because each of these sources is a unique "behavior environment," inputs and outcomes are by no means consistent with each other. An important aspect of modern organization design is the cultivation of such consistency. Those feedback sources which exist within the organization have at least the possibility for coherence and consistency, since they encourage adherence to the system's values and goals. The trade union is a different system, whose values and goals relate to the quality of life of workers across many organizations. Thus, behavior which may gain a man status and prestige in a trade union will by no means also call forth an equivalent reward from the work system.

There is also likely to be variability in individual responses to inputs. The predisposition to seek certain kinds of outcomes rather than others, and to engage in the behaviors which will achieve the desired outcomes, lies at the core of individual personality. People tend to respond differentially to any given range of inputs, according to the dictates of their individual predispositions or personalities. Thus the man whose internal value for research enquiry and acquisition of knowledge exceeds his value for power and authority may refuse promotion to a higher managerial position, even though such a promotion is a commonly accepted form of organizational reward.

This illustration reminds us that if a feedback loop is to function effectively, it is important that the information transmitted have the same meaning for the recipient as it has for the sender. All too often an outcome input may be experienced as pleasing by one person and displeasing by another. For instance, a supervisor may reward successful behavior by increasing the autonomy of the person concerned. If the person welcomes autonomy so much the better, but if he experiences this change as a decrease in the extent to which his needs for dependence and affiliation are met, what was intended as a reward may well turn out to be a penalty.

This brings us to the crucial role of the supervisor. We have noted above that each person has a characteristic pattern of differential response to a given range of inputs—what we call his personality. It is at the interface between supervisor and subordinate that such variations in individual personality need to be taken into account, for it seems likely that a higher proportion of information inputs are transmitted to the individual from the supervisor than from any other single source. Similarly the supervisor generates or influences many of the subsequent pleasing or displeasing outcomes.

The appropriateness of behavior outcomes depends on (1) the range of outputs within the supervisor's power of influence—for instance, does he have the authority to grant pay raises or time off; and (2) the extent to which he is able to perceive the meanings of different inputs to different people, and to select and provide those which have greatest potential for pleasure, reward, or satisfaction to each person. It is this need for interpersonal sensitivity and responsiveness which constitutes the all-important social content of supervisory work activities, and which at present is so inaccessible to theoretical analysis, due to both the amount of time and effort required and to limitations of current theory.

We have deliberately used the phrase "activities at work" rather than "work activities" because by no means all the things which people do in a work setting fall within the range of formal work activities. Burns and Stalker (1961) have commented on the personal needs, and the methods of satisfying them, which men and women bring with them into an organization:

. . . in pursuing these private purposes which are irrelevant to the working organization, individuals affiliate themselves to groups, and seek to bind others in association. They acquire commitments . . . which involve some surrender of personal autonomy, not this time in a bargain with employers and in return for money and other benefits, but in the hope of further material or non-material rewards or in order to avoid discomfort, embarrassment, or loss (perhaps of self-esteem). Commitments to others involve loss of autonomy in that the right to spontaneous divergent action outside the group is surrendered in respect of the objectives the private combination has been constituted to attain. . . .

. . . The activities within a concern of private combinations and individuals in pursuit of ends distinct from those of the working organization account for the distortions and frustrations recurrently or chronically experienced by working organizations. . . .[3]

INTRAPERSONAL STATES AND CAPABILITIES

As we have said before, more than motivation is needed if a person is to produce the behavior necessary to the accomplishment of a particular task. He must also be capable of doing so: that is, he must possess the technical and social skills and knowledge needed, and also be in a state of sufficient physical and psychological health to carry it out. These states and competences constitute the variables within the individual's personal system which must be maintained in an appropriate steady state if work behavior is to occur. We conceptualize them as (1) a state of health, both physical and psychological, (2) technical-educational competence and (3) social-interpersonal competence.[4]

These variables are interactive in at least two significant ways. Problems in one area are often linked with problems or symptoms in another. For example, physical ailments are frequently associated with mood changes such as feelings of depression or withdrawal, psychological problems may often be expressed somatically, and poor health of either kind is likely to hamper the acquisition and exercise of technical and social competence. There may also be a degree of choice about the way in which a problem is expressed, one which is dictated in part by the

---

[3] Burns, Tom and G. M. Stalker. *The Management of Innovation* (London: Tavistock, 1961), p. 100.

[4] An interaction also clearly occurs between "motivation" and intrapersonal states and competences which is easily observable in any work system. For example, a person who is highly motivated to the accomplishment of a particular task is likely to persist in working at it in a state of ill health which, given another task, would lead him to go home and take to his bed. Conversely, some people are on the whole much more predisposed to carry out work activities for which they possess the necessary competences than those for which they do not.

personality of the individual, and in part by the culture of the organization. Thus individuals learn what kind of symptomatology the organization regards as acceptable expressions of problems (no matter what or where their origin), as Hill and Trist found (1962).

SYSTEM INTERVENTION

Work systems vary tremendously in the extent to which they perceive different intrapersonal states and capabilities as relevant to successful work behavior; in the extent to which they attempt to influence these states and capabilities; and in the ways in which they exercise such influence. Broadly speaking, three alternatives are open: (1) they can leave everything to the supervisor, or to other relevant role-holders such as a personnel manager; (2) they can purchase specialist help from sources beyond the system, such as hospitals and local colleges; (3) they can provide some degree of specialist help within the system. Whether these activities focus on the maintenance and improvement of physical or mental health, technical or social competence, they function to adjust and maintain the intrapersonal characteristics of people in a way which will foster their ongoing adaptive success at work. Because they exist to promote such adaption, we shall hereon refer to them as *proadaptive activities*. We see the development of such activities as involving two elements: (1) the sophistication of the organization in appropriately conceptualizing the nature of its people-problems; (2) the differentiated structures and processes which the organization develops, as a result of its conceptions.

PROADAPTIVE ACTIVITIES: STRUCTURE

The provision of internal service involves the differentiation of a specific role structure. This may occur at one of three different levels.

Combined Roles. Differentiation within an existing role occurs when a role-holder expands his activities to encompass some aspect of human behavior with which he has not previously been concerned. The doctor or nurse who spends more and more time in psychological counseling, or the technical instructor who fits himself to provide T-group training are examples of such combined activities. Sometimes activities are combined in a formal explicit way. Sometimes they are simply the result of an individual role-holder's perception that the service is needed, and of his willingness to provide it. Higher levels of management within an organization may not always be aware that an activity is going on. Conversely some combination of activities may be regarded as normative for a particular role. Thus personnel managers are almost invariably expected to provide career and personal counseling.

Combination of functions in a single role may have other implications. A function may not be wholly accepted within the system as a legitimate industrial activity. For example, a training manager might well test the acceptability of sensitivity training by including it among existing programs before deciding whether or not to hire a trainer. Or combined roles may simply be an effective use of existing talents, as in smaller companies whose populations do not justify a fulltime practitioner. Sometimes a combination of activities may reflect a resistance within the system to further differentiation. Many a doctor and personnel manager know how extensive are problems of mental and emotional ill health, simply because they have to deal with them in systems where the introduction of formal psychiatric service would meet intensive opposition.

Differentiated Roles. The differentiation of full-fledged roles occurs either when there is sufficient volume of work to warrant them, or when the task is of such a kind that it cannot be performed by an existing member of the system—perhaps because it required a trained professional such as a physician or nurse. When only a single role is differentiated, the role-holder may be part of another subsystem—the counselor who works in a personnel department, or the technical instructor who is part of a production group.

Differentiated Subsystems. Where the need for service is extensive or its nature complex, the activity may be differentiated at the level of a complete subsystem; that is, a department or group with two or more reporting levels, such as the medical department with a physician and subordinate nurses, or the training department with a manager and instructors.

The presence of formal structure indicates that the organization perceives the management of a particular aspect of behavior to be a legitimate organizational need—although to some degree it may also reflect organizational predilection for certain forms of structural response.

Though the development of differentiated structures and processes has the advantage of fostering more profound understanding of problems and more sophisticated means of dealing with them, it has the major disadvantage of splitting human problems into different and not necessarily connected aspects. This highlights the reciprocal need for some form of integration around the individual.

PROADAPTIVE ACTIVITIES: PROCESS

Inputs. Inputs to proadaptive structures consist of information and people. Information consists of requests for information and/or service. The people involved are those to be evaluated and served.

Inputs may originate (*a*) from potential inputs to the system as a whole, e.g., requests for evaluation of the health of job candidates; (*b*) from the system's existing population, e.g., requests for retraining of employees before transfer to different work; or (*c*) from system output, e.g., requests for pre-retirement counseling.

Input or Subsystem Boundary Control. The nature of the input request will be determined by the perceptions of the person making the demand about the nature of the problem and about the function of the proadaptive subsystem or role. Such perceptions may not always be accurate. Consequently the subsystem must determine the fit of the input demand with its own definition of function. Input control may constitute part of a role activity within a subsystem, such as the nurse who screens all employees asking to see a doctor.

Process.   Proadaptive process is of two kinds: (*a*) condition-determining process and (*b*) condition-changing process.

(*a*) *Condition-determining process* consists of evaluating the present state of a person in relation to internal subsystem and system values and other external (e.g. professional) criteria, in order to determine the presence and nature of existing or potential inadequacies in intra-personal states-competences, and the extent to which such inadequacies contribute to present or predicted problems. Origins of adaptive problems may be individual-centered or environment-centered. It is interesting to note that at present attention is far more readily paid to correcting environment-centered problems which stem from the physical nature of the work situation than from its psychological character.

The subsystem provides feedback to the source of the request as to the nature and origin of the problems, and gives recommendations for solution, namely the extent to which change in individual or environment, or both, is desirable and possible. Where the solution involves some degree of change in the individual, the subsystem is likely to indicate whether it is willing to act as change agent. If unwilling or unable, it may recommend referral elsewhere, or it may simply dump the problem back in the individual's or supervisor's lap.

(*b*) *Condition-changing process* has two aspects, though the process may be similar in each case. It may be *remedial*—change which enables the individual to regain a lost ability to supply needed behavior. Or it may be *preventive*—change which maintains the individual's ability to supply appropriate behavior, possibly in the face of altering work demands. Process may be provided in-plant, or be supplied by an appropriate local agency, such as hospital, clinic, school, or even outside consultant. In all cases, process depends on the willingness of the individual to participate, which in turn is likely to be influenced by local organizational values. We think of one organization whose normative response to psychological problems was to suppress or deny all knowledge of them; a second system in which the norm was to acknowledge the physiological aspect of the problem by consulting the doctor, but to discourage attention to psychological aspects; and a third where the norm was to consult a counselor on problems with psychological components.

The extent to which complex work organizations have a social responsibility to influence and change the behavior of their members is presently a matter of public debate. Certainly employers are today regarded as having responsibility for more than the mere provision of work activities and equitable payment, and it cannot be disputed that membership in a large work organization does effect change in individual attitudes and behavior.

We have all too little information about the extent to which work systems attempt to bring about such change deliberately through the creation of explicit proadaptive roles and task activities. In this short paper we have presented a systems-theoretic approach to the problem as a first step toward developing a framework in terms of which such activities are capable of observation, measurement, and comparison. To some extent we have relabelled concepts already existing in other fields. This allows us to bring these concepts into relation with systems theory, and so view both individual and organization as systems in their own right, albeit at different levels. More important, we can then consider the interaction within and between them in a single set of terms. The value of this approach lies in the power of systems theory to encompass entities whose study previously occurred in the traditionally separate fields of sociology and psychology. Observers can thus look systematically at ways in which an organization acts toward its members, and members of an organization can hopefully make more effective choices about the extent to which they will devote limited resources to these processes.

REFERENCES

Burns, Tom and G. M. Stalker. *The Management of Innovation.* London, Tavistock Publications, 1961.

Hill, J. M. and E. L. Trist. "Industrial Accidents, Sickness and Other Absences." *Tavistock Pamphlets,* No. 4, Tavistock Institute of Human Relations, London, 1962.

Homans, George C. *The Human Group.* New York, Harcourt Brace, 1950.

Kahn, Robert L., Donald M. Wolfe, Robert P. Quinn and J. Diedrich Snoek. *Organizational Stress: Studies in Role Conflict and Ambiguity,* New York, Wiley, 1964.

Lawrence, Paul R. and Jay W. Lorsch. *Organization and Environment.* Boston, Harvard University Press, 1967.

Vroom, Victor. *Work and Motivation.* New York, Wiley, 1964.

Ralph M. Stogdill

9. Basic Concepts for a Theory of Organization

CLASSICAL THEORIES of organization have been concerned with principles of departmentation and the structure of responsibility and authority relationships. Behavioral theories emphasize the effects of interpersonal relations upon member satisfaction and organizational productivity. Mathematical models have been developed for plant operations and decision processes. All of these approaches are concerned with organization, but deal with different subsets of variables. There is an obvious need for a general theory that will weld the different subjects onto a coherent system.

The accomplishment of the classical theorists in developing highly rationalized structures of departments and positions for the accomplishment of different organizational objectives is not to be minimized. However, these theories overlook the fact that their high degree of rationalization is made possible by the existence in any society of biological and psychosocial substructures that are directly transferable into formalized organization structures. A general theory should take account of, and should be based upon, these basic substructures. Allee [1] reports that all social aggregates among mammals and the social insects exhibit differentiated structures in which at least a few individuals perform specialized functions. Tuckman [1] analyzed the results of some 60 studies dealing with developmental trends in initially unstructured groups of strangers. The following stages of development were found to be common to groups engaged in therapy, role playing, sensitivity training, and problem solving.

Stage 1. *Forming*—testing and dependence; development of role structures and interpersonal dependencies.

Stage 2. *Storming*—intragroup conflict and hostility; competition for position; emotional expression; group drive.

From *Management Science*, June 1967, vol. 13, no. 10, pp. B-666–B-676. Reprinted with permission of the author and the publisher.

Stage 3. *Norming*—development of group cohesion, norms, and standards; pressures toward conformity.

Stage 4. *Performing*—productive task activity; stable role structure and interdependencies channel group energy into task performance.

The same sequence is not observed in all groups. The stages alternate in some groups, and occur in cycles in other groups. Some groups progress rapidly through the four stages. But if the task is difficult to face (as in therapy), if there is difference of opinion about objectives or methods, if feelings run high, or if there is competition for leadership, a group may fluctuate between the different stages over a long period of time. A group cannot engage in more than sporadic task performance until a role structure has been stabilized, drive channelized, and regulative norms and cohesiveness attained.

It is possible to draw a chart which specifies in advance the role structure of an organization. This procedure averts the necessity of developing structure out of the rigors and uncertainties of face-to-face interaction. However, it does not necessarily solve the problem of channelizing emotion and drive, or that of developing norms and cohesiveness. The subgroups of a newly formed organization must resolve these problems in much the same manner as that observed in problem solving groups. The classical theories do not address themselves to such problems. But these are the kinds of problems with which the behavioral sciences are concerned.

Davis [4] suggests that organization comes into being because of its ability to acquire, create, preserve, or distribute values. Simon [9] has developed different models for the firm (with profit as a primary objective) and the organization (with survival as a primary objective). The firm can measure profits in terms of the dollar value of inputs and outputs. But the firm represents only a small percentage of the total number of organizations that exists in a society. A theory of organization should apply to organizations in general, whether or not they deal with material or monetary values. In fact, the concept of organization can be more clearly developed if physical and monetary values are treated as specific examples of more general psychosocial values.

Since organizations differ widely in objectives and operations, a general theory must of necessity be formulated in terms of quite general concepts. Such a theory should be capable not only of describing the basic dimensions of organization structure and process; it should also account accurately for the research findings.

GENERAL STRUCTURE OF THE THEORY

An organization, in the real world, can be regarded as an input-process-output system. In order to account for such a system, a theory of organization should also be designed as an input-output system. The present paper will attempt merely to state a new theory in outline form.

A detailed discussion of the system and experimental evidence in its support are set forth in a prior publication [10].

The theory here presented is built upon three sets of variables: (1) inputs, (2) mediators (processors), and (3) outputs, as shown in Table 1. The three sets of concepts are assumed to characterize any organization regardless of it size, and whether or not it operates upon any form of material or monetary input. The variables will be defined in the discussion that follows.

Each of the ten variables shown in Table 1 and Figure 1 is assumed

TABLE 1

Input, Processing, and Output Variables

1. *Inputs*	2. *Processors*	3. *Outputs*
Actions (A)	Operations (O)	Product (P)
Interactions (I)	Interpersonnel (L)	Drive (D)
Expectations (E)	Structure (S)	Cohesiveness (C)
Task materials (T)		

FIGURE 1

Direction of Effects in the System

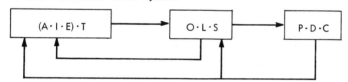

to be complex in structure. Action (A), for example, represents a complex set of performances exhibited by all the members of the organization. Since the most routine performance of a single member exhibits considerable variability, the action input (A) for the total organization will be complex indeed. It is further assumed that actions, interactions, and expectations are interdependent, and are determined in part by task objectives and materials. Similarly, the processors (operation, interpersonnel, and structure) are interdependent, as are the outputs (product, drive, and cohesiveness).

It is assumed that the inputs are affected by feedback effects from the mediators and outputs, and that the mediators are influenced by feedback from the outputs, as shown in Figure 1. An organization also engages in exchange relationships with its environment at both the input and output ends of its operations. The theory under discussion, while recognizing the interactions between organization and environment, is concerned solely with the internal organization as a structured operational system.

THE INPUT VARIABLES

The theory requires three sets of input variables identified as member *actions* (performances), *expectations*, and *interactions*.[1] Performance is defined as any action by an individual that identifies him as a member of the group or organization being observed. Expectation is defined as a member's readiness for reinforcement by the system of which he is a member. Interaction is defined as an action-reaction sequence in which the actions of any one member of a social system are reactions to the performances of one or more other members, and these in turn are reacted to by other members of the system. If the members do nothing but talk with each other, the inputs are numerous, varied, and complex in nature.

The *actions*, or *performances*, of the members are of primary importance in carrying out the operations of the organization. Members may work essentially alone (as in operating a lathe) or they may work in interaction with other members (as in discussing a plan). Performances are member contributions to the organization.

Member *interactions* are basic to cooperative effort. An organization, in its most elementary form, can be defined as a social interaction system in which the structuring of member expectations accounts for the differentiation of positions and roles. Such organization grows out of the interaction process itself. It is of course possible to draw an organization chart before recruiting members to fill the positions specified on the chart. But the drawing of such a chart is almost always preceded by the formation of a small group of individuals who interact, acknowledge a role structure, and agree to cooperate in the development of a larger organization. It is suggested, then, that organization is defined by the differentiation of positions and roles that grows out of interaction and the mutual confirmation of intermember expectations.

Member *expectations* are also contributions to the organization. As such, they define the role that the individual believes he and others should play on behalf of the organization. Members bring into the organization various expectations that it may or may not be able to satisfy. In general, members are able to perform their tasks more effectively when they know what they are expected to do. They tend to be better satisfied with an organization when their expectations relative to pay, advancement, hours, recognition, and treatment by supervisors are fulfilled by the organization. Many organizations thus find it advantageous to use job descriptions, policy manuals, and the like, as instruments for structuring member expectations along lines that are capable of being satisfied.

Some sort of task material is utilized as an input in any kind of

---

[1] The inputs specified for this system are similar to those used by Parsons [8] to explain a societal system and by Homans [6] to explain the social group. Parsons and Homans built their theories on the basis of actions, interactions, and sentiments as input variables which are more broadly defined than are the inputs for the system presented here. Katz and Kahn [7] also regard individual actions and expectations as human inputs in any kind of organization.

organization. Pig iron, coal, lumber, clay, leather, petroleum, and so on, are used as input materials in manufacturing organizations. In a discussion group, the input materials may be the words and sentences used in discussions. If the objective of the organization is to play a game, the input materials may be balls, bats, gloves, and the rules of the game.

Actions, expectations, and interactions are human inputs. If the group does not employ any material or financial inputs, then its structure, operations, and outputs are accounted for entirely by its human inputs. Many kinds of organization, of course, utilize money, materials, and the services of other organizations, as inputs. Such inputs will influence the size, structure, operations, and outputs of the organizations that use them. An organization can exist without such inputs, but cannot exist without the inputs of human performance, expectation, and interaction. For this reason, only human inputs have been utilized in the development of the proposed theory of organization.

MEDIATING (PROCESSING) VARIABLES

Not all the inputs of an organization are transformed directly into outputs. Some portion of the inputs must be utilized in the development and maintenance of operations, structure, and interpersonnel. The mediating and processing variables that transform inputs into outputs are listed in Table 2.

TABLE 2

The Processing Variables

Operations:
    Human performances
    Technical processes
Interpersonnel:
    Interpersonal relations (superior, subordinate, peer)
    Personal—organizational exchange relations
    Reinforcement, satisfactions and dissatisfactions
    Internal—external relations (reference groups)
    Subgroup norms and pressures
    Subsystem interrelations
Structure:
    Positions (status and function)
    Roles (responsibility, authority, delegation)
    Formal and informal subsystems
    Departmentation

The *operations* of an organization are carried out by the performances of the members individually and in interaction with others. Task performances may consist of talking, thinking, using tools, operating machines, handling materials, writing letters, selling merchandise, rendering a service, and the like. Social, educational, religious, governmental, military, financial, mercantile, and manufacturing organizations differ in the kinds of operations they perform. What is common to all of them is that operations are carried out by the task performances of their members. Even automated plants must be supplied, programmed and monitored by the performances of responsible members.

The nature of group *structure* is best understood by examining newly created groups that have no predetermined structure. Bales and Strodtbeck [2] and Heinicke and Bales [5] have shown that experimental groups assigned the task of solving problems by discussion tend to make little progress toward task accomplishment until a role structure has been differentiated and stabilized. The work of the group remains at a standstill as long as the leadership of the group remains in doubt or in contest. However, the emergence of a leader who maintains his position of leadership tends to reinforce the expectations of the group members that he will continue to act as leader, and they look to him for direction and the resolution of differences among the members. Once a structure of positions has been established, the group is free to engage in task performance. Research of the sort described above indicates that group structure facilitates operations.

The prior structuring of a group by those members who organize it and recruit new members for it tends to increase the likelihood that the structure of positions will be acknowledged and accepted by the members. The position accepted by a member defines his status and function in relation to other members and in relation to the operational tasks of the organization. The status of a position defines the stable expectations made upon the occupant of the position in relation to his accountability to and for other members of the organization. The functions of a position define the stable expectations that are made upon the occupant of the position in relation to the tasks he is to perform on behalf of the organization. The status and functions of a position tend to remain constant regardless of the person who occupies the position.

Despite the fact that the status and functions of a position remain relatively constant over time, no two occupants of the position exhibit identical behaviors. One occupant may exercise a higher degree of authority than his predecessor, and may delegate more (or less) to his subordinates. He may assume responsibility for certain functions that his predecessor disregarded, and he may assign responsibility for other functions to his subordinates. In other words, no two occupants of the same position ever play the same role in the organization. The above considerations suggest that responsibility and authority should be regarded as characteristics of role performance. Whereas status and function are stable characteristics of positions, authority and responsibility are variable aspects of role performance. Authority is highly related to status, and responsibility is highly related to function. The higher the level of a member's position, the greater his area of freedom for making decisions that affect the organization as a whole. The more clearly defined his functions, the greater the likelihood that he will assume responsibility for them.

The concept *interpersonnel* was introduced by the author [12] to account for a dimension that has not been heretofore logically integrated into a theory of organization. Interpersonnel refers to a large complex of behaviors and relationships that account for exchanges between member and member, and between member and organization. These are intraorganizational exchanges.

A member pays certain costs for belonging to any organization. For some organizations, he may invest only a small amount of money in the form of paying dues. But for others, he may make heavy investments of time, effort, skill, expecation, and endurance of interpersonal stress. All of these are costs to the individual member, and inputs into the organization. In exchange for these contributions, the individual expects certain returns in the form of rewards and satisfaction of expectations from other members and from the organization as a whole.

Members generally tend to expect that those members in status levels one or two steps above their own positions will act as mediators between themselves and the organization. They look to those persons in positions of leadership for clear definition of tasks, provision of freedom for task performance, and recognition of their contributions to, and support of, the organization. The relationships between members and their supervisors tend to affect the outputs of subgroups and of the organization as a whole.

Individual members differ markedly in their expectations, ability to interact harmoniously with other members, and ability to derive satisfaction from work, social interaction, and rewards. Their identifications with other groups and organizations in the external environment tend to condition their expectations and loyalties within the organization. On the other hand, there is a tendency for individuals to incorporate, and to be governed by, the norms of the subgroups in which they have membership in the organization. The norms of subgroups tend to regulate member performance, condition their loyalties, and affect group outputs. These factors, in turn, affect the operations and outputs of the organization as a whole.

One of the most critical exchange relationships existing between a member and the organization is that involving expectations. There are numerous reasons, some having their origins in ancient customs and even in superstitions, why individuals and organizations exhibit reluctance to make their expectations clearly known. If expectations are not communicated at the beginning of a relationship, it becomes increasingly difficult as time goes on to explain respects in which a member fails to satisfy the expectations of the organization and for a member to explain his disappointments with the organization. The author has shown [13] that member satisfaction with the company and subgroup loyalty to the company are highly related to a form of supervisory behavior that lets employees know what is expected of them and what they can expect of the organization. Thus, working in return for a wage is not the only important exchange relationship that exists between a member and an organization. The personnel manager of a large organization is concerned with a wide variety of exchanges. The union-company contract specifies a variety of exchange obligations. But there remains a set of exchanges (between member and member, member and subgroup, member and organization) that is not subject to negotiation or to administration by the personnel director, and this set is the one that is most likely to affect individual and subgroup performance and organizational outputs.

The confirmation and satisfaction of member expectations, in accordance with Cyert and March [3], should be regarded as an inter-member and intraorganizational exchange process. The organization expects certain contributions and loyalties from its members. The members expect task definition, freedom for performance, and an adequate return for their contributions. Members may perform to the best of their abilities without the satisfaction of these expectations, but are likely to feel frustrated and ill at ease as a result. Extremely high degrees of satisfaction or dissatisfaction may be associated with raised or lowered norms in the subgroups. Such norms consist of mutually acknowledged expectations among the members of a subgroup, and tend to regulate the performances and loyalties of all its members even though some members may dislike the norms. When member satisfaction or dissatisfaction is translated into raised or lowered subgroup norms throughout an organization, the outputs of the organization are likely to be affected as a result. Under conditions of prevailing dissatisfaction, it is more likely that organizational cohesiveness will be seriously affected than that productivity will decline. The result may be no less costly in that the organization may find that it does not have the support of its members when it needs their loyalty in some critical situation.

In summary, the concepts *operations, structure,* and *interpersonnel* have been used to effect an integration of classical theories of organization with behavior science conceptions of organization, and to generate some new insights into the differences between positions and roles. Status and function are shown to be characteristics of positions, while authority and responsibility are shown to be characteristics of role performance. The concept *interpersonnel* accounts for a great variety of exchange relationships that exist between member and member, member and subgroup, member and organization, and subgroup and organization.

THE OUTPUT VARIABLES

Inputs, in business organizations, are usually measured in terms of the dollar value of materials, man-hours, and support facilities. Output is measured as a ratio between the dollar value of the end product or service and the cost of the inputs, using a base period of time as a standard for later comparisons. The very survival of a large business organization depends upon the adequacy with which these factors are measured and controlled. But there are many kinds of organizations (religious, educational, fraternal, recreational, and the like) for which these measures have little or no relevance. Even for business organizations, we believe that there are valuable outputs other than product.

When one considers the inputs (performances, interactions, and expectations) and the mediating variables (structure, operations, and interpersonnel) used in this theoretical system, it seems reasonable to believe that they might generate a complex output. Such is found to be the case. What then is the nature of the outputs? Research in small groups indicates that groups develop cohesiveness and drive as they engage in task performance. Cohesiveness and drive may vary in degree in different groups, or in the same group over periods of time. Cohesiveness is variously defined in the literature as intermember harmony, mutual liking and acceptance, group resistance to disruptive influences, and ability to function as an integrated unit. Drive is variously defined as group freedom of action, morale, task enthusiasm, application of pressure upon an objective, member satisfaction, and spirited action. In social science literature, these two variables are considered to condition (increase or decrease) group productivity, or product. Intensive work with a great variety of models convinced the writer [10] that a logically consistent system cannot be developed when cohesiveness and morale are treated as conditioners of productivity. Nor can apparent discrepancies in experimental results be reconciled by such a model. However, if cohesiveness, morale, and productivity are all treated as outputs, a logically consistent system emerges and the previous discrepancies are reconciled.

In order to satisfy the system, the outputs must be defined in reference to the input and mediating variables. Assuming that there are no material or monetary inputs, *productivity* is then defined as the relative changes in the goal expectancy and goal achievement values of the organization. Goal expectancy sets a standard in terms of which the rate or degree of progress toward goal achievement is evaluated. Various members of the organization may evaluate productivity differently, depending upon their perceptions of the goals to be achieved and of the progress made toward goal achievement. If an organization operates upon a material input, it is both convenient and legitimate to use units of product or dollar value of product in the computation of productivity indexes.

Drive is here defined as organizational morale or freedom of action. Drive thus defined is a characteristic of organizations and of subgroups within organizations. The morale (drive) of an organization may be channeled toward task performance or toward some activity that bears little relation to the primary task of the organization. It can be diverted into competing or contradictory activities. Thus, it may not be advantageous to increase drive unless it can be utilized in strengthening the cohesiveness or productivity of the organization.

Cohesiveness is defined as the maintenance of, or capacity to maintain, structure and operations under stress. Cohesiveness characterizes the organization as a whole, as well as the subgroups within the organization. Cohesiveness in the subgroups may, or may not, add to the cohesiveness of the total organization. The latter is likely to suffer when member loyalties are to the subgroups rather than to the organization, and when the norms of the subgroups are such that they refuse to support the organization when support is needed. Cohesiveness is of importance to the survival of an organization in times of stress and emergency.

The three output variables (product, cohesiveness, and morale) are interrelated, but the correlations between them are seldom very high. A survey (10) of the research on small groups and large organizations suggested the following general trends.

1. Productivity and drive tend to be positively related.
2. Productivity and cohesiveness tend to be negatively related.
3. Drive and cohesiveness may be either positively or negatively related.

The three hypotheses stated above are based on empirical findings that were never satisfactorily explained by the assumption that cohesiveness and morale are conditioners of productivity. However, the findings appear to be logically consistent when the three variables are regarded as interrelated outputs.

The author recently completed a study in which the productivity, cohesiveness (harmony), and drive of 1267 work groups in 26 industrial and governmental organizations were rated by the superiors of the supervisors of the work groups. Correlations between the three variables are in accord with the general expectation that productivity and drive will be positively correlated, productivity and cohesiveness will be negatively correlated, and drive and cohesiveness may be either positively or negatively correlated. The number of positive and negative correlations between the three variables were as follows: volume of output and drive, 22 plus and 4 minus; volume and cohesiveness, 4 plus and 22 minus; drive and cohesiveness, 15 plus and 11 minus. Although only 17 of the 78 correlation coefficients were significantly larger than zero, the direction of correlation tended to be in the direction hypothesized by the theory.

Exceptions to the above hypotheses are found. But it seems reasonable to expect that some portion of increased group drive (morale) would be reflected in increased productivity. It also seems reasonable that as time and effort are devoted to strengthening cohesiveness, productivity would tend to decrease; and that as efforts are made to increase productivity, either the internal cohesiveness of subgroups or their support of the organization would tend to decrease. In regard to the relationship between drive and cohesiveness, the experimental literature suggests that morale (drive) tends to be expended in strengthening cohesiveness when groups are under threat. However, increased drive is associated with reduced cohesiveness when subgroups are divided into cliques or when differences in loyalties are involved. Thus, drive can contribute either to increased or decreased cohesiveness. Input and mediating variables interact in complex combinations to condition the relationships between productivity, drive, and cohesiveness. Extensive research will be required to determine the nature of these combinations.

The relationships hypothesized above are the ones most frequently encountered under stabilized operating conditions and with constant inputs. Under such conditions, an increase in one output variable tends to be obtained at the cost of another. In order to increase all outputs simultaneously, it should be necessary to increase the inputs. The author [11] found support for this hypothesis in a study of six football games. Teams are able to invest intense inputs of effort, skill, and motivation during one hour of play. Such high input levels cannot be expected in organizations that require eight hours of work, day after day, throughout the year. But an intensification of inputs appears to provide one of the conditions under which all outputs can be increased at the same time.

In summary, it is suggested that any organization, regardless of its size or the nature of its material inputs or technical operations, generates three distinct outputs: product, drive, and cohesiveness. The outputs are interrelated. Material inputs and outputs were disregarded in this formulation, not because they are unimportant, but because they are present only in specific types of organizations.

DISCUSSION

The proposed theory can be said to be new in the following respects. (1) It is based on a set of newly defined input variables. (2) The mediating variables for the first time in any theory, establish a logical differentiation between (a) the status and function of a position and (b) the responsibility and authority of a role. The concept *interpersonnel* accounts for a great complex of exchange relationships that exist within an organization, but are usually treated as separate from a theory of organization. (3) The theory generates a new set of output variables. Whereas it is generally assumed that productivity is the only output measure, the new theory requires that productivity, cohesiveness, and drive be regarded as outputs of organization. (4) The theory reconciles findings that have heretofore been regarded as contradictory. (5) The theory has provided a basis for new research designs, the results of which are in conformity with the theory, and has stimulated the investigation of variables and relationships that would have gone unnoticed without the theory.

It should be noted that a theory of organization is not a theory of management. Organization theory is concerned only with what exists in organization; not with what ought to exist. Management theory, on the other hand, is concerned with decisions designed to optimize the attainment of values through the instrumentality of organization. An adequate theory of organization should provide an advantage in the understanding and management of organizations even though it generates no rules for effective management.

REFERENCES

1. Allee, W. C. *Cooperation among Animals, with Human Implications.* Henry Schuman, New York, 1951.
2. Bales, R. F. and Strodtbeck, F. L. "Phases in Group Problem-Solving." *Journal of Abnormal Social Psychology,* 46, 1951, pp. 485–95.
3. Cyert, R. M. and March, J. G. *A Behavioral Theory of the Firm.* Prentice-Hall, Englewood Cliffs, N.J., 1963.

4. Davis, R. C. *The Fundamentals of Top Management.* Harper, New York, 1951.
5. Heinicke, C. and Bales, R. F. "Developmental Trends in the Structure of Small Groups." *Sociometry, 16,* 1953, pp. 7–38.
6. Homans, G. C. *The Human Group.* Harcourt, Brace, New York, 1950.
7. Katz, D. and Kahn, R. L. *The Social Psychology of Organizations.* Wiley, New York, 1966.
8. Parsons, T. *The Social System.* Free Press, New York, 1951.
9. Simon, H. A. *Models of Man.* Wiley, New York, 1957.
10. Stogdill, R. M. *Individual Behavior and Group Achievement.* Oxford University Press, New York, 1959.
11. ——— *Team Achievement under High Motivation.* Ohio State University, Bureau of Business Research, Columbus, Ohio, 1963.
12. ——— "Dimensions of Organization Theory," in J. D. Thompson, *Approaches to Organizational Design.* University of Pittsburgh Press, 1965.
13. ——— *Managers, Employees, Organizations.* Ohio State University, Bureau of Business Research, Columbus, Ohio, 1965.
14. Tuckman, B. W. "Developmental Sequence in Small Groups." *Psychological Bulletin, 63,* 1965, pp. 384–99.

Section III

ORGANIZATIONAL ENVIRONMENTS

To CONCEPTUALIZE an organization as an open system is to emphasize the importance of its environment, upon which the maintenance, survival, and growth of an open system depend. Emery and Trist present the view that a comprehensive understanding of organizational behavior requires a consideration not only of the internal activities of an organization, but also of the across-the-boundaries transactions between the organization and its environment and the external processes characterizing relationships among the various parts of the environment. They offer a typology of four "ideal types" of environment and discuss the effect of differing environmental conditions upon an organization existing in each type of environment.

Shirley Terreberry extends and elaborates the ideal types of environment conceptualized by Emery and Trist. Her paper argues that evolutionary processes occur in the environments of organizations and that recent literature gives evidence that organizations experience decreasing autonomy and increasing interdependence. Two hypotheses are offered: (1) that organizational change is increasingly externally induced; (2) that organizational adaptability is a function of the ability to learn and to perform according to changes in the environment.

Lawrence and Lorsch (1967), in a pioneering study, developed a systemic model for studying the relationship between environment and organizations. In a paper especially prepared for this volume, Gabarro reviews this research of Lawrence and Lorsch and its findings, and then reports his attempt to apply their organizational systems framework. Lawrence and Lorsch compared business organizations in different environments, focusing on the processes of differentiation and integration within these systems. Gabarro also focused on differentiation and integration but studied these variables as they relate to the problem of adaptation of school organizations within a single environment in the process of change.

Organizational systems develop specialized roles in conducting input and output transactions with their environment. In his paper, Thompson offers a typology of output roles characterizing boundary-spanning transactions related to the distribution of the products of organizational operations. Transaction structures are particularly relevant in terms of linking the organization to conditions existing in a dynamic environment. This paper provides some insight into the kind of organizational adaptation which may be called for in meeting the varied environmental contingencies described in the other papers in this section.

In the last paper in this section, Brown discusses the role of organizational boundaries as these boundaries control the impact of environmental forces. He makes use of a general systems framework in exploring selected aspects of organizational boundary problems, especially those concerning information flows. Brown offers a general schema based on the classification of decision making in the firm for bringing various types of information into an organization.

F. E. Emery
E. L. Trist

10. The Causal Texture
of Organizational Environments

IDENTIFICATION OF THE PROBLEM

A MAIN PROBLEM in the study of organizational change is that the environmental contexts in which organizations exist are themselves changing, at an increasing rate, and towards increasing complexity. This point, in itself, scarcely needs labouring. Nevertheless, the characteristics of organizational environments demand consideration for their own sake, if there is to be an advancement of understanding in the behavioral sciences of a great deal that is taking place under the impact of technological change, especially at the present time. This paper is offered as a brief attempt to open up some of the problems, and stems from a belief that progress will be quicker if a certain extension can be made to current thinking about systems.

In a general way it may be said that to think in terms of systems seems the most appropriate conceptual response so far available when the phenomena under study—at any level and in any domain—display the character of being organized, and when understanding the nature of the interdependencies constitutes the research task. In the behavioural sciences, the first steps in building a systems theory were taken in connection with the analysis of internal processes in organisms, or organizations, when the parts had to be related to the whole. Examples include the organismic biology of Jennings, Cannon, and Henderson; early Gestalt theory and its later derivatives such as balance theory; and the classical theories of social structure. Many of these problems could be represented in closed-system models. The next steps were taken when wholes had to be related to their environments. This led to open-system models.

A great deal of the thinking here has been influenced by cybernetics and information theory, though this has been used as much to extend

From *Human Relations*, 1965, vol. 18, pp. 21–32. Reprinted with permission of the authors and the publisher.

the scope of closed-system as to improve the sophistication of open-system formulations. It was von Bertalanffy (1950) who, in terms of the general transport equation which he introduced, first fully disclosed the importance of openness or closedness to the environment as a means of distinguishing living organisms from inanimate objects. In contradistinction to physical objects, any living entity survives by importing into itself certain types of material from its environment, transforming these in accordance with its own system characteristics, and exporting other types back into the environment. By this process the organism obtains the additional energy that renders it 'negentropic'; it becomes capable of attaining stability in a time-independent steady state—a necessary condition of adaptability to environmental variance.

Such steady states are very different affairs from the equilibrium states described in classical physics, which have far too often been taken as models for representing biological and social transactions. Equilibrium states follow the second law of thermodynamics, so that no work can be done when equilibrium is reached, whereas the openness to the environment of a steady state maintains the capacity of the organism for work, without which adaptability, and hence survival, would be impossible.

Many corollaries follow as regards the properties of open systems, such as equifinality, growth through internal elaboration, self-regulation, constancy of direction with change of position, etc.—and by no means all of these have yet been worked out. But though von Bertalanffy's formulation enables exchange processes between the organism, or organization, and elements in its environment to be dealt with in a new perspective, it does not deal at all with those processes in the environment itself which are among the determining conditions of the exchanges. To analyse these an additional concept is needed— *the causal texture of the environment*—if we may re-introduce, at a social level of analysis, a term suggested by Tolman and Brunswik (1935) and drawn from S. C. Pepper (1934).

With this addition, we may now state the following general proposition: that a comprehensive understanding of organizational behaviour requires some knowledge of each member of the following set, where L indicates some potentially lawful connection, and the suffix 1 refers to the organization and the suffix 2 to the environment:

$$L_{1\,1}, \; L_{1\,2}$$
$$L_{2\,1}, \; L_{2\,2}$$

$L_{1\,1}$ here refers to processes within the organization—the area of interdependencies; $L_{1\,2}$ and $L_{2\,1}$ to exchanges between the organization and its environment—the area of transactional interdependencies, from either direction; and $L_{2\,2}$ to processes through which parts of the environment become related to each other—i.e. its causal texture—the area of interdependencies that belong within the environment itself.

In considering environmental interdependencies, the first point to which we wish to draw attention is that the laws connecting parts of the environment to each other are often incommensurate with those connecting parts of the organization to each other, or even with those which govern the exchanges. It is not possible, for example, always to reduce organization-environment relations to the form of 'being included in'; boundaries are also 'break' points. As Barker and Wright (1949), following Lewin (1936), have pointed out in their analysis of this problem as it affects psychological ecology, we may lawfully connect the actions of a javelin thrower in sighting and throwing his weapon; but we cannot describe in the same concepts the course of the javelin as this is affected by variables lawfully linked by meteorological and other systems.

THE DEVELOPMENT OF ENVIRONMENTAL CONNECTEDNESS (CASE I)

A case history, taken from the industrial field, may serve to illustrate what is meant by the environment becoming organized at the social level. It will show how a greater degree of system-connectedness, of crucial relevance to the organization, may develop in the environment, which is yet not directly a function either of the organization's own characteristics or of its immediate relations. Both of these, of course, once again become crucial when the response of the organization to what has been happening is considered.

The company concerned was the foremost in its particular market in the food-canning industry in the U.K. and belonged to a large parent group. Its main product—a canned vegetable—had some 65 percent of this market, a situation which had been relatively stable since before the war. Believing it would continue to hold this position, the company persuaded the group board to invest several million pounds sterling in erecting a new, automated factory, which, however, based its economies on an inbuilt rigidity—it was set up exclusively for the long runs expected from the traditional market.

The character of the environment, however, began to change while the factory was being built. A number of small canning firms appeared, not dealing with this product nor indeed with others in the company's range, but with imported fruits. These firms arose because the last of the post-war controls had been removed from steel strip and tin, and cheaper cans could now be obtained in any numbers—while at the same time a larger market was developing in imported fruits. This trade being seasonal, the firms were anxious to find a way of using their machinery and retaining their labour in winter. They became able to do so through a curious side-effect of the development of quick-frozen foods, when the company's staple was produced by others in this form. The quick-freezing process demanded great constancy at the growing end. It was not possible to control this beyond a certain point, so that quite large crops unsuitable for quick freezing but suitable for canning became available—originally from another country (the United States) where a large market for quick-frozen foods had been established. These surplus crops had been sold at a very low price for animal feed. They were now imported by the small canners—at a better but still comparatively low price, and additional cheap supplies soon began to be procurable from underdeveloped countries.

Before the introduction of the quick-freezing form, the company's own canned product—whose raw material had been specially grown at additional cost—had been the premier brand, superior to other varieties and charged at a higher price. But its position in the product spectrum now changed. With the increasing affluence of the society, more people were able to afford the quick-frozen form. Moreover, there was competition from a great many other vegetable products which could substitute for the staple, and people preferred this greater variety. The advantage of being the premier line among canned forms diminished, and demand increased both for the not-so-expensive varieties among them and for the quick-frozen forms. At the same time, major changes were taking place in retailing; supermarkets were developing, and more and more large grocery chains were coming into existence. These establishments wanted to sell certain types of goods under their own house names, and began to place bulk orders with the small canners for their own varieties of the company's staple that fell within this class. As the small canners provided an extremely cheap article (having no marketing expenses and a cheaper raw material), they could undercut the manufacturers' branded product, and within three years they captured over 50 percent of the market. Previously, retailers' varieties had accounted for less than 1 percent.

The new automatic factory could not be adapted to the new situation until alternative products with a big sales volume could be developed, and the scale of research and development, based on the type of market analysis required to identify these, was beyond the scope of the existing resources of the company either in people's or in funds.

The changed texture of the environment was not recognized by an able but traditional management until it was too late. They failed entirely to appreciate that a number of outside events were becoming connected with each other in a way that was leading up to irreversible general change. Their first reaction was to make an herculean effort to defend the traditional product, then the board split on whether or not to make entry into the cheaper unbranded market in a supplier role. Group H.Q. now felt they had no option but to step in, and many upheavals and changes in management took place until a 'redefinition of mission' was agreed, and slowly and painfully the company re-emerged with a very much altered product mix and something of a new identity.

FOUR TYPES OF CAUSAL TEXTURE

It was this experience, and a number of others not dissimilar, by no means all of them industrial (and including studies of change problems in hospitals, in prisons, and in educational and political organizations), that gradually led us to feel a need for re-directing conceptual attention to the causal texture of the environment, considered as a quasi-independent domain. We have now isolated four 'ideal types' of causal texture, approximations to which may be thought of as existing simultaneously in the 'real world' of most organizations—though, of course, their weighting will vary enormously from case to case.

The first three of these types have already, and indeed repeatedly, been described—in a large variety of terms and with the emphasis on an equally bewildering variety of special aspects—in the literature of a number of disciplines, ranging from biology to economics and including military theory as well as psychology and sociology. The fourth type, however, is new, at least to us, and is the one that for some time we have been endeavouring to identify. About the first three, therefore, we can be brief, but the fourth is scarcely understandable without reference to them. Together, the four types may be said to form a series in which the degree of causal texturing is increased, in a new and significant way, as each step is taken. We leave as an open question the need for further steps.

Step One

The simplest type of environmental texture is that in which goals and noxiants ('goods' and 'bads') are relatively unchanging in themselves and randomly distributed. This may be called the *placid, randomized environment.* It corresponds to Simon's idea of a surface over which an organism can locomote: most of this is bare, but at isolated, widely scattered points there are little heaps of food (1957, p. 137). It also corresponds to Ashby's limiting case of no connection between the environmental parts (1960, S15/4); and to Schutzenberger's random field (1954, p. 100). The economist's classical market also corresponds to this type.

A critical property of organizational response under random conditions has been stated by Schutzenberger: that there is no distinction between tactics and strategy, 'the optimal strategy is just the simple tactic of attempting to do one's best on a purely local basis' (1954, p. 101). The best tactic, moreover, can be learnt only by trial and error and only for a particular class of local environmental variances (Ashby, 1960, p. 197). While organizations under these conditions can exist adaptively as single and indeed quite small units, this becomes progressively more difficult under the other types.

Step Two

More complicated, but still a placid environment, is that which can be characterized in terms of clustering: goals and noxiants are not randomly distributed but hang together in certain ways. This may be called the *placid, clustered environment,* and is the case with which Tolman and Brunswik were concerned; it corresponds to Ashby's 'serial system' and to the economist's 'imperfect competition.' The clustering enables some parts to take on roles as signs of other parts or become means-objects with respect to approaching or avoiding. Survival, however, becomes precarious if an organization attempts to deal tactically with each environmental variance as it occurs.

The new feature of organizational response to this kind of environment is the emergence of strategy as distinct from tactics. Survival becomes critically linked with what an organization knows of its environment. To pursue a goal under its nose may lead it into parts of the field fraught with danger, while avoidance of an immediately difficult issue may lead it away from potentially rewarding areas. In the clustered environment the relevant objective is that of 'optimal location,' some positions being discernible as potentially richer than others.

To reach these requires concentration of resources, subordination to the main plan, and the development of a 'distinctive competence,' to use Selznick's (1957) term, in reaching the strategic objective. Organizations under these conditions, therefore, tend to grow in size and also to become hierarchical, with a tendency towards centralized control and coordination.

Step Three

The next level of causal texturing we have called the *disturbedreactive environment*. It may be compared with Ashby's ultra-stable system or the economist's oligopolic market. It is a type 2 environment in which there is more than one organization of the same kind; indeed, the existence of a number of similar organizations now becomes the dominant characteristic of the environmental field. Each organization does not simply have to take account of the others when they meet at random, but has also to consider that what it knows can also be known by the others. The part of the environment to which it wishes to move itself in the long run is also the part to which the others seek to move. Knowing this, each will wish to improve its own chances by hindering the others, and each will know that the others must not only wish to do likewise, but also know that each knows this. The presence of similar others creates an imbrication, to use a term of Chein's (1943), of some of the causal strands in the environment.

If strategy is a matter of selecting the 'strategic objective'—where one wishes to be at a future time—and tactics a matter of selecting an immediate action from one's available repertoire, then there appears in type 3 environments to be an intermediate level of organizational response—that of the *operation*—to use the term adopted by German and Soviet military theorists, who formally distinguish tactics, operations, and strategy. One has now not only to make sequential choices, but to choose actions that will draw off the other organizations. The new element is that of deciding which of someone else's possible tactics one wishes to take place, while ensuring that others of them do not. An operation consists of a campaign involving a planned series of tactical initiatives, calculated reactions by others, and counter-actions. The flexibility required encourages a certain decentralization and also puts a premium on quality and speed of decision at various peripheral points (Heyworth, 1955).

It now becomes necessary to define the organizational objective in terms not so much of location as of capacity or power to move more or less at will, i.e., to be able to make and meet competitive challenge. This gives particular relevance to strategies of absorption and parasitism. It can also give rise to situations in which stability can be obtained only by a certain coming-to-terms between competitors, whether enterprises, interest groups, or governments. One has to know when not to fight to the death.

Step Four

Yet more complex are the environments we have called *turbulent fields*. In these, dynamic processes, which create significant variances for the component organizations, arise from the field itself. Like type 3 and unlike the static types 1 and 2, they are dynamic. Unlike type 3, the dynamic properties arise not simply from the interaction of the component organizations, but also from the field itself. The 'ground' is in motion.

Three trends contribute to the emergence of these dynamic field forces:

1. The growth to meet type 3 conditions of organizations and linked sets of organizations, so large that their actions are both persistent and strong enough to induce autochthonous processes in the environment. An analogous effect would be that of a company of soldiers marching in step over a bridge.
2. The deepening interdependence between the economic and the other facets of the society. This means that economic organizations are increasingly enmeshed in legislation and public regulation.
3. The increasing reliance on research and development to achieve the capacity to meet competitive challenge. This leads to a situation in which a change gradient is continuously present in the environmental field.

For organizations, these trends mean a gross increase in their area of *relevant uncertainty*. The consequences which flow from their actions lead off in ways that become increasingly unpredictable: they do not necessarily fall off with distance, but may at any point be amplified beyond all expectation; similarly, lines of action that are strongly pursued may find themselves attenuated by emergent field forces.

THE SALIENCE OF TYPE 4 CHARACTERISTICS (CASE II)

Some of these effects are apparent in what happened to the canning company of case I, whose situation represents a transition from an environment largely composed of type 2 and type 3 characteristics to one where those of type 4 began to gain in salience. The case now to be presented illustrates the combined operation of the three trends described above in an altogether larger environmental field involving a total industry and its relations with the wider society.

The organization concerned is the National Farmers Union of Great

Britain to which more than 200,000 of the 250,000 farmers of England and Wales belong. The presenting problem brought to us for investigation was that of communications. Headquarters felt, and was deemed to be, out of touch with county branches, and these with local branches. The farmer had looked to the N.F.U. very largely to protect him against market fluctuations by negotiating a comprehensive deal with the government at annual reviews concerned with the level of price support. These reviews had enabled home agriculture to maintain a steady state during two decades when the threat, or existence, of war in relation to the type of military technology then in being had made it imperative to maintain a high level of homegrown food without increasing prices to the consumer. This policy, however, was becoming obsolete as the conditions of thermonuclear stalemate established themselves. A level of support could no longer be counted upon which would keep in existence small and inefficient farmers—often on marginal land and dependent on family labour—compared with efficient medium-size farms, to say nothing of large and highly mechanized undertakings.

Yet it was the former situation which had produced N.F.U. cohesion. As this situation receded, not only were farmers becoming exposed to more competition from each other, as well as from Commonwealth and European farmers, but the effects were being felt of very great changes which had been taking place on both the supply and marketing sides of the industry. On the supply side, a small number of giant firms now supplied almost all the requirements in fertilizer, machinery, seeds, veterinary products, etc. As efficient farming depended upon ever greater utilization of these resources, their controllers exerted correspondingly greater power over the farmers. Even more dramatic were the changes in the marketing of farm produce. Highly organized food processing and distributing industries had grown up dominated again by a few large firms, on contracts from which (fashioned to suit their rather than his interests) the farmer was becoming increasingly dependent. From both sides deep inroads were being made on his autonomy.

It became clear that the source of the felt difficulty about communications lay in radical environmental changes which were confronting the organization with problems it was ill-adapted to meet. Communications about these changes were being interpreted or acted upon as if they referred to the 'traditional' situation. Only through a parallel analysis of the environment and the N.F.U. was progress made towards developing understanding on the basis of which attempts to devise adaptive organizational policies and forms could be made. Not least among the problems was that of creating a bureaucratic elite that could cope with the highly technical long-range planning now required and yet remain loyal to the democratic values of the N.F.U. Equally difficult was that of developing mediating institutions—agencies that would effectively mediate the relations between agriculture and other economic sectors without triggering off massive competitive processes.

These environmental changes and the organizational crisis they induced were fully apparent two or three years before the question of

Britain's possible entry into the Common Market first appeared on the political agenda—which, of course, further complicated every issue.

A workable solution needed to preserve reasonable autonomy for the farmers as an occupational group, while meeting the interests of other sections of the community. Any such possibility depended on securing the consent of the large majority of farmers to placing under some degree of N.F.U. control matters that hitherto had remained within their own power of decision. These included what they produced, how and to what standard, and how most of it should be marketed. Such thoughts were anathema, for however dependent the farmer had grown on the N.F.U. he also remained intensely individualistic. He was being asked, he now felt, to redefine his indenity, reverse his basic values, and refashion his organization—all at the same time. It is scarcely surprising that progress has been, and remains, both fitful and slow, and ridden with conflict.

VALUES AND RELEVANT UNCERTAINTY

What becomes precarious under type 4 conditions is how organizational stability can be achieved. In these environments individual organizations, however large, cannot expect to adapt successfully simply through their own direct actions—as is evident in the case of the N.F.U. Nevertheless, there are some indications of a solution that may have the same general significance for these environments as have strategy and operations for types 2 and 3. This is the emergence of *values that have overriding significance for all members of the field.* Social values are here regarded as coping mechanisms that make it possible to deal with persisting areas of relevant uncertainty. Unable to trace out the consequences of their actions as these are amplified and resonated through their extended social fields, men in all societies have sought rules, sometimes categorical, such as the ten commandments, to provide them with a guide and ready calculus. Values are not strategies or tactics; or Lewin (1936) has pointed out, they have the conceptual character of 'power fields' and act as injunctions.

So far as effective values emerge, the character of richly joined, turbulent fields changes in a most striking fashion. The relevance of large classes of events no longer has to be sought in an intricate mesh of diverging causal strands, but is given directly in the ethical code. By this transformation a field is created which is no longer richly joined and turbulent but simplified and relatively static. Such a transformation will be regressive, or constructively adaptive, according to how far the emergent values adequately represent the new environmental requirements.

Ashby, as a biologist, has stated his view, on the one hand, that examples of environments that are both large and richly connected are not common, for our terrestrial environment is widely characterized by being highly subdivided (1960, p. 205); and, on the other, that, so far as they are encountered, they may well be beyond the limits of human adaptation, the brain being an ultrastable system. By contrast the role here attributed to social values suggests that this sort of environment may in fact be not only one to which adaptation is possible, however difficult, but one that has been increasingly characteristic of the human condition since the beginning of settled communities. Also, let us not forget that values can be rational as well as irrational and that the rationality of their rationale is likely to become more powerful as the scientific ethos takes greater hold in a society.

MATRIX ORGANIZATION AND INSTITUTIONAL SUCCESS

Nevertheless, turbulent fields demand some overall form of organization that is essentially different from the hierarchically structured forms to which we are accustomed. Whereas type 3 environments require one or other form of accommodation between like, but competitive, organizations whose fates are to a degree negatively correlated, turbulent environments require some relationship between dissimilar organizations whose fates are, basically, positively correlated. This means relationships that will maximize cooperation and which recognize that no one organization can take over the role of 'the other' and become paramount. We are inclined to speak of this type of relationship as an *organizational matrix*. Such a matrix acts in the first place by delimiting on value criteria the character of what may be included in the field specified—and therefore who. This selectivity then enables some definable shape to be worked out without recourse to much in the way of formal hierarchy among members. Professional associations provide one model of which there has been long experience.

We do not suggest that in other fields than the professional the requisite sanctioning can be provided only by state-controlled bodies. Indeed, the reverse is far more likely. Nor do we suggest that organizational matrices will function so as to eliminate the need for other measures to achieve stability. As with values, matrix organizations, even if successful, will only help to transform turbulent environments into the kinds of environment we have discussed as 'clustered' and 'disturbed-reactive'. Though, with these transformations, an organization could hope to achieve a degree of stability through its strategies, operation, and tactics, the transformations would not provide environments identical with the originals. The strategic objective in the transformed cases could no longer be stated simply in terms of optimal location (as in type 2) or capabilities (as in type 3). It must now rather be formulated in terms of *institutionalization*. According to Selznick (1957) organizations become institutions through the embodiment of organizational values which relate them to the wider society.[2] As Selznick has stated in his analysis of leadership in the modern American corporation, 'the default of leadership shows itself in an acute form

---

[2] Since the present paper was presented, this line of thought has been further developed by Churchman and Emery (1964) in their discussion of the relation of the statistical aggregate of individuals to structured role sets:

"Like other values, organizational values emerge to cope with relevant uncertainties and gain their authority from their reference to the requirements of larger systems within which people's interests are largely concordant."

when *organizational* achievement or survival is confounded with *institutional* success' (1957, p. 27). ' . . . the executive becomes a statesman as he makes the transition from administrative management to institutional leadership' (1957, p. 154).

The processes of strategic planning now also become modified. In so far as institutionalization becomes a prerequisite for stability, the determination of policy will necessitate not only a bias towards goals that are congruent with the organization's own character, but also a selection of goal-paths that offer maximum convergence as regards the interests of other parties. This became a central issue for the N.F.U. and is becoming one now for an organization such as the National Economic Development Council, which has the task of creating a matrix in which the British economy can function at something better than the stop-go level.

Such organizations arise from the need to meet problems emanating from type 4 environments. Unless this is recognized, they will only too easily be construed in type 3 terms, and attempts will be made to secure for them a degree of monolithic power that will be resisted overtly in democratic societies and covertly in others. In the one case they may be prevented from ever undertaking their missions; in the other one may wonder how long they can succeed in maintaining them.

An organizational matrix implies what McGregor (1960) has called Theory Y. This in turn implies a new set of values. But values are psycho-social commodities that come into existence only rather slowly. Very little systematic work has yet been done on the establishment of new systems of values, or on the type of criteria that might be adduced to allow their effectiveness to be empirically tested. A pioneer attempt is that of Churchman and Ackoff (1950). Likert (1961) has suggested that, in the large corporation or government establishment, it may well take some ten to fifteen years before the new type of group values with which he is concerned could permeate the total organization. For a new set to permeate a whole modern society the time required must be much longer—at least a generation, according to the common saying—and this, indeed, must be a minimum. One may ask if this is fast enough, given the rate at which type 4 environments are becoming salient. A compelling task for social scientists is to direct more research onto these problems.

SUMMARY

(a) A main problem in the study of organizational change is that the environmental contexts in which organizations exist are themselves changing—at an increasing rate, under the impact of technological change. This means that they demand consideration for their own sake. Towards this end a redefinition is offered, at a social level of analysis, of the causal texture of the environment, a concept introduced in 1935 by Tolman and Brunswik.

(b) This requires an extension of systems theory. The first steps in systems theory were taken in connection with the analysis of internal processes in organisms, or organizations, which involved relating parts to the whole. Most of these problems could be dealt with through closed-system models. The next steps were taken when wholes had to be related to their environments. This led to open-system models, such as that introduced by Bertalanffy, involving a general transport equation. Though this enables exchange processes between the organism, or organization, and elements in its environment to be dealt with, it does not deal with those processes in the environment itself which are the determining conditions of the exchanges. To analyse these an additional concept—the causal texture of the environment—is needed.

(c) The laws connecting parts of the environment to each other are often incommensurate with those connecting parts of the organization to each other, or even those which govern exchanges. Case history I illustrates this and shows the dangers and difficulties that arise when there is a rapid and gross increase in the area of relevant uncertainty, a characteristic feature of many contemporary environments.

(d) Organizational environments differ in their causal texture, both as regards degree of uncertainty and in many other important respects. A typology is suggested which identifies four 'ideal types', approximations to which exist simultaneously in the 'real world' of most organizations, though the weighting varies enormously:

1. In the simplest type, goals and noxiants are relatively unchanging in themselves and randomly distributed. This may be called the placid, randomized environment. A critical property from the organization's viewpoint is that there is no difference between tactics and strategy, and organizations can exist adaptively as single, and indeed quite small, units.

2. The next type is also static, but goals and noxiants are not randomly distributed; they hang together in certain ways. This may be called the placid, clustered environment. Now the need arises for strategy as distinct from tactics. Under these conditions organizations grow in size, becoming multiple and tending towards centralized control and coordination.

3. The third type is dynamic rather than static. We call it the disturbed-reactive environment. It consists of a clustered environment in which there is more than one system of the same kind, i.e., the objects of one organization are the same as, or relevant to, others like it. Such competitors seek to improve their own chances by hindering each other, each knowing the others are playing the same game. Between strategy and tactics there emerges an intermediate type of organizational response—what military theorists refer to as operations. Control becomes more decentralized to allow these to be conducted. On the other hand, stability may require a certain coming-to-terms between competitors.

4. The fourth type is dynamic in a second respect, the dynamic properties arising not simply from the interaction of identifiable components systems but from the field itself (the 'ground'). We call these environments turbulent fields. The turbulence results from the complexity and multiple character of the causal inter-connec-

tions. Individual organizations, however large, cannot adapt successfully simply through their direct interactions. An examination is made of the enhanced importance of values, regarded as a basic response to persisting areas of relevant uncertainty, as providing a control mechanism, when commonly held by all members in a field. This raises the question of organizational forms based on the characteristics of a matrix.

(*e*) Case history II is presented to illustrate problems of the transition from type 3 to type 4. The perspective of the four environmental types is used to clarify the role of Theory X and Theory Y as representing a trend in value change. The establishment of a new set of values is a slow social process requiring something like a generation—unless new means can be developed.

REFERENCES

Ashby, W. Ross (1960). *Design for a Brain.* London: Chapman & Hall.

Barker, R. G. & Wright, H. F. (1949). Psychological Ecology and the Problem of Psychosocial Development. *Child Development* 20, 131–43.

Bertalanffy, L. von (1950). The Theory of Open Systems in Physics and Biology. *Science* 111, 23–9.

Chein, I. (1943) Personality and Typology. *J. Soc. Psychol.* 18, 89–101.

Churchman, C. W. & Ackoff, R. L. (1950). *Methods of Inquiry.* St. Louis: Educational Publishers.

Churchman, C. W. & Emery, F. E. (1964). On various approaches to the study of organizations. Proceedings of the International Conference on Operational Research and the Social Sciences, Cambridge, England, 14–18 September 1964. To be published in book form as *Operational Research and the Social Sciences.* London: Tavistock Publications, 1965.

Heyworth, Lord (1955). *The Organization of Unilever.* London: Unilever Limited.

Lewin, K. (1936). *Principles of Topological Psychology.* New York: McGraw-Hill.

Lewin, K. (1951). *Field Theory in Social Science.* New York: Harper.

Likert, R. (1961). *New Patterns of Management.* New York, Toronto, London: McGraw-Hill.

McGregor, D. (1960). *The Human Side of Enterprise.* New York, Toronto, London: McGraw-Hill.

Pepper, S. C. (1934). The Conceptual Framework of Tolman's Purposive Behaviorism. *Psychol. Rev.* 41, 108–33.

Schutzenberger, M. P. (1954). A Tentative Classification of Goal-Seeking Behaviours. *J. Ment. Sci.* 100, 97–102.

Selznick, P. (1957). *Leadership in Administration.* Evanston, Ill.: Row Peterson.

Simon, H. A. (1957). *Models of Man.* New York: Wiley.

Tolman, E. C. & Brunswik, E. (1935). The Organism and the Causal Texture of the Environment. *Psychol. Rev.* 42, 43–77.

Shirley Terreberry

11. The Evolution of Organizational Environments

DARWIN PUBLISHED *The Origin of Species by Means of Natural Selecttion* in 1859. Modern genetics has vastly altered our understanding of the variance upon which natural selection operates. But there has been no conceptual breakthrough in understanding *environmental* evolution which, alone, shapes the direction of change. Even today most theorists of change still focus on *internal* interdependencies of systems—biological, psychological, or social—although the external environments of these systems are changing more rapidly than ever before.

INTRODUCTION

Von Bertalanffy was the first to reveal fully the importance of a system being open or closed to the environment in distinguishing living from inanimate systems.[1] Although von Bertalanffy's formulation makes it possible to deal with a system's exchange processes in a new perspective, it does not deal at all with those processes in the environment *itself* that are among the determining conditions of exchange.

Emery and Trist have argued the need for one additional concept, "the causal texture of the environment."[2] Writing in the context of formal organizations, they offer the following general proposition:

That a comprehensive understanding of organizational behaviour requires some knowledge of each member of the following set, where L indicates some potentially lawful connection, and the suffix 1 refers to the organization and the suffix 2 to the environment:

$$L_{11} \ L_{12}$$
$$L_{21} \ L_{22}$$

From *Administrative Science Quarterly*, 1968, vol. 12, pp. 590–613. Reprinted with permission of the author and the publisher.

[1] Ludwig von Bertalanffy, General System Theory, *General Systems*, 1 (1956), 1–10.

[2] F. E. Emery and E. L. Trist, The Causal Texture of Organizational Environments, *Human Relations*, 18 (1965), 21–31.

L_{11} here refers to processes within the organization—the area of internal interdependencies; L_{12} and L_{21} to exchanges between the organization and its environment—the area of transactional interdependencies, from either direction; and L_{22} to processes through which parts of the environment become related to each other—i.e., its causal texture—the area of interdependencies that belong within the environment itself.[3]

We have reproduced the above paragraph in its entirety because, in the balance of this paper, we will use Emery and Trist's symbols (i.e., L_{11}, L_{21}, L_{12}, *and* L_{22}) to denote intra-, input, output, and extra-system interdependencies, respectively. Our purpose in doing so is to avoid the misleading connotations of conventional terminology.

Purpose

The theses here are: (1) that contemporary changes in organizational environments are such as to increase the ratio of externally induced change to internally induced change; and (2) that *other* formal organizations are, increasingly, the important components in the environment of any focal organization. Furthermore, the evolution of environments is accompanied—among viable systems—by an increase in the system's ability to learn and to perform according to changing contingencies in its environment. An integrative framework is outlined for the concurrent analysis of an organization, its transactions with environmental units, and interdependencies among those units. Lastly, two hypotheses are presented, one about organizational *change* and the other about organizational *adaptability;* and some problems in any empirical test of these hypotheses are discussed.[4]

Concepts of Organizational Environments

In Emery and Trist's terms, L_{22} relations (i.e., interdependencies within the environment itself) comprise the "causal texture" of the field. This causal texture of the environment is treated as a quasi-independent domain, since the environment cannot be conceptualized except with respect to some focal organization. The components of the environment are identified in terms of that system's actual and *potential* transactional interdependencies, both input (L_{21}) and output (L_{12}).

Emery and Trist postulate four "ideal types" of environment, which can be ordered according to the degree of *system connectedness* that exists among the components of the environment (L_{22}). The first of these is a "placid, randomized" environment: goods and bads are relatively unchanging in themselves and are randomly distributed (e.g., the environments of an amoeba, a human foetus, a nomadic tribe). The second is a "placid, clustered" environment: goods and bads are

[3] Ibid., 22.

[4] I am particularly grateful to Kenneth Boulding for inspiration and to Eugene Litwak, Rosemary Sarri, and Robert Vinter for helpful criticisms. A Special Research Fellowship from the National Institutes of Health has supported my doctoral studies and, therefore, has made possible the development of this paper.

relatively unchanging in themselves but clustered (e.g., the environments of plants that are subjected to the cycle of seasons, of human infants, of extractive industries). The third ideal type is "disturbed-reactive" environment and constitutes a significant qualitative change over simpler types of environments: an environment characterized by similar systems in the field. The extinction of dinosaurs can be traced to the emergence of more complex enviroments on the biological level. Human beings, beyond infancy, live in disturbed-reactive environments in relation to one another. The theory of oligopoly in economics is a theory of this type of environment.[5]

These three types of environment have been identified and described in the literature of biology, economics, and mathematics.[6] "The fourth type, however, is new, at least to us, and is the one that for some time we have been endeavouring to identify."[7] This fourth ideal type of environment is called a "turbulent field." Dynamic processes "arise from the *field itself*" and not merely from the interactions of components; the actions of component organizations and linked sets of them "are both persistent and strong enough to induce autochthonous processes in the environment."[8]

An alternate description of a turbulent field is that the accelerating rate and complexity of interactive effects exceeds the component systems' capacities for prediction and, hence, control of the compounding consequences of their actions.

Turbulence is characterized by complexity as well as rapidity of change in causal interconnections in the environment. Emery and Trist illustrate the transition from a disturbed-reactive to a turbulent-field environment for a company that had maintained a steady 65 percent of the market for its main product—a canned vegetable—over many years. At the end of World War II, the firm made an enormous investment in a new automated factory that was set up exclusively for the traditional product and technology. At the same time postwar controls on steel strip and tin were removed, so that cheaper cans were available; surplus crops were more cheaply obtained by importers; diversity increased in available products, including substitutes for the staple; the quick-freeze technology was developed; home buyers became more

---

[5] The concepts of ideal types of environment, and one of the examples in this paragraph, are from Emery and Trist, op. cit., 24–26.

[6] The following illustrations are taken from Emery and Trist, ibid.: For random-placid environment see Herbert A. Simon, *Models of Man* (New York: John Wiley, 1957), p. 137; W. Ross Ashby, *Design for a Brain* (2d ed.; London: Chapman and Hall, 1960), Sec. 15/4; the mathematical concept of random field; and the economic concept of classical market.

For random-clustered environment see Edward C. Tolman and Egon Brunswik, The Organism and the Causal Texture of the Environment, *Psychological Review*, 42 (1935), 43–72; Ashby, op. cit., sec. 15/8; and the economic concept of imperfect competition.

For disturbed-reactive environment see Ashby, op. cit., sec. 7; the concept of "imbrication" from I. Chein, Personality and Typology, *Journal of Social Psychology*, 18 (1943), 89–101; and the concept of oligopoly.

[7] Emery and Trist, op. cit., 24.

[8] Ibid., 26.

affluent; supermarkets emerged and placed bulk orders with small firms for retail under supermarket names. These changes in technology, international trade, and affluence of buyers gradually interacted (L_{22}) and ultimately had a pronounced effect on the company: its market dwindled rapidly. "The changed texture of the environment was not recognized by an able but traditional management until it was too late."[9]

Sociological, social psychological, and business management theorists often still treat formal organizations as closed systems. In recent years, however, this perspective seems to be changing. Etzioni asserts that interorganizational relations need intensive empirical study.[10] Blau and Scott present a rich but unconceptualized discussion of the "social context of organizational life."[11] Parsons distinguishes three distinct levels of organizational responsibility and control: technical, managerial, and institutional.[12] His categories can be construed to parallel the intraorganizational (i.e., technical or L_{11}), the interorganizational (i.e., managerial or L_{21} and L_{12}), and the extra-organizational levels of analysis (i.e., the institutional or L_{22} areas). Perhaps in the normal developmental course of a science, intrasystem analysis necessarily precedes the intersystem focus. On the other hand, increasing attention to interorganizational relations may reflect a real change in the phenomenon being studied. The first question to consider is whether there is evidence that the environments of formal organizations are evolving toward turbulent-field conditions.

Evidence for Turbulence

Ohlin argues that the sheer rapidity of social change today requires greater organizational adaptability.[13] Hood points to the increasing complexity, as well as the accelerating rate of change, in organizational environments.[14] In business circles there is growing conviction that the future is unpredictable. Drucker[15] and Gardner[16] both assert that the kind and extent of present-day change precludes prediction of the future. Increasingly, the rational strategies of planned-innovation and long-range planning are being undermined by unpredictable changes. McNulty found no association between organization adaptation and the

[9] Ibid., 24.

[10] Amitai Etzioni, New Directions in the Study of Organizations and Society, *Social Research*, 27 (1960), 223–28.

[11] Peter M. Blau and Richard Scott, *Formal Organizations* (San Francisco: Chandler, 1962), pp. 194–221.

[12] Talcott Parsons, *Structure and Process in Modern Societies* (New York: Free Press, 1960), pp. 63–64.

[13] Lloyd E. Ohlin, Conformity in American Society Today, *Social Work*, 3 (1958), 63.

[14] Robert C. Hood, Business Organization as a Cross Product of Its Purposes and of Its Environment," in Mason Haire (ed.), *Organizational Theory in Industrial Practice* (New York: John Wiley, 1962), p. 73.

[15] Peter F. Drucker, The Big Power of Little Ideas, *Harvard Business Review*, 42 (May 1964), 6–8.

[16] John W. Gardnere, *Self-Renewal* (New York: Harper & Row, 1963), p. 107.

introduction of purposeful change in a study of 30 companies in fast-growing markets.[17] He suggests that built-in flexibility may be more efficient than the explicit reorganization implicit in the quasi-rational model. *Dun's Review* questions the effectiveness of long-range planning in the light of frequent failures, and suggests that error may be attributable to forecasting the future by extrapolation of a noncomparable past. The conclusion is that the rapidity and complexity of change may increasingly preclude effective long-range planning.[18] These examples clearly suggest the emergence of a change in the environment that is suggestive of turbulence.

Some writers with this open-system perspective derive implications for interorganizational relations from this changing environment. Blau and Scott argue that the success of a firm increasingly depends upon its ability to establish symbiotic relations with other organizations, in which extensive advantageous exchange takes place.[19] Lee Adler proposes "symbiotic marketing."[20] Dill found that the task environments of two Norwegian firms comprised four major sectors: *customers*, including both distributors and users; *suppliers* of materials, labor, capital, equipment, and work space; *competitors* for both markets and resources; and *regulatory groups*, including governmental agencies, unions, and interfirm associations.[21] Not only does Dill's list include many more components than are accommodated by present theories, but all components are themselves evolving into formal organizations. In his recent book, Thompson discusses "task environments," which comprise the units with which an organization has input and output transactions (L_{21} and L_{12}), and postulates two dimensions of such environments: homogeneous-heterogenous, and stable-dynamic. When the task environment is *both* heterogeneous and dynamic (i.e., probably turbulent), he expects an organization's boundary-spanning units to be functionally differentiated to correspond to segments of the task environment and each to operate on a decentralized basis to monitor and plan responses to fluctuations in its sector of the task environment.[22] He does not focus on other organizations as components of the environment, but he provides a novel perspective on structural implications (L_{11}) for organizations in turbulent fields.

Selznick's work on TVA appears to be the first organizational case study to emphasize transactional interdependencies.[23] The next study was Ridgway's 1957 study of manufacturer-dealer relationships.[24]

[17] James E. McNulty, Organizational Change in Growing Enterprises, *Administrative Science Quarterly,* 7 (1962), 1–21.

[18] Long Range Planning and Cloudy Horizons, *Dun's Review,* 81 (Jan. 1963), 42.

[19] Blau and Scott, op. cit., p. 217.

[20] Lee Adler, Symbiotic Marketing, *Harvard Business Review,* 44 (November 1966), 59–71.

[21] W. R. Dill, Environment as an Influence on Managerial Autonomy, *Administrative Science Quarterly,* 2 (1958), 409–43.

[22] James D. Thompson, *Organizations in Action* (New York: McGraw-Hill, 1967), pp. 27–28.

[23] Philip Selznick, *TVA and the Grass Roots* (Berkeley: University of California, 1949).

[24] V. F. Ridgway, Administration of Manufacturer-Dealer Systems, *Administrative Science Quarterly,* 2 (1957), 464–83.

Within the following few years the study by Dill[25] and others by Levine and White,[26] Litwak and Hylton,[27] and Elling and Halebsky[28] appeared, and in recent years, the publication of such studies has accelerated.

The following are examples from two volumes of the *Administrative Science Quarterly* alone. Rubington argues that structural changes in organizations that seek to change the behavior of "prisoners, drug addicts, juvenile delinquents, parolees, alcoholics [are] . . . the result of a social movement whose own organizational history has yet to be written."[29] Rosengren reports a similar phenomenon in the mental health field whose origin he finds hard to explain: "In any event, a more symbiotic relationship has come to characterize the relations between the [mental] hospitals and other agencies, professions, and establishments in the community."[30] He ascribes changes in organizational goals and technology to this interorganizational evolution. In the field of education, Clark outlines the increasing influence of private foundations, national associations, and divisions of the federal government. He, too, is not clear as to how these changes have come about, but he traces numerous changes in the behavior of educational organizations to interorganizational influences.[31] Maniha and Perrow analyze the origins and development of a city youth commission. The agency had little reason to be formed, no goals to guide it, and was staffed by people who sought a minimal, no-action role in the community. By virtue of its existence and broad province, however, it was seized upon as a valuable weapon by other organizations for the pursuit of their own goals. "But in this very process it became an organization with a mission of its own, in spite of itself."[32]

Since uncertainty is the dominant characteristic of turbulent fields, it is not surprising that emphasis in recent literature is away from algorithmic and toward heuristic problem-solving models;[33] that optimizing models are giving way to satisficing models;[34] and that rational decision making is replaced by "disjointed incrementalism."[35] These

---

[25] Dill, op. cit.

[26] Sol Levine and Paul E. White, Exchange as a Conceptual Framework for the Study of Interorganizational Relationships, *Administrative Science Quarterly*, 5 (1961), 583–601.

[27] Eugene Litwak and Lydia Hylton, Interorganizational Analysis: A Hypothesis on Coordinating Agencies, *Administrative Science Quarterly*, 6 (1962), 395–420.

[28] R. H. Elling and S. Halebsky, Organizational Differentiation and Support: A Conceptual Framework, *Administrative Science Quarterly*, 6 (1961), 185–209.

[29] Earl Rubington, Organizational Strain and Key Roles, *Administrative Science Quarterly*, 9 (1965), 350–69.

[30] William R. Rosengren, Communication, Organization, and Conduct in the "Therapeutic Milieu," *Administrative Science Quarterly*, 9 (1964), 70–90.

[31] Burton R. Clark, Interorganizational Patterns in Education, *Administrative Science Quarterly*, 10 (1965), 224–37.

[32] John Maniha and Charles Perrow, The Reluctant Organization and the Aggressive Environment, *Administrative Science Quarterly*, 10 (1965), 238–57.

[33] Donald W. Taylor, "Decision Making and Problem Solving," in James G. March (ed.), *Handbook of Organizations* (Chicago: Rand McNally, 1965), pp. 48–82.

[34] James G. March and Herbert A. Simon, *Organizations* (New York: John Wiley, 1958), pp. 140–41.

[35] David Braybrooke and C. E. Lindblom, *A Strategy of Decision* (Glencoe: The Free Press, 1963), especially ch. 3, 5.

trends reflect *not* the ignorance of the authors of earlier models, but a change in the causal texture of organizational environments and, therefore, of appropriate strategies for coping with the environment. Cyert and March state that "so long as the environment of the firm is unstable —and predictably unstable—the heart of the theory [of the firm] must be the process of short-run adaptive reactions."[36]

In summary, both the theoretical and case study literature on organizations suggests that these systems are increasingly finding themselves in environments where the complexity and rapidity of change in external interconnectedness (L_{22}) gives rise to increasingly unpredictable change in their transactional interdependencies (L_{21} and L_{12}). This seems to be good evidence for the emergence of turbulence in the environments of many formal organizations.

INTERORGANIZATIONAL ENVIRONMENT

Evidence for Increasing Dependence on Environment

Elsewhere the author has argued that Emery and Trist's concepts can be extended to *all* living systems; furthermore, that this evolutionary process gives rise to conditions—biological, psychological, and social—in which the rate of evolution of environments exceeds the rate of evolution of component systems.[37]

In the short run, the openness of a living system to its environment enables it to take in ingredients from the environment for conversion into energy or information that allows it to maintain a steady state and, hence, to violate the dismal second law of thermodynamics (i.e., of entropy). In the long run, "the characteristic of living systems which most clearly distinguishes them from the nonliving is their property of progressing by the process which is called evolution from less to more complex states of organization."[38] It then follows that to the extent that the environment of some living system X is comprised *of other living systems*, the environment of X is *itself* evolving from less to more complex states of organization. A major corollary is that the evolution of environments is characterized by an increase in the ratio of externally induced change over internally induced change in a system's transactional interdependencies (L_{21} and L_{12}).

For illustration, let us assume that at some given time, each system in some set of interdependent systems is equally likely to experience an internal (L_{11}) change that is functional for survival (i.e., improves its L_{21} or L_{12} transactions). The greater the number of other systems in that set, the greater the probability that some system other than X will experience that change. Since we posit interdependence among

[36] Richard M. Cyert and James G. March, *A Behavioral Theory of the Firm* (Englewood Cliffs, N.J.: Prentice-Hall, 1963), p. 100.

[37] Shirley Terreberry, "The Evolution of Environments" (mimeographed course paper, 1967), pp. 1–37.

[38] J. W. S. Pringle, On the Parallel Between Learning and Evolution, *General Systems*, 1 (1956), 90.

members of the set, X's viability over time depends upon X's capacity (L_{11}) for adaptation to environmentally induced (L_{22}) changes in its transactive position, or else upon control over these external relations.

In the case of formal organizations, disturbed-reactive or oligopolistic environments require some form of accommodation between like but competitive organizations whose fates are negatively correlated to some degree. A change in the transactional position of one system in an oligopolistic set, whether for better or worse, automatically affects the transactional position of all other members of the set, and in the opposite direction (i.e., for worse or better, as the case may be).[39] On the other hand, turbulent environments require relationships between dissimilar organizations whose fates are independent or, perhaps, positively correlated.[40] A testable hypothesis that derives from the formal argument is that the evolution of environments is accompanied, in viable systems, by an increase in ability to learn and to perform according to changing contingencies in the environment.

The evolution of organizational environments is characterized by a change in the important constituents of the environment. The earliest formal organizations to appear in the United States (e.g., in agriculture, retail trade, construction, mining)[41] operated largely under placid-clustered conditions. Important inputs, such as natural resources and labor, as well as consumers, comprised an environment in which strategies of optimal location and distinctive competence were critical organizational responses.[42] Two important attributes of placid-clustered environments are: (1) the environment is itself *not* formally organized; and (2) transactions are largely initiated and controlled by the organization (i.e., L_{12}).

Later developments, such as transport technology and derivative overlap in loss of strength gradients, and communication and automation technologies that increased economies of scale, gave rise to disturbed reactive (oligopolistic) conditions in which similar formal organizations become the important actors in an organization's field. They are responsive to its acts (L_{12}) *and* it must be responsive to theirs (L_{21}). The critical organizational response now involves complex operations, requiring sequential choices based on the calculated actions of others, and counteractions.[43]

When the environment becomes turbulent, however, its constituents are a multitude of other formal organizations. Increasingly, an organization's markets consist of other organizations; suppliers of material, labor, and capital are increasingly organized; and regulatory groups are more numerous and powerful. The critical response of organizations under these conditions will be discussed later. It should be noted that *real* environments are often mixtures of these ideal types.

[39] Assuming a nonexpanding economy, in the ideal instance.

[40] Emery and Trist argue that fates, here, are positively correlated. This writer agrees, if an expanding economy is assumed.

[41] Arthur L. Stinchcombe, "Social Structure and Organizations," in March (ed.), op. cit., p. 156.

[42] Emery and Trist, op. cit., 29.

[43] Ibid., 25–26.

The evolution from placid-clustered environments to turbulent environments[44] can be summarized as a process in which formal organizations evolve: (1) *from* the status of systems within environments not formally organized; (2) *through* intermediate phases (e.g., Weberian bureaucracy); and (3) *to* the status of subsystems of a larger social system.

Clark Kerr traces this evolution for the university in the United States.[45] In modern industrial societies, this evolutionary process has resulted in the replacement of individuals and informal groups by organizations as *actors* in the social system. Functions that were once the sole responsibility of families and communities are increasingly allocated to formal organizations; child-rearing, work, recreation, education, health, and so on. Events which were long a matter of chance are increasingly subject to organizational control, such as population growth, business cycles, and even the weather. One wonders whether Durkheim, if he could observe the current scene, might speculate that the evolution from "mechanical solidarity" to "organic solidarity" is now occurring on the *organizational level*, where the common values of organizations in oligopolies are replaced by functional interdependencies among specialized organizations.[46]

Interorganizational Analysis

It was noted that survival in disturbed-reactive environments depends upon the ability of the organization to anticipate and counteract the behavior of similar systems. The analysis of interorganizational behavior, therefore, becomes meaningful only in these and more complex environments. The interdependence of organizations, or any kind of living systems, at less complex environmental levels is more appropriately studied by means of ecological, competitive market, or other similar models.

The only systematic conceptual approach to interorganizational analysis has been the theory of oligopoly in economics. This theory clearly addresses only disturbed-reactive environments. Many economists admit that the theory, which assumes maximization of profit and perfect knowledge, is increasingly at odds with empirical evidence that organizational behavior is characterized by satisficing and bounded rationality. Boulding comments that "it is surprisingly hard to make a really intelligent conflict move in the economic area simply because of the complexity of the system and the enormous importance of side effects and dynamic effects."[47] A fairly comprehensive search of the

---

[44] The author does not agree with Emery and Trist, that *formal* (as distinct from social) organization will emerge in placid-random environments.

[45] Clark Kerr, *The Uses of the University* (New York: Harper Torchbooks, 1963).

[46] Emile Durkheim, *The Division of Labor in Society,* trans. George Simpson (Glencoe: The Free Press, 1947).

[47] Kenneth E. Boulding, "The Economies of Human Conflict," in Elton B. McNeil (ed.), *The Nature of Human Conflict* (Englewood Cliffs, N.J.: Prentice-Hall, 1965), p. 189.

literature has revealed only four conceptual frameworks for the analysis of interorganizational relations outside the field of economics. These are briefly reviewed, particular attention being given to assumptions about organization environments, and to the utility of these assumptions in the analysis of interorganizational relations in turbulent fields.

William Evan has introduced the concept of "organization-set," after Merton's "role-set."[48] Relations between a focal organization and members of its organization-set are mediated by the role-sets of boundary personnel. "Relations" are conceived as the flow of information, products or services, and personnel.[49] Presumably, monetary, and legal, and other transactions can be accommodated in the conceptual system. In general, Evan offers a conceptual tool for identifying transactions at a given time. He makes no explicit assumptions about the nature of environmental dynamics, nor does he imply that they are changing. The relative neglect of interorganizational relations, which he finds surprising, is ascribed instead to the traditional intraorganizational focus, which derives from Weber, Taylor, and Barnard.[50] His concepts, however, go considerably beyond those of conventional organization and economic theory (e.g., comparative versus reference organizations and overlap in goals and values). If a temporal dimension were added to Evan's conceptual scheme, then, it would be a very useful tool for describing the "structural" aspects of transactional interdependencies (L_{21} and L_{12} relations) in turbulent fields.

Another approach is taken by Levine and White who focus specifically on relations among community health and welfare agencies. This local set of organizations "may be seen as a system with individual organizations or system parts varying in the kinds and frequencies of their relationships with one another."[51] The authors admit that interdependence exists among these local parts only to the extent that relevant resources are not available from *outside* the local region, which lies beyond their conceptual domain. Nor do we find here any suggestion of turbulence in these local environments. If such local sets of agencies are increasingly interdependent with other components of the local community and with organizations outside the locality, as the evidence suggests, then the utility of Levine and White's approach is both limited and shrinking.

Litwak and Hylton provide a third perspective. They too are concerned with health and welfare organizations, but their major emphasis is on coordination.[52] The degree of interdependence among organizations is a major variable; low interdependence leads to *no* coordination and high interdependence leads to merger; therefore they deal only with conditions of moderate interdependence. The type of coordinating

---

[48] William M. Evan, "The Organization-Set: Toward a Theory of Interorganizational Relations," in James D. Thompson (ed.), *Approaches to Organizational Design* (Pittsburgh, Pa.: University of Pittsburgh Press, 1966), pp. 177–80.

[49] Ibid., pp. 175–76.

[50] Ibid.

[51] Levine and White, op. cit., 586.

[52] Litwak and Hylton, op. cit.

mechanism that emerges under conditions of moderate interdependence is hypothesized to result from the interaction of three trichotomized variables: the *number* of interdependent organizations; the degree of their *awareness* of their interdependence; and the extent of *standardization* in their transactions. The attractive feature of the Litwak and Hylton scheme is the possibility it offers of making different predictions for a great variety of environments. Their model also seems to have predictive power beyond the class of organizations to which they specifically address themselves. If environments are becoming turbulent, however, then increasingly fewer of the model's cells (a $3 \times 3 \times 3$ space) are relevant. In the one-cell turbulent corner of their model, where a large number of organizations have low awareness of their complex and unstandardized interdependence, "there is little chance of coordination,"[53] according to Litwak and Hylton. If the level of awareness of interdependence increases, the model predicts that some process of arbitration will emerge. Thus the model anticipates the interorganizational implications of turbulent fields, but tells us little about the emerging processes that will enable organizations to adapt to turbulence.

The fourth conceptual framework available in the literature is by Thompson and McEwen.[54] They emphasize the interdependence of organizations with the larger society and discuss the consequences that this has for goal setting. "Because the setting of goals is essentially a problem of defining desired relationships between an organization and its environment, change in either requires review and perhaps alteration of goals."[55] They do not argue that such changes are more frequent today, but they do assert that reappraisal of goals is "a more constant problem in an unstable environment than in a stable one," and also "more difficult as the 'product' of the enterprise becomes less tangible."[56]

Thompson and McEwen outline four organizational strategies for dealing with the environment. One is competition; the other three are subtypes of a cooperative strategy: bargaining, co-optation, and coalition. These cooperative strategies all require direct interaction among organizations and this, they argue, increases the environment's potential control over the focal organization.[57] In bargaining, to the extent that the second party's support is necessary, that party is in a position to exercise a veto over the final choice of alternative goals, and thus takes part in the decision. The co-optation strategy makes still further inroads into the goal-setting process. From the standpoint of society, however, co-optation, by providing overlapping memberships, is an important social device for increasing the likelihood that organizations related to each other in complicated ways will in fact find compatible goals.

[53] Ibid., 417.

[54] James D. Thompson and William J. McEwen, Organizational Goals and Environment, *American Sociological Review*, 23 (1958), 23–31.

[55] Ibid., 23.

[56] Ibid., 24.

[57] Ibid., 27.

Co-optation thus aids in the integration of heterogeneous parts of a complex social system. Coalition refers to a combination of two or more organizations for a common purpose and is viewed by these authors as the ultimate form of environmental conditioning of organization goals.[58]

The conceptual approaches of Levine and White and of Litwak and Hylton therefore appear to be designed for nonturbulent conditions. Indeed, it may well be that coordination *per se,* in the static sense usually implied by that term, is dysfunctional for adaptation to turbulent fields. (This criticism has often been leveled at local "councils of social agencies."[59]) On the other hand, Evan's concept of organization-set seems useful for describing static aspects of interorganizational relations in either disturbed-reactive *or* turbulent-field environments. Its application in longitudinal rather than static studies might yield data on the relationship between structural aspects of transactional relations and organizational adaptability. Lastly, Thompson and McEwen make a unique contribution by distinguishing different *kinds* of interorganizational relations.

As an aside, note that Evan's extension of the role-set concept to organizations suggests still further analogies, which may be heuristically useful. A role is a set of acts prescribed for the occupant of some position. The role accrues to the position; its occupants are interchangeable. If formal organizations are treated as social actors, then one can conceive of organizations as occupants of positions in the larger social

FIGURE 1

Structure of Living Systems Such as a Formal Organization

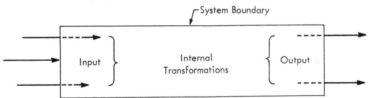

system. Each organization has one or more roles in its behavioral repertoire (these are more commonly called functions or goals). The organization occupants of these social positions, however, are also interchangeable.

---

[58] Ibid., 25–28.

[59] Examples include: Robert Morris and Ollie A. Randall, Planning and Organization of Community Services for the Elderly, *Social Work*, 10 (1965), 96–103; Frank W. Harris, A Modern Council Point of View, *Social Work*, 9 (1964), 34–41; Harold L. Wilensky and Charles N. Lebeaux, *Industrial Society and Social Welfare* (New York: Russell Sage Foundation, 1958), especially pp. 263–65.

INTEGRATIVE FRAMEWORK

Model

It is assumed that the foregoing arguments are valid: (1) that organizational environments are increasingly turbulent; (2) that organizations are increasingly less autonomous; and (3) that other formal organizations are increasingly important components of organizational environments. Some conceptual perspective is now needed, which will make it possible to view any formal organization, its transactional interdependencies, and the environment itself within a common conceptual framework. The intent of this section is to outline the beginnings of such a framework.

A formal organization is a system primarily oriented to the attainment of a specific goal, which constitutes an output of the system and which is an input for some other system.[60] Needless to say, the output of any living system is dependent upon input into it. Figure 1 schematically illustrates the skeletal structure of a living system. The input and output regions are partially permeable with respect to the environment, which is the region outside the system boundary. Arrows coming into a system represent input and arrows going out of a system represent output. In Figure 2, rectangles represent formal organizations and circles represent individuals and *non*formal social organizations. Figure 2 represents the *statics* of a system X and its turbulent environment. Three-dimensional illustration would be necessary to show the *dynamics* of a turbulent environment schematically. Assume that a third, temporal dimension is imposed on Figure 2 and that this reveals an increasing number of elements and an increasing rate and complexity of change in their interdependencies over time. To do full justice to the concept of turbulence we should add other sets of elements even in Figure 2 above, although these are not yet linked to X's set. A notion that is integral to Emery and Trist's conception of turbulence is that changes outside of X's set, and hence difficult for X to predict and impossible for X to control, will have impact on X's transactional interdependencies in the future. The addition of just one link at some future time may not affect the supersystem but may constitute a system break for X.

This schematization shows only one-way directionality and is meant to depict energic inputs (e.g., personnel and material) and output (e.g., product). The organization provides something in exchange for the inputs it receives, of course, and this is usually informational in nature —money, most commonly. Similarly the organization receives money for its product from those systems for whom its product is an input. Nor does our framework distinguish different kinds of inputs, although the analysis of interorganizational exchange requires this kind of taxonomic device. It seems important to distinguish energic inputs and outputs

---

[60] Talcott Parsons, "Suggestions for a Sociological Approach to the Theory of Organizations," in Amitai Etzioni (ed.), *Complex Organizations* (New York: Holt, Rinehart, and Winston, 1962), p. 33.

FIGURE 2

Illustration of System *X* in Turbulent Environment

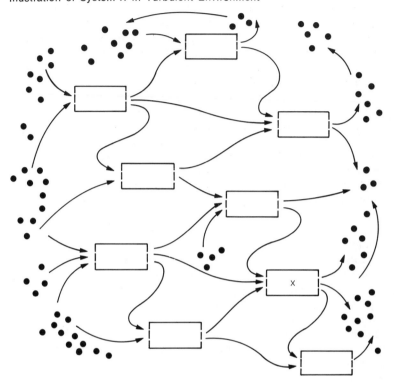

from informational ones. Energic inputs include machinery, personnel, clientele in the case of service organizations, electric power, and so on. Informational inputs are not well conceptualized although there is no doubt of their increasing importance in environments which are more complex and changeable. Special divisions of organizations and whole firms devoted to information collecting, processing, and distributing are also rapidly proliferating (e.g., research organizations, accounting firms, the Central Intelligence Agency).

An input called "legitimacy" is popular in sociological circles but highly resistant to empirical specification. The view taken here is that legitimacy is mediated by the exchange of other resources. Thus the willingness of firm *A* to contribute capital to *X*, and of agency *B* to refer personnel to *X* and firm *C* to buy *X*'s product testifies to the legitimacy of *X*. This "willingness" on the part of organizations *A*, *B*, and *C*, however, can best be understood in terms of informational exchange. For example, *A* provides *X* with capital on the basis of *A*'s information about the market for *X*'s product. Or *B* refuses to refer skilled workmen to *X* since *B* has information on *X*'s discriminatory employment practices and also knows of consequences to itself from elsewhere if it is party to *X*'s practice. Technology is also sometimes treated as an input to organizations. We use the term, however, to refer to the complex set of interactions among inputs which takes place in the internal region shown in Figure 1. It is technology which transforms the inputs of the system into the output of the system. Transportation and communication technologies, however, are of a uniquely different order; the former constitutes an energic and the latter an informational transcendence of space-time that enabled the evolution of the more complex environments (L_{22}) which concern us here. Automation and computer technologies are roughly equivalent (i.e., energic and informational, respectively) but on an intraorganizational (L_{11}) level.

Our attention to "legitimacy" and "technology" was tangential to our main theme, to which we now return. Our simplistic approach to an integrative framework for the study of organizations (L_{11}), their transactional interdependencies $(L_{21}$ and $L_{12})$, and the connectedness within their environments (L_{22}), gives the following conceptual ingredients: (1) units that are mainly formal organizations, and (2) relationships between them that are the directed flow[61] of (3) energy and information. The enormous and increasing importance of informational transaction has not been matched by conceptual developments in organization theory. The importance of information is frequently cited in a general way, however, especially in the context of organizational change or innovation. Dill has made a cogent argument on the need for more attention to this dimension.[62]

The importance of communication for organizational change has been stressed by Ohlin, March and Simon, Benne, Lippitt, and others.[63] Diversity of informational input has been used to explain the creativity of individuals as well as of social systems.[64] The importance of boundary positions as primary sources of innovative inputs from the environment has been stressed by March and Simon[65] and by Kahn et al.[66] James Miller hypothesizes that up to a maximum, which no living system has yet reached, the more energy a system devotes to information processing (as opposed to productive and maintenance activity), the more likely the system is to survive.[67]

---

[61] Dorwin Cartwright, "The Potential Contribution of Graph Theory to Organization Theory," in Mason Haire (ed.), *Modern Organization Theory* (New York: John Wiley, 1959), pp. 254–71.

[62] William R. Dill, "The Impact of Environment in Organizational Development," in Sidney Mailick and Edward H. Van Ness (eds.), *Concepts and Issues in Administrative Behavior* (Englewood Cliffs, N.J.: Prentice-Hall, 1962), pp. 94–109.

[63] Ohlin, op. cit., 63; March and Simon, op. cit., pp. 173–83; Kenneth D. Benne, "Deliberate Changing as the Facilitation of Growth," in Warren G. Bennis et al. (eds.), *The Planning of Change* (New York: Holt, Rinehart, and Winston, 1962), p. 232; Ronald Lippitt, *The Dynamics of Planned Change* (New York: Harcourt, Brace, and World, 1958), p. 52.

[64] For example: Floyd H. Allport, *Theories of Perception and the Concept of Structure* (New York: John Wiley, 1955), p. 76; William F. Ogburn and Meyer F. Nimkoff, *Sociology* (4th ed.; Boston: Houghton Mifflin, 1964), pp. 662–70.

[65] March and Simon, op. cit., pp. 165–66, 189.

[66] Robert L. Kahn et al., *Organizational Stress* (New York: John Wiley, 1964), pp. 101–26.

[67] James G. Miller, Toward a General Theory for the Behavioral Sciences, *The American Psychologist,* 10 (1955), 530.

Evolution on the biological level is accompanied by improvement in the ability of systems to discover and perform according to contingencies in their environments. The random walk which suffices in a placid-randomized environment must be replaced by stochastic processes under placid-clustered conditions, and by cybernetic processes in disturbed-reactive fields. Among biological/psychological systems, only man appears to have the capacity for the purposeful behavior that may permit adaptation to or control of turbulent environments. There is some question, of course, as to whether man actually *has* the capacity to cope with the turbulence that he has introduced into the environment.

Analogous concepts are equally applicable to the evolution of social systems in general and to formal organizations in particular. The capacity of *any* system for adapting to changing contingencies in its environment is inversely related to its dependence upon instinct, habit, or tradition. Adaptability exists, by definition, to the extent that a system (L_{11}) can survive externally induced (L_{22}) change in its transactional interdependencies $(L_{21}$ and $L_{12})$; therefore viability equals adaptability.

Hypotheses

Hypothesis 1. Organizational change is largely externally induced.

Any particular change may be adaptive or maladaptive, and it may be one of these in the short run and the other in the long run. There is *no* systematic empirical evidence on the relative influence of internal versus environmental antecedents to organizational change. The empirical task here is to identify organizational changes, and the internal or external origins of each change.

It is crucial to distinguish change on the organizational level from the multitude of changes that may occur in or among subsystems, only some of which give rise to change on the system level. Many social psychologists, for example, study change in individuals and groups *within* organizations, but with no reference to variables of organizational level. Likert's book is one noteworthy exception.[68] The important point is that change on the organizational level is analytically distinct from change on other levels.

Organizational change means any change in the kind or quantity of output. Ideally, output is treated as a function of inputs and of transfer functions (i.e., intraorganizational change is inferred from change in input-output relations). Haberstroh illustrates the use of these general system concepts in the organization context.[69] An excellent discussion of the efficiency and effectiveness of organizations, in an open-systems framework, is given in Katz and Kahn.[70]

However, the input-output functions in diversified industries and the outputs of many service organizations are resistant to objective specifica-

---

[68] Rensis Likert, *New Patterns of Management* (New York: McGraw-Hill, 1961).

[69] Chadwick J. Haberstroh, "Organization Design and Systems Analysis," in March (ed.), op. cit., pp. 1171–1211.

[70] Daniel Katz and Robert L. Kahn, *The Social Psychology of Organizations* (New York: John Wiley, 1966), especially pp. 149–70.

tion and measurement. An empirical test of this hypothesis, with presently available tools, may have to settle for some set of input and internal change that seems to be reasonably antecedent to output change.

The identification of the origin of change is also beset by difficulties. An input change may indeed have external antecedents, but external events may also be responses to some prior internal change in the focal organization. And internal change may be internally generated, but it may also be the result of an informational input from external sources. Novel informational inputs, as well as novel communication channels, often derive from change in personnel inputs. Increasingly, organizations seek personnel who bring specialized information rather than "manpower" to the organization. The presence of first, second, and higher order causation poses a problem for any empirical test of this hypothesis.

Hypothesis 2. System adaptability (e.g., organizational) is a function of ability to learn and to perform according to changing environmental contingencies.

Adaptability exists, by definition, to the extent that a system can survive externally induced change in its transactional interdependencies in the long run. Diversity in a system's input (L_{21}) and output (L_{12}) interdependencies will increase adaptability. The recent and rapid diversification in major industries illustrates this strategy. Flexible structure (L_{11}, e.g., decentralized decision making) will facilitate adaptation. Beyond this, however, adaptability would seem to be largely a function of a system's perceptual and information-processing capacities.[71] The following variables appear crucial: (1) *advance information* of impending externally induced (L_{22}) change in L_{21} or L_{12} transactions; (2) *active search* for, and activation of, more advantageous input and output transactions; and (3) *available memory store* (L_{11}) of interchangeable input and output components in the environment.

Advance information and active search might be empirically handled with Evan's concept of the role-sets of boundary personnel, along with notions of channel efficiency. For example, overlapping memberships (e.g., on boards) would constitute a particularly efficient channel. Likewise, direct communication between members of separate organizations, while less effective than overlapping memberships, would be a more efficient channel between agencies *A* and *B* than instances where their messages must be mediated by a third agency, *C*. Efficiency of interorganizational communication channels should be positively associated with access to advance information, and be facilitative of search, for example. The members of an organization's informational set may become increasingly differentiated from its energic set. Communication channels to research and marketing firms, universities, governmental agencies and other important information producing and distributing

---

[71] Igor Ansoff speaks of the "wide-open windows of perception" required of tomorrow's firms, and offers a perspective on the future that is fully compatible with that presented here; see The Firm of the Future, *Harvard Business Review*, 43 (September 1965), 162.

agencies would be expected to increase long-run viability. The third variable, memory store, is probably a function of the efficiency of past and present informational channels, but it involves internal (L_{11}) information processing as well.

Lastly, *any* internal change that improves an organization's transactional advantage (e.g., improved technology) will also be conducive to adaptability. Since organizational innovation is more often imitation than invention,[72] these changes are usually also the product of informational input and can be handled within the same integrative framework.

SUMMARY

The lag between evolution in the real world and evolution in theorists' ability to comprehend it is vast, but hopefully shrinking. It was only a little over one hundred years ago that Darwin identified natural selection as the mechanism of evolutionary process. Despite Darwin's enduring insight, theorists of change, including biologists, have continued to focus largely on internal aspects of systems.

It is our thesis that the selective advantage of one intra- or inter-organizational configuration over another cannot be assessed apart from an understanding of the dynamics of the environment itself. It is the environment which exerts selective pressure. "Survival of the fittest" is a function of the fitness of the environment. The dinosaurs *were* impressive creatures, in their day.

---

[72] Theodore Levitt, Innovative Imitation, *Harvard Business Review*, 44 (September 1966), 63–70.

John J. Gabarro

12. Organizational Adaptation to Environmental Change

THE PROBLEM of organizational adaptation to change has become central to the survival of organizations in contemporary society. Urban school systems undergoing rapid increases in minority enrollments are a particular example of organizations in which an urgent need exists for concepts which can be used by administrators to make these systems more adaptive to the needs of their changing pupil populations (Schrag, 1967; Gittell, 1967). Although their frequent lack of adaptiveness has been seen by some as a problem of community power, several recent studies have identified organizational factors as also being important, particularly in terms of administrative organization at the system-wide level (Gittell, 1967; Rogers, 1968; Watson, 1967).

This paper reports the findings of a study which explored the problem of adaptation by urban school systems to the needs of increasing minority enrollments. The study set out to apply a set of systems-based concepts to see if they were useful in identifying organizational variables bearing on a school system's ability to adapt effectively. Within this framework school systems were viewed as organizations dealing with changing environments.

The relationship between organization and environment has been the subject of a growing body of research and theory. Much of this research has conceived of organizations, either explicitly or implicitly, as open systems (Dill, 1962; Rice, 1963; Emery and Trist, 1965; Lawrence and Lorsch, 1967; Thompson, 1967; and Terreberry, 1968). The dominant theme in these studies is that some patterns of organization and behavior seem to be more appropriate to certain environmental and task characteristics than others (Burns and Stalker, 1961; Woodward, 1965; Perrow, 1967; Lawrence and Lorsch, 1967; and Thompson, 1967).

Building on several studies of organizational-environment "fit," Lawrence and Lorsch (1967) developed a systemic model for studying the relationship between environment and complex organizations, i.e.,

This material was prepared especially for this volume.

organizations comprised of several interrelated major subunits. By comparing business firms in different environments, they were able to relate effective patterns of organization to external environmental variables. The obvious implication of the work by Lawrence and Lorsch and others is that there is no single "best" prescription of organization appropriate to all situations, but rather that appropriate patterns of organization vary, depending on task and environmental demands. A second implication, which is particularly salient to the problem of change, is that if an environment changes in significant ways, the organizational patterns needed within a system must also change. Even if an organizational pattern were initially well suited to its environment, the pattern would cease to be viable if major environmental changes occurred.

The study reported here attempted to take the concepts developed by Lawrence and Lorsch in their comparison of organizations in different environments and apply them to the problem of adaptation by organizations within a single environment in the process of change. The two organizations studied were school systems trying to adapt to the needs of increasing minority enrollments—a major environmental change. By studying school systems, this research also set out to explore the applicability of the Lawrence and Lorsch concepts to the study of change in human services organizations. Their theory, like most of the work done by organizational systems theorists, is based on research in business organizations. Increasingly, however, the most critical need for such concepts is in non-business organizations, especially those dealing with human services. Urban school systems undergoing rapid increases in minority enrollments provide a compelling example of human services organizations trying to adapt to a changing environment.

The Lawrence and Lorsch Model

Before describing the design of this study, it may be useful to briefly review the Lawrence and Lorsch research and its findings. Conceptually, Lawrence and Lorsch viewed organizations as open systems capable of internal differentiation.[1] Observing that organizational environments are often comprised of a wide diversity of issues, they postulated that organizations segment themselves into subgroups, each subgroup concentrating on one aspect of the organization's task and environment.

Based on these assumptions, Lawrence and Lorsch theorized that segmentation into subgroups had two consequences: (1) segmentation required that the efforts of the various segmented parts be integrated to make the organization viable; and (2) it resulted in cognitive dif-

---

[1] Several others have also pointed out that differentiation by organizations facing complex environments is a precondition to survival and effectiveness. Working from both an open system model and a Parsonian frame of reference, Katz and Kahn (1966) have theorized that to survive, an organization differentiates itself into several subsystems which include technical, productive, maintenance, supportive, and adaptive subsystems. Rice (1963) and his associates at the Tavistock Institute have theorized that organizations differentiate themselves to perform the "primary task," i.e., the task which the organization must do to survive.

ferentiation among members of the different parts of the organization. By differentiation they meant differences in *attitudes and behavior,* not simply division of labor or specialization of knowledge. They operationally defined differentiation as the differences between major subgroups in terms of their cognitive and attitudinal orientations. They operationalized integration as the state of collaboration existing among subgroups as perceived by members of those subgroups.

Using these constructs, they postulated, and later showed empirically, that the greater the differentiation among parts, the more difficult it was to bring about integration of effort.[2] They also found that organizations dealing with highly diverse environments were more differentiated than those dealing with less diverse environments because a greater range of orientations was needed to deal with the wider variety of external issues. Operationally they defined environmental diversity as the degree to which the subenvironments corresponding to the various subgroups differed in terms of their uncertainty of information. They measured environmental uncertainty along three dimensions: uncertainty of information concerning causal relationships in the subenvironment; rate of change in the subenvironment; and the time span of subenvironment feedback on actions taken by the organization.

Lawrence and Lorsch found that in the more diverse environments, the more effective organizations (in terms of economic criteria such as product introduction, growth, etc.) had attained higher states of both differentiation and integration among subgroups than the less effective organizations. Greater differentiation was needed to deal with the diversity of demands posed by the total environment, while a high level of integration was needed to bring together these differentiated subgroups.

By comparing organizations in more diverse environments with those in less diverse environments they identified several structural and behavioral characteristics which enabled organizations in diverse environments to achieve integration despite their greater differentiation. They found that organizations in the more diverse environments developed a number of integrative devices such as integrative roles, cross-functional teams, and integrative departments which have as their purpose the achievement of collaboration and integration of effort among subgroups. They also found that the managerial hierarchy, as well as the rules, procedures, and the "paper" system, functioned as basic integrating devices which existed in all organizations.

By comparing more effective with less effective firms within each environment, they also identified a number of determinants of effective

---

[2] Differentiation between subunits makes integration of effort difficult because it increases the potential conflict between subunits. Walton and Dutton (1967) have made this point in a review of the literature on inter-departmental conflict. Similarly, March and Simon (1958) have explained the tendency which each subunit has to pursue its own subgoals at the expense of other organizational goals as a necessary consequence of the cognitive limitations of humans as decision-makers. Organizational differentiation, as Lawrence and Lorsch have defined it, is a more extensive elaboration of the same concept because it also includes the consequences of differences in work styles and other orientations, which are sources of difficulty in interunit relationships.

conflict resolution, including the influence attributed to administrators in integrating roles, and the prevalent modes used for resolving conflicts. They found that the more effective organizations were able to maintain both differentiation and integration because people in key integrating roles had relatively higher influence with which to facilitate integration than integrators in the less effective organizations. They also discovered that the more effective organizations were able to achieve needed integration through the use of more open and confronting conflict-resolving behavior.

THE RESEARCH APPROACH

The Lawrence and Lorsch framework was applied to studying adaptation by urban school systems. In using these concepts, the study focused on two central research questions:

1. How did the changing environment, in terms of increasing minority enrollments, affect the need for differentiation and integration among subgroups?[3]
2. What patterns of organization appeared most effective for dealing with these changing environmental requirements?

The approach taken to explore these questions was a comparative case study of two urban school systems trying to adapt to the needs of increasing minority enrollments. The two systems were of similar size and were serving similar cities and communities in terms of socio-economic context variables, minority populations, and rates of change of these variables over the most recent ten-year period.[4] The two school systems also enjoyed reputations of having quality administrations and progressive boards of education, and had been characterized by outside professionals as having tried to adapt. The premise in matching the two systems on these criteria was that any comparison of organizational variables would be of questionable validity unless both were dealing with relatively similar socioeconomic environments (Coleman, 1969), and the quality and intentions of administrations and school boards were relatively similar.

The two systems studied were also chosen because they did in fact differ in the effectiveness with which they adapted. Effectiveness of adaptation was based on the relative improvement or decline of several performance indicators[5] over the most recent ten-year period of increasing minority enrollments.

---

[3] It should be noted that the term "minority enrollments" has been used in a very specific way to include only black, Puerto Rican, and other Spanish-speaking children; other ethnic minorities, white or nonwhite, are not included in this definition.

[4] The two cities had populations of between 125,000 and 200,000 people, with pupil populations of between 20,000 and 40,000. Minority enrollments of pupils in the two systems had increased by 17 percent over the most recent 5 years at a nearly identical rate of change in each system.

[5] Performance indicators were based on achievement test scores compared on a national norm basis, aggregate daily attendance, quality of placement, drop-out

The research questions were explored in three phases paralleling the design of the Lawrence and Lorsch inquiry. First, hypotheses were developed from interviews about the degree of differentiation and integration required to meet the needs posed by increasing minority enrollments. This phase of the study suggested that adaptation required increased integration among subgroups of the two school systems and that it was important to maintain differentiation.

Second, a comparison was made of the actual states of differentiation and integration attained by the two systems. The more adaptive system was found to have attained higher states of both differentiation and integration, lending support to the hypotheses developed in the first phase of the study.

The third phase set out to determine whether the organizational variables identified by Lawrence and Lorsch as being related to an organization's ability to achieve integration were also salient within school systems. This phase also identified a number of other possible factors bearing on a system's ability to achieve differentiation and integration and, more directly, its ability to adapt.

The Concepts as Applied to School Systems

For purposes of the study, *environment* was thought of in terms of the Levine and White (1961) definition of an organization's task domain, i.e., the claims which an organization "stakes out for itself" in terms of clients, tasks, services, and important groups, whether inside or outside of the organization. This is a broader treatment of environment than that of Lawrence and Lorsch, who conceived of it as uncertainty of information. It is also more inclusive than definitions which include only external groups and organizations such as those used by Dill (1958), Thompson (1967), and Evan (1966). It was felt appropriate to conceive of environment in intentionally broad terms, given the exploratory nature of the study, and the notion of task domain was inclusive enough to cover a wide array of possibly relevant factors, including those considered by Lawrence and Lorsch and others.

The focal unit of *organization* was the total school system rather than individual schools. The two school systems studied differed considerably in organization from the business firms studied by Lawrence and Lorsch. For example, no counterparts were found in the school systems for the production, sales, and research subsystems of the industrial organizations which they studied. The two school systems were

---

rate, reported incidents of violence per 100 pupils, and the number of days cancelled because of disturbance.

It must be acknowledged that the performance indicators used cover only a small part of the relevant criteria for effectiveness of adaptation. Their choice was limited by the availability of comparable data in the two systems. For a more detailed discussion of the development and choice of these indicators, their limitations, and the context factors used in choosing the two school systems see: John J. Gabarro, "School System Organization and Adaptation to a Changing Environment," unpublished doctoral dissertation, Harvard University Graduate School of Business Administration, 1971, pp. II-1 to 4; pp. IV-1 to 38.

comprised of territorially dispersed schoolhouses and a number of centralized support groups, some of which had resident professionals in the schoolhouses.

Three major subgroups were identified as being central to the school system's primary task of teaching and socializing a community's children: *schoolhouses, curriculum services,* and *pupil services.* These subgroups were chosen because they were described in pilot interviews with several superintendents as being the subgroups most important to adaptation, in addition to comprising most of the school system's professional staff. These three subgroups are defined in more detail by the following descriptions:

Schoolhouses: the individual schools and the professionals who comprise their staffs, i.e., teachers, principals, and central administrators directly in charge of principals.
Curriculum Services: centralized personnel and staff concerned with providing curriculum assistance to teachers and principals in the schools, e.g., subject supervisors, curriculum development groups, consultants, special curriculum coordinators, resource teachers, media resources, etc.
Pupil Services: specialists and staff groups providing services to the individual pupil, e.g., guidance, psychological services, school social workers, school psychologists, specialists in teaching the emotionally and physically handicapped, etc.

Differentiation was defined as the differences between subgroups along four cognitive orientations or attributes:

Time orientation: time horizon of problems most typically worked on by the individual.
Goal orientation: priority ordering of organizational goals by an individual in performing his job.
Interpersonal orientation: style of work interaction most preferred by the individual in his job, i.e., task-centered interaction as compared to socially centered interactions (after Fiedler).
Formality of structure: the degree of structure characteristic of the subgroup's organization, in terms of reporting procedures, span of control, and levels in the hierarchy (after Hall, Woodward, Evan, and Burns and Stalker).

These four dimensions were used by Lawrence and Lorsch and were felt to be applicable to the school system setting. In addition, field data suggested that a subgroup's orientation to change would also be a relevant dimension of differentiation:

Orientation to change: the degree to which a subgroup's work involves the changing of methods and programs.

The addition of this fifth dimension was based on the premise that environmental change would affect the subgroups differently, so that some subgroups would be more comfortable with, or better able to deal with, change than other subgroups. It was assumed that differences between subgroups along this dimension might create an added basis for differentiation and conflict among subgroups.

It was expected that the *need* for differentiation would be reflected by two indicators of environmental diversity: (1) the degree to which the subenvironments differed in terms of their environmental uncertainty as defined by Lawrence and Lorsch; and (2) differences among subenvironments in terms of the groups and organizations which comprised them as indicated by a time allocation instrument developed by Allen (1969). However, observations revealed that factors other than environmental diversity were related to the need for differentiation and they were considered in the analysis.

Integration among subgroups was defined as the state of perceived collaboration existing among them. It was expected that the *need* for integration would be reflected by the degree of interdependence which principals reported on the various support groups, including curriculum services and pupil services.

DEMANDS OF THE CHANGING ENVIRONMENT: THE NEED FOR DIFFERENTIATION AND INTEGRATION POSED BY INCREASING MINORITY ENROLLMENTS

The two school systems studied were conceived of as systems interacting with larger environments, and their changing pupil compositions were concomitantly thought of as major environmental changes. Interviews with administrators, teachers, and community people in the two systems suggested that the increasing minority enrollments posed a number of needs which the two systems had to meet if they were to adapt effectively. In particular, three major needs stood out as recurrent themes in interviews: a need for major curriculum change; a need for changes in teaching staff background and preparation; and a need for increased pupil services of an individual nature.

The need for major curriculum change was one of the adaptation requirements most often mentioned. In both cities, minority groups and other constituencies were major forces pressing for changes in curriculum and methods. Declining achievement scores, student disturbances, and other problems were offered as manifestations of why more relevant curricula were needed. Both systems had attempted to respond by initiating changes in programs and methods. Interviews with administrators and teachers suggested that much of the curricula of the early sixties (when both systems were predominantly nonminority) were either inappropriate or irrelevant to the needs of minority children. The increasing number of Puerto Rican children in the two systems posed an additional need for curriculum change since English was a second or unknown language to many. Both systems were in the process of developing and expanding programs at the elementary and secondary level in which Spanish-speaking pupils could continue to study such courses as arithmetic and geography in Spanish while taking English as a second language.

The need for more carefully selected and trained staff was related to the need for curriculum change. Teachers who may have successfully taught middle-class or ethnic white pupils in the past were often unable to relate to or understand the needs of increasing numbers of black and Puerto Rican students. In addition, minority parents and students were demanding more minority teachers and administrators. As a result of these pressures, the recruitment of minority staff had taken on prime importance in both school systems.

Although the impact of these needs was felt most keenly by schoolhouses, the need for change in curriculum, methods, and staffing also had major implications for curriculum services in both systems. Interviews with administrators suggested that implementing the changes needed in the schools required considerable assistance from curriculum services, particularly in terms of assistance from special resource teachers, subject area supervisors, and curriculum development specialists. In order to meet the needs for curriculum change, curriculum services had to become more concerned with innovation and program development in those areas where a single schoolhouse did not have the necessary staff or resources. Such major curriculum changes as the development of an intensive reading curriculum or an early childhood program or a new career field were all given as examples of major changes which no one single school could plan and implement. Meeting the needs of the changing environment also required that curriculum services be able to assist schoolhouse personnel in acquiring special resources or materials needed for curriculum changes.

The increased availability of state and federal funds during the sixties was an additional environmental change which had an impact on curriculum services. The acquisition and effective use of these funds required that curriculum services become more involved in program development. In both systems a large part of the task of converting these funds into effective curriculum changes fell within the purview of the curriculum services subgroup.

Similarly, interviews suggested that increasing minority enrollments created a need for more extensive pupil services. Administrators in the two systems explained that as an individual schoolhouse's minority enrollment increased, individual pupil problems of a physical or emotional nature, stemming from an environment of poverty and deprivation, also increased. Dealing with these problems required services such as those provided by nurses, guidance counselors, special education teachers, school social workers, and school psychologists.

The Need for Subgroup Integration

Interviews and observation suggested that meeting these needs had implications for the relationships among *schoolhouses, curriculum services,* and *pupil services,* particularly the integration of effort required among these subgroups. Interviews with administrators in the two systems indicated that in order to meet these needs, schoolhouses became more dependent on central staff services in general and curriculum and pupil services in particular. Other environmental changes, particularly the increased availability of special state and federal funding, also acted to increase schoolhouse dependence on central staff functions because the acquisition and allocation of those special resources were largely administered by "downtown" departments.

These interview data were supported by questionnaire responses of principals in the two systems which showed a significant association between the school's rate of increase in minority enrollments and the principal's perceived dependence on central staff support in general, and pupil services in particular. Table 1 summarizes the responses of princi-

TABLE 1

Rate of Increase in Minority Enrollment and Perceived Dependence on Central Staff Departments

	Percent Reporting High Dependence		
Rate of Increase in Minority Enrollment	*On All Central Departments*	*On Curriculum Services*	*On Pupil Services*
Slowly changing schoolhouses	27	44	27
Rapidly changing schoolhouses	73	55	73

N=22. Pooled sample of principals' responses from both school systems. Cramer's Phi for rate of change and dependence on all central departments is .45 (significant at the .05 level); Cramer's Phi for rate of change and dependence on curriculum services is .10 (not significant); Cramer's Phi for rate of change and dependence on pupil services is .45 (significant at the .05 level). Respondents' scores for dependence on each subgroup were dichotomized at the median into either high or low. Rates of change for the sample schoolhouses were dichotomized at the median into either stable or changing. Median rate was 9.6 percent change in minority enrollment over the most recent three years. Individual rates of change varied from 1.0 percent to 28.5 percent.

pals in both systems, arranged according to rate of change in the schools. As can be seen, principals in schools with rapidly changing enrollments were more dependent on all staff functions in general, and pupil services in particular, than principals in slowly changing schools. Similarly, they tended to report higher dependence on curriculum services (although the association between rate of change and dependence on curriculum services was not significant). A similar pattern was found when principal responses were compared by minority composition of schoolhouses rather than rate of change. Principals of schoolhouses with compositions above the sample median reported significantly higher dependence (P < .05) on pupil services than principals below the sample median. Thus, the questionnaire data tended to support interview data which suggested that as a total system's minority enrollment increased, the need for integration among subgroups also increased. These data provided a basis for hypothesizing that effective adaptation to increasing minority enrollments required increased integration among schoolhouses and the two major support groups, curriculum services and pupil services.

The Need for Differentiation

Although the environments of the two systems were in a state of change, field observation and questionnaire data suggested that the subenvironments corresponding to schoolhouses, curriculum services, and pupil services did not substantially differ from each other. The subenvironments corresponding to these subgroups were not found to differ significantly on any of the three dimensions of environmental uncertainty used by Lawrence and Lorsch (Kruskal-Wallis one-way analysis of variance); nor did the groups comprising these subenvironments differ markedly. Administrators in all three subgroups identified essentially the same groups of internal and external organizations as being important to their tasks (although the amount of time spent with these groups did vary according to subgroup). Environmental diversity, per se, then, did not provide a basis for hypothesizing that differentiation was needed in order to adapt. Indeed had the study not included observation and interviews, the study might have concluded that differentiation among subgroups was not an especially important requirement for adapting to the environmental changes described earlier.

The Need to Adapt While Coping

Filed observations and interviews suggested that the nature of the demands posed by the changing environment, not just its diversity, were relevant to the degree of differentiation required, and that meeting these needs required that differentiation exist among subgroups. As described earlier, each system had to bring about a number of changes and provide a number of specialized services if it was to effect long-term adaptation. At the same time, however, field observation suggested that in order to remain viable in the short term, each system also had to cope with a number of immediate problems of an urgent and sometimes crisis nature. Observations of the two systems suggested a distinction between *adapting* to environmental change and *coping* with it. Coping with environmental change is therefore used here to describe the process of dealing with the consequences of environmental change on a day-to-day basis so that the organization is kept going. Adaptation is used to describe the process of providing the services and changes needed to meet changing needs. Interviews suggested that these two sets of responses, adapting and coping, were to some extent in conflict with each other. The immediate coping problems were so compelling that they tended to receive more attention than the longer term or more specialized activities *unless* enough differentiation existed so that some subgroups could focus on these longer term, change oriented, or more specialized activities.

Several themes stood out in interviews with administrators in the two systems concerning the need to cope: (1) As an individual schoolhouse's minority enrollment increased, the number of urgent problems of a coping nature also increased, especially if corresponding changes had not occurred in curricula. (2) As the total system's enrollment increased, the coping problems faced by the system as a whole increased. (3) The need to cope while bringing about change was real and could not be ignored without serious consequences to individual schools or the system as a whole. (4) The immediate needs facing schoolhouses were so compelling that a tendency existed for support groups to orient themselves to these needs at the expense of other more specialized, longer-term, or more change-oriented concerns. (5) Many of the coping problems faced by the two systems resulted from and were manifestations of the system's past inability to bring about change.

Interviews with administrators in schools of varying compositions suggested that as a school's minority enrollment increased, the number and urgency of such coping problems as riots, students needing medical attention, major disturbances, student protests, and parent boycotts also increased. These interview and field observations were borne out by questionnaire responses of elementary principals in the pooled sample. These responses showed that principals in schools above the sample's median minority enrollment reported spending significantly more time on problems which had to be solved within a day's time than principals of schools below the sample's median. When a broader measure of a principal's concern with short-term problems was used, the same pattern was found. An immediacy-of-time-orientation score was computed for each principal[6] which reflected his allocation of time on problems of varying time horizons, e.g., a week, month, quarter, etc. Again, the relationship between composition and time orientation was seen. The immediacy-of-time-orientation scores were negatively correlated with the minority composition of schoolhouses. (Spearman Rank Order Correlation Coefficient $= -.44$, $P < .05$.) In other words, the greater the minority enrollment of a schoolhouse, the more immediate was the principal's time orientation.

A number of explanations can be offered for the greater number of problems of a coping nature which accompanied increases in minority enrollments. One explanation has already been offered, i.e., that the problems were themselves manifestations of the need for change and of client dissatisfaction with the system's programs. For example, many of the disciplinary or disruptive problems facing principals were due to irrelevancy in programs. Minority students found the old curriculum irrelevant or inappropriate, did not become engaged in the school's program, and created a number of disciplinary problems.

However, whatever the source of these coping problems, the problems themselves were real, and could not be ignored without serious consequences to the short-term viability of individual schools and the system itself. In this respect they "had" to be dealt with. As an example of the nondeferability of such problems, the investigator observed the administrative staff of one of the high schools spend most of a morning trying to prevent a fight from occurring in the school cafeteria that noon. Preventing this fight required that the administrators talk with the students involved, explore the reasons for the conflict, and finally get the students to talk about their problems with each other and

---

[6] The immediacy-of-time-orientation scores were computed based on responses to the Lawrence and Lorsch time orientation instrument. It should be noted that the correlation existed for *elementary* school principals only. Scores for high school principals were not included because their larger administrative staffs enabled them to be less immediacy oriented than elementary principals. The correlation is not significant if they are included in the sample.

agree not to fight. The prospect of such a fight in the school's cafeteria was an urgent problem since it presented the potential for violence, widespread injury, damage to property, and a possible aftermath of fear and anxiety on the part of pupils and staff.

Dealing with such an urgent problem would likely take precedence (and many would argue *should* take precedence) over a report which might be due at the superintendent's office that afternoon. Many less dramatic but nonetheless urgent problems were observed which suggested that these coping problems had to be dealt with as they occurred. Interviews also suggested that it was quite difficult for principals to deal with requirements of a short-term coping nature and simultaneously plan and implement longer-term change. This difficulty may in part explain their increased dependence on central office support groups. The two sets of demands—*coping,* i.e., reacting to immediate problems, and *adapting,* that is, bringing about longer-term change, were to a certain extent incompatible.

The urgency of these schoolhouse coping problems was also found to have a pervasive influence on the curriculum services and pupil services subgroups. Administrators from these two support groups commented on the pressures they felt to respond to immediate school-house needs at the expense of longer-term or more specialized activities. The urgency of these problems made it difficult for support groups to maintain orientations which were longer term, more change oriented, or specialized—orientations needed to provide the planning and specialized services necessary for the system as a whole to adapt in the longer term. For example, in the absence of pressures to the contrary, a tendency existed for principals in troubled schools to use such pupil service specialists as guidance personnel or school social workers to check on absences, perform record-keeping or administrative chores, or, in extreme cases, to patrol halls—functions clearly inappropriate to the goals of pupil services. Oftentimes this resulted in tensions and conflicts between principals and pupil services administrators. In some cases, involvement by support groups took the very direct form of dispatching support staff personnel to schools experiencing or attempting to prevent disturbances.

These field observations and interviews provided a basis for hypothesizing that differentiation was needed if the systems were to adapt while coping. Specifically, it is possible to suggest that differentiation was needed among the three subgroups so that curriculum services and pupil services could focus on some of the change oriented or specialized tasks which administrators and teachers in schoolhouses could not perform because of the many coping demands put on them by the environment. Differentiation was needed so that curriculum services could develop and maintain orientations consistent with initiating and facilitating program changes, and pupil services could maintain orientations needed to provide specialized services of an individual pupil nature. Otherwise, all subgroups would become involved in the urgent, coping needs of the schoolhouse, and other functions necessary for effective adaptation would not be performed.

A COMPARISON OF THE TWO SCHOOL SYSTEMS

The two school systems were compared in terms of the differentiation and integration they had attained at the time of the study. In addition, they were also compared in terms of other organizational variables which may have had a bearing on either adaptation effectiveness or their ability to attain differentiation or integration.

Differentiation among Subgroups

Given the analysis of environmental demands described above, one would expect that the more adaptive system would have attained higher states of both differentiation and integration (assuming, of course, that the states achieved were not superoptimal for dealing with the environment).[7]

A differentiation score was determined for each pairing of the three major subgroups as reported in Table 2. A comparison of the two school systems shows that the more adaptive system (in terms of the performance indicators described earlier) had attained a higher degree of

TABLE 2

Differentiation Scores for the Three Major Subgroups of Each System

	*Differentiation Score**			
Pairing	*Less Adaptive System*		*More Adaptive System*	
Schoolhouses and curriculum services	6	(4)†	20	(10)†
Schoolhouses and pupil services	13	(6)	14	(9)
Curriculum services and pupil services	12	(6)	17	(9)
Average differentiation score for each school system‡	10.3	(5.3)‡	17.0	(9.3)‡

° The higher the score, the greater the differentiation.
† Numbers in parentheses are the differentiation scores computed using the Garrison and Lawrence and Lorsch conventions.
‡ It is not possible to make a statistical comparison of the average differentiation scores reported for the two systems since the individual differentiation scores reported for each of the three pairings of subgroups are not independent.

---

[7] Since the hypotheses developed earlier pertain to the longitudinal process of adaptation, the reader must be cautioned by two *caveats*. First, the data reported on actual states of differentiation and integration are not of a longitudinal nature. The data reflect states at the time of the study and say nothing about how actual differentiation and integration among subgroups may have changed over time as the two systems tried to adapt. All that can be said is that at the time of the study, one organization appeared to be doing a better job of maintaining differentiation or achieving integration than the other.

Second, there is no basis for saying how "appropriate" the states achieved by either system might be. Lawrence and Lorsch were able to make judgments of this kind because they had studied the requirements of a number of different environments and had a basis for saying that some environments required higher states of differentiation or integration than others. It is not possible to do this with the two school systems studied since there are no data on other school system environments against which to make comparisons. Had school systems dealing with stable environments also been studied, there would have been some basis for making such judgments.

differentiation among subgroups. There is no way of knowing how the patterns of differentiation may have changed in each system over time as it tried to adapt to the needs of its increased minority enrollments. It is possible to speculate that the need for more specialized central support services may have resulted in increased differentiation among subgroups as the two systems tried to adapt. However, it is also possible to argue that the increasing number of problems of a coping nature may have resulted in reducing differentiation over time. The one generalization which can be made on the basis of interviews is that both systems were having difficulty maintaining differentiation because of the many coping demands they were facing. The data reported above do provide a basis for saying that the more effective system was doing a better job of maintaining differentiation at the time of the study. In this regard, the actual comparison of the two systems is not inconsistent with a hypothesis that differentiation was needed to deal with the changing environment.

It is not possible to directly compare the differentiation scores reported in Table 2 with scores reported by Lawrence and Lorsch for industrial organizations because of the inclusion in our index of an added dimension of differentiation—*orientation to change*—and because the norms used in assigning differentiation scores for each dimension were based on the range of attributes reported by subgroups of the two school systems rather than those developed by Garrison (1966) and reported by Lawrence and Lorsch (p. 258). The decision to develop different norms was based on the judgment that norms for an industrial setting might not be appropriate to the school system setting. However, a set of differentiation scores for the two school systems was also computed using the Garrison norms. These alternative scores do not include the *orientation to change* dimension and are therefore comparable to those reported by Lawrence and Lorsch if one assumes that the Garrison norms can be applied to school systems. This alternative set of differentiation scores is given in parentheses in Table 2. Again, a comparison shows greater differentiation being reported by the more adaptive system.

Differences in Subgroup Orientations

A comparison of subgroup scores showed that in the more adaptive system the subgroup orientations differed from each other significantly on all orientations except for concern with external issues (one of the goal orientations). In the less adaptive system, subgroup scores differed significantly on all orientations except for interpersonal orientation, and two of the goal orientations (concern with external issues, and concern with teaching problems). Table 3 reports the significance levels of these differences for the two systems. These data are consistent with the higher differentiation scores reported for the more adaptive system since the differences were in all cases at least as significant, if not more so, in the more adaptive system than in the less adaptive one.

In both systems, schoolhouses reported the shortest time orientation

TABLE 3

Differences among Subgroups in the Two School Systems

Orientation	*Level of Significance of Differences among Subgroups (Kruskal-Wallis One-Way Analysis of Variance)*	
	More Adaptive System	*Less Adaptive System*
Orientation to change	.05	.05
Interpersonal orientation	.10	NS
Time orientation	.01	.02
Goal orientation		
Concern with pupils as individuals versus concern with programs	.001	.001
Concern with external groups and issues	NS	NS
Concern with teaching problems	.02	NS
Formality of structure	Indices not possible to test for significance.	

and the greatest concern with goals related to the teaching process; curriculum services reported the greatest concern with programs and the greatest orientation to change, while pupil services reported the most concern with pupils as individuals, and the least concern with change. The patterns were not the same in both systems on all dimensions however. In the more adaptive system, curriculum services had the longest time orientation, while in the less adaptive system, curriculum had a shorter time orientation than pupil services (although not as short as schoolhouses).

Degree of Integration

The actual integration scores reported for the two school systems indicate that they had attained varying levels of integration among their different subgroups. Integration scores were determined for the relationships existing among schoolhouses and curriculum services, pupil services, and a third subgroup, special programs and planning. Special programs and planning was a comparatively small group in each of the two systems but one which provided special planning and fund-seeking support and thus had a bearing on adaptation. Table 4 reports the averages of the ratings given to each of these three relationships by administrators in each system. In both systems, the tightest integration was reported between schoolhouses and curriculum services. For all three pairings, the scores reported for the more adaptive system are higher than those reported for the less adaptive system, although the difference between systems was not significant for the schoolhouses–curriculum services pairing. The systematic occurrence of these differences for all pairings provides a basis for suggesting that, as a total system, the more adaptive system was more tightly integrated than the

TABLE 4

Degree of Integration among Pairings of Schoolhouses and Support Groups[a]

	More Adaptive System	Less Adaptive System
Schoolhouses and curriculum services[b]	4.6	4.5
Schoolhouses and pupil services[c]	4.3	4.2
Schoolhouses and special programs[d]	4.3	3.6

[a] The higher the score, the greater the integration.
[b] Differences between systems is not significant (P < .47), N=76, Mann Whitney U Test.
[c] Difference between systems is significant at P < .06, N=79.
[d] Difference between systems is significant at P < .07, N=63.

less adaptive system.[8] If the average of integration scores assigned to all pairings is looked at by individual subgroups rather than by pairings, the same pattern is found with all subgroups of the more adaptive system reporting higher average integration scores than their counterparts in the less adaptive system (p <.10, Mann Whitney U Test).

Again, it is not possible to know how the state of integration may have changed over time in the two school systems since the data reported in Table 4 apply only to the time of the study. The data do show, however, that at the time of the study, the more adaptive system was doing a better job of providing integration among subgroups. If the assumption is made that current differences in integration scores are representative of historical differences, then the more adaptive system's ability to achieve integration may have been an important factor underlying its relatively more effective adaptation.

Thus, the actual comparison of school systems also tends to support the hypotheses developed earlier that adapting to the needs of minority enrollments required that the systems be able to increase integration while maintaining differentiation.

ACHIEVING DIFFERENTIATION AND INTEGRATION

If differentiation and integration are important variables in adaptation, as the environmental analysis and actual comparison of systems would suggest, then a second set of questions arises concerning those organizational characteristics which enabled the more adaptive system to attain greater integration while maintaining a higher level of differentiation. A comparison of the two systems showed a number of ways in which the more adaptive system differed from the less adaptive system.

In terms of integrating devices, the more adaptive system had developed program office organizations for secondary and elementary education which had as one of their purposes the coordination and planning of programs as they applied to schools. These offices served

[8] Integration scores for relationships among support groups, i.e., curriculum services with pupil services, etc., did not differ significantly for the two systems, although in all cases the scores were higher for the more adaptive system.

to facilitate integration among schoolhouses and the two support groups. The two systems also differed in terms of their managerial hierarchies, the most basic integrating device identified by Lawrence and Lorsch. In the more adaptive system, schoolhouses and curriculum services both came together under the directors of elementary and secondary education, one level removed from principals and curriculum services administrators. In the less adaptive system, the two subgroups reported through separate hierarchies, coming together at the superintendent's level, three levels removed from principals and two levels removed from curriculum services administrators.

In terms of influence attributed to integrators, middle level administrators, whose purpose it was to link schoolhouses with central office support groups, were attributed with having more influence in the more adaptive systems than in the less adaptive system ($p < .05$). This greater influence attributed to administrators in integrating roles was one of the determinants identified by Lawrence and Lorsch as facilitating integration. The two systems were not found to differ significantly, however, in the extent to which they used problem solving as their prevalent mode of conflict resolution ($p < .34$), another important determinant identified by Lawrence and Lorsch.

OTHER DIFFERENCES AND SIMILARITIES

The more adaptive system was also found to differ along several variables not considered by the Lawrence and Lorsch framework but which may have had a bearing on integration or adaptation. One of these variables was the number of schoolhouses in each system. Although the less adaptive system had a somewhat smaller pupil population, it had a greater number of schools than the more adaptive system, resulting in a greater span of control within the schoolhouses subgroup. It is possible to speculate that the greater span of control may have been a factor making integration of effort intrinsically more difficult in the less adaptive system and limiting some of the organizational choices available to it.

Another difference between systems was the size of support staffs. The less adaptive system had 35 percent fewer curriculum services professionals and 18 percent fewer pupil services professionals on a per pupil basis than the more adaptive system.[9] Given the increasing schoolhouse dependence on these services described earlier, this difference may have had a very direct bearing on system adaptation, as well as an indirect influence on each system's ability to achieve differentiation and integration.

A number of similarities were also identified which may have had a bearing on adaptation. Both systems responded to the ten-year period

---

[9] The more adaptive system's larger support staffs may have been related to larger per pupil expenditures over the most recent ten years, and a growing tax base per pupil over that period, two contextual factors not controlled in the initial site selection.

of increasing minority enrollments by expanding the size and activities of curriculum services and pupil services, as well as by creating special fund-seeking roles. Both systems also dealt with the resulting increased internal complexity by adding middle level administrators and by enlarging the superintendent's office through the addition of a deputy or associate's position. It is possible to suggest that many of these structural changes may have been functional for adaptation since in both systems the rates of decline in attendance and achievement scores had leveled off within the most recent three-year period (despite constant increases in minority enrollments) while drop-out rates and quality of placement had actually improved.

CONCLUSIONS

The findings of this exploratory study suggest that the Differentiation-Integration model can be usefully applied to understanding the organizational requirements of adapting to a changing environment. An important implication of the findings is that a need for differentiation may exist within organizations in changing environments even if those environments are not particularly diverse. In this sense the present study extends the Lawrence and Lorsch findings which related the need for differentiation to environmental diversity. The changing environments described in this paper not only posed requirements for longer-term change, but also posed a number of problems of an urgent and immediate nature, many of which were consequences of past system maladaptiveness. Nonetheless, to ignore these problems would have threatened the short-term viability of the systems. Because these coping problems were so compelling, a tendency existed for the organizations to develop orientations and patterns of behavior appropriate for coping but not for longer-term adaptation unless some parts of the system were differentiated enough to focus on these issues. In these terms, differentiation is needed if a system is to simultaneously cope while adapting to environmental change.

The need for differentiation in meeting these two sets of needs is also consistent with recent findings of cognitive psychologists. Work by Schroder, Driver, and Streufert (1967) and Driver and Streufert (1969) suggests that persons who are dealing with urgent or threatening situations or are overloaded with problems of an immediate nature are not psychologically disposed to performing complex tasks such as planning or searching. Seen in these terms, differentiation allows subgroups to focus on different aspects of environmental demands. In the school systems studied, differentiation was needed so that support groups could focus on some of the change oriented or planning tasks which schoolhouse administrators could not perform because of the many coping demands put on them by the environment.

Finally, these findings suggest that the differentiation-integration concepts can be applied in human services organizations to better understand the consequences of environment for organization even though the setting is quite different from the business organizations originally studied by Lawrence and Lorsch and most other systems theorists.

REFERENCES

Allen, Stephen A., III. "Managing Organizational Diversity," Unpublished doctoral dissertation, Harvard University Graduate School of Business Administration, 1969.

Burns, Tom and Stalker, G. M. *The Management of Innovation.* London: Tavistock Publications Limited, 1961.

Coleman, James S. "A Brief Summary of the Coleman Report." *Equal Educational Opportunity,* Editorial Board, *Harvard Educational Review.* Cambridge: Harvard University Press, 1969.

Dill, William. "The Impact of Environment on Organizational Development." *Concepts and Issues in Administrative Behavior.* Edited by Sydney Marlick and E. H. Van Ness. Englewood Cliffs, N.J.: Prentice–Hall, 1962.

Dutton, John M. and Walton, Richard E. "Interdepartmental Conflict and Cooperation: Two Contrasting Studies." *Human Organization,* Volume 25, November 2, Fall, 1966.

Emery, R. E. and Trist, E. L. "The Causal Texture of Organizational Environments." *Human Relations,* Volume 18, Number 1, February, 1965.

Evan, William. "The Organization-Set: Toward A Theory of Interorganizational Relations." *Approaches to Organizational Design.* Edited by James Thompson. Pittsburgh: The University of Pittsburgh, 1966.

Gabarro, John J. "School System Organization and Adaptation to a Changing Environment." Unpublished doctoral dissertation. Harvard University Graduate School of Business Administration, 1971.

Garrison, James S. "Organizational Patterns and Industrial Environments." Unpublished doctoral dissertation, Harvard University Graduate School of Business Administration, 1966.

Gittell, Marilyn. *Participants and Participation.* New York: Frederick A. Praeger, Publishers, 1967.

Katz, Daniel and Kahn, Robert L. *The Social Psychology of Organizations.* New York: John Wiley & Sons, Inc., 1966.

Lawrence, Paul R. and Lorsch, Jay W. *Organization and Environment.* Boston: Division of Research, Harvard University Graduate School of Business Administration, 1967.

Lorsch, Jay W. and Allen, Stephen A., III. *Managing Diversity and Interdependence.* Boston: Division of Research, Harvard University Graduate School of Business Administration, 1973.

March, James G. and Simon, Herbert A. *Organizations.* New York: John Wiley & Sons, Inc., 1958.

Miller, E. J. "Technology, Territory and Time: The Internal Differentiation of Complex Production Systems." *Human Relations,* Volume 12, Number 3.

Perrow, Charles. "A Framework for the Comparative Analysis of Complex Organizations." *American Sociological Review,* Volume 32, Number 2, April 1967.

Pondy, Lewis R. "Organization Conflict: Concepts and Models." *Administrative Science Quarterly*, Volume 12, Number 2, September 1967.

Rice, A. K. *The Enterprise and its Environment.* London: Tavistock Publications Limited, 1963.

Rogers, David. *110 Livingston Street.* New York: Random House, Inc., 1968.

Schrag, Peter. *Village School Downtown.* Boston: Beacon Press, 1967.

Schroder, Harold M., Driver, Michael J., and Streufert, Siegfried. *Human Information Processing.* New York: Holt, Rinehart, and Winston, Inc., 1967.

Starbuck, William H. "Organizational Growth and Development." *Handbook of Organizations.* Edited by James G. March. Chicago: Rand McNally & Company, 1965.

Streufert, Siegfried, and Driver, Michael J. "Integrative Complexity: An Approach to Individuals and Groups as Information-Processing Systems." *Administrative Science Quarterly*, Volume 14, Number 2, June 1969.

Terreberry, Shirley. "The Evolution of Organization Environments." *Administrative Science Quarterly*, Volume 12, Number 4, March 1968.

Thompson, James D. *Organizations in Action* New York: McGraw-Hill Book Company, 1967.

Walton, Richard E., and Dutton, John. "Management of Interdepartmental Conflict: Model and Review." *Administrative Science Quarterly*, Volume 12, Number 3, December 1967.

Watson, Goodwin. "Toward a Conceptual Architecture of a Self-Renewing School System." *Change in School Systems.* Edited by Goodwin Watson. Washington: National Training Laboratories, 1967.

Woodward, Joan. *Industrial Organization: Theory and Practice.* London: Oxford University Press, 1965.

James D. Thompson

13. Organizations and Output Transactions[1]

COMPLEX PURPOSIVE organizations receive inputs from, and discharge outputs to, environments, and virtually all such organizations develop specialized roles for these purposes. *Output roles*, designed to arrange for distribution of the organization's ultimate product, service, or impact to other agents of the society thus are *boundary-spanning* roles linking organization and environment through interaction between member and non-member.

Organizational output roles are defined in part by reciprocal roles of non-members. Teacher, salesman, and caseworker roles can only be understood in relation to pupil, customer, and client roles. Both member and non-member roles contain the expectation of closure or completion of interaction, leading either to the severance of interaction or bringing the relationship into a new phase.[2] Each output role, together with the reciprocating non-member role, can be considered as built into a *transaction structure*.

Because output roles exist in structures that span the boundaries of the organization, they may be important sources of organization adaptation to environmental influences. Empirical studies reflect this fact more than do theories of organization.

From *The American Journal of Sociology*, 1962, vol. 68, pp. 309–24. Reprinted with permission of the author and the publisher. © 1962 by The University of Chicago. All rights reserved.

[1] I am indebted to my colleagues—Robert Avery, Carl Beck, Richard Carlson, Joseph Eaton, Robert Hawkes, Axel Leijonhufvud, Morris Ogul, and C. Edward Weber—for helpful reactions to earlier versions of this paper.

[2] Role theory has devoted much more attention to structure-maintaining behavior than to transaction behavior, in spite of the variety and quantity of transaction in modern societies. See, however, William J. Goode, "A Theory of Role Strain," *American Sociological Review*, XXV (August, 1960), 483–96; George C. Homans, *Social Behavior: Its Elementary Form* (New York: Harcourt, Brace & Co., 1961), and John W. Thibaut and Harold H. Kelley, *The Social Psychology of Groups* (New York: John Wiley & Sons, 1959).

Classic bureaucratic theory is preoccupied with behavioral relations ordered by a single, unified authority structure from which the client is excluded,[3] and only recently has an explicit correction for this one-sided approach been introduced by Eisenstadt's theory of debureaucratization.[4] Another strain of organization theory, following Chester Barnard, clouds the significance of input-output problems by lumping investors, clients, suppliers, and customers, as members of the "cooperative system."[5] The developing inducements-contributions theory of March and Simon has so far been directed primarily at the problem of recruiting and motivating members or employees.[6]

One purpose of this paper will be to focus theoretical attention on boundary-spanning behavior by way of output roles. A second will be to indicate that there are several types of transaction structures, each having peculiar significance for the comparative analysis of organizations. A third aim will be to indicate that transaction processes can be studied profitably through sequential analysis. Finally, some larger consequences of output relationships will be suggested.

Consideration will be limited to those transaction structures that call for face-to-face interpersonal interaction between member and non-member, thus ignoring the cigarette "salesman" who may periodically load a vending machine without seeing his customers or knowing who they are, and the soldier who may deliver destruction to an enemy he neither sees nor could identify. Consideration will also be limited to those cases in which the output role is occupied by an employed agent of the organization.

CHARACTERISTICS OF TRANSACTION STRUCTURES

For any transaction structure there appear to be three possible transaction outcomes: (1) *completion* of a transaction as defined by organizational norms, (2) *abortion,* in which interaction is terminated without completion of the transaction, or (3) *side transaction,* in which member and non-member complete an exchange not desired or approved by the organization.[7] Which of these three outcomes emerges will in part be determined by the desires, attitudes, and actions of the two parties involved. But the likely paths from initiation of interaction to termination, and the branching points which lead to one or another outcome, are largely defined by the type of transaction structure.

[3] Max Weber, *The Theory of Economic and Social Organization* (Glencoe: Free Press, 1957), and Robert K. Merton, *Social Theory and Social Structure* (rev. ed.; Glencoe: Free Press, 1957), chap. vi.

[4] S. N. Eisenstadt, "Bureaucracy, Bureaucratization, and Debureaucratization," *Administrative Science Quarterly,* IV (1959), 302–20.

[5] Chester Barnard, *The Functions of the Executive* (Cambridge, Mass.: Harvard University Press, 1938).

[6] James G. March and Herbert A. Simon, *Organizations* (New York: John Wiley & Sons, 1958).

[7] This concept was suggested by that of "side payment" as developed by R. M. Cyert and J. G. March, "A Behavioral Theory of Organizational Objectives," in Mason Haire (ed.), *Modern Organization Theory* (New York: John Wiley & Sons, 1959).

ELEMENTS OF A TYPOLOGY

The organization cannot predict in advance of any specific encounter just what desires, attitudes, or actions the non-member will bring to the transaction structure, but the organization can estimate in advance with reasonable accuracy two things: (1) the extent to which it has armed its agents with routines, and (2) the extent to which the non-member is compelled to participate in the relationship.

The first dimension will be labeled one of *specificity of control over member*. Undoubtedly this forms a continuum, but it will be discussed here only in its extremes. At one extreme the member is equipped with a single, complete program—a standard procedure which supposedly does not vary, regardless of the behavior of the non-member. At the other extreme, the member's behavior is expected to be guided primarily by the behavior of the non-member, although always in relation to some organizational target or goal.[8] The supermarket check-out clerk approximates the *programmed* role, while the social caseworker illustrates the *heuristic* variety.

The second dimension, also a continuum but here dichotomized, will be labeled *degree of non-member discretion*. At one extreme the non-member finds interaction *mandatory*, at the other it is *optional*. It may be presumed that the prisoner, for example, finds interaction with the guard mandatory. To be sure, the prisoner may evade interaction by escape or by behavior which results in transfer to another cell block or prison. But short of these extremes, the prisoner cannot choose whom he will interact with or whether to interact; the relationship is mandatory.

The optional state is exemplified by the salesman-customer relationship under conditions of "perfect competition," where the prospect has a wide choice of salesmen. Not only does the customer have discretion over whether to interact, but he also may terminate it at will, before completion of a transaction.

When these two dichotomized dimensions are combined, four types of output structures emerge:

	Specificity of Organizational Control	
Degree of Non-Member Discretion	*Member Programmed*	*Member Heuristic*
Interaction mandatory..............	I	III
Interaction optional................	II	IV

Temptation to label each cell is strong, since there are familiar categorizations readily available which appear to correspond to each.

---

[8] March and Simon make a similar distinction by referring to programs specifying activities (means) and programs specifying product or outcome (ends). They observe that the latter allow discretion to the individual (op. cit., p. 147).

Cell I, for example, might be considered "clerical," Cell II "commercial," Cell III "semi-professional," and Cell IV "professional." The temptation has been resisted, however, because casual, traditional categorizations may in fact hide some of the distinctions which this typology attempts to bring out. Thus, transactions which might typically be lumped under the term "commercial" may, in fact, appear in any of the cells defined here.

OCCURRENCE OF TRANSACTION STRUCTURES

Organizations which develop elaborate programs for those in output roles appear to be those that either (1) provide services for large numbers of persons and, therefore, face many non-members relative to each member at the output boundary, or (2) employ a mechanized production technology which places a premium on large runs of standardized products, attaches heavy costs to retooling, and, therefore, depend on a large volume of standardized transactions per member at the output boundary.[9] The first condition seems appropriate for clerical activities, such as the issuance of licenses or permits by a government bureau, and is especially likely when the organization holds a monopoly position, as a government often does. This seems to correspond to Cell I in the typology above, and to classic bureaucratic theory.

The second condition seems to describe commercial transactions of mass-produced products under competitive conditions and corresponds to Cell II in the typology. When competition is removed, as in the seller's market for automobiles after World War II, the role of salesman can be redefined into a clerical role, with the salesman merely writing orders and adding names to the waiting list—and perhaps adding a side transaction to give the customer a priority rank in exchange for private payment.[10] Under competitive conditions, however, the salesperson expresses one or more organizational programs governing size, style, color, price, terms, delivery schedules, and so on, and the customer either takes it or leaves it to search among other organizations for a more suitable program.

Neither of these transaction structures is appropriate for the organization which must "tailor" its output, for here the exigencies make it impractical if not impossible to develop standard programs in advance. Instead the organization must rely on the judgment of the member at the output boundary. Such roles tend to be assigned to professionally trained or certified persons, for it is believed that the professional type of education qualifies individuals to make judgments or exercise discretion in situations appropriate to their specialization.

When non-member participation is mandatory the transaction structure corresponds to Cell III in the typology above. Examples would include the therapy-oriented prison and the military hospital, the public

---

[9] I am indebted to Axel Leijonhufvud for contributing this insight from economic theory.

[10] When competition returned to this field, dealers complained frequently that their "salesmen" had forgotten how to "sell."

school, and the *public* (as distinguished from "voluntary") welfare agency. In each case, the non-member (prisoner, patient, pupil, or applicant) is obliged to participate in the structure, and in each case the member is expected to vary his behavior to suit the particular condition of the non-member.

When interaction of the heuristic variety is optional for the non-member, the transaction structure corresponds to Cell IV in the typology. This would encompass the "voluntary" (non-governmental) welfare agency and the voluntary hospital. Many of the services which fit this category are dispensed by private entrepreneurial arrangements—by private practitioners—rather than in large-scale organizational contexts. It appears that this reflects the relative complexity of the process in this kind of transaction structure, which makes it especially difficult when subjected to the additional constraints of an organizational context.

It is suggested that in the order listed above, and numbered in the typology, the four types of transaction structures are increasingly difficult to operate.

TRANSACTION PROCESSES

The same three types of outcomes—successful transaction, abortion, and side transaction—are available for all of the transaction structures, but significant contrasts appear in the possible courses of interaction. The final state, and the paths to it, *depend in each case on contingencies and on responses to contingencies.*

Our conceptual apparatus for analyzing contingent interaction is ill-developed, since interaction theory has been preoccupied with structure-maintaining behavior, that is, with behavior conforming to stable norms and with social control mechanisms to correct deviation. The contingencies and possible paths of interaction, therefore, will be shown in flow diagrams. These are offered as hypothetical, for literature search did not reveal sufficient data to "test" them, but illustrative citations will be made.

For illustration, the possibilities for Type I, where the member is programmed and the non-member finds interaction mandatory, are shown in Figure 1. Once contact is made, the non-member may respond in one of three ways, and each, in turn, presents several response possibilities to the member. If, following initial contact, the non-member responds appropriately by offering necessary information, the member may routinely complete the transaction. Instead of responding appropriately, however, the non-member may resist or offer a bribe. If the non-member resists, the member applies punishments, and the non-member may respond by increasing resistance, by cooperating, or by offering a bribe. The bribe attempt thus may be made immediately upon initial contact or following an exchange of activity. In either case, however, the member may elect one of three alternative responses to the bribe offer: accept it, apply additional punishments, or explore the pros and cons of the bribe possibility. And so on, as is indicated in

FIGURE 1

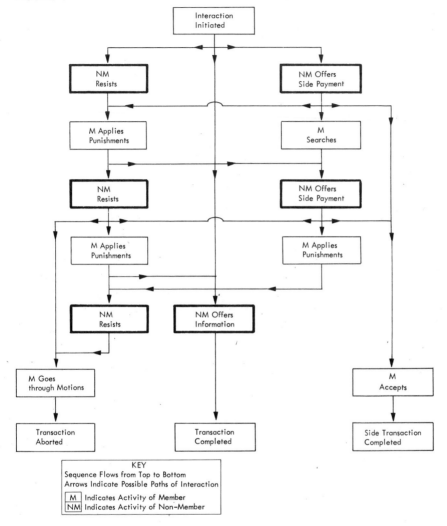

Figure 1. In each flow diagram we have attempted to depict as the central pattern that course of interaction most desired by the organization. In each, however, there are several possibilities for the interaction to digress from the desired path, swinging either to the right or left. In the more complex transaction structures, the course of interaction may swing from one stream to the other several times before the final outcome is reached, and one of the major problems for the member is to counter each digression so that it returns to the central path and ultimately results in a successful transaction.

Why members and non-members elect one route rather than another at each switching point is a subject for microanalysis and is beyond the scope of this paper.

The number of "ifs" or contingencies—even in this simplest type of transaction—is impressive when diagrammed, but too large to make verbal reproduction possible. The following discussion will focus, therefore, on certain implications of contingencies for behavior within each type of transaction structure.

INTERACTION IN TYPE I STRUCTURES

Here the non-member's participation is mandatory, the member is programmed, and the organization expects a large number of output transactions from the member (see Fig. 1).

The course of interaction is simplest if the non-member chooses to complete the transaction speedily, for he appears to the member to be cooperative. The member scores a completed transaction in a minimum of time, which contributes toward maximizing the number of transactions in serial and not only puts the member in line for organizational rewards but may provide the personal satisfaction of a "job well done." Blau, for example, noted that employment interviewers in a government agency avoided operations that took time without helping to improve their statistical records. Since the records served as the basis for evaluating performance, concentration on this goal made interviewers "unresponsive to requests from clients that would interfere with its attainment."[11]

If the non-member chooses to seek concession from the member, he must offer personal satisfactions sufficient to induce the member to risk penalties that the organization can apply if deviation from programs is detected. Blau found that at times the social reward of the client's appreciation led governmental employment interviewers to offer extra help,[12] and students of prisons have reported favored treatment by guards of prisoners who could exercise (or withhold) informal control over other inmates.[13]

That the opposite result—penalties for seeking too much help—can occur is suggested by the analysis of another employment office. Francis and Stone note that the client's eligibility for unemployment compensation is in effect determined by the interviewer's report, and that this report reflects the interviewer's *judgment* as to whether the claimant has given an accurate or exaggerated account of his history and present circumstances.[14] Blau also reports the use of punitive measures to vent *antagonism against aggressive clients.*[15]

---

[11] Peter M. Blau, *The Dynamics of Bureaucracy* (Chicago: University of Chicago Press, 1955), pp. 43, 70.

[12] *Ibid.*, p. 70.

[13] See summary in Lloyd E. Ohlin, *Sociology and the Field of Corrections* (New York: Russell Sage Foundation, 1956), pp. 20–21.

[14] Roy G. Francis and Robert C. Stone, *Service and Procedure in a Bureaucracy* (Minneapolis: University of Minnesota Press, 1956), pp. 83–84.

[15] *Op. cit.*, pp. 24, 87.

The bribe attempt involves risk for both member and non-member. If the inducements offered by the non-member are insufficient, the member may regard him as unco-operative and therefore apply penalties. If, on the other hand, the non-member's offer is sufficiently inducing, the member must somehow hide his deviation from the program, and runs the risk of being caught. The bribe attempt, therefore, often calls for a rather delicate "sounding out process."[16]

If neither party is able to marshal enough power to bring a transaction or a side transaction to a conclusion, this mandatory relationship is likely to settle into a going-through-the-motions, with each party seeking to maximize his rewards or minimize his costs.[17] From the standpoint of official organizational goals, then, the transaction is aborted. If interaction must be sustained, as in the prison, it is likely to become a struggle for control. That the struggle for control can be most subtle is brought out by Gresham Sykes' study of the corruption of authority in the maximum security prison.[18]

INTERACTION IN TYPE II STRUCTURES

In this case the non-member's participation is optional, the member is programmed, and the organization expects a large number of output transactions from the member (see Fig. 2).

Since the relationship is not mandatory, the organization usually develops an *array* of programs, one to appeal to each class or category of potential customer. The non-member, in interacting with this organization, is foregoing interaction with another, and hence seeks to determine rapidly whether a satisfactory program is available. The "gambit," or "opening move," is therefore a crucial issue in this interaction process. Lombard reports some of the difficulties salesgirls faced in "sizing up" prospective customers. In a children's clothing department, for example, "each customer presented . . . a different set of values that determined her taste in the clothes she bought for her children." Knowledge of style was not alone sufficient to make a sale, for the salesgirl could not express this in a way that criticized the customer's taste. With respect to price, the store suggested that "unless a customer made some other request, a salesgirl would do well to show her clothes in a middle price range."[19]

---

[16] For analysis in another context of the "sounding out process" see James D. Thompson and William J. McEwen, "Organizational Goals and Environment," *American Sociological Review,* XXIII (February 1958), 23–30.

[17] Thomas C. Schelling, *The Strategy of Conflict* (Cambridge, Mass.: Harvard University Press, 1960), reports an unusual case. This involved a club for motorists, and the membership card identified the holder as a person who would keep quiet if his bribe offer was accepted by a policeman. Schelling also reports, however, that if the police could identify card-carrying motorists by sight, they could concentrate arrests on card-carrying drivers, threatening a ticket unless payment were received! (See pp. 140–41.)

[18] "The Corruption of Authority and Rehabilitation," in Amitai Etzioni (ed.), *Complex Organizations* (New York: Holt, Rinehart & Winston, 1960), pp. 191–97.

[19] George F. Lombard, *Behavior in a Selling Group* (Boston: Harvard Business School, 1955), pp. 176–80.

FIGURE 2

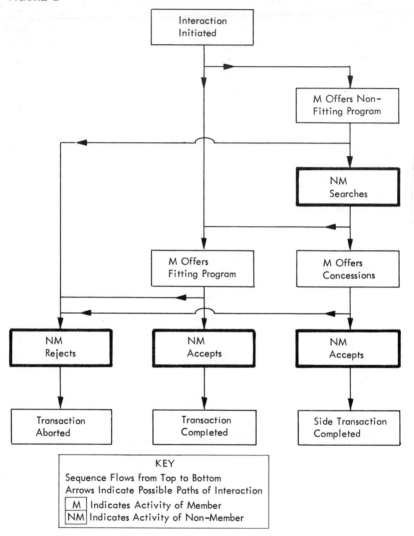

The simplest situation in this output transaction occurs when the gambit is successful, either because the non-member offers enough accurate information or the member "sizes up" the non-member correctly. On either basis the most suitable program is offered and the non-member accepts or rejects. The transaction has been completed successfully in short order, or has been aborted clearly and speedily, permitting the member to devote attention to other prospects and the non-member to seek a more satisfactory supplier.

A complication arises when the member selects from his repertoire of programs an unsuitable one because of inaccurate "size up," incorrect interpretation of the information given by the non-member, or false or misleading information offered by the non-member. The result may be (*a*) further exploration until a fitting program is identified, which is relatively costly to both parties, (*b*) withdrawal of the non-member from the output structure, in which case a potential transaction is lost to a competitor, or (*c*) continued interaction on the basis of false optimism, with mounting frustration for one or both parties as they find the investment mounting and the possibility of satisfaction dwindling.[20] In the latter case, the eventual abortion of the transaction is likely to be unpleasant, reducing the possibility of future interaction between member and non-member. It is in this frustrating situation that the member is most likely to deviate from approved programs, misrepresenting the product, offering inferior substitutes, or finding ways of offering arrangements not approved by the organization, such as special deliveries or price reductions by shaving commissions. Another possible outcome of this situation is for the non-member to reduce his standards—accepting another color, style, or settling for a different size, delivery date, and so on. When this occurs, the transaction is completed, although perhaps at the expense of future transactions.

Because prolonged search behavior is "costly," requiring investment of the time and energy where the outcome is problematic, ability to "close the deal" as early as possible becomes almost as crucial as the gambit.

INTERACTION IN TYPE III STRUCTURES

Here the non-member's participation is mandatory, the member must tailor his service to the particular needs of the non-member, and the organization must therefore judge the member's performance in terms of results rather than in terms of conformity to prescribed procedures (see Fig. 3).

The straightforward and relatively easy case occurs when the non-member elects to complete the transaction with least effort and responds appropriately. The ability of the non-member to take *active* part in the heuristic process seems frequently to be important, and especially so in comparison with the Type I output structure. Nurses were found to prefer an active patient who could make his needs and wants known, if for no other reason than to help his own treatment. In contrast, ward aides, whose duties defined the patient as "something like a necessary evil, preferred the passive patient above all others."[21]

When the non-member's responses are appropriate, both parties achieve a satisfactory outcome at minimum cost, and the member has the additional reward of evidence that his skills were instrumental—

---

[20] *Ibid.*, p. 177. When "merchandise which a customer liked was not available in the desired color or size, the situation could quickly become most difficult for both salesgirl and customer."

[21] Reported in Leonard Reissman and John H. Rohrer (eds.), *Change and Dilemma in the Nursing Profession* (New York: G. P. Putnam's Sons, 1957), p. 143.

FIGURE 3

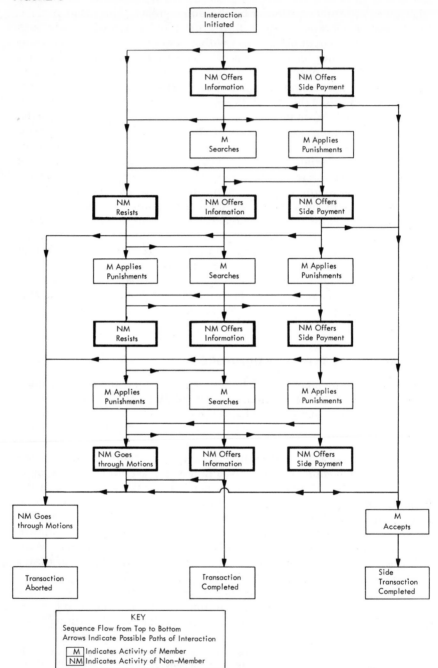

that he did a job well. But what happens when the response of the non-member is inappropriate?[22] The inappropriate response might come either from a non-member with a sincere desire to cooperate or from a non-member who is resisting. Cooperation by the non-member does not guarantee his appropriate response, for he may be incapable of articulating appropriate information, may not understand which responses are desired by the member, or may zealously offer a flood of irrelevant information. Extreme status differences between member and non-member, often found in the "helping professions," can lead to this situation. Fanshel concludes that casework help is tailored more for the verbal, communicative group than it is for "the significantly large group of clients who find difficulty in expressing their feelings and their basic ideas about the problems that bring them to the agency."[23]

When an inadequate or inappropriate response is made, therefore, the member must form a judgment (or leap to a conclusion) about the reason. If he defines the non-member as uncooperative he may apply punishments, while if he defines the non-member as sincere but inept, he may search for relevant information by seeking to educate the non-member to the appropriate role. This educational activity under such circumstances is clearly brought out by Israeli responses to large numbers of immigrants who had yet to learn client roles,[24] but there is no reason to believe that it does not occur in more "normal" or less striking situations.

Whether the member will search for information or, rather, punish the non-member for inappropriate responses, will depend largely on the member's conclusions regarding the reasons for inappropriate responses. If the non-member's search for the proper role is unsuccessful, or if he responds to punishment by resisting (or increasing resistance), the interaction process in this output structure is likely to degenerate into a struggle for control. This would appear, perhaps, in the relationship between highschool students and teachers, in the therapy-oriented prison, and in the therapy-oriented mental hospital.[25] Often the non-member will resist "help" but will "go through the motions," appearing to cooperate

---

[22] This dilemma frequently arises in treatment-oriented organizations which also have a custodial responsibility. Oscar Grusky notes that if an inmate in a traditional prison system violates the rules, the guard simply writes up a "ticket" and the inmate is punished by a central disciplinary court or a disciplinary officer; the transaction fits our Type I. However, if the same violation occurs in a treatment-oriented prison, it complicates the guard's response and creates conflict, for he must decide whether he ought to write up a ticket or whether, for treatment reasons, he ought to let the inmate "express his emotions" (see "Role Conflict in Organization," *Administrative Science Quarterly*, vol. III [March, 1959]).

[23] David Fanshel, "A Study of Caseworkers' Perceptions of Their Clients," *Social Casework*, XXXIX (December 1958), 543–51.

[24] Elihu Katz and S. N. Eisenstadt, "Some Sociological Observations on the Response of Israeli Organizations to New Immigrants," *Administrative Science Quarterly*, vol. v (June 1960).

[25] Morris S. Schwartz and Gwen Tudor Will describe "mutual withdrawal" in a mental hospital ward, with the withdrawal of the nurse perpetuating the withdrawal of the patient and the withdrawal of the patient reinforcing the nurse's withdrawal. Withdrawal in this case was described as affective, communicative, and physical (see "Low Morale and Mutual Withdrawal on a Mental Hospital Ward," *Psychiatry*, XVI [November 1953], 337–53).

to escape punishments. Eaton has suggested that for the social case-worker the client may effectively control the relationship by "presenting a challenge," by "showing appreciation," or by promising to change his behavior at some future time—provided that the caseworker keep trying.[26]

The abortive struggle for control can also occur when the non-member offers inadequate bribes for a side transaction. A bribe offer that is rejected, however, gives the member added leverage in avoiding the stand-off and, instead, completing a transaction successfully. Blau reports that being offered a bribe constituted a special tactical advantage for the field agent of a law enforcement agency. A non-member who had violated one law was caught in the act of compounding his guilt by violating another one. He could no longer claim ignorance or inadvertence as an excuse for his violations, and agents exploited this situation to strengthen their position in negotiations.[27]

INTERACTION IN TYPE IV STRUCTURES

Now the interaction is heuristic, the relationship is optional for the non-member, and the member must attract clients and maintain interaction before he can complete a transaction (see Fig. 4).

Whether interaction is initiated by the member or by the nonmember, making the initial contact is likely to be a difficult experience. Lack of knowledge of appropriate role behavior may indeed lead many individuals who need "professional help" to avoid interaction with appropriate professionals. In a sample survey seeking to determine "normal" American attitudes toward mental health, "lack of knowledge about means" of seeking professional help was a major reason given by those who felt they could have used help but did not seek it.[28] A study of "well-trained" life insurance underwriters concluded that "anxiety over intrusion on prospect privacy" was a major deterrent to making contacts,[29] and it has been reported that "a salesman of accident and health insurance is expected to make at least 36 prospect contacts by cold canvass each week. If he can sell four of these, he is regarded as performing very well. Thus he must steel himself to an average of 32 rejections 50 weeks in the year."[30]

Initial contact, at best, simply sets the stage for further exploration. Friedson concludes that the first visit by urban patients to a medical practitioner is often tentative, a tryout. Whether the physician's prescription will be followed, and whether the patient will come back, seem to rest at least partly on his retrospective assessment of the professional

---

[26] Joseph W. Eaton, private conversation.

[27] Blau, op. cit., p. 151.

[28] Gerald Gurin, Joseph Veroff, and Sheila Feld, *Americans View Their Mental Health* (New York: Basic Books, 1960), chap. xi.

[29] Herbert E. Krugman, "Salesman in Conflict," *Journal of Marketing*, XXIII (July 1958), 59–61.

[30] Robert N. McMurry, "The Mystique of Super-Salesmanship," *Harvard Business Review*, XXXIX (March–April 1961), 113–22.

FIGURE 4

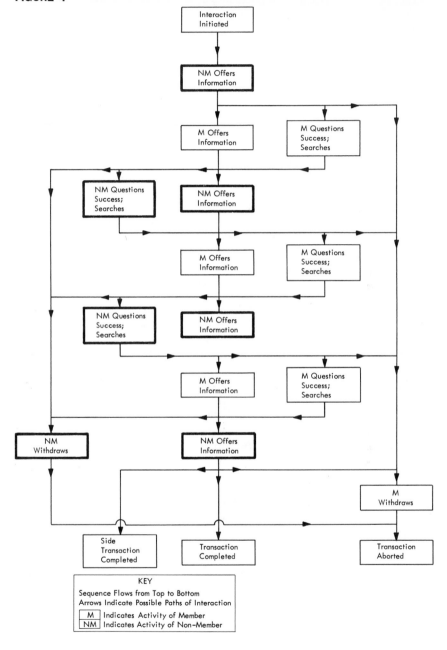

consultation.[31] Kadushin reports a considerable amount of shopping around by individuals undertaking psychotherapy.[32] The organizational counterpart of shopping around is exercised, for example, by the intake committee of the psychiatric clinic that, on the basis of initial exploration, may select those most likely to succeed as patients.[33]

Nevertheless, both parties enter this transaction structure with a measure of uncertainty, because the behavior of each must be keyed to information possessed by the other. The potential client seeks a personalized transaction for which there may be no standard pattern, and hence, no readily available schedule of costs or of probabilities of satisfactory outcome. The member's diagnosis and prognosis depend on the unfolding of information during the interaction process; hence the member cannot at the outset guarantee success, nor can he predict the complete costs to the non-member. The member must hold out enough hope of results to maintain interaction in the face of uncertainty—but he also needs "escape clauses." The client likewise must disclose enough information and motivation to enable the member to make estimates—but without irrevocable commitment to the transaction.

If the non-member becomes fully committed to the transaction, he can make few choices and exercises little control. Friedson notes that when the patient penetrates into "organizational practice" such as found in hospitals, he is well into the professional referral system, which involves maximal restriction on client choice of services, and his efforts at control are most likely to take the form of evasion.[34] In other words, when the client becomes fully committed to the transaction, the transaction structure is converted into a Type III structure.

Short of this conversion, however, there is a fully developed sounding-out process. In the early stages of interaction, a large proportion of behavior of both parties is oriented toward establishment of rapport and a small proportion to the substance of the transaction relationship. As the interaction process nears successful achievement the proportions are reversed, but en route the joint search may yield information that changes the prognosis of one or both parties, either making costs appear higher than anticipated or making success appear less likely. When the unfolding of the relationship in this way changes the conditions for the client, the transaction may be aborted. Each successive time the relationship is thus challenged by the unfolding of information which bears on costs and rewards, rapport must be reestablished. Eaton notes that an evaluation of the actual results of treatment is often impossible until a good deal of time has elapsed. "Along the way, during this protracted search for happiness, both practitioner and client want to know if there is evidence of progress, and often a judgment is made with an implied

[31] Eliot Friedson, "Client Control and Medical Practice," *American Journal of Sociology*, LXV (January 1960), 374–82.

[32] Charles Kadushin, "Individual Decisions To Undertake Psychotherapy," *Administrative Science Quarterly*, III (December 1958), 379–411 (see also his "Social Distance between Client and Professional," *American Journal of Sociology*, LXVII [March 1962], 517–31.

[33] Kadushin, "Individual Decisions To Undertake Psychotherapy," op. cit.

[34] Op. cit., p. 381.

presumption that the *quality of the therapeutic relationship* is predictive of its outcome."[35]

Each successive time that the relationship is questioned, additional investment has been made and thus the magnitude of the dilemma is greater. This may be true for the member as well as for the non-member, for crisis over rapport is likely to be personally dissatisfying to the practitioner as well as threatening to his organizational rewards; thus, he may be reluctant to alter early diagnosis and prognosis (in a direction more costly to the client) so long as there appears to be a possibility of success on the original terms. This would operate, then, to diminish the heuristics of the interaction process; the early prognosis becomes a program not easily disrupted.

The possibility of a side-transaction occurs when the member has exclusive access to something desired by the non-member, such as drugs that only a licensed physician can prescribe, or when the non-member seeks some form of attention rather than technical treatment. In medicine, under these circumstances, the placebo may satisfy the patient's need for attention and maintain the physician-patient relationship. Friedson notes that whether their motive be to heal the patient or to survive professionally, physicians dependent on the lay referral system will feel pressure to accept or manipulate lay expectations by administering harmless placebos or by giving up unpopular drugs.[36]

Because of the uncertainties of the heuristic transaction process and the difficulties of establishing and maintaining rapport, it would appear that the distinction between a process leading toward side transaction often is difficult to distinguish from one leading toward transaction. For this reason, Figure 4 does not distinguish a separate flow for transactions and side transactions.

The theme of the Type IV transaction structure thus appears to be rapport, and members must not only be able to establish and maintain rapport but, it would seem, to "cool out" the non-member if rapport cannot be reestablished at one of the crisis points;[37] that is, when the transaction is aborted, the member needs some means of convincing the disappointed non-member that the attempt was a reasonable one.

DYNAMICS OF OUTPUT RELATIONS

Because disposition of its product is imperative for any complex, purposive organization, the transaction structure is crucial. It may dictate or place significant constraints on (*a*) the acquisition of necessary

---

[35] Joseph W. Eaton, "The Client-Practitioner Relationship as a Variable in the Evaluation of Treatment Outcome," *Psychiatry*, XXII (May 1959), 189.

[36] Op. cit., p. 378.

[37] The "cooling out" concept appears to have been first introduced into sociological literature by Erving Goffman in his analysis of confidence games, in "On Cooling the Mark Out: Some Aspects of Adaptations to Failure," *Psychiatry*, XV (November 1952), 451–63. The concept appears to be relevant to other types of activities where, acting in good faith, the practitioner must disappoint the client and faces the need to "let him down gently." For the application of the cooling-out concept in the junior college see Burton R. Clark, "The 'Cooling-Out' Function in Higher Education," *American Journal of Sociology*, LXV (May 1960), 569–76.

inputs by the organization, (b) the internal arrangements for allocation and coordination of resources and activities, and (c) the political or "institutional" requirements of the organization. The pivotal nature of the output relationship thus claims the attention not only of those in output roles but also of those at the several administrative levels.

SIGNIFICANCE OF SUPERVISION

At the technical level of the organization is found not only the output role but, inevitably, a supervisory role—charged with direct responsibility for evaluating, facilitating, or modifying the behavior of those in output roles vis-à-vis non-members. The foregoing analysis has suggested differences in the role of the supervisor for each type of transaction structure. For the Type I structure, supervisory responsibility revolves around enforcement of conformity to program and detection of deviation. The supervisory focus on the Type II structure is on subordinates' skills in gambits and in closing transactions. For the Type III structure, the supervisory problem centers on the maintenance of heuristics—and preventing the development of *ad hoc* programs. For the supervisor of a Type IV relationship, the concern is with balancing rapport and treatment.

Though the content of the supervisory role differs from one type of structure to another, its significance is common to all; the supervisory role reinforces the organizational dimension in the definition of the output role. Since the non-member also is important in defining the role of member, the stage is set for a three-person game, with the supervisor having to make sure that the coalition always is between himself and the member in the output role, and not between member and non-member, that is, a side transaction.

Even when the coalition includes the supervisor, however, satisfactory disposal of output is not assured, for performance at the output boundary always is subject to the constraints of the transaction structure. If the structure departs from one of the "pure types" discussed above, members at the technical level can only adapt to those impurities, but administrators at other levels of the organization may have the resources necessary to reinforce or "purify" the structures. If it is to the organization's advantage to convert from one type of structure to another, that too must result from administrative decision and action. Finally, if the organization faces a changing environment for which existing transaction structures are inadequate, the necessary adaptation can come only from administrative action.

MIXED STRUCTURES AND MANAGERIAL ACTION

Poetic license was exercised earlier to deal only with the polar extremes of two variables that in reality are continuous. The four pure types yielded by this maneuver may indeed describe the "ideal models" toward which organizations strive—but often only approximate. Impurities may stem from the situation of the non-member or of the member.

Non-members may be *more or less* under compulsion to interact with members, rather than at the mandatory or optional poles. Even where the government has a monopoly on licensing certain activities, the potential non-member may choose to risk illegal activity rather than to negotiate for a license. The member in the output role and his supervisor are powerless to combat this. If the transaction structure is to be reinforced in this case, it requires managerial policies and commitment of resources to find and punish evaders. Prisoners may be unable to effect escape, but the fact that they are *batched*[38] makes rioting possible, and the fact that they and their guards engage in a prolonged series of transactions gives prisoners a certain subtle bargaining power that can place limits on the complete arbitrariness of the prison.[39] If the transaction structure is to be reinforced in this case, it requires managerial policies and commitment of resources to provide for segmentation of prisoners or rotation of guards.

Potential customers may, of course, "do without" or find alternative sources of supply, but there are inconveniences of time and place involved in both alternatives. Potential clients may have a choice of suffering without professional help or seeking out other professionals, which often is difficult for the layman, and there are costs attached to these alternatives also. If the transaction structure is to be purified in these cases where the non-member's direction is limited, policies stressing service or professional norms must be adopted and emphasized by managers.

Impurities can be introduced into the transaction structure from the organization's side also. The programmed member can be furnished with such a large number of programs that he must, in fact, behave heuristically. If this situation is to be purified, it must be through managerial action to reduce the number of programs or, more likely, to departmentalize, with non-members initially screened and routed to the appropriate department. Scarcity of prospects, in the Type II structure, also leads to impurities, for this structure operates on a statistical rather than an absolute basis, that is, a percentage of interactions are expected to abort. When prospects are scarce, members in the output role may be reluctant to permit an abortion, seeking instead to complete a side transaction. If this situation is to be purified, it is likely to be through managerial action to increase the flow of prospects. Advertising, extension of credit, or emphasis on convenience are among the tactics employed.

An overload on the member in the structure calling for heuristics also introduces impurities by limiting the time available to engage in heuristics and encouraging the use of *ad hoc* programs. Here again, if impurities are to be removed, managerial action is required. In the Type IV

---

[38] See Goffman, "On the Characteristics of Total Institutions," *Proceedings of the Symposium on Preventive and Social Psychiatry* (Washington: Walter Reed Army Institute of Research, 1957).

[39] For a discussion of the substitution of functional diffuseness and particularism for functional specificity and universalism under similar conditions, see Peter B. Hammond, "The Functions of Indirection in Communication," in James D. Thompson et al. (eds.), *Comparative Studies in Administration* (Pittsburgh: University of Pittsburgh Press, 1959), pp. 183–94.

structure managerial policies can turn away prospective clients, but in the Type III structure the most likely course is to increase the size of the output staff.

MANAGEMENT OF ORGANIZATIONAL POSTURE

Managerial or administrative influence over transaction structures rests less on authority to dictate standards or procedures than on ability to negotiate changes in *organizational posture,* that is, a relationship between the organization and its relevant environment resulting from the joint action of both.[40]

Hawkes has shown, for the psychiatric hospital case, how the contractual negotiations of the administrator can significantly influence transactions at the boundaries. Both patient input and patient output are affected by the administrator's activities outside the hospital.[41] Levine and White have indicated that administrative negotiations can affect the *domain* of a welfare agency within the larger welfare system, and hence affect what are here termed transaction structures.[42] Economic evidence clearly indicates that administrative negotiation can arrange for monopoly or cartel systems, which in turn have important implications for output relationships.

While most of the illustrations given above were of actions that rectify some imbalance caused either by (*a*) a positive advantage held by the prospective non-member or (*b*) a disadvantage imposed on the member, administrative maneuvers can sometimes result in a posture more favorable to the organization than to the non-member. It can be hypothesized that organizations faced with Types II or IV transaction structures will attempt to reduce the freedom of prospective non-members, thus converting the structures into Types I or III. Achievement of a monopoly, of course, accomplishes this immediately.

While monopoly is illegal for certain types of American organizations, many others are not subject to such constraints. Governments usually operate monopolistic agencies, and public school systems are monopolistic or quasi-monopolistic. Hospitals, clinics, and voluntary social welfare organizations are monopolistic in many non-metropolitan settings. In larger cities they may achieve monopoly or near-monopoly by actions of the referral system, if referring agencies recognize the organization's claim to exclusive domain or jurisdiction for certain types of cases.

Even in the commercial sphere where legal codes prohibit economic monopoly, social-psychological monopoly can be achieved or approached through extension and control over credit purchasing, through brand-

[40] For a discussion of the concept of organizational postures see James D. Thompson, "Organizational Management of Conflict," *Administrative Science Quarterly,* V, No. 4 (March 1960), 389–409.

[41] Robert W. Hawkes, "The Role of the Psychiatric Administrator," *Administrative Science Quarterly,* VI (June 1961), 89–106.

[42] See Sol Levine and Paul E. White, "Exchange as a Conceptual Framework for the Study of Interorganizational Relationships," *Administrative Science Quarterly,* V (March 1961), 583–601.

name advertising, or by the achievement of *extrinsic prestige* by the organization.[43]

CONCLUSION

It is hoped that this paper has called attention to the need for further consideration by organization theorists of the output relationship, by showing that it occurs in transaction structures built in part of elements not within the organization and hence that it cannot be authoritatively dictated.

Finally, it is hoped that a case has been made for the necessity of developing or adapting analytic tools—such as flow charts—for the investigation of contingent interaction in social roles.

---

[43] For the concept of extrinsic prestige and its implications for output relationships see Charles Perrow, "Organizational Prestige: Some Functions and Dysfunctions," *American Journal of Sociology*, LXVI (January 1961), 335–41.

Warren B. Brown

14. Systems, Boundaries, and Information Flow

THE SYSTEMS CONCEPT

WE WILL FIRST briefly establish some concepts and vocabulary about systems since these will provide us with some tools. Although any definition of a particular system is somewhat unique and arbitrary, various definitions of the general systems concept itself will invariably contain many common elements that can be applied to all types of systems.[1]

Some System Characteristics

Here we use the term system to refer to a group or complex of parts (such as people, machines, etc.) interrelated in their actions towards some goals. Typically it is the interrelationships among these system elements, and between them and their environments, that is of most interest in system design both for human organizations and for purely physical systems. Although these interrelationships often are difficult to describe fully and precisely for human organizations, we nonetheless can anticipate that a systems approach will provide a useful framework for examining organizational phenomena, and allow us to fill in whatever detail we find appropriate.

In our attempts to obtain a good description of how a system operates, we shall use a variety of system characteristics. The first three[2] are rather broad but useful; these are: (*a*) flows (of information, materials,

From *Academy of Management Journal,* 1966, Vol. 9, pp. 318–27. Reprinted with permission of the author and the publisher.

[1] Some examples are: Johnson, Kast, and Rosenzweig, *The Theory and Management of Systems* (New York: McGraw-Hill, 1963); S. Beer, *Cybernetics and Management* (New York: John Wiley & Sons, 1959); W. R. Ashby, *Cybernetics* (New York: Wiley Science Editions, 1963); *The Yearbook of the Society for General Systems,* Vols. 1,2,3; G. E. Briggs, "Engineering Systems Approaches to Organizations," *New Perspectives in Organization Research,* ed. Cooper, Leavitt and Shelly (New York: John Wiley & Sons, 1964), pp. 479–92; C. J. Haberstroh, "Organization Design and Systems Analysis," *Handbook of Organizations,* ed. J. G. March (Chicago: Rand McNally, 1965). pp. 1171–1211.

[2] R. C. Hopkins, "A Systematic Approach for System Development," *IRE Trans. on Eng. Mgt.,* EM-8 (June 1961).

money, etc.); (*b*) structure (referring to physical and geographic aspects, organizational design, etc.); and (*c*) procedures (the pre-planned activities which affect the flows and structure). Interest here focuses on complex, formal organizations; where the structure and flows in the organization are purposeful and planned, i.e., rational systems as opposed to self-organizing or natural ones. Organizational change and adaptation, of course, are not ruled out.

Another general system characteristic is that of control. This centers on the prevention and correction of deviations in a system's behavior from those standards which are specified at a given time.

Closed and Open Systems

In a self-regulating system, the key element for a control procedure is feedback. This refers to a circularity of the flows among two or more parts of the system structure such that the output of the system (or subsystem) is maintained at a desired level. Typically this procedure involves a constant monitoring of the system output, a comparison of the output values with the standards, evaluation of any discrepancies, and a flow of information concerning the degree of deviation back to the other elements in the system structure so that the procedures may be changed if necessary.

Example of a closed-loop system with feedback:

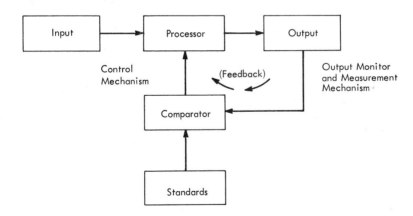

For fully mechanistic systems, with only deterministic elements in the structure and with all control procedures consisting of direct coupling and feedback mechanisms, we can obtain complete automation. Human organizations of course present more variability.

Systems are also characterized as closed or open. A closed system, as indicated in the above sketch, is called information-tight.[3] This means that the information loop is closed, and that the control mechanism has the capability to affect the processor so that the desired output is

---

[3] W. R. Ashby, "General Systems Theory as a New Discipline" in *General Systems Yearbook*, Vol. 3 (1958).

achieved. Commonly used examples are the Watts governor for the control of engine speeds, and the household thermostat for temperature. We should note that information-tight systems may be open in terms of other flows, such as materials and energy, through the input-output processes.

Open systems do not have the information-tight control units which are characteristic of closed systems. Instead, the relations among the elements of the system, and between the system and its environment are often unknown, and the precise causes of system changes may be a mystery. Examples are an individual's behavior, a nation's economy, and a rocket system where there is no further control once the rocket has been fired.

However, even where one does not have complete knowledge or control of the internal workings of a system, it often still is possible to make many inferences about the interrelationships of the system elements. The technique used is called the "Black Box" approach,[4] wherein the system analyst tries to deduce the nature of the internal workings by manipulating the inputs to the system and observing the consequent outputs. For large, complex systems this approach obviously presents many practical difficulties.

Example of an open system:

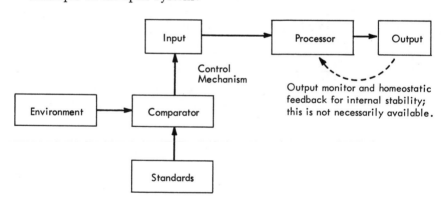

Our interest focuses on human systems that are open to their environments. Such systems are capable of bringing in resources by which they can modify their own internal flows, structures, and procedures.

This behavior prevents the entropic process which is well known in physical and biological systems: if there are no counteracting forces, entropy (a measure of the unavailable energy in a system) increases toward a maximum, and the system elements become more randomized and less differentiated.[5] The result of the entropic process is that the system runs down and achieves an equilibrium at the most probable distribution of the elements.

---

[4] For a good discussion of this concept see Beer, op. cit., Ch. 6.

[5] D. Katz, and R. L. Kahn, *The Social Psychology of Organizations* (New York: John Wiley & Sons, 1966) p. 21.

Open systems counteract this phenomenon by obtaining from their environments more energy and resources than they expend in their operations. They can store resources for possible future crises, and otherwise utilize these additional resources for deliberate system changes, such as adaptation and growth, which are in accord with the system goals.

Open systems also have the characteristic of striving to maintain a dynamic balance (homeostasis).[6] As the environment changes, an open system will receive signals to this effect, and then try to adapt itself to the change in keeping with its character and goals. We note that over time this is a dynamic balance, rather than a static one.

ORGANIZATIONAL BOUNDARIES

Once we have described a system, we necessarily have separated those elements belonging to the system from other elements in its environment. However, the separation is rarely absolute, i.e., some of the elements in the system typically will interact with the system environment. For human organizations the amount of interaction can be thought of in terms of the permeability of the organization's boundaries. This refers to the flow of both people and information across the boundary; these two factors may be interrelated or separate. The degree and kind of permeability will affect the adaptation of the organization to its environment.

Permeability of Boundaries

Some organization boundaries are easily penetrated,[7] for example those of voluntary civic organizations and political groups. With these rather extreme boundary conditions, persons can enter or leave such organizations largely at their own will, and consequently boundary control of membership is minimal. The other extreme occurs when a person has no choice about his membership in an organization, for example when a citizen is drafted by a society into its army. Business firms tend to lie between these extremes, with an individual's hiring and firing often occurring in a semivoluntary manner due to practical complications like union restrictions, seniority rights, and so on.

Since people and information coming into an organization will bring knowledge and values from the external environment, the relative permeability of an organization's boundaries affects the degree to which the organization's members are exposed to such influences. If the boundaries effect a tight screening of both information and individuals so that only those people and pieces of information which reinforce the current internal values are allowed in, then the organization may facilitate internal agreement among its members. However, at the same time it may lose its position in its environment, such as when a firm gets into a rut with its technology and product line. An alert management often can

---

[6] Katz and Kahn, op. cit., pp. 23–4.
[7] Katz and Kahn, op. cit., pp. 59–62, 122–24.

select the degree of boundary permeability which seems to best fit the needs of the organization.

Impact of Environmental Factors

As an aid in understanding the impact of environmental factors on behavior in organizations, Dill[8] has argued that the environment can be conceptualized as the flow of information which crosses the organizational boundaries. This study illustrated how differences in autonomy exercised by various high-level management groups resulted from the different environments that they faced. One interesting factor concerned the reception points for information crossing the boundary. Those who had relatively exclusive access to pertinent inputs were better able to manipulate the information flow to their peers or other organization members, thus gaining a measure of internal control. Such a position can further be used to avoid unsolicited advice which otherwise could be pressed in an informal manner to affect their decisions.

While the focus in Dill's study was on rather high-level decision makers, the activities of many other organization members of course may also involve the crossing of organizational boundaries. For example, persons concerned with the sales and purchasing functions in business firms routinely will deal with outsiders.

Three Levels of Decision-Making

However, before we examine another organizational level in this regard, we will first distinguish among three traditional levels of decision-making in organizations:[9,10] (a) the institutional level for strategic planning, i.e., those concerned with general company objectives and the broad problems of the position of the organization in its environment; (b) the managerial level which focuses on the gathering, coordinating, and allocating of resources for the organization, e.g., planning budgets, formulating personnel practices, and deciding on routine capital expenditures; and, (c) the technical level, involving the acquisition and utilization of technical knowledge for operational controls, e.g., production scheduling, inventory controls, and decisions concerning the measurement of workers' efficiency. Although some descriptions regarding the levels of decision-making are more restrictive, here we will argue that all three basic levels involve boundary-spanning activities.

In a descriptive study of purchasing agents, Strauss[11] found that they

---

[8] W. R. Dill, "Environment as an Influence on Managerial Autonomy," *Administrative Science Quarterly* (March 1958); reprinted in *Comparative Studies in Administration*, ed. J. D. Thompson et al. (Pittsburgh: Univ. of Pittsburgh Press, 1959).

[9] T. Parsons, *Structure and Process in Modern Societies* (New York: Free Press, 1960).

[10] R. N. Anthony, *Planning and Control Systems* (Boston: Division of Research, Harvard Business School, 1965).

[11] G. Strauss, "Tactics of Lateral Relationship: The Purchasing Agent," *Administrative Science Quarterly* (Sept. 1962).

too were concerned with controlling information flows across organizational boundaries in order to enhance their influence. He found that many of their tactics involved influence based on information exchanges, though of course other techniques also were used, such as an imaginative interpretation of the rules, and the involvement of allies both within and outside the organizational boundaries to bolster their positions.

In a broader analysis, Thompson[12] has examined face-to-face interactions across organizational boundaries and set them into four types of transaction structures, involving two dichotomized dimensions: (a) the specificity of organizational control over members; and, (b) the degree of non-member discretion.

The first of these dimensions is of particular interest; it refers to the degree to which the behavior of an organization member at the boundary is pre-set by standard operating procedures. At one extreme the member performs highly routine (or programmed) activities, while at the other end of this dimension a member's activities are guided only by general guidelines or heuristics. The programmed positions involve routine flows of information, e.g., data concerning current deliveries of raw materials, while the non-programmed activities often involve special information searches. Within a business organization the less programmed activities generally are associated with higher positions in the authority hierarchy, i.e., with those who set corporate policies and long-range strategies, though this is not always true.

INFORMATION FLOWS

The manner in which an organization determines the kind and amount of information allowed to cross its boundaries can be thought of as a coding process. This process refers to both mechanical and human transmission of information, and it reflects the norms, values, and subculture of the organization. Those persons who have accepted membership in an organization presumably accept the coding process—indeed they are part of it—and consequently they will perceive and interpret various pieces of information differently from non-members. While many factors affect one's perceptions,[13] we simply note that a coding process can (and often does) lead to distortions, elaborations, and other transformations of the information. Further, such effects occur not only when the information crosses an organization's boundaries, but also when it is passed among various members with differing organizational roles.

Adjustment Processes for Overload

It is not only the coding process that can lead to problems in information transmission across organization boundaries, but also the adjustment processes that an organization can evoke when its information system is overloaded. This condition occurs when a system which is ra-

---

[12] J. D. Thompson, "Organizations and Output Transactions," *American Journal of Sociology*, No. 3 (Nov. 1962).

[13] S. S. Zalkind and T. W. Costello, "Perception: Some Recent Research and Implication for Administration," *Administrative Science Quarterly* (Sept. 1962).

tionally designed to operate effectively under certain fairly constant conditions suddenly is hit by changes in its inputs. Such unanticipated changes tend to overload one or more parts of the total system.

Organizational responses to such overload conditions are quite varied. Miller[14] has classified them into seven categories: (1) omission, simply not processing the information; (2) error, processing information correctly and not making necessary adjustments; (2) queuing, delaying responses during peak demand periods and catching up during lulls; (4) filtering, systematic omission of certain categories of information according to some priority scheme; (5) approximation, less accurate responses because there is no time to be precise; (6) multiple channels, using parallel transmission subsystems that can do comparable tasks at the same time as in the special case of decentralization; and, (7) escape, leaving a situation entirely or effectively cutting off the flow of information.

A similar listing of adjustment processes has been noted by Meier,[15] based on his study of an undergraduate university library during a period when there was an increasing demand for books. Briefly, his listing includes: queuing; priorities in queues; destruction of low priority inputs (filtering); omission; reduction of processing standards (approximation); decentralization; formation of independent organizations near the periphery (multiple channels); mobile reserves (multiple channels); rethinking procedures; redefining the boundaries of the system, e.g., instituting customer self-service; escape; retreat to formal, ritualistic behavior; and lastly, the possibility of dissolving the larger system with salvage of its assets into several new, smaller units.

As science and technology constantly change the environments of many organizations, this problem of information overloads and changing inputs will probably become more common. However, we must note that in the lists above, the responses vary considerably in long-range usefulness. Some (e.g., the error response) are simply very temporary and often costly adaptations. Other responses, such as decentralization, may involve the deliberate restructuring of the organization on a long-range basis. Similarly, queuing and filtering can become a deliberate feature of the organizational coding process which can be routinely evoked whenever an overload condition seems imminent.

Choice of Appropriate Response

The particular selection of one or more of these responses depends at least in part on the perceived costs to the organization. The costs will involve factors such as the availability of necessary or desirable resources; time constraints; survival threats; and the like. These costs will be weighed against the amount of resources stored or readily available (the

---

[14] J. C. Miller, "Information Input, Overload, and Psychopathology," *American Journal of Psychiatry* (116), 695–704. Cited in Katz and Kahn, op. cit., p. 231.

[15] R. L. Meier, "Information Input Overload: Features of Growth in Communications—Oriented Institutions," *Libri*, Vol. 13, No. 1, pp. 1–44.

negative entropy in the system). Then the rational organization will strive for a response suitable to the organization's objectives, and which also maintains its dynamic balance.

A dynamic system also must have mechanisms for obtaining from the environment the various kinds of information it needs to maintain both its internal operations and its position of dynamic balance within its environment. Commenting on organization controls, Roberts[16] has stated, "The boundaries of a management control system design study must not be drawn to conform with organizational structure merely because of that structure. System boundaries cannot ignore vital feedback channels of information and action if the system is to be effective." Consequently, in terms of our earlier comments, the organization's boundaries deliberately must be made permeable to facilitate the flow of a variety of information into the organization.

Levels and Kinds of Information

Although there are several ways of classifying information needed by organizations,[17] we use here a scheme of separating the information according to the level of decision-making for which it is needed, i.e., strategic, managerial, and technical.

Strategic information refers to such knowledge as environmental trends concerning political, economic, and social factors, and the impact of the organization on its environment. Managerial information is typically used to focus on an evaluation of the organization's activities, assessing the effectiveness of the institution's past performance, and programs against other similar—possibly competing—organizations. Technical or operational information is primarily obtained for use with organizational control functions, and consequently it becomes the critical feedback needed for small closed-loop subsystems within the total organization. If this activity is highly programmed, then there may be little new information flow across the organization boundary. However, even seemingly specific functions at the organization's boundary can be greatly modified by the role occupants, thus stimulating additional information flow as we noted earlier in the example of the purchasing function.

Related Intelligence Mechanisms

The need for these three basic kinds of information necessitates a corresponding need for three intelligence structures whereby an organization can regularly search its environment for new information. Each intelligence structure must also provide an appropriate coding procedure, since most information channels are not multi-purpose. For example,

---

[16] E. B. Roberts, "Industrial Dynamics and the Design of Management Control Systems," *Management Controls,* ed. Bonini, Jaedicki, and Wagner (New York: McGraw-Hill, 1964), p. 124.

[17] For example, see D. R. Daniel, "Management Information Crisis," *Harvard Business Review* (Sept.–Oct. 1961).

an information channel constructed to handle technical data will not be useful for transmitting information about new companies entering a firm's traditional markets, or community relations, or about new tax laws affecting business which are proposed in the state legislature.

We shall call these three types of intelligence mechanisms filters, screens, and sieves. In general, the filters are to provide the technical information from the environment, the screens will bring in managerial level information, and the sieves should provide information for the strategy or policy-making level. We must note in passing that although such information nets can be made virtually identical with an organization's authority structure, for most complex organizations specialized structures are evolved for both the authority and information flows.[18]

Each of these intelligence mechanisms must be allocated enough organizational resources to carry out its own function. If the intelligence function is assigned as a secondary mission to a subpart of the organization, it is not surprising if much relevant information never gets into the appropriate channel. For example, if a market research group is assigned the additional responsibility of giving to management an overall evaluation of the firm's environment, it is not likely to provide an adequate substitute for the sieve function mentioned above. Such a group may well generate a thorough evaluation of consumer reactions to the firm's current products, but it probably will not have the resources or expertise to evaluate such things as the impact of new technical developments upon the organization's formal and informal structures.

Proposed Information Agency

In addition to the three kinds of intelligence mechanisms, we propose the creation of an information agency which would formally be placed in the information flow pattern. This group, providing a kind of clearing house for information flows, would have representatives from all major divisions of the organization. It would have as its primary responsibility the integration, evaluation, and comparison of the various kinds of information coming into the organization.

In discharging this responsibility, the agency will have its main impact in its recommendations to the policy level of the firm, though it would not be restricted to communications at that level. In fact, it is important that it have access to all three basic levels of information flow and decision-making if it is to carry out its mission effectively. While processing the selected parts of the total information inputs, this agency also can provide an antibias function by making adjustments in its evaluations based on the perceived distortions existing at the various levels in the coding processes.

Further, such an information agency would aid in providing standards concerning the operations of the filter, screen, and sieve functions, e.g., stating the desired specificity of an individual's behavior when it in-

---

[18] H. Guetzkow, "Communications in Organizations," *Handbook of Organizations*, ed. J. G. March (Chicago: Rand McNally, 1965).

volves boundary-spanning activities. When overload problems occur, the type of action taken by the organization would also be determined by this group. In addition, this agency would be intimately concerned with the kinds of information that the intelligence mechanisms are generating; if some types of information for evaluations appear to be lacking, this agency could suggest an appropriate change in the search and coding processes for that level. Conceivably it could have the formal authority to enforce such changes, though basically it serves in an advisory capacity.

In the power structure of the organization we suggest that this group be placed between the policy and managerial levels. It is not meant to replace decision-making at these levels by the regular persons involved, but to aid their efforts and thus provide a vital link for an open, human system in its efforts to maintain a dynamic balance vis-à-vis its environment.

Some selected aspects of information flows:

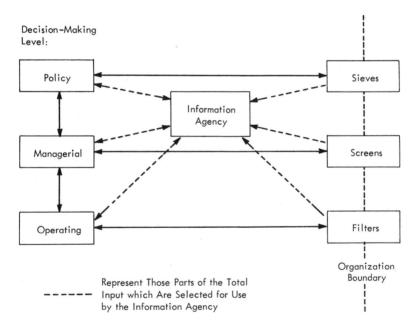

Decision-Making Level:

Policy — Sieves

Information Agency

Managerial — Screens

Operating — Filters

Organization Boundary

– – – – – Represent Those Parts of the Total Input which Are Selected for Use by the Information Agency

CLOSING COMMENT

This discussion was designed to indicate the usefulness of a systems framework as an aid in analyzing at least some types of organizational problems. Specifically, we focused on problems related to the flow of information across organizational boundaries, where the boundaries involve both human and physical variables that buffer the organization's interaction with its environment. After noting several types of problems we evolved a general, three-level scheme for bringing various kinds of information into the organization, based on a classification of decision-making within the firm. In addition, we proposed an internal information agency, and explained its functions.

As a final note we want to add that for specific companies and specific formats of information flow, there are good possibilities of using a computer to simulate these organizational properties. In this manner a company could test out in advance the impact of proposed changes in its information channels, as well as testing the efficiency of any particular pattern over a range of possible demands on the system.

Section IV

INTRAORGANIZATIONAL SYSTEMS

How DOES VIEWING the organization as an open system affect one's approach to intraorganizational processes? The aim of this section is to give the reader some understanding of systems approaches to internal organizational structures and dynamics.

In Britain, workers at the Tavistock Institute of Human Relations have been very active in developing system models of complex organizations. The first two papers in this section describe an approach developed by the Tavistock group in their studies of different types of industrial organizations which views the enterprise as an open sociotechnical system. The "sociotechnical" approach emphasizes the importance of considering not only the social and psychological components of an organization, but the technological system as well. The focus here is on the way in which the technological system interacts with the social system to affect the ability of the internal organization to set tasks and accommodate to environmental stresses. Emery and Trist outline the method of sociotechnical systems analysis using examples from the "longwall" coal-mining investigation, the action research which Trist and others used as a basis for their development of models of the organization as an open system.

Eric J. Miller defines the three dimensions of territory, technology, and time as intrinsic to the structure of the task of any production system. He uses examples from complex industrial production to organizations to illustrate the way in which the subsystems of a complex organizational system are differentiated along these three dimensions.

In relating to complex, rapidly changing environments, organizations experience particular internal problems in establishing effective internal linkage mechanisms. Lynton describes four different ways that decision makers of a system assess environmental requirements for change, and discusses the results of the particular kind of assessment made on system differentiation and linkage. The linkage mechanisms adopted have associated costs and differ in the way that they interact on the total system. He asserts that in a turbulent environment, a system must develop subsystems and also linkage mechanisms which are clearly differentiated yet are supported by integration into the system as a whole. If systems support is inadequate, linkage problems tend to provoke a multiplication of innovative subsystems.

In the next paper, Rosen focuses on an upholstering department as a subsystem within a large furniture manufacturing organization. He reports a field experimental study conducted with this organization which tested and found support for the three explicit hypotheses derived from the application of open systems theory to formal organizations. He finds evidence that a stable system or subsystem receiving negative feedback from its environment, along with a substantial manipulation of a central internal system variable, experiences disequilibrium and attempts corrective mobilization of energy. Rosen's findings also show that the system is assisted in adjusting itself and establishing a new equilibrium by internal system feedback mechanisms. The system-component relationships, after temporary disruption, then tend to restabilize in a form closely resembling its initial equilibrium state.

Obviously, key components or subsystems of an organizational system are comprised of individuals and groups and their behaviors. In the spirit of the effort by general systems theorists to find analogies and isomorphisms which apply across different levels of living systems, the last paper in this section attempts to apply system theory developed at the organization level to processes at the individual and group levels. This paper is one of the last published of the late A. K. Rice, a key member of the Tavistock research group. Rice (1958) followed up the studies conducted by Trist and other Tavistock researchers in the British mining industry with action research in the textile mills of India. In this paper he summarizes his version of a theory of the organization as an open system and then discusses the concepts of boundary control functions and task systems derived from analysis of organizational enterprises as they apply to the individual, the group, and the intergroup.

F. E. Emery
E. L. Trist

15. Socio-Technical Systems

THE ANALYSIS of the characteristics of enterprises as systems would appear to have strategic significance for furthering our understanding of a great number of specific industrial problems. The more we know about these systems the more we are able to identify what is relevant to a particular problem and to detect problems that tend to be missed by the conventional framework of problem analysis.

The value of studying enterprises as systems has been demonstrated in the empirical studies of Blau (4), Gouldner (6), Jaques (8), Selznick (15) and Lloyd Warner (21). Many of these studies have been informed by a broadly conceived concept of bureaucracy, derived from Weber and influenced by Parsons and Merton:

They have found their main business to be in the analysis of a specific bureaucracy as a complex social system, concerned less with the individual differences of the actors than with the situationally shaped roles they perform. (6)

Granted the importance of system analysis there remains the important question of whether an enterprise should be construed as a 'closed' or an 'open system,' i.e., relatively 'closed' or 'open' with respect to its external environment. Von Bertalanffy (3) first introduced this general distinction in contrasting biological and physical phenomena. In the realm of social theory, however, there has been something of a tendency to continue thinking in terms of a 'closed' system, that is, to regard the enterprise as sufficiently independent to allow most of its problems to be analysed with reference to its internal structure and without reference to its external environment. Early exceptions were Rice and Trist (11) in the field of labour turnover and Herbst (7) in the analysis of social flow systems. As a first step, closed system thinking has been fruitful, in psychology and industrial sociology, in directing attention to the existence of structural similarities, relational determina-

From *Management Sciences: Models and Techniques*, vol. II, 1960, edited by C. W. West, Churchman and M. Verhulst. Reprinted with permission of the authors and Pergamon Press, Ltd.

tion and subordination of part to whole. However, it has tended to be misleading on problems of growth and the conditions for maintaining a 'steady state.' The formal physical models of 'closed systems' postulate that, as in the second law of thermo-dynamics, the inherent tendency of such systems is to grow toward maximum homogeneity of the parts and that a steady state can only be achieved by the cessation of all activity. In practice the system theorists in social science (and these include such key anthropologists as Radcliffe-Brown) refused to recognize these implications but instead, by the same token, did "*tend* to focus on the statics of social structure and to neglect the study of structural change" (10). In an attempt to overcome this bias, Merton suggested that "the concept of dysfunction, which implies the concept of strain, stress and tension on the structural level, provides an analytical approach to the study of dynamics and change" (10). This concept has been widely accepted by system theorists but while it draws attention to sources of imbalance within an organization it does not conceptually reflect the mutual permeation of an organization and its environment that is the cause of such imbalance. It still retains the limiting perspectives of 'closed system' theorizing. In the administrative field the same limitations may be seen in the otherwise invaluable contributions of Barnard (2) and related writers.

The alternative conception of 'open systems' carries the logical implications that such systems may spontaneously reorganize toward states of greater heterogeneity and complexity and that they achieve a 'steady state' at a level where they can still do work. Enterprises appear to possess at least these characteristics of 'open systems.' They grow by processes of internal elaboration (7) and manage to achieve a steady state while doing work, i.e., achieve a quasi-stationary equilibrium in which the enterprise as a whole remains constant, with a continuous 'throughput,' despite a considerable range of external changes (9, 11).

The appropriateness of the concept of 'open system' can be settled, however, only by examining in some detail what is involved in an enterprise achieving a steady state. The continued existence of any enterprise presupposes some regular commerce in products or services with other enterprises, institutions and persons in its external social environment. If it is going to be useful to speak of steady states in an enterprise, they must be states in which this commerce is going on. The conditions for regularizing this commerce lie both within and without the enterprise. On the one hand this presupposes that an enterprise has at its immediate disposal the necessary material supports for its activities—a workplace, materials, tools and machines—and a work force able and willing to make the necessary modifications in the material 'throughput' or provide the requisite services. It must also be able, efficiently, to utilize its material supports and to organize the actions of its human agents in a rational and predictable manner. On the other hand, the regularity of commerce with the environment may be influenced by a broad range of independent external changes affecting markets for products and inputs of labour, materials and technology. If we examine the factors influencing the ability of an enterprise to maintain a steady state in the face of these broader environmental influences we find that:

a) The variation in the output markets that can be tolerated without structural change is a function of the flexibility of the technical productive apparatus—its ability to vary its rate, its end product or the mixture of its products. Variation in the output markets may itself be considerably reduced by the display of distinctive competence. Thus the output markets will be more attached to a given enterprise if it has, relative to other producers, a distinctive competence—a distinctive ability to deliver the right product to the right place at the right time.

b) The tolerable variation in the 'input' markets is likewise dependent upon the technological component. Thus some enterprises are enabled by their particular technical organization to tolerate considerable variation in the type and amount of labour they can recruit. Others can tolerate little.

The two significant features of this state of affairs are:—

a) That there is no simple one-to-one relation between variations in inputs and outputs. Depending upon the technological system, different combinations of inputs and different 'product mixes' may be produced from similar inputs. As far as possible an enterprise will tend to do these things rather than make structural changes in its organization. It is one of the additional characteristics of 'open systems' that while they are in constant commerce with the environment they are selective and, within limits, self-regulating.

b) That the technological component, in converting inputs into outputs, plays a major role in determining the self-regulating properties of an enterprise. It functions as one of the major boundary conditions of the social system of the enterprise in thus mediating between the ends of an enterprise and the external environment. Because of this the materials, machines and territory that go to making up the technological component are usually defined, in any modern society, as 'belonging' to an enterprise and excluded from similar control by other enterprises. They represent, as it were, an 'internalized environment.'

Thus the mediating boundary conditions must be represented amongst "the open system constants" (3) that define the conditions under which a steady state can be achieved. The technological component has been found to play a key mediating role and hence it follows that the open system concept must be referred to the sociotechnical system, not simply to the social system of an enterprise.

It might be justifiable to exclude the technological component from the system concept if it were true, as many writers imply, that it plays only a passive and intermittent role. However, it cannot be dismissed as simply a set of limits that exert an influence at the initial stage of building an enterprise and only at such subsequent times as these limits are overstepped. There is, on the contrary, an almost constant accommodation of stresses arising from changes in the external environment; the technological component not only sets limits upon what can be done, but also in the process of accommodation creates demands that must be reflected in the internal organization and ends of an enterprise.

Study of a productive system therefore requires detailed attention to both the technological and the social components. It is not possible to understand these systems in terms of some arbitrarily selected single aspect of the technology such as the repetitive nature of the work, the coerciveness of the assembly conveyor or the piecemeal nature of the task. However, this is what is usually attempted by students of the enterprise. In fact:

It has been fashionable of late, particularly in the 'human relations' school, to assume that the actual job, its technology, and its mechanical and physical requirements are relatively unimportant compared to the social and psychological situation of men at work. (5)

Even when there has been a detailed study of the technology this has not been systematically related to the social system but been treated as background information. (21)

In our earliest study of production systems in coal mining it became apparent that "So close is the relationship between the various aspects that the social and the psychological can be understood only in terms of the detailed engineering facts and of the way the technological system as a whole behaves in the environment of the underground situation." (19)

An analysis of a technological system in these terms can produce a systematic picture of the tasks and task interrelations required by a technological system. However, between these requirements and the social system there is not a strictly determined one-to-one relation but what is logically referred to as a correlative relation.

In a very simple operation such as manually moving and stacking railway sleepers ('ties') there may well be only a single suitable work relationship structure, namely, a cooperating pair with each man taking an end of the sleeper and lifting, supporting, walking and throwing in close coordination with the other man. The ordinary production process is much more complex and there it is unusual to find that only one particular work relationship structure can be fitted to these tasks.

This element of choice and the mutual influence of technology and the social system may both be illustrated from our studies, made over several years, of work organization in British deep seam coal mining. The following data are adapted from Trist and Murray. (20)

Thus Table I indicates the main features of two very different forms of organization that have both been operated economically within the same seam and with identical technology.

The conventional system combines a complex formal structure with simple work roles: the composite system combines a simple formal

TABLE 1

Same Technology, Same Coalseam, Different Social Systems

	A Conventional Cutting Long-Wall Mining System.	A Composite Cutting Long-Wall Mining System.
Number of men	41	41
No. of completely segregated task groups	14	1
Mean job variation for members:		
Task groups worked with	1.0	5.5
Main tasks worked	1.0	3.6
Different shifts worked	2.0	2.9

structure with complex work roles. In the former the miner has a commitment to only a single part task and enters into only a very limited number of unvarying social relations that are sharply divided between those within his particular task group and those who are outside. With those 'outside' he shares no sense of belongingness and he recognizes no responsibility to them for the consequences of his actions. In the composite system the miner has a commitment to the whole group task and consequently finds himself drawn into a variety of tasks in cooperation with different members of the total group; he may be drawn into any task on the coal-face with any member of the total group.

That two such contrasting social systems can effectively operate the same technology is clear enough evidence that there exists an element of choice in designing a work organization.

However, it is not a matter of indifference which form of organization is selected. As has already been stated, the technological system sets certain requirements of its social system and the effectiveness of the total production system will depend upon the adequacy with which the social system is able to cope with these requirements. Although alternative social systems may survive in that they are both accepted as "good enough" (17), this does not preclude the possibility that they may differ in effectiveness.

In this case the composite systems consistently showed a superiority over the conventional in terms of production and costs.

This superiority reflects, in the first instance, the more adequate coping in the composite system with the task requirements. The constantly changing underground conditions require that the already complex sequence of mining tasks undergo frequent changes in the relative magnitudes and even the order of these tasks. These conditions optimally require the internal flexibility possessed in varying degrees by the composite systems. It is difficult to meet variable task requirements with any organization built on a rigid division of labour. The only justification for a rigid division of labour is a technology which demands specialized non-substitute skills and which is, moreover, sufficiently superior, as a technology, to offset the losses due to rigidity. The conventional longwall

TABLE 2

Production and Costs for Different Forms of Work Organization with Same Technology

	Conventional	*Composite*
Productive achievement*	78	95
Ancilliary work at face (hrs. per man-shift)	1.32	0.03
Average reinforcement of labour (percent of total face force)	6	—
Percent of shifts with cycle lag	69	5
No. consecutive weeks without losing a cycle	12	65

    ° Average percent of coal won from each daily cut, corrected for differences in seam transport.

cutting system has no such technical superiority over the composite to offset its relative rigidity—its characteristic inability to cope with changing conditions other than by increasing the stress placed on its members, sacrificing smooth cycle progress or drawing heavily upon the negligible labour reserves of the pit.

The superiority of the composite system does not rest alone in more adequate coping with the tasks. It also makes better provision to the personal requirements of the miners. Mutually supportive relations between task groups are the exception in the conventional system and the rule in the composite. In consequence, the convenional miner more frequently finds himself without support from his fellows when the strain or size of his task requires it. Crises are more likely to set him against his fellows and hence worsen the situation.

Similarly, the distribution of rewards and statuses in the conventional system reflects the relative bargaining power of different roles and task groups as much as any true differences in skill and effort. Under these conditions of disparity between effort and reward any demands for increased effort are likely to create undue stress.

The following table indicates the difference in stress experienced by miners in the two systems.

These findings were replicated by experimental studies in textile mills in the radically different setting of Ahmedabad, India. (12)

TABLE 3

Stress Indices for Different Social Systems

	Conventional	*Composite*
Absenteeism (Percent of possible shifts)		
Without reason	4.3	0.4
Sickness or other	8.9	4.6
Accidents	6.8	3.2
Total:	20.0	8.2

However, two possible sources of misunderstanding need to be considered:

1. Our findings do not suggest that work group autonomy should be maximized in all productive settings. There is an optimum level of grouping which can be determined only by analysis of the requirements of the technological system. Neither does there appear to be any simple relation between level of mechanization and level of grouping. In one mining study we found that in moving from a hand-filling to a machine-filling technology, the appropriate organization shifted from an undifferentiated composite system to one based on a number of partially segregated task groups with more stable differences in internal statuses.

2. Nor does it appear that the basic psychological needs being met by grouping are workers' needs for friendship on the job, as is frequently postulated by advocates of better 'human relations' in industry. Grouping produces its main psychological effects when it leads to a system of work roles such that the workers are primarily related to each other by way of the requirements of task performance and task interdependence. When this task orientation is established the worker should find that he has an adequate range of mutually supportive roles (mutually supportive with respect to performance and to carrying stress that arises from the task). As the role system becomes more mature and integrated, it becomes easier for a worker to understand and appreciate his relation to the group. Thus in the comparison of different composite mining groups it was found that the differences in productivity and in coping with stress were not primarily related to differences in the level of friendship in the groups. The critical prerequisites for a composite system are an adequate supply of the required special skills among members of the group and conditions for developing an appropriate system of roles. Where these prerequisites have not been fully met, the composite system has broken down or established itself at a less than optimum level. The development of friendship and particularly of mutual respect occurs in the composite systems, but the friendship tends to be limited by the requirements of the system and not to assume unlimited disruptive forms such as were observed in conventional systems and were reported by Adams (1) to occur in certain types of bomber crews.

The textile studies (12) yielded the additional finding that *supervisory roles* are best designed on the basis of the same type of socio-technical analysis. It is not enough simply to allocate to the supervisor a list of responsibilities for specific tasks and perhaps insist upon a particular style of handling men. The supervisory roles arise from the need to control and coordinate an incomplete system of men-task relations. Supervisory responsibility for the specific parts of such a system is not easily reconcilable with responsibility for overall aspects. The supervisor who continually intervenes to do some part of the productive work may be proving his willingness to work but is also likely to be neglecting his main task of controlling and coordinating the system so that the operators are able to get on with their jobs with the least possible disturbance.

Definition of a supervisory role presupposes analysis of the system's requirements for control and coordination and provision of conditions that will enable the supervisor readily to perceive what is needed of him and to take appropriate measures. As his control will in large measure rest on his control of the boundary conditions of the system—those activities relating to a larger system—it will be desirable to create 'unified commands' so that the boundary conditions will be correspondingly easy to detect and manage. If the unified commands correspond to natural task groupings, it will also be possible to maximize the autonomous responsibility of the work group for internal control and coordination, thus freeing the supervisor for his primary task. A graphic illustration of the differences in a supervisory role following a socio-technical reorganization of an automatic loom shed (12) can be seen in the following two figures: Figure 1 representing the situation before and Figure 2 representing the situation after change.

This reorganization was reflected in a significant and sustained improvement in mean percentage efficiency and a decrease in mean percentage damage.

FIGURE 1

Management Hierarchy before Change

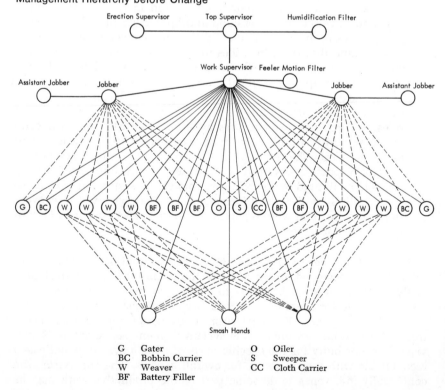

G	Gater	O	Oiler
BC	Bobbin Carrier	S	Sweeper
W	Weaver	CC	Cloth Carrier
BF	Battery Filler		

FIGURE 2

Management Hierarchy after Change

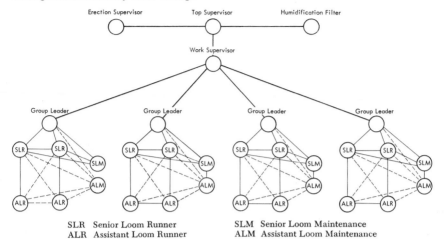

SLR Senior Loom Runner   SLM Senior Loom Maintenance
ALR Assistant Loom Runner   ALM Assistant Loom Maintenance

The significance of the difference between these two organizational diagrams does not rest only in the relative simplicity of the latter (although this does reflect less confusion of responsibilities) but also in the emergence of clearly distinct areas of command which contain within themselves a relatively independent set of work roles together with the skills necessary to govern their task boundaries. In like manner the induction and training of new members was recognized as a boundary condition for the entire shed and located directly under shed management instead of being scattered throughout subordinate commands. Whereas the former organization had been maintained in a steady state only by the constant and arduous efforts of management, the new one proved to be inherently stable and self-correcting, and consequently freed management to give more time to their primary task and also to manage a third shift.

Similarly, the primary task in managing the enterprise as a whole is to relate the total system to its environment and is not in internal regulation per se. This does not mean that managers will not be involved in internal problems but that such involvement will be oriented consciously or unconsciously to certain assumptions about the external relations of the enterprise.

This contrasts with the common postulate of the structural-functional theories that "the basic need of all empirical systems is the maintenance of the integrity and continuity of the system itself." (14) It contrasts also with an important implication of this postulate, namely, that the primary task of management is "continuous attention to the possibilities of encroachment and to the forestalling of threatened aggressions or deleterious consequences from the actions of others." (14)

In industry this represents the special and limiting case of a management that takes for granted a previously established definition of its primary task and assumes that all they have to do, or can do, is sit tight and defend their market position. This is, however, the common case in statutorily established bodies and it is on such bodies that recent studies of bureaucracy have been largely carried out.

In general the leadership of an enterprise must be willing to break down an old integrity or create profound discontinuity if such steps are required to take advantage of changes in technology and markets. The very survival of an enterprise may be threatened by its inability to face up to such demands, as for instance, switching the main effort from production of processed goods to marketing or from production of heavy industrial goods to consumer goods. Similarly, the leadership may need to pay 'continuous' attention to the possibilities of making their own encroachments rather than be obsessed with the possible encroachments of others.

Considering enterprises as 'open socio-technical systems' helps to provide a more realistic picture of how they are both influenced by and able to act back on their environment. It points in particular to the various ways in which enterprises are enabled by their structural and functional characteristics ('system constants') to cope with the 'lacks' and 'gluts' in their available environment. Unlike mechanical and other inanimate systems they possess the property of 'equifinality'; they may achieve a steady state from differing initial conditions and in differing ways. (3) Thus in coping by internal changes they are not limited to simple quantitative change and increased uniformity but may, and usually do, elaborate new structures and take on new functions. The cumulative effect of coping mainly by *internal* elaboration and differentiation is generally to make the system independent of an increasing range of the predictable fluctuations in its supplies and outlets. At the same time, however, this process ties down in specific ways more and more of its capital, skill and energies and renders it less able to cope with newly emergent and unpredicted changes that challenge the primary ends of the enterprise. This process has been traced out in a great many empirical studies of bureaucracies. (4, 10, 15)

However, there are available to an enterprise other aggressive strategies that seek to achieve a steady state by transforming the environment. Thus an enterprise has some possibilities for moving into new markets or inducing changes in the old; for choosing differently from amongst the range of personnel, resources and technologies offered by its environment or training and making new ones; and for developing new consumer needs or stimulating old ones.

Thus, arising from the nature of the enterprise as an open system, management is concerned with 'managing' both an internal system and an external environment. To regard an enterprise as a closed system and concentrate upon management of the 'internal enterprise' would be to expose the enterprise to the full impact of the vagaries of the environment.

If management is to control internal growth and development it must in the first instance control the 'boundary conditions'—the forms of exchange between the enterprise and its environment. As we have seen, most enterprises are confronted with a multitude of actual and possible exchanges. If resources are not to be dissipated the management must select from the alternatives a course of action. The causal texture of competitive environments is such that it is extremely difficult to survive on a simple strategy of selecting the best from among the alternatives immediately offering. Some that offer immediate gain lead nowhere, others lead to greater loss; some alternatives that offer loss are avoidable, others are unavoidable if long run gains are to be made. The relative size of the immediate loss or gain is no sure guide as to what follows. Since also the actions of an enterprise can improve the alternatives that are presented to it, the optimum course is more likely to rest in selecting a strategic objective to be achieved in the long run. The strategic objective should be to place the enterprise in a position in its environment where it has some assured conditions for growth—unlike war, the best position is not necessarily that of unchallenged monopoly. Achieving this position would be the *primary task* or overriding mission of the enterprise.

In selecting the primary task of an enterprise, it needs to be borne in mind that the relations with the environment may vary with (*a*) the productive efforts of the enterprise in meeting environmental requirements; (*b*) changes in the environment that may be induced by the enterprise and (*c*) changes independently taking place in the environment. These will be of differing importance for different enterprises and for the same enterprises at different times. Managerial control will usually be greatest if the primary task can be based on productive activity. If this is not possible, as in commerce, the primary-task will give more control if it is based on marketing than simply on foreknowledge of the independent environmental changes. Managerial control will be further enhanced if the primary task, at whatever level it is selected, is such as to enable the enterprise to achieve *vis-à-vis* its competitors, a *distinctive competence.* Conversely, in our experience, an enterprise which has long occupied a favoured position because of distinctive productive competence may have grave difficulty in recognizing when it is losing control owing to environmental changes beyond its control.

As Selznick has pointed out, (16) an appropriately defined primary task offers stability and direction to an enterprise, protecting it from adventurism or costly drifting. These advantages, however, as he illustrates (16), may be no more than potential unless the top management group of the organization achieves solidarity about the new primary task. If the vision of the task is locked up in a single man or is the subject of dissension in top management, it will be subject to great risk of distortion and susceptible to violent fluctuations. Similarly, the enterprise as a whole needs to be reoriented and reintegrated about this primary task. Thus, if the primary task shifts from heavy industrial goods to durable consumer goods, it would be necessary to ensure that there is a corresponding shift in values that are embodied in such sections as the sales force and design department.

REFERENCES

1. Adams, S. 'Status Congruency as a Variable in Small Group Performance.' *Social Forces*, 1953, 32, 16–22.

2. Barnard, C. I. *The Functions of the Executive.* Cambridge, Mass.: Harvard University Press, 1948.

3. Bertalanffy, L. v. 'The Theory of Open Systems in Physics and Biology.' *Science*, 1950, *111*, 23–29.

4. Blau, P. *The Dynamics of Bureaucratic Structure: A Study of Interpersonal Relations in Two Government Agencies.* Chicago, University of Chicago Press, 1955.

5. Drucker, P. F. 'The Employee Society.' *American Sociological Review*, 1952, *58*, 358–63.

6. Gouldner, A. W. *Patterns of Industrial Bureaucracy.* London: Routledge, Kegan Paul, 1955.

7. Herbst, P. G. 'The Analysis of Social Flow Systems.' *Human Relations*, 1954, 7, 327–36.

8. Jaques, E. *The Changing Culture of a Factory.* London: Tavistock, 1951.

9. K. Lewin. *Field Theory in Social Science*, Harper, 1951.

10. R. K. Merton. *Social Theory and Social Structure*, Free Press, 1949.

11. A. K. Rice and E. L. Trist. 'Institutional and sub-institutional determinants of change in labour turnover (The Glacier Project-VIII).' *Human Relations*, vol. 5 (1952), 347–72.

12. A. K. Rice. *Productivity and Social Organization: The Ahmedabad Experiment*, Tavistock, 1958.

13. M. P. Schutzenberger. 'A tentative classification of goal-seeking behaviors.' *Journal of Mental Science*, vol. 100 (1954) 97–102.

14. P. Selznick. 'Foundations of the theory of organization.' *American Sociological Review*, vol. 13 (1848), 25–35.

15. P. Selznick, *TVA and the Grass Roots*, University of California Press, 1949.

16. P. Selznick. *Leadership in Administration.* Row Peterson, 1957.

17. H. A. Simon. *Models of Man.* Wiley, 1957.

18. E. C. Tolman and E. Brunswik. 'The organism and the causal texture of the environment', *Psychological Review*, vol. 42 (1935), 43–77.

19. E. L. Trist and K. W. Bamforth. 'Some social and psychological consequences of the longwall method of coal-getting.' *Human Relations*, vol. 4 (1951).

20. E. L. Trist and H. Murray, 'Work organization at the coal face: A comparative study of mining systems.' *Tavistock Institute of Human Relations, Doc.*, no. 506 (1948).

21. W. L. Warner and J. O. Low. *The Social System of the Modern Factory*, Yale University Press, 1947.

Eric J. Miller

16. Technology, Territory, and Time: The Internal Differentiation of Complex Production Systems[1]

INTRODUCTION

THE CONCEPT OF a production system as a socio-technical system was introduced by Trist (13), was further developed by Rice in his recent book (10), and has now been reviewed by Emery (1). Of particular relevance here is Rice's concept of operating and managing systems,[2] summarized below:

> Any production system may be defined by reference to what is imported into and exported from it . . . Complex industrial systems import raw materials, convert them into a variety of products, and export the products by selling them . . .
>
> The process of changing import into export may require the carrying out of either sequential or simultaneous operations, or of both. When different operations are carried out discretely, a production system may be differentiated into *operating systems* . . .
>
> In the most general case production systems may be differentiated into three kinds of operating systems concerned with . . . import . . . , conversion . . . [and] export . . .
>
> In a complex organization there may be more than one system of each kind. In a simple organization there may be incomplete differentiation, more than one operation being carried out in the same operating system.
>
> When a production system is differentiated into discrete operating systems, its management cannot be contained in any one operating system, and

---

[1] I am indebted to my colleagues in the Tavistock Institute of Human Relations for many helpful comments on and criticisms of earlier drafts of this paper.

[2] In an earlier paper Rice and Trist used the term 'governing system' (11).

From *Human Relations*, 1959, vol. 12, pp. 243–72. Reprinted with permission of the author and the publisher.

a system external to the operating systems is required to control and service their activities. This is the *managing system* (10, pp. 41–42).

The managing system may contain differentiated *service and control functions*. Typical examples in industry are planning, research and development, cost control and accounting, engineering, personnel, etc., all of which help to service and control the operating systems.

Rice goes on to point out that in a complex hierarchical production system the main or first-order operating systems may be differentiated into second-order operating and managing systems, the second-order operating systems into third-order systems, and so on, until one reaches the *primary production system*, within which management is internally structured and is not a discrete function.

This set of concepts, in conjunction with the concept of the *primary task* as the task that a system is created to perform (10, pp. 32 ff.), provides a useful framework for analysing an industrial or business organization.

The present paper explores the principles of differentiation of operating units within a complex system. Part I describes these principles or dimensions of differentiation in the context of transition from a primary or simple production system to a complex system. An analysis is made of the forces that may act on a simple system to transform it into a complex system containing differentiated operating and managing systems. There is discussion of what happens inside the system when such a transition takes place. In Part II the dimensions are examined in relation to the order of differentiation in a multi-level, multi-shift complex system. Finally, Part III deals with the patterns of relationships— and therefore the problems of management—which are inherent in the structure of certain typical production systems, according to the basis of internal differentiation into subsystems and the nature and degree of sub-system interdependence.

In Lewinian terms, we are dealing here with the 'foreign hull' of the life space of the individual—'facts which are not subject to psychological laws but which influence the state of the life space' (8, p. 216; see also pp. 68–74).

PART I: THE TRANSITION FROM SIMPLE
TO COMPLEX SYSTEM

The typical simple production system in industry is the primary workgroup. Elsewhere it appears in the small workshop, the retail shop, the service station, and so forth. The essential feature of such a system is that management is inherent in relationships within the group: either there is no recognized leader at all (as is the case in some mining groups [5, 14]), or, if there is one, he spends all or most of his time working alongside the other members of the group on tasks comparable to theirs. His contribution to the output of the group tends to be directly productive rather than indirect and facilitative.

Herbst has described certain characteristics of simple and complex behaviour systems (3). He finds that one significant criterion of a simple system is that the relationship between input, size, and output is linear. (Input and output are here measured in money rather than goods.) In small retail shops, for example, the total amount paid in wages (i.e., input) increases at a linear rate with sales turnover (output) achieved, while sales turnover increases at a linear rate with the size of the shop, measured by the number of persons employed. In a complex system the relationship is non-linear: 'The presence of an administrative unit concerned with ongoing activities increases the rate at which sales turnover increases with size of the organization, and . . . the loss incurred by withdrawing personnel from production tasks decreases as the organization becomes larger' (3, p. 344). Herbst has this to say about the transition from a simple to a complex system:

As the size of the simple system increases, and depending also on the extent of both its internal and external linkages, more and more work has to be carried out on the co-ordination of component functioning, so that a critical boundary value with respect to size is reached, beyond which intrinsic regulation breaks down. An increase in size beyond this point will become possible by differentiating out a separate integrating unit, which takes over the function of both control and co-ordination of component units, thus leading to a transition from a simple to a complex system. The point at which intrinsic regulation breaks down will be determined by the effectiveness of the organizational structure. The less efficient the organizational structure happens to be the earlier the point at which intrinsic regulation breaks down (3, pp. 337–38).

In other words, three critical factors in the transition are size, complexity, and efficiency. Herbst here seems to be implying that the efficiency of a simple system is measured by the extent to which the system can tolerate increased size and complexity without throwing up a differentiated management function. However, if efficiency is measured by the ratio of output to size—a ratio that Herbst himself uses elsewhere in the same article—then it would appear that this assumption might not always be justifiable. If, for example, a simple system of given size could secure a greater output by becoming reorganized as a complex system of the same size then its persistence as a simple system would be relatively inefficient. 'Resilience' might therefore be a more appropriate term than 'efficiency' to describe the capacity of a simple system to withstand pressures—both external and internal—towards transformation into a complex system. This would be an omnibus term embracing a number of factors of small group functioning which counter the effects of increased size and complexity. Some of these factors are considered later in this section.

Size by itself is not a critical factor. Apart from the pair, groups of six to twelve are often said to be the most stable, in both psychotherapeutic and other situations (10, pp. 36–37). Fissiparous forces tend to develop in groups outside this optimum range, but there is no known maximum number beyond which emergence of a full-time management function is inevitable. Much depends on the need for differentiation that is intrinsic in the task of the group and in the way the task has to be, or is being, performed. Herbst notes, for example, that in independent retail shops a differentiated administrative function tends to appear when the staff numbers around five, whereas shops that belong to a retail chain retain the characteristics of simple systems until the size reaches about nine. Certain services are supplied to the latter by the larger organization to which they belong. In the Durham coalfield, autonomous groups of 41 have been shown to be effective (14). They have internally structured controls and services and lack any overtly recognized and titled leader. These groups are further discussed below.

Complexity may be considered in terms of Rice's import-conversion-export formulation. Imports into the system and exports from it may become more diverse. Complexity of the conversion process is likely to increase through diversification of input or output or both, or through a change in the techniques or rates of production.

Before considering these factors of size and complexity in more detail, it seems necessary to stress that *an essential preliminary to differentiation of a managing system is the formation of subsystems with discrete subtasks within the simple system.* Role-relationships cluster around the subtasks; such clusters of relationships become potential subsystems; and areas of less intensive relationships become potential boundaries between subsystems. Clustering may be functional for subtask performance, but the associated discontinuities between clusters may be dysfunctional for integrated performance of the total task. It becomes a function of a differentiated managing system to compensate for these discontinuities. Management mediates relationships among the lower-order systems, which constitute the higher-order system, in such a way as to ensure that the subtasks performed by the subunits add up to the total task of the whole unit.

If the principles of differentiation of subsystems can be identified, then the effects of changes of size and complexity can be more clearly understood; and furthermore the notion of 'resilience'—the capacity to withstand, without sacrifice of efficiency, the pressures towards creation of a differentiated managing system—will become less vague.

It is postulated here that there are three possible bases for clustering of role-relationships and thus for the internal differentiation of a production system. These are technology, territory, and time. Whenever forces towards differentiation operate upon a simple production system, it is one or more of these dimensions that will form the boundaries of the emergent subsystems and will provide the basis for the internal solidarity of the groups associated with them.

'Technology' here is given a broad meaning. It refers to the material means, techniques, and skills required for performance of a given task. Differentiation of the import, conversion, and export systems (the purchasing, manufacturing, and selling of an industrial unit) is in this sense a technological differentiation; so also is differentiation of phases of the conversion operation (successive manufacturing processes), or specialization in buying or selling particular commodities. The greater the diversity of technologies used within a group, the stronger the forces towards differentiation of fully-fledged subsystems—especially when the skills of some members are so specialized that others cannot aspire to have them or even comprehend them, and interchange of roles between members of the total group becomes impracticable.[3] Increase in technological complexity or diversity tends to have this effect even though the quantum of input and output remains unchanged. It may even occur where the size of the system, in terms of the number of roles, is reduced.[4]

The dimension of territory is straightforward: it relates to the geography of task performance. An increase in the staff of a retail shop from three persons to five may not precipitate formation of a differentiated management function. If, however, the two extra persons are employed to start a branch store—if, in other words, two potential subsystems are formed, spatially separated from one another—then the forces towards differentiation will be greatly increased. Physical separation is not essential to produce this result, but a sharp physical boundary of some kind is probably necessary before territory by itself can become a basis of subsystem differentiation within a simple production system. Identification of the group with its territory is of course a basic feature of all human societies and is found too among many of the higher mammals. Even boundaries that are imperceptible to an external observer may have highly charged emotional significance for the members of the groups they divide—especially when territorial differentiation is reinforced by technological differentiation. Technology, indeed, seems to seek the support of territory, and only seldom stands by itself as a differentiating factor. (In most parts of India, castes which are differentiated from one another by their traditional occupations are also segregated spatially, living in different parts of the village or in different villages.)

The third dimension—time—is more commonly relevant in increasing the levels of differentiation in an already differentiated complex system, but may also reinforce an increase in size in bringing about the transition from a simple system to a complex one. Forces towards differentiation probably begin to develop when the requirements of task performance are such that the length of the working day or working week of the group exceeds the working period of any individual member. This factor of time is of course most pronounced in multi-shift systems. As in the case of territorial separation, subsystems tend to emerge with well-defined boundaries, which in this case are based on time separation.

The subsystems and associated groupings described in the preceding paragraphs are those that are intrinsic to the structure of the task. Task structure is assumed to be inseparable from the type of technology and

---

[3] The obverse point was made by Rice (10, pp. 37–39), who postulated that small work-groups in modern mechanized industry usually require sufficient variety of roles (implying some technological differentiation) as to need some internal structuring, but should not have so much specialization as would lead to formation of inflexible subgroups.

[4] In her study of industrial firms in south Essex, Joan Woodward noted that 'the number of levels of authority in the management hierarchy increased with technical complexity,' while 'the span of control of the first-line supervisor . . . decreased' (15, p. 16).

specialization involved, from the geography of the territory in which the task is performed, and from the time-scale of task performance—though within these limiting factors alternative structures may be possible.

Among the persons manning the roles of a production system other groupings may occur, based on propinquity, sex, age, religion, race, and many other principles of association, and on occasion these groupings and related cleavages, perhaps by their coincidence with task-oriented groupings, may accelerate differentiation; or, if they cut across these groupings, they may retard it. It is the task-oriented subsystems themselves, however, which are relevant to task performance. These seem invariably to be differentiated by technology, territory, time, or some combination of these.[5] Production systems can probably not be satisfactorily broken down into subsystems on any other basis.[6]

If territory, technology, and time, singly or in combination, provide the basis for differentiation into task-relevant subsystems, the capacity of a simple system to tolerate growth and remain efficient without becoming transformed into a complex system is apparently related to two main factors: (*a*) mobility or fluidity, and (*b*) subsystem interdependence.

If individual members move frequently from one subsystem to another, so that there are no permanent subgroups of workers coinciding with the task subsystems, then the simple system will have greater capacity to tolerate an increase in size or complexity. Such movement compensates for discontinuities between subsystems. Secondly, the more immediately and directly performance of the task of each subsystem depends upon the performance of all the other subsystems, the more likely is the total simple system to remain viable in the face of forces towards differentiation. (Without some task interdependence it is, of

[5] Since drafting this paper I have seen Gulick's five-fold classification of the ways in which work in an organization can be grouped: by purpose (cf. by product), by process, by clientele, by place, and by time (2). My 'technological' category would embrace the first three of these. There seems, however, to be a conceptual distinction between the grouping of work and the differentiation of socio-technical systems, and in several respects the similarity between the 'scientific-management' theories of Gulick, Urwick, et al. and the aspect of socio-technical-system theory elaborated in this paper is only superficial. One difference is that (as, for example, March and Simon [9] have pointed out) Gulick's models do not take human motivations into account: he is dealing with management of technical systems, not of *socio*-technical systems. In Gulick's scheme, principles of association extraneous to work are obstructions to be overcome in establishing an essentially static and inflexible organization; in the more dynamic socio-technical-system theory they are a relevant part of the total reality situation. In general, the Gulick type of organizational theory does not lend itself to predictions about behaviour and relationships.

[6] The women's services in the forces may at first sight appear to be an exception to this rule. Closer examination shows, however, that these do not constitute production systems within which the members are interrelated and integrated by performance of a specific task. The production systems in the forces are the training and fighting units to which members of the women's services are attached (in much the same way as medical and signals personnel are attached to an infantry battalion) and to the task of which they contribute. In terms of the tasks of these units, the women may do jobs which are similar or dissimilar to jobs of men in the same unit; but in either event it is the technological specialization which is relevant to the total task, and their sex is incidental.

course, not a production system but an assembly or aggregate of individuals.)

Exceptionally large simple production systems occur in longwall
coal-mining, in a form of composite working described by Trist and
Murray (14) and Higgin (5). As mentioned earlier, some of the composite groups have forty-one members, working over three shifts. Both
'resilience' factors operate strongly in these groups, which are internally
differentiated by both technology and time. Although the individual
subsystems have well-defined tasks, mobility between the subsystems
allows many or all of the members to view intersubsystem relationships
from the perspective of the total system rather than from the perspective
of the subsystems they happen to belong to at any one time. Close
reciprocal interdependence, necessary in these mining groups for achieving the total task, evidently helps to reinforce this global perspective.

It may well be that it is not the number of persons that limits the
maximum size of a simple production system, but the number of subsystems. (A subsystem may consist of either an individual or a subgroup.) Certainly, complexity in task structuring can actually contribute
to the cohesion of large simple systems. Where there is a number of
subsystems interdependent in more than one direction, the complex
conditions of equilibrium can be a substitute for a differentiated management function. It is the very lack of such complexity built in to the
task that helps to lower the threshold of 'resilience' in less structured
simple production systems. Internal structuring for which the primary
task does not cater is sought in other groupings (based on age, sex,
etc.), implying involvement in other tasks that to a greater or lesser
extent conflict with the primary task for which the system was constituted. In some cases it may be possible to use these factors of resilience and to restructure roles in such a way as to postpone the
emergence of a differentiated managing system.

It can be inferred that, in any expanding or changing system in which
no such restructuring has occurred, there is an optimum or 'natural'
stage for creating a new level of management. This is applicable equally
to the initial transition from a simple system to a complex system and to
the addition of a new level to an already complex hierarchical system.

Premature differentiation is uneconomic because the cost of adding
a specialized administrative function is greater than the gain from any
increase in efficiency that results. As subsystems have not yet been
crystallized by task differentiation, government is more efficiently contained as an undifferentiated internal function. Indeed, extrinsic government, if imposed prematurely, may tend to be more destructive than
integrative. (This is the kind of situation in which the internal collaborative relationships, which before the change have been used constructively for task performance, are likely to be mobilized destructively
against the imposed external management (cf. Trist and Bamforth
[13].)

Postponement of differentiation of the management function beyond
the optimum stage also leads to a decline in the efficiency of the system,
but for a different reason. The energies of group members, instead of

being devoted to the primary task, are increasingly diverted to the task of holding the group together in the face of the fissiparous forces of subgroup formation and of differentiation. This is especially likely to happen if there is imbalance in the pattern of subsystem interdependence. Individuals experience conflict between identification with an emergent subgroup and identification with the total group. Only the creation of a new level of management which allows the subsystems to become fully explicit simple systems and which reintegrates them as parts of a higher-order system permits the energies of the members to revert to primary task performance.

Herbst uses the input-size-output relationship as an index for measuring the level of behaviour systems and for diagnosing whether a given system is simple or complex (3). The reverse approach may also be useful. If a production system, which is known to have the structural characteristics of a simple system, increases in size, and if this expansion is unaccompanied by a linear increase in output, then (other things being equal) it is worth investigating whether the system has passed the optimum stage for differentiation—either because the subsystems are in a stage of disequilibrium or because of emergence of subgroups unrelated to the primary task of the system. The same possibility may exist if a simple system, remaining constant in size, shows over a period of time a declining output. Equally, if a structural transition from a simple to a complex system is not accompanied by the kind of change in size-output ratio predictable from a Herbst-type formula for systems of that kind, then it is possible that differentiation of the managing system has been premature.

PART II: STRUCTURE OF COMPLEX PRODUCTION SYSTEMS

The forces towards transforming a simple system into a complex system, or towards increasing the levels of differentiation in a system that is already complex, are not only of theoretical interest to social scientists, but also of practical interest to those concerned with management. It has already been suggested, for instance, that working efficiency and cost are likely to be adversely affected if the timing of a change in response to the accumulating forces towards differentiation is not opportune. A second cause of inefficiency may lie in an inappropriate choice of the basis of differentiation into subunits.

In the initial transition from a simple to a complex system, the basis of differentiation is usually directly traceable to the forces leading to differentiation. Consider the example of a small privately owned workshop that manufactures simple components, all of the same kind, for the automobile industry. Raw materials are delivered and the finished products removed by the company it supplies. Administration takes up little of the owner's time and he himself works at a bench alongside his employees. This is the typical simple production system. Let us imagine that demand grows, and because of lack of space for expansion the owner acquires two more workshops in the vicinity. If the three workshops are sufficiently far apart, the owner is likely to spend less of his time at the bench and to take a nearly full-time managerial role. The three workshops then become three simple operating systems within a complex production system. In other words, territorial expansion has led to differentiation and it is territorially demarcated subunits that are explicitly recognized.

Alternatively, the expansion might have been achieved by adding two more shifts in the original workshop. The shifts would then become the recognized subunits, and, because of the need for control and co-ordination over the twenty-four hours, the owner would again take a full-time managerial role.

We now have to consider what happens when additional forces towards differentiation operate on a production system that has already become complex and there is the prospect of extending the hierarchy by further differentiation. Here again the forces themselves will dictate the new basis for differentiation, but not necessarily the level at which it will occur.

Reverting to our example, let us now suppose that further expansion requires all three workshops to run on three shifts. Each shift in each workshop is now likely to develop into a simple subsystem, and sooner or later the owner-manager will be compelled to realize that there are nine workshop-shifts to be managed, instead of merely three workshops on one shift, or three shifts in one workshop (Figure 1).

FIGURE 1

Sub-units Differentiated by Both Territory and Time

Time Differentiation

	Workshop A Shift I	Workshop A Shift II	Workshop A Shift III
Territorial Differentiation	Workshop B Shift I	Workshop B Shift II	Workshop B Shift III
	Workshop C Shift I	Workshop C Shift II	Workshop C Shift III

Increase in the number of subunits does not, of course, necessarily lead to further differentiation and to an increase in the number of levels. If, for example, output had been tripled by expanding from three workshops to nine, instead of by adding more shifts, the additional simple production systems so created could have become explicit without necessarily overextending the span of the overall manager's command. Even in the present example, it might be practicable to maintain direct control of the nine subunits, perhaps by employing additional staff for time keeping and recording production—that is, by increasing the size of the managing system without adding to the number of levels in the hierarchy. However, since the subsystems in this case are differentiated and interdependent along two dimensions (territory and time) that cut across each other, and therefore have to be coordinated along these two dimensions, it is likely that an additional level of management will be interposed.

The owner is now faced with a choice. He may introduce the new level by managing the three territories (workshops) through three fore-men, delegating to the foreman in each workshop the task of co-ordinating the three shifts within it (Figure 2). Alternatively, he may

FIGURE 2

First-Order Differentiation by Territory
Second-Order Differentiation by Time

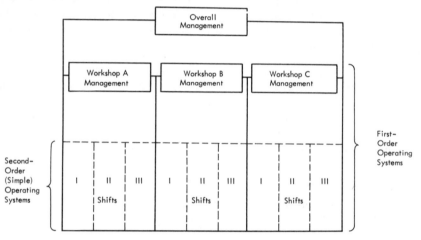

elect to undertake coordination of the three shifts himself by appointing three shift foremen, each of whom is responsible for the work on one shift in all three workshops (Figure 3).[7] The fact that territorial dif-ferentiation preceded the addition of shifts by no means presupposes that, in the management hierarchy, territorial differentiation need occur at a higher level than differentiation by time.

It is now necessary to consider this choice in more detail. In fact, it is a real choice only in so far as territory and time are equally salient in differentiating the nine simple systems from one another. In terms of task relationships, this is so only when one shift in one workshop is equally interdependent with other shifts in the same workshop and with the corresponding shift in other workshops. Workshop A Shift I (A I) belongs then to two larger systems: it is part of the 'A' system, within which the other systems are A II and A III, and it is part of the 'I' system, within which the other systems are B I and C I (cf. Figure 1). In the situation of equal interdependence,

---

[7] It was A. K. Rice who first drew my attention to this kind of choice, to which he also refers in his recent book (10, pp. 177 and 200–201).

FIGURE 3

First-Order Differentiation by Time
Second-Order Differentiation by Territory

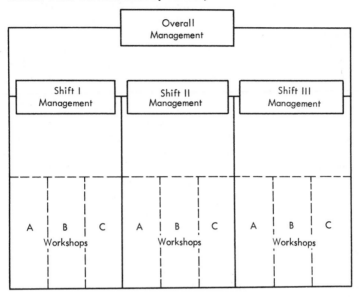

$$R(A \text{ I}, A \text{ II}, A \text{ III}) = R(A \text{ I}, B \text{ I}, C \text{ I}),$$

where R is a measure of task interrelatedness between the simple systems. Such an equilibrium may make it possible for the nine workshop shifts to be managed directly without interposing a new level of differentiation.

We have seen that the formation of subsystems with discrete subtasks is a necessary preliminary to transition from a simple to a complex system. Similarly, in an expanding complex system, the clustering of subsystems precipitates an additional level of differentiation, in which the clusters are acknowledged as explicit systems of a higher order than the constituent subsystems.

When two dimensions of differentiation are involved, with two implicit sets of systems cutting across one another, it is seldom that they actually have equal salience. Task relationships generally draw the basic units into the orbit of one system more strongly than into the other, and so dictate the lines of higher-order and lower-order differentiation. Furthermore, even if task-oriented interrelations themselves do have equal salience, other factors may tend to tilt the balance one way or the other. Persons who share a compact territory over three shifts, for example, may feel more strongly identified than those who share the same shift-timing over dispersed territories. Alternatively, if the dispersal is limited, going to work at the same time, and hence sharing free time, may lead to closer identification.

Failure to differentiate on the appropriate basis will create stress in relationships, because the natural groupings inherent in the structure of task performance will run counter to the groupings dictated by the formal organization. Formal boundaries will cut through these natural groupings. This will inhibit development of solidarity in the formal units, with consequent lowering of work satisfaction and morale. In general, we can suggest that to the extent that the formal structuring deviates from the reality of the task situation, whether in the basis for differentiation or in the boundaries of the formal subunits, to that extent will the management function itself have to multiply and become 'top-heavy' in order to deal with the resultant dysphoria. Additional controls will have to be imposed. This tendency will increase in proportion to the interdependence of the formal units. If on the other hand a unit is appropriately subdivided in relation to total task performance —if it is cut, so to speak, with the grain and not against it—both the internal management of the constituent subunits and the overall integration of the total task are likely to require less effort.

Flexibility is not entirely lacking. Imposition of a managing system itself helps to crystallize the selected basis and boundaries of differentiation of operating systems. Therefore, provided that the salience of two dimensions is not too unequal, differentiation at the higher level along the dimension of lower salience may increase the salience of that dimension to a point where it exceeds that of the other. This would not appreciably increase the difficulties of management. Similarly, if prior clustering of subunits is not too strong, the emergent boundaries can be supplanted by formal boundaries that do not necessarily coincide with them. Such flexibility, however, occurs only in marginal cases.

So far, instances of only two orders of differentiation have been discussed—by territory and time. We have seen that there is, subject to certain limiting factors, a choice between:

a)   first-order differentiation by territory and second-order differentiation by time (Figure 2); and
b)   first-order differentiation by time and second-order differentiation by territory (Figure 3).

A third possibility, provided the salience of the two dimensions is roughly equal, is to accept only one order of differentiation, operating systems being differentiated simultaneously by the time dimension in one direction and by the territorial dimension in the other. This was illustrated in Figure 1, which shows three time subdivisions and three territorial subdivisions, making nine subunits in all. Theoretically there is yet another way of compressing differentiation by two dimensions into one level. This occurs when the two dimensions, instead of operating at right angles, coincide and reinforce one another. Time and territory coincide in this way when shift working is used in highly mechanized road construction. A piece of mobile equipment—the common technology—is operated by one team in one stretch of road on Shift I, by a second team on a fresh territory on Shift II, and so on. In longwall coal-getting, time and technology coincide as differentiating dimensions, territory being undifferentiated: a different technology is used on each of three different shifts on one coalface. Both these combinations are fairly rare in industry, where it is territory and technology that most frequently coincide as reinforcing dimensions: In manufacturing operations, more often than not, each of a group of technologically differentiated subunits has its own discrete territory of task performance as well.

When all three dimensions of differentiation occur (if, in the example of the workshop, several products are manufactured in each of the three workshops on three shifts), the theoretical choice of order of differentiation is greatly increased. Assuming that differentiation occurs only once along each dimension, there are six combinations of three levels of differentiation, six more of two, and one of one level. These are listed in Figure 4. It should be noted that in the seven combinations (Nos.

FIGURE 4

Internal Differentiation of a Complex Production System by Territory, Technology, and Time (assuming that differentiation occurs once, and only once, along each dimension)

	Orders of Differentiation	
1st	*2d*	*3d*
1. Territory	Time	Technology
2. Territory	Technology	Time
3. Technology	Territory	Time
4. Technology	Time	Territory
5. Time	Technology	Territory
6. Time	Territory	Technology
7. Territory and time	Technology	—
8. Technology	Territory and time	—
9. Territory and technology	Time	—
10. Time	Territory and technology	—
11. Technology and time	Territory	—
12. Territory	Technology and time	—
13. Territory, technology, and time	—	—

7–13) involving differentiation by more than one dimension at one level, the simultaneous differentiation may be either (*a*) cross-cutting or (*b*) coincident and reinforcing.

Where there are more than three levels, at least one dimension will become the basis of differentiation at more than one level. In a large manufacturing concern, for example, there may be first-order differentiation into purchasing, manufacturing, and sales (technology); second-order differentiation of manufacturing into product units (technology, probably reinforced by territory); third-order differentiation of the product units into departments responsible for various phases of the process (again technology plus territory); and so forth. Time differentiation will occur in a multi-shift concern, but a twenty-four-hour command is narrowed down into eight-hour commands at only one level in any segment of a hierarchy. It may nevertheless occur at different levels in different segments of the same total hierarchy.

Very often the internal structure of a large organization is the cumulative result of many small local changes. Adherence to a particular pattern of differentiation adopted in response to one change may limit the possible responses to subsequent changes. In so far as the enterprise is a system, a change in one area will affect other areas. Accordingly, any organizational change must be planned in the context of the total structure, to ensure that it provides for the most efficient performance of the primary task of the enterprise.

Theoretically there are very many ways of differentiating a large organization. Depending on the size of commands and the amount of delegation, the number of levels can vary, and the basis of differentiation at each level offers many possible combinations. As we have seen, however, the inherent structure of the task imposes limitations on choice. It is now necessary to discuss in more detail some of the factors that make one choice more appropriate than others.

The highest level at which differentiation by time can take place in a multi-shift organization is to a large extent determined by throughput time. This is the intrinsic time dimension of the process itself—in effect, the interval between input and output. A series of consecutive processes places the associated task-roles into a temporal order that follows the flow of material, even though the occupants of these roles are all working simultaneously and are therefore not differentiated from one another along the time dimension. Some consequences of this temporal order of tasks are examined in Part III. What is relevant in the present context is that the head of a shift command cannot be held fully accountable for his shift's performance if the throughput time of the operation exceeds the duration of the shift.

In a composite spinning and weaving mill, for example, it takes two or three weeks for a piece of raw cotton to emerge as a length of woven cloth. During that period, if it is a multi-shift concern, the cotton goes through the entire shift cycle many times as it undergoes one process after another. In such a mill, first-order differentiation by time would be impractical. Figure 5 illustrates this point. A manager in charge of a shift that extended longitudinally through all these processes would be only partially and intermittently responsible for what happened to a given piece of raw material in the various stages of manufacture. For two-thirds of the time it would be outside his control. Responsibility for the total operation would be diffused among all three shift managers, and it would seldom be possible to attribute faults in the end-product to one of them more than to another. In relation to such a long throughput time, in other words, the shift is no more than a category: it is not a task-relevant grouping. Technological differentiation would be necessary at a higher level to narrow down the command to groups of processes and then to individual processes. Only at this point could the twenty-four-hour command be broken down into three shift commands. (It may happen of course that throughput time still exceeds shift time even when no further differentiation by technology or territory is pos-

FIGURE 5

Representation of a Continuous Three-Shift Manufacturing Operation (showing the route through various processes and shifts of a piece of material entering Process A in Shift 1)

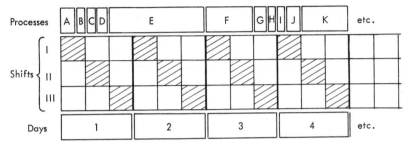

Throughput time of Process A is less than one shift; throughput time of Process E is four shifts; etc.

sible. If a multi-shift command at this level cannot constitute a simple production system, then it will be necessary to tolerate shift commands which do not allow full responsibility to be taken by the shift head.)

In contrast to textiles, electrical generation is a good example of short throughput time. Coal is transformed into steam and steam into electricity almost instantaneously. Consequently it is possible to differentiate by time at a high level: shift engineers can take charge of boilers, turbines, switchgear, etc. Indeed, so intimate and immediate is the connection between the various operations that it would be unrealistic to differentiate technologically between, say, boilers and turbines at a higher level than by time. The people performing all the operations on one shift constitute a more coherent system than the people performing one operation over several shifts. Thus, the highest level at which time differentiation can occur by the criterion of throughput time is also probably the optimum level for such differentiation. Slow throughput time emphasizes the territorial and technological groupings and cleavages; swift throughput emphasizes those that pertain to time.

In some multi-shift concerns, there is ostensibly first-order technological differentiation into departments and second-order differentiation into shifts, implying that the department heads have a twenty-four-hour responsibility. In addition, however, the overall manager may appoint second and third shift managers to coordinate the departments during his own absence. (Similarly, a naval vessel has a chief engineer, a chief gunnery officer, etc., in charge of their respective 'departments' over twenty-four hours, but there is also an executive officer of the watch who is in full charge of shipboard operations during his period of duty.) Shift heads within a department are thus subordinates in certain ways both to their department head and to their shift manager. In this situation the responsibility and authority of the two superior roles need careful clarification. In some cases it may prove necessary to make explicit first-order differentiation by time and second-order differentiation by technology: Here, as in the power-station example, departments would be only one shift 'deep' along the time dimension and department heads would have only an eight-hour responsibility. In other cases the shift managers do not represent an additional level of authority. They are part of the first-order managing system—extensions of the overall manager himself to secure twenty-four-hour control. Department heads retain twenty-four-hour accountability, but existence of shift managers reduces its scope, particularly in the area of maintaining coordination with other departments. This is what happens in naval vessels, and often too in coal-mining, and such an extension of the managing system seems not inappropriate in hazardous operations when someone is needed to take immediate charge in an emergency. Elsewhere, the resultant problems of confused authority and impaired accountability may outweigh any gain in interdepartmental coordination.

Short throughput time calls for a high level of differentiation along the time dimension largely because any disturbance in a part of the system has almost immediate repercussions in the rest of it, and swift corrective action is therefore required. More generally, the greater the extent to which a system's survival demands rapid adaptation to changes, whether in its external environment or internally, the greater the amount of self-containment it needs.[8] This implies that the system requires a clear-cut operational goal and the necessary resources to attain it.

Even where the interdependence of subsystems within the larger system places substantial limits on the self-determination of the subsystems, it is still desirable to break down the total organization in such a way that each component system has an operational goal or 'whole' task. This not only allows the head of each component command to be held accountable for performance, but also helps people to derive greater satisfaction from their work (cf. 10, p. 34 *et passim*). If task relationships within a particular command are less intensive than relationships that cross the boundaries of the command, there is a strong indication that the command does not have a whole task. The type of differentiation, the level at which it occurs, or the boundaries of differentiation may be discrepant.

To determine the subdivisions at each level that are most realistic in relation to the groupings and cleavages inherent in task performance, and to break the organization down into commands which allow accountability, is often a matter of common sense. In continuous manufacturing operation, for instance, adjacent processes have an affinity with one another that requires them to be differentiated only at a low level. Adjacence is determined by the flow of work and their successional dependence.

Major differences in technology can, however, obscure appropriate differentiation. In cotton textile manufacture, bleaching, dyeing, and

---

[8] March and Simon make much the same point (9, p. 193; also pp. 159 and 169). They also emphasize that 'if there were not boundaries to rationality, or if the boundaries varied in a rapid and unpredictable manner, there could be no stable organizational structure' (pp. 170–71).

other processes after weaving are for the most part chemical, while the processes up to and including weaving are not. Sizing, which occurs just before weaving, is an exception. In a composite mill performing all these operations it is not uncommon to group sizing with the post-weaving chemical processes, because of the technological affinity. But the flow of work is from warping to sizing to weaving to bleaching to dyeing, etc., and the grouping of nonconsecutive operations (for example, sizing with bleaching) into one command does not provide a whole task (see Figure 6). On these grounds it is contraindicated.

FIGURE 6(a)

"Unnatural" Technological Grouping of Non-consecutive Processes in a Continuous Manufacturing Operation

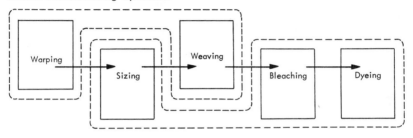

FIGURE 6(b)

"Natural" Grouping of Consecutive Processes in a Continuous Manufacturing Operation

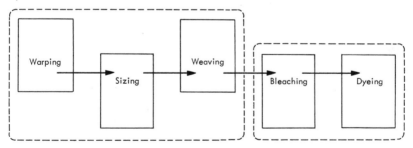

Territory is an important variable here. Usually each process has its own territory. Grouping of nonconsecutive processes means that the territory of the group may be discontinuous, other processes forming enclaves within it. When the layout is planned in such a way that material flows smoothly from one operation to the next, with minimum handling, then grouping of consecutive processes into one command creates a consolidated territory. The sharp territorial boundary—for example, between one building and another—is generally more relevant than any technological affinity between processes that are territorially remote and not sequential. Poorly planned layouts are not only inefficient from the point of view of material handling but also create organizational difficulties. Suppose, for example, that there are three consecutive processes, P1, P2, and P3; and that P1 and P3 are located in one building while P2, the intervening process, is in another building some way off. Whereas the order of operations inherent in the task indicates a closer relationship between P1 and P2 and between P2 and P3 than between P1 and P3, propinquity in this case may well modify the task relationships. Geography here is to some extent incompatible with effective task performance. Inclusion of P1 and P3 in one command that excluded P2 would be consistent with the territorial breakdown but not with the sequence of task performance: It would correspond to the 'unnatural' grouping of Figure 6(a). Combination of P1 and P2 in one command and exclusion of P3 would conform to the operational sequence but would conflict with territorial affinity. As we have seen, a common territory is a most compelling factor in group solidarity, and when it is not in harmony with the operational sequence, organizational difficulties almost invariably arise.

Re-examination of an existing complex organization often shows that the higher levels of differentiation do not always correspond with the deepest 'natural' cleavages and reveals alternatives that have previously been overlooked. One not uncommon choice occurs when a sequence of processes is duplicated in two neighbouring factories. The question is whether to give precedence to technological or to territorial differentiation. In textiles, technological precedence would imply combining a given group of processes in both mills into one command, inter-mill differentiation occurring at a lower level. A simplified example is given in Figure 7(a). If Mill *A* Spinning feeds only Mill *A* Weaving, and Mill *B* Spinning feeds only Mill *B* Weaving, the lack of interdependence between the two mills calls for territorial differentiation at a higher level.

FIGURE 7

Territorial and Technological Differentiation

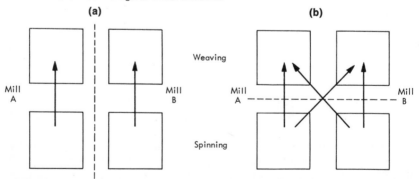

Mill *A* Spinning feeds only Mill *A* Weaving, Mill *B* Spinning feeds only Mill *B* Weaving; differentiation is territorial.

Mill *A* Spinning feeds *A* and *B* Weaving, which are also both fed by Mill *B* Spinning; technological differentation is indicated.

When, on the other hand, as in Figure 7(b), some cross-feeding also occurs, the cleavage between the two mills may be less significant than the cleavage between $A + B$ Spinning and $A + B$ Weaving, in which case technological differentiation could take precedence. The more spatially separated the two mills are, the greater the extent to which territory, rather than technology, will become the unifying factor. (And, of course, separation reduces the practicability of cross-feeding.)

Another common high-level choice is between two kinds of technological differentiation—the so-called 'functional' differentiation into purchasing, manufacturing, and marketing, or differentiation into product units, each unit managing its own functions. This is represented in Figure 8. Self-contained product units are more appropriate when there

FIGURE 8(a)

Self-contained Product Units

FIGURE 8(b)

"Functional" Differentiation

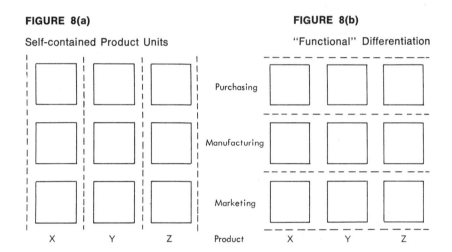

is little overlap between them, especially in the manufacturing processes and in markets. Separate factories reinforce such an organization by adding a territorial component to the technological dimension.

Management of a command is easier if the sizes of subcommands within it are roughly equal.[9] Inequality creates disequilibrium between the sub-commands; this has to be compensated by strengthening the managing system. The need to equalize commands may involve com-

[9] The criterion of size is not simply the number of persons in the command, but the total responsibility or level of work required of the manager. On 'level of work,' see Jaques (7, pp. 44 ff.). For the purposes of this argument it is unnecessary to examine the difficulties that have been encountered in developing effective techniques for job evaluation. We are concerned here with the more glaring differences about which there is likely to be intuitive agreement among those concerned. Disagreement or uncertainty about the relative levels of work of two managerial jobs is often presumptive evidence that they are sufficiently equal to be included at the same organizational level in one command.

bining two or more operating systems into one. If, for example, there are four factories, *A*, *B*, *C*, and *D*, and *C* and *D* are much smaller than *A* or *B*, it may be possible to include *C* and *D* in one command.

Compromises for the sake of equalization may also entail mixed methods of differentiation in the same command. Figures 8(a) and 8(b) illustrate two ways of differentiating an organization within which three products are manufactured and sold. Figure 9 shows a third

FIGURE 9

Example of a Mixed Basis of Differentation (two kinds of techno-logical differentation in one command)

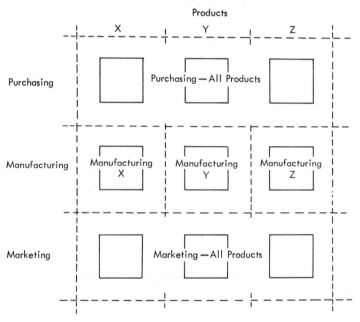

possibility: a common purchasing unit feeding three separate manufac-turing units, which supply a common marketing unit.

Wherever commands can be equalized only by creating boundaries that do not conform to the task structure, management has to choose between two sources of reduced efficiency. The amount of additional coordination and control required by each alternative—unequal com-mands and unrealistic boundaries—has to be assessed.

A comparable difficulty often arises where an organization contains a large number of technologically similar units which are territorially differentiated from one another and between which there is little or no direct interdependence. Shops in a retail chain or the 'beats' of travelling salesmen are typical examples. The less the interdependence and inter-action among them, the larger the number of such units that can be included in a single command, and thus the flatter the hierarchy.

Delegation is a variable here; but though the actual amount of delegation can vary considerably, it is clear that maximum delegation is possible where interdependence is least. However, unless delegation is carried to the extreme of abdication, there is at some stage a limit to the number of subordinates who can be directly accountable to one superior. No matter how much staff assistance he has, some form of direct communication between the superior and each of his immediate subordinates must be maintained—and the superior is one person, with limited time. Short lines of communication are generally more efficient, but not if they are allowed to become too tenuous, and there comes a point where a thicker line through an intermediate level is more efficient than a tenuous direct line.

The decision to establish an intervening level of management immediately creates the problem of how to group the units into larger commands. Each unit has a clear-cut whole task and the overall—perhaps nation-wide—organization has a clear-cut task, but there may be no intermediate grouping whether intrinsic to task performance or not. Propinquity is not enough. Often in fact such intermediate commands are not 'systems' at all, but merely aggregates. They have no unique 'whole' task to integrate them. Boundaries are dictated by extrinsic administrative convenience. Interaction between individual units—or absence of interaction—is just as pronounced across the arbitrary territorial boundaries as it is within them. The intermediate managers will frequently be perceived by their superior as a barrier between himself and their subordinates, and by the subordinates as a barrier between themselves and their proper superior.[10]

To the extent that the interposed commands have some task validity, this problem is diminished. (Organizational devices which are not really relevant to the task, and which are designed merely to give the boundaries an illusion of validity, are unlikely to be very effective.) In the case of the retail chain, for example, a group of stores might be linked to a regional supply depot and distribution system. Regional commands in a marketing organization might be reinforced by genuine regional differences in patterns of consumption, and therefore in the techniques of selling. Alternatively, the administrative need for the intermediate level may be eliminated by reorganization elsewhere. If the primary selling unit, for instance, is changed from an individual salesman to an internally led group of salesmen covering a larger territory, the number of units to be managed is greatly reduced.

To sum up, therefore, any production system, complex or simple, can be defined along the dimensions of territory, technology, and time. A large system is broken down into progressively smaller systems along

---

[10] The decision to create a new level of management involves, of course, not only the appointment of additional managers, but re-examination and possibly redeployment of control and service functions and probably changes in the quantum of delegation. However, the tendency for the intervening level to be experienced as a barrier to communication and to effective task performance is directly related to its irrelevance to the task structure and is not primarily a function of any changes in delegation and control that may have occurred at the time the level was interpolated.

one or more of these dimensions at each level. The smallest systems are sometimes coextensive along one or even two dimensions with the overall system, but more often in a manufacturing organization they are shorter along all three dimensions. Each component system, however, has boundaries that serve to separate it from parallel systems, and also boundaries that form part of the higher-order system's boundaries. Work-oriented relations crossing the former boundaries should be more intensive than those that cross the latter; if not, it can be inferred that an inappropriate basis of differentiation has been adopted and that the efficiency of the total system is less than optimal.

PART III: INTERNAL DIFFERENTATION AND PROBLEMS OF MANAGEMENT

Where a complex production system is differentiated into subsystems, the total task is also broken down into subtasks associated with these subsystems. As Rice has pointed out, such a hierarchy of tasks may often lead to situations where 'decisions taken within one component system which are consistent with its primary task may appear irrelevant or even harmful in a system of a different order' (10, p. 228). Differentiation into subsystems therefore throws up a managing system which has the reintegrative function of seeing that the constituent tasks of the subsystems are so performed that they add up to the overall task of the system as a whole.

It is suggested here that the way in which a task is broken down—in terms of the dimensions along which the subsystems are differentiated and in terms of the intrinsic interdependence between them—is a major determinant of the kind and quality of management required, including the kinds of control mechanism that will be appropriate. Fundamentally, of course, the dimension along which the system is differentiated at a given level is the dimension along which the major controls have to be exercised to secure reintegration.

Differentiation by territory, technology, and time, singly and in combination, can at any one level take seven different forms—three one-dimensional, three two-dimensional, and one three-dimensional. These are set out in Figure 10. Multi-dimensional differentiation can be reinforcing, cross-cutting, or mixed, though the examples given in Figure 10 are all of reinforcing differentiation: that is, at the level of differentiation in question, each component system is differentiated from every other along both the named dimensions. Examples of cross-cutting and mixed differentiation could also be added.

Types of task dependence have been classified in some detail by Herbst (4) and Emery (1). For present purposes it is relevant to consider the extent to which, at a given level of differentiation, the component systems of a larger system are *codependent* on supplies, equipment, and services, and *interdependent* for the attainment of the end-result or goal of the larger system. One or both of these types of dependence may be present. Emery points out that interdependence may be simultaneous or successional, and that successional dependence may

FIGURE 10

The Seven Basic Forms of Differentiation at One Organizational Level. The Examples of Two- and Three-dimensional Differentation Given Are of 'Reinforcing' Type. Brief Notes on These Examples Are Given in the Text.

Differentiated Dimensions	*Undifferentiated Dimensions*	*Examples*
1. Territory	Technology and time	(a) Separate sections within a factory, or separate factories, making same product
		(b) Marketing organization; chain of retail stores
2. Technology	Time and territory	Shipbuilding
3. Time	Territory and technology	Typical and multi-shift structure in process and other industries
4. Territory and technology	Time	(a) Quasi-independent product units
		(b) Consecutive manufacturing operations
5. Technology and time	Territory	Longwall coal-mining
6. Time and territory	Technology	Mechanized road-making with shift working
7. Territory, technology, and time	—	Milk: collection; processing and bottling; and delivery

be further classified as cyclic, convergent, or divergent. Distinctions can also be drawn between simple and complex dependence and between reciprocal and non-reciprocal.

If the differentiation variables were separately considered in relation to all the dependency variables, the resultant number of combinations would be enormous. Here it will be sufficient to examine the three basic differentiation variables in a little more detail and to discuss a few models that occur fairly frequently in industry. From these the implications of other models can be inferred.

There is one other respect in which the present discussion is restricted. While the basis on which subsystems are differentiated and the nature of their dependencies are the internal system elements that create a particular pattern of demands on management, it is also a function of management to mediate in certain ways between the system and its environment (which may include successively larger systems of which it is a part), and environmental factors will inevitably impose certain other demands. Such factors, for example, may call for additional control mechanisms within the system. The more complex and diverse these environmental factors are, the greater the number and variety of control and service functions likely to be differentiated within the managing system, and the greater the consequent complexity of intrasystem relationships. Here, however, environmental factors are held constant and attention is focused on internal factors relevant to the relationships of a manager with his immediate subordinate group.

Differentiation by Territory

It is characteristic of operating systems differentiated from one another only along the territorial dimension that the output of the total system to which they belong is the added sum of the outputs of the constituent systems. Output from one system can be high, low, or even absent without directly affecting output from the others. In other words, where differentiation is only territorial, interdependence is minimal.

The extent to which the systems are codependent—on a single source of supplies, for example, or on centralized service functions—can vary considerably. Spatial segregation can be an important factor here, though not necessarily a determining one. To take the examples given in Figure 10, if the territorially differentiated units are neighbouring sections in the same factory—for example, the series of groups of workers on groups of looms in the textile mills described by Rice (10)—they are likely to draw their input from the same source and to be jointly dependent on a number of centralized services. If, however, the units are separate factories making identical products in different parts of the country, their codependence may well be less. Canneries and other food-processing plants are often dispersed in this way in order to be close to agricultural sources of supply. Decentralized control over input is practicable in such cases but is less appropriate where the factories (perhaps dispersed to be close to their markets) share a common and limited source of supply. Codependence may also extend to output: the smaller the fluctuation of output permissible in the total system, the greater the centralized control over the outputs of the constituent systems.

Putting it in another way, we can say that where a unit is differentiated into territorial subunits, the individual subunits and the total unit are the same 'length' along the input-output dimension. The constraints on procurement of input and on disposal of output that operate on the whole unit will place upper limits on the autonomy that can be given to the subunits. The stronger these external constraints, the greater the codependence of the subunits.

Some of the problems that arise when territorially differentiated subunits have had to be created only because of the size of the total command have been discussed in Part II. However, in other cases, so long as the territorial boundaries conform to the reality of the task structure and so long as subunit performance can be measured separately, this is one of the easiest kinds of command to manage, especially if the subunits are roughly equal in size. Because the operations of his subordinates are not interdependent, the superior is not concerned with maintaining collaborative relations between them. Indeed, competitive relations are often more appropriate. Their homogeneity makes comparisons straightforward and a highly productive subunit can be used as an example and pace-setter for the others. Subject to the external constraints on autonomy, substantial delegation is possible, which means that a fairly large number of units can be included in one command, producing a flat hierarchy.

One practical difficulty that sometimes arises in such a command, however, is that the competitive situation gets out of hand. The superior may become so involved in resolving problems of real or imagined incomparability between the subordinates that he loses sight of the primary task of the system. The subordinates for their part are liable to seek short-term competitive advantages that may be detrimental in the long run; or alternatively they may go into collusion to protect themselves from competitive stress by establishing safely attainable norms. The common restrictive practices in industry and commerce are special cases of this form of organization.

There is another management problem that may occur in manufacturing units. This is the tendency for the subunits to develop an 'individuality' that is based on more than their territorial differentiation from one another. Here we are not concerned with the general tendency of groups to develop a structure and culture that apparently transcend what is needed for attainment of their overt goals. We are concerned more specifically with a tendency to supplement territorial differentiation by technological differentiation. This is pace-setting of a special kind. In a manufacturing operation such as weaving, identical machinery may be used to turn out several varieties of one product. Even though all varieties are spread equitably among all subunits, individual subunits may develop a special proficiency in some. They acquire what Selznick has called a 'distinctive competence' (12). This distinctive competence may be encouraged, perhaps almost accidentally, by assigning more of these varieties to the subunits in question. Such specialization is the beginning of technological differentiation. Management needs to be alert to such incipient changes and to recognize their implications. It is not simply a question of deciding whether the gains from specialization —probably in improved efficiency and quality—outweigh the disadvantages of reduced flexibility in production planning. Different methods of management are required: competition ceases to be an appropriate control mechanism when the subunits become heterogeneous. In the extreme situation, the varieties, by ceasing to be interchangeable, acquire the status of separate products and the territorial differentiation becomes secondary to what is, in effect, technological differentiation between product units. Management of such units is discussed in the next section.

Differentiation by Technology

In cases of differentiation by technology, the notion of distinctive competence is very much present. The organization is built up around clusters of specialized skills and often specialized equipment too. Members of a subunit that is differentiated from others along the technological dimension derive solidarity from their distinctive competence, often by exaggerating its distinctiveness. Preservation of that distinctiveness may become the primary task of the subunit. Management of a unit in which the subunits are differentiated, and therefore have to be reintegrated, along the technological dimension, involves using the specialized contributions of the subunits to perform the primary task of the whole. To achieve this, the solidarity that the subunits derive from distinctive skills should be sufficient for them to maintain their viability as separate systems, but not so great that they lose sight of the total task of the larger unit. To strike such a balance is no easy task. Perceived threats to the integrity and distinctiveness of subunit skills mobilize the energies of subunit members towards preservation of the subunit at the expense of the unit as a whole. Closed-shop movements in departments of automobile factories and demarcation disputes in shipyards are familiar examples.

Operating systems are seldom differentiated from one another by technology alone. Perhaps the nearest approximation is in enterprises such as ship-building where what is being made is also the territory of task performance. Even in ship-building, however, there is some supplementary differentiation by territory and time: certain jobs have to be done elsewhere in the yard and certain jobs on the ship itself cannot be started until preceding jobs are complete. The occupational groups at work on the ship at any one time have shifting and overlapping territorial boundaries and it is along the technological dimension that they have primarily to be coordinated. Conventional longwall coal-getting involves differentiation by both territory and time (5, 13, 14). The team working on a particular section of the coal-face over a 24-hour period is subdivided into shifts that are distinguished from one another both by the times they work and by the kinds of task they do. Reinforcing differentiation by technology, territory, and time may occur in a milk business: milk is collected from the farms in the afternoon and evening and brought to the central depot; there it is processed and bottled during the night; and next morning it goes out on the delivery rounds.

In industry, technological differentiation is commonly accompanied by territorial differentiation. (The word 'department,' for example, often carries both connotations.) Where the two are combined in this way, the former distinction always seems to be primary: territorial differentiation supplements and reinforces the technological. To some extent the combination also facilitates coordination by giving the technological groupings the security of a clear-cut physical boundary—contrasted with the vague and shifting boundaries of the shipyard.

Where the technological differentiation is primary, the nature and extent of codependence and interdependence among the subunits concerned can vary. It will illuminate the kinds of problem that can occur in management if we consider two extreme—but nevertheless common —models, one with little or no interdependence and one in which interdependence is substantial.

The self-contained product units of Figure 8(a) are an example of the former. Codependence tends to be minimal when the units are spatially remote, and not merely manufacture different products but market them separately as well. Their heterogeneity and their lack of interconnectedness make it difficult for the superior to weld the heads of the subordinate units into a group and to compare their performance. Centralization of controls and services increases their codependence, but probably reduces the extent to which the individual units can exploit their distinctive competence. Decentralization permits the units to respond more quickly and creatively to relevant environmental changes.[11] This capacity is encouraged to the full if the manager of the total command delegates really substantial authority to the product units, letting them operate virtually, or actually, as separate companies. Comparisons can then be made in terms of profits. In other words, a common denominator is found at a higher level of abstraction, which makes the technological distinction less relevant. In general, the fewer the intrinsic or imposed dependencies between technologically differentiated subunits, the greater the extent to which efforescence of subunit distinctive competence can be not merely permitted but encouraged as contributing to overall task performance of the total unit.

Where units manufacturing different products share common organizations for purchasing and marketing, there may be first-order differentiation into purchasing, manufacturing, and marketing, and second-order differentiation of the manufacturing system into product units. Alternatively, the two levels of technological differentiation can be compressed into one (cf. Figure 9). The purchasing unit, the marketing unit, and, say, three product units all become coordinate systems within the larger system. This is illustrated in Figure 11. There is no direct interdependence between the three manufacturing units, *A, B,* and *C,* but there is substantial codependence between them resulting

FIGURE 11

Differentation by Technology [and Territory]

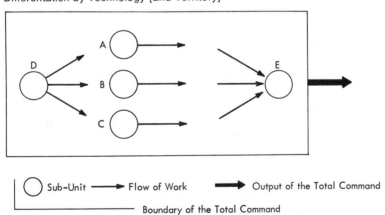

Codependent manufacturing operations (A, B, and C), each separately interdependent with common purchasing (D) and marketing (E) operations. (two dimensions of technological differentiation: cf. Figure 9)

---

[11] March and Simon suggest that interdependence and codependence restrain innovative activity: 'The greater the interdependence among subunits and the higher the dependence of line units on auxiliary units, the less vigorous will be the innovative activity of the line units' (9, p. 198).

from the pattern of successional interdependence: $D \rightarrow A \rightarrow E$, $D \rightarrow B \rightarrow E$, and $D \rightarrow C \rightarrow E$. The effects of this relatively complex pattern on relationships are not always predictable. The manufacturing units may be drawn closer together, into a manufacturing subgroup; but against this conflicts may arise between the manufacturing units over relationships with purchasing and marketing. The head of one manufacturing unit, for example, may try to 'capture' the head of purchasing. The management problem is to control these groupings and cleavages and to use them constructively. Pressures from the manufacturing subgroup may make purchasing and marketing more effective; on the other hand, if there is too much pressure backwards from marketing the manufacturing units may go into collusion to keep performance standards down.

The organization just described is in effect a combination of our two extremes of technological differentiation. The minimally interdependent product units have already been discussed, and we now have to consider the model of high successional interdependence. This is illustrated in Figure 12. *A* feeds *B; B* feeds *C;* and *C's* output is the

FIGURE 12

Differentation by Technology [and Territory];
Successional Interdependence

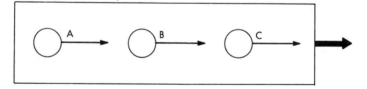

output of the total command. Although the tasks are carried out simultaneously, the flow of material places them in a temporal order. This is an extremely common organization of production systems at many levels. In manufacturing operations, for example, it occurs where material passes consecutively through a series of processes. At a higher level, the *A, B,* and *C* of the diagram may be the purchasing, manufacturing, and selling operations. What is said here primarily concerns manufacturing operations, but much of it is relevant to other kinds of operation besides.

There are five conditions under which management of this kind of command is facilitated. First, the unit as a whole should have clearcut territorial boundaries and if possible the subunits within it should be territorially, as well as technologically, differentiated from one another. Secondly, if the unit as a whole is technologically differentiated from neighbouring units, the technologies of its constituent subunits should have more in common with one another than with units outside. Thirdly, the output of the unit as a whole (and if possible the output of the individual subunits) should be clearly measurable. Fourthly, the subunits should be reasonably equal as commands. Finally, the sequence of operations should be unbroken: that is to say, material should not pass from *A* to *B* and then undergo a further process outside the organization before going on to operation *C* (cf. Figure 6). Interdependence is to some extent reduced if there are banks of part-processed materials between *A* and *B* and between *B* and *C*: a breakdown in one subunit does not then immediately affect following subunits.

Even when all these optimal conditions obtain, there remain certain management problems inherent in this kind of organization. One characteristic is the existence of overlapping consecutive pair relationships among the subordinates. Each, except the last, supplies unfinished material to—and has power over—his successor; each, except the first, is dependent on his predecessor in the command. The last man is in the unhappy position of having within the unit no successor to whom he can transfer his unfinished goods. Dependence always tends to give rise to conscious or unconscious hostility and resentment. *B* may express in two kinds of ways his resentment at dependence on *A*: He can complain about the quality of the material that *A* feeds to him; and, more seriously, he may unconsciously deal with his grudge by doing poor work and so sabotaging the output of his successor. In other words, he compensates for his dependence by misusing his power. The last man, in order to compensate for his lack of power over a successor within the command, may even unconsciously sabotage the output of the group as a whole. It is therefore the major task of the manager in this kind of unit to create an integrated group, within which the individual subordinate feels not only that he is responsible for the task of his subunit, but that he shares in the responsibility for the total task of the group.

Differentiation by Time

In the ordinary multi-shift situation, where the subunits are differentiated from one another only by time and share a common territory and a common technology, their codependence is considerable. For example, maintenance failures on one shift affect the others. Generally this codependence is accomplished by a circular form of successional interdependence: each shift not only completes certain operations, exporting the material outside the total unit, but also passes on some semi-finished material to the next shift for completion. This three-shift pattern is depicted in Figure 13. Throughput time is a major determinant of interdependence. The longer the throughput time, the higher the proportion of semi-finished to finished material at the end of each shift. Also, the less likely it is that individual shift performance, in terms of quantity and quality, can be measured precisely. Continuous operations of process industries provide an obvious example, but the production lines of the engineering industry also contain at any one time components in various stages of completion. Another factor that reduces the clear-cut self-containment of the shifts in the most highly automated industries, where shift working is most prevalent, is that the functions of so-called 'production workers' have increasingly been taken over by the machines themselves. The task of the workers is to monitor

FIGURE 13

Differentiation by Time; Territory and Technology
Undifferentiated; Circular Interdependence

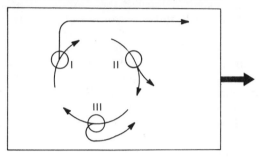

and maintain, and the consequences of things they do or fail to do are often not immediately and clearly visible: The benefits or otherwise may fall upon succeeding shifts.

Furthermore, in most industries—indeed in the society at large—night-work is considered unnatural; a certain stigma attaches to it. Men who work while the rest of the world is asleep tend to feel cut off from society—and no doubt some select night-work for this reason, and may even become neurotically addicted to it. This is not the place for a discussion of the psychology of shift work: The point to be emphasized is that night-shifts often have a distinctive 'atmosphere' of their own.[12] This is particularly true where a group of workers is permanently on night-shift. Night-shifts are less differentiated in this particular respect in enterprises such as chemical plants, steel plants, or power stations, where continuous operations are dictated by the basic nature of the technology, and also where all shifts rotate.

It is clear that differentiation by time calls for positive managing skills to maintain the tempo and quality of work and to prevent the circular dependence from becoming a deteriorating cycle. The management problems inherent in this model make it important to eliminate avoidable complexities. Many of these stem from a failure on the part of management to conceptualize second and third shifts as discrete systems. Outside industries where continuous operations are intrinsic to the technology, the second and third shifts have generally been introduced in order to supplement production from single-shift working without increasing capital investment; and the notion that they are supplementary tends to be perpetuated not only in management attitudes but also in organization. Rice has given a good example of this kind of situation in a textile mill[13] and also indicated that acceptance

---

[12] Often, too, the level of attention is lower and mistakes are more numerous. Accidents may be fewer: cf. Hill and Trist (6).

[13] '. . . Supervision on the third shift had never been as good as on the other shifts. It had been unpopular with management as well as with workers. In consequence, the most junior supervisors had been posted to it. Not unnaturally the

of the organizational consequences of three-shift working can lead to higher productivity and improved quality (10).

An avoidable complication occurs, for example, when the overall head of the three shifts has himself the additional role of first-shift supervisor. The twenty-four-hour responsibility of the overall head naturally cannot be discharged if he is regularly tied for eight hours to one shift only. A separate first-shift supervisor is therefore necessary. Related to this is a tendency to confuse first-shift control and service functions with headquarters functions, usually because office hours more nearly coincide with first-shift hours than with those of other shifts. The first-shift supervisor may be given responsibility for such functions as pertain to all shifts, or alternatively—and less frequently —certain services that are decentralized to the second- and third-shift supervisors are, for the first shift, retained under headquarters control. It is appropriate either to centralize such functions fully under the head of the total command or to decentralize them equally to his three subordinates, but not to delegate them to one or two subordinates only (cf. 10, p. 46). Difficulties of coordination are also increased if one shift supervisor—commonly the third shift—has an operating command that is smaller than the other two. Equalization of shift commands, by allowing the heads of the three shifts to collaborate as equals, may reduce the load on the managing system to an extent that more than offsets the cost of increased third-shift working. (This is not possible, of course, where there are wide fluctuations in the load—for example, in some engineering firms—and a 'spill-over' night-shift is required irregularly in order to absorb these fluctuations and to maintain a steady day-shift load.)

The head of this kind of command therefore has to take specific precautions appropriate to the pattern of differentiation and interdependence: he needs to be aware of his twenty-four-hour responsibility, to attend shift handovers as often as possible, to avoid delegating either too much or too little to the first-shift head, and to avoid giving too small a command to the third-shift head. Meetings of the superior with his subordinates as a group help to emphasize the complementary contributions of the shifts to the total task of the command. Meeting the subordinates only individually makes it more difficult to ensure that all three shifts work together coherently. There are possibly advantages in a form of shift rotation which periodically alters the order of dependence of the shifts.

Where there are only two shifts, although the general problems are very much the same as in the three-shift situation—especially the sharing of territory and equipment—the reciprocity makes equilibrium easier to sustain because dependence and power balance each other. There is one drawback in having only two shift heads reporting to one superior: it is too small a command. Coordination and control of two subordinates generally give the superior too little to do. He may tend to bypass

combination of least experienced workers and most junior supervisors had succeeded in proving correct those who believed that third shifts were not worth while' (10, p. 176).

his immediate subordinates, withdrawing authority and responsibility from them. Consequently it may prove desirable to combine at the same level differentiation by time with cross-cutting differentiation by territory and/or technology. As was pointed out in Part II, however, it is unlikely that the task structure will be such that interrelatedness along the time dimension will be equal to interrelatedness along the territorial/technological dimension. In all cases of cross-cutting differentiation, where two dimensions of differentiation are compressed into one level, formation of subgroups is to be expected along one dimension or the other. It has to be realized, however, that such groupings have no formal identity in this kind of structure, so that controls and services must be either fully centralized or else fully decentralized to the individual subunits.

Though the few models discussed here only touch the fringe of all the possible variations, they serve to indicate the different kinds of demand placed on management according to the types of boundary that separate the subunits and according to the type and degree of dependence between them. Consideration of these factors may be relevant to the selection, training, and placement of managers. Though it is probably a little far-fetched to suggest that management of territorially differentiated units requires a special kind of person, it is certainly clear that techniques of management appropriate in that situation cannot effectively be transplanted into a situation where the units are differentiated along other dimensions and the patterns of codependence and interdependence are more complex.

SUMMARY

Any production system can be defined along the three dimensions of territory, technology, and time, which are intrinsic to the structure of the task of the system. Task performance is impaired if subsystems within a system are differentiated along any other dimensions than these.

In a simple production system, management is inherent in internal relationships. A prerequisite of transition from a simple to a complex system, with a differentiated managing system, is the emergence of subsystems with discrete subtasks. Simple systems vary in their 'resilience' —the capacity to resist forces towards transformation into a complex system, while remaining viable and efficient. The resilience of a simple system is increased if there is mobility of individuals between subsystems associated with subtasks and if the subsystems themselves are closely interdependent. For any expanding or changing simple system, however, there is an optimum stage for transition to a complex system: Task performance suffers if the transition is premature or belated.

A large, complex production system, such as a multi-shift manufacturing concern, is broken down into progressively smaller systems along all three dimensions of territory, technology, and time. There are several levels of differentiation. Differentiation along some dimensions occurs at more than one level. Differentiation along more than one dimension can also occur at a single level. The dimensions may then be either at right angles to one another ('cross-cutting'), in which case each component subsystem is interdependent with—and has to be coordinated with—some subsystems along one dimension and some along the other; or the dimensions may be coincident and mutually reinforcing (for example, departments in a factory which are differentiated from one another technologically and also occupy separate territories).

In such a complex system there is sometimes a choice: differentiation along one dimension could occur at a higher level than differentiation along another; or at a lower level. However, the basis of differentiation must not violate the task structure: boundaries should be so located as to associate each command with a 'whole' task, for which the head of the command can be held accountable. If at any level the formal organization is such that task relationships which cross the boundaries of a command are more intensive than relationships within it, there will be a loss of efficiency and/or an expansion of the managing system at that level.

The appropriate level of differentiation by time is related to the throughput time of the process (the interval between input and output). Time differentiation occurs at a low level in an organization with a long throughput time and at a high level when throughput time is short. When a sequence of operations is involved, consecutive operations that are also territorially adjacent are appropriately differentiated at a low level, even though technologically they may be heterogeneous. Operations that are technologically homogeneous but spatially and sequentially separated can seldom be combined into a viable system.

The need to ensure that subcommands within a command are approximately equal in size and that each has a 'whole' task may entail combining more than one system into one command or adopting mixed methods of differentiation. These needs, however, sometimes conflict.

Differentiation implies reintegration, to ensure that the subtasks of subsystems add up to the total task of the whole system. Within any command, the way in which the task is broken down—i.e. the dimensions along which the constituent systems are differentiated and the interdependence between the systems—largely determines the kind of management required. Organizational models drawn from industry are used to illustrate this point.

REFERENCES

1. Emery, F. E. *Characteristics of Socio-Technical Systems.* Tavistock Institute of Human Relations, Document No. 527, January 1959 (mimeo).

2. Gulick, L. 'Notes on the Theory of Organization, with special reference to Government in the United States.' In *Papers on the Science of Administration,* edited by L. Gulick and L. Urwick. New York: Institute of Public Administration, 1937.

3. Herbst, P. G. 'Measurement of Behaviour Structures by Means of Input-Output Data.' *Hum. Relat.,* Vol. X, No. 4, 1957, pp. 335–46.

4. Herbst, P. G. 'Task Structure and Work Relations.' T.I.H.R., Document No. 528, January 1959 (mimeo).

5. Higgin, G. W. 'Studies in Work Organization at the Coal Face—I.' *Hum. Relat.*, forthcoming.

6. Hill, J. M. M., and Trist, E. L. 'Changes in Accidents and other Absences with Length of Service.' *Hum. Relat.*, Vol. VIII, No. 2, 1955, pp. 121–52.

7. Jaques, E. *Measurement of Responsibility.* London: Tavistock Publications Ltd., 1956; Cambridge, Mass.: Harvard University Press.

8. Lewin, K. *Principles of Topological Psychology.* New York and London: McGraw-Hill Book Company, 1936.

9. March, J. G., and Simon, H. A. *Organizations.* New York: John Wiley & Sons; London: Chapman and Hall Ltd., 1958.

10. Rice, A. K. *Productivity and Social Organization: The Ahmedabad Experiment.* London: Tavistock Publications, 1958.

11. Rice, A. K., and Trist, E. L. 'Institutional and Sub-Institutional Determinants of Change in Labour Turnover.' *Hum. Relat.*, Vol. V, No. 4, 1952, pp. 347–71.

12. Selznick, P. *Leadership in Administration: A Sociological Interpretation.* Evanston, Ill.: Row, Peterson and Company, 1957.

13. Trist, E. L., and Bamforth, K. W. 'Some Social and Psychological Consequences of the Longwall Method of Coal-Getting.' *Hum. Relat.*, Vol. IV, No. 1, 1951, pp. 3–38.

14. Trist, E. L., and Murray, H. 'Work Organization at the Coal Face: a Comparative Study of Mining Systems.' Forthcoming.

15. Woodward, J. 'Management and Technology.' Dept. of Scientific and Industrial Research: Problems of Progress in Industry, No. 3. London: H.M.S.O., 1958.

Rolf P. Lynton

17. Linking an Innovative Subsystem into the System

> Innovation . . . is not so much the adoption of objects by individuals as it is the acceptance of ideas by (people in) an organization. (Andrews and Greenfield, 1966:81)

> Any research process . . . has an inbuilt tendency towards the formation of a relatively closed system, in which self-generated intakes crowd out intakes from the external environment. . . . A major problem is to steer between . . . creative research activities that are irrelevant and . . . relevant research activities that are uncreative. (Miller and Rice, 1967: 159–60.)

FORMAL ORGANIZATIONAL devices for facilitating change in social institutions, particularly institutions in environments characterized by increasing differentiation and complexity, are largely integrative devices. It is on these integrative devices that this paper will concentrate.[1]

Integrative devices provide linkage in two crucial areas: (1) between the institution and its specific publics of consumers and suppliers, as well as its wider public of political and social sanction, which determines the relevance of the institution; and (2) between different parts within the institution, which determines the extent to which the institution operates as a unit.

Both these linkages link subsystems and are commonly themselves subsystems. This paper focuses on systems providing a specified kind of service or material goods which have outgrown the possibility of attaining optimal integration through interpersonal, spontaneous, face-to-face contact, and therefore require formal integrative devices.

In classical organization theory the dominant criterion for this classification was size: a system greater than a given size required formal devices, a smaller system did not. Recent empirical research findings indicate that the degree of differentiation, and therefore the need for

From *Administrative Science Quarterly*, 1969, vol. 14, pp. 398–416. Reprinted with permission of the author and the publisher.

[1] Particular thanks are due to Prof. Louis R. Pondy, Business Administration and Community Health Sciences, Duke University.

integration, correlates much more closely with the degrees of uncertainty in technology and markets than with absolute size (Downs, 1967; Harvey, 1968; Lorsch, 1965). The focus of this paper therefore is on systems (1) that need to cope with high degrees of uncertainty, and (2) that have developed subsystems to develop innovations for the system. In industry, organizations manufacturing electronic apparatus and certain chemicals are examples of such a system; and research and development departments are an example of such a subsystem.

ASSESSMENT OF UNCERTAINTIES

The accuracy with which decision makers in the system assess uncertainties for the system guides formal differentiation within the system as well as the design of integrating devices. There are two basic models for this assessment. One model treats uncertainties as a succession of discrete stimuli to which the system needs to respond with appropriate innovations. In this model differentiation and integration are functions of the frequency and force of the expected stimuli. It is a model that has affinities with familiar stimulus-response models in learning theory and with the early formulations of an administrator's functions in terms of weight and frequency of decision making. The integrative devices appropriate to this assessment are linkages of minimal complexity and duration. As stimuli for innovative responses increase in frequency, temporary and informal devices may become permanent and formal, but the kinds of devices and their limitations will not change unless the model of assessment changes. The other model treats uncertainties as a continuous general state requiring a continuously varying response from the whole system. This model treats the system and its environment as in continuous interaction in uncertainty and as aiming to achieve and maintain a steady state in and through this interaction. Emery uses the word "turbulence" to describe the environment in this model (1967:225).

The spontaneous response to a turbulent environment is to reduce the turbulence. The first model aims at reducing the complexity and uncertainty by segmentation or dissociation, which are essentially the defense mechanisms that lead to passive adaptation. The second model

FIGURE 1

Two Models of Assessing Environmental Uncertainties
and Characteristic Linkages

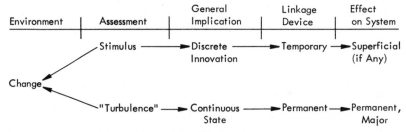

aims at establishing hierarchies of goals, which may then serve as guides to behavior, i.e., to allocating attention and other resources to causal strands on the basis of established priorities. Through the development of such codes, the field is no longer perceived as turbulent. Questions about the values themselves, and of the transformation of values into priorities and codes, will not be pursued in this paper, though the later discussion of institutionalized linkages could be analyzed in terms of the values that could emerge from the interactions they encourage (Emery, 1967:225, 229). In the first model, the cost of the responses is a permanent ceiling on the rate of innovation; in the second model, costs occur from the slower rate of innovation due to various rigidities associated with the acceptance and revision of codes.

This paper treats the environment as turbulent and explores the kind of appropriate response by the system, particularly the design and operation of linkages devices. Figure 1 illustrates the basic two-directional schema for this paper.

The empirical data to be introduced and analyzed at some length in later sections of the paper show a heavy incidence of failure for *ad hoc* and temporary linkage devices. They refer to situations in which linkages were designed to perform some specific function(s) and failed to do so. Actually an unrecorded kind of failure may be more serious, namely the system rigidities that the introduction of the inadequate linkages reinforced. The basic weakness of the discrete stimulus-innovation model shows up most sharply when the intended linkage *succeeds:* For there is no evidence that the successful integration of two or three subsystems increases the degree of integration of the system as a whole; on the contrary, commonly, a well-integrated subsystem (or a set of subsystems) may become isolated, and then it and the rest of system become rigid separately, in defense against one another. Katz and Kahn (1966:390) state that "the major error is to disregard the systemic properties of the organization and to confuse individual change with modifications in organizational variables." Very little of the growing literature on innovation and its diffusion has been attentive to this problem. The prevailing focus of attention is on the individual innovator when he adopts an innovation, why, etc. (Rogers, 1962), not on the organizational setting in which innovation takes place. The early sociological role studies may have unwittingly supported this orientation. In the rural economy the typical adopter was an individual farmer rather than a collectivity, such as an organization. Even in later studies the influence of the larger social setting on the farmer-innovator tended to be underestimated. Studies of innovation in educational systems, like schools of various kinds and school districts, either show a lack of available knowledge about the functioning of organizations and communities or suffer from a kind of great-man orientation (Ross, 1958; Carlson, 1964). In anthropological studies too, the properties of a particular innovation and its diffusion across systems and integration within systems have been overemphasized without corresponding attention to the dynamics and functioning of the receiving organization. Gallagher (1963) discussed the power structure in innovation receiving systems, the actual prestige of advocates of the innovation, and other matters influencing if and how an innovation would be integrated into the local organization. But his primary focus too is on the substance of the particular innovation, taking the local system itself as a kind of unmodifiable ground against which the innovation showed up starkly. The currently widespread emphasis on the importance of "dissemination of research findings" seems to continue the popular view that the content or demonstrated efficacy of a particular innovation as such is crucial in determining whether it will be adopted and used effectively. In short, to use an image from Gestalt psychology, specific planned change attempts have most typically been "in figure," occupying the focus of attention, while the organization itself has remained "the ground." That this model commonly persists in the face of growing experience of its inadequacy indicates its long antecedents in history and also the deep reluctance many decision makers seem to feel about making major system changes that may be required to interact effectively with a turbulent environment. They prefer to assess the uncertainty as small and the response to it as routine, whereas it might be more useful to take as a primary target the improvement of system dynamics, specifically the linkage design. To continue the analogy then, this paper treats linkage design as "figure," and innovative needs as "ground."

In Emery's formulation (1967), turbulent environments require the linkage of dissimilar subsystems whose goals correlate positively, i.e., subsystems which are not in competition, cannot take over the role of the other. Therefore linkages between them will tend to maximize cooperation. As Thompson (1967) has shown, such a matrix can take several shapes, but all provide for two aspects: (1) that which links the system to the environment, through which broad social sanctioning is secured and the system is attuned to the needs of the environment and (2) that linkage which enables subsystems to engage in effective joint search for common ground rules, while retaining the degree of privacy, protection, and autonomy they need to carry out their distinctive functions.

The strategic objective then has to be formulated in terms of institutionalization. As institutionalization becomes a prerequisite for achieving a steady state in a turbulent environment, then subsystem goals have to be found and formulated which accord with system goals and which offer a maximum convergence between its subsystems.

SYSTEM DIFFERENTIATION

Effective linkage presupposes appropriate differentiation, which is in turn required by new task complexities. As these occur, subsystems need to be built to deal with primary tasks. Over time, successive levels of differentiation can be established, each reaching primary production systems. The primary task differentiates one subsystem from other subsystems at the same organizational level, and also from subsystems at higher and lower levels of the hierarchy (Lorsch, 1965:151; Lawrence and Lorsch, 1967:213; Miller and Rice, 1967:157; Thompson, 1967).

Appropriate differentiation has three possible dimensions: technology, territory, and time; and Miller (1959) states: "Task performance is impaired if subsystems . . . are differentiated along any other dimension than these." He indicates that at any one organizational level, these three dimensions can have seven possible combinations.

Types of task dependence have been classified in some detail by Herbst (1958), Emery and Trist (1960), and Thompson (1967). It is relevant here to consider how, at a given level of differentiation, the subsystems of a larger system are all dependent on supplies, equipment, and services, and are interdependent for the attainment of the goal of the larger system. Emery's and Trist's study (1960) classifies interdependence into simultaneous or successional, and classifies successional dependence further into cyclic, convergent, or divergent. They also distinguish between simple and complex dependence and between reciprocal and nonreciprocal dependence. Thompson (1967) presents a simpler typology of interdependence: "pooled," "sequential," and "reciprocal." Lorsch and Lawrence (1968) used Thompson's typology to distinguish between organizations having different degrees of differentiation and using different linkage devices.

While subsystems have in common "an inherent centrifugal tendency (Parsons, 1956), Miller (1959) noted some particular tendencies associated with differentiation by technology, territory, or time. Differentiation by technology tended to make the preservation of distinctiveness a primary task of the subsystem (for example, demarcation disputes between skilled craftsmen). Differentiation by territory led subsystems to develop a special climate and culture, a particular proficiency. The most complex tendencies were associated with differentiation by time, e.g., tasks with different time cycles or organized on different shifts (Trist *et al.*, 1963; Lawrence and Lorsch, 1967).

Three of Litwak's (1961) mechanisms of segregation seem to overlap with Miller's for determining points of development at which a shift from one form of structure to another is to be made. His "transferral occupations" arise in the linkage subsystem which this paper examines.

RESEARCH VERSUS OPERATIONS

Since linkage devices between research and development subsystem(s) and operating subsystem(s) in industrial systems best illustrate the main issues, most of the data for this paper were drawn from studies in industry. Research scientists and operating personnel differ sharply in all three major dimensions of differentiation. In technology, the researcher deals with nonuniform processes, the man in operations with uniform or nonuniform, but programmed, processes (Perrow, 1967); in territory, the two systems are usually kept physically apart, often to avoid conflict (Litwak, 1961); in time, research scientists have longer-range concerns and expect much later gratification from feedback about the results of their work than operating personnel (Lawrence and Lorsch, 1967: 35, 36).

The differences are extreme in organizations in which research scientists have great freedom in their choice of problems and procedures. In these cases, the norms and organization of the research department approximate those of an academic environment (Litwak, 1961:183). Burns and Stalker (1961:225, 229) maintain that the "prima donna" scientist is actually a more familiar figure in the electronics industry than in the academic world. The characteristic norms of innovative subsystems include: high energy devoted to "novel, significant, focused, internalized, shared goals"; "esprit de corps," "mutual identification with peers"; "high autonomy and spontaneity, with freedom for creative experimentation"; and high involvement and commitment (Miles, 1964:655). The researcher's commitment tends to be to the task or to the discipline, whereas the operating personnel's commitment is to the organization; he therefore responds to the sender's recognized research status, whereas operating personnel responds to the level of administrative authority (Smith, 1966:58). He tends to be a younger person than persons of equivalent rank in operations and to have more years of formal education than they (Dalton, 1950). The differentiation between research and operations is marked further by a series of interlocking signs and differences of conduct, for example, different working hours and differences in dress and leisure pursuits (Burns and Stalker, 1961:186).

The differentiation between research and operating subsystems in industry seems then to be institutionalized at all levels: image and self-image (personal), face-to-face relations (interpersonal), organizational structure (intersubsystem). Moreover, although differentiated groupings tend to be self-perpetuating, special technological factors reinforce this tendency in the case of research and operating subsystems. For research generates further problems for research, and these tend to crowd out problems on which other subsystems would prefer the researchers to work. "Research institutions have therefore a natural tendency to become increasingly divorced from their (organizational) environments and their boundaries to become increasingly impermeable" (Miller and Rice, 1967:157). Yet, although researchers prefer to work on their own problems, they may need access to data and certainly do need resources from other subsystems. These needs leave important aspects of control and power in the hands of other kinds of people and other kinds of subsystems. In short, technological factors inherent in the research process exaggerate the centrifugal tendencies of research subsystems (Havelock, 1967: 23–24), and feedback loops that tend to develop between individuals in subsystems and restrict communication in self-confirming, stabilizing ways are especially tight in this case (Miles, 1964:644). In such sharply separated subsystems, members have even greater difficulty than usual in understanding the language of the members of other subsystems (Burns and Stalker, 1961:155, 174; Kahn *et al.*, 1964:133; Kast and Rosenzweig, 1962:325–26), for each perceives and internalizes those aspects of any new situation that relate specifically to the activities and goals of his subsystem (Dearborn and Simon, 1958:149–144; Barnes, 1960:3–4). In these circumstances, misunderstandings are a common occurrence and tensions are normal. In an atmosphere of tension, joint decision processes between research and operating subsystems then tend to be characterized by bargaining rather then problem solving (Pondy, 1967:319). Some of the characteristics of this bargaining are "careful rationing of information and its deliberate distortion; rigid, formal, circumscribed relations; (and) suspicion, hostility and disassocation" among the subsystems (Pondy, 1967:319; Burns and Stalker, 1961: 192, 194; Kahn *et al.*, 1964:134; Litterer, 1963:405–407; Chandler, 1963:154). Shepard (1956) summarized the differences in outlook between the researcher and the operations man along seven dimensions (Table 1).

TABLE 1

Researcher versus Operations Man: Differential Perceptions of Seven Items*

	Researcher	*Operations Man*
Research	Future-oriented, perhaps distant future	Focused on present, or just beyond
Identification of researcher	His profession	The company
Research results	Published achievements	Guarded secrets
Research budget	Based on research requirements	Based on company needs
Authority for research	Shared among scientists	Delegated by management
Organization of research	Project groups	Functional groups
Complexity of research management	Desirable fulfillment of managerial competence	Undesirable regression into poor administrative practice

° After Shepard (1956).

These rigidities and negative attitudes provide the potential for conflict. Joan Woodward found that "hostility between research and production personnel . . . probably exaggerated in turn the differentiation based on task differences. This was a noticeable feature of all the large batch and unit production firms studied" (1965:149–51). Studies in a variety of settings show that conflict and tension add strong impetus to the centrifugal tendencies of differentiation. In this atmosphere the parties each have a set to expect conflict and therefore may perceive conflict even when none exists (Pondy, 1967:319). This separation involves high cost. Miller (1959) found that the higher the degree of tension, the smaller the energy directly invested in the pursuit of the objectives of subsystems or of those of the system as a whole (see also Hermann, 1963).

HYPOTHESES

It is now possible to focus more systematically on the main thesis of this paper. In a turbulent environment, institutions must innovate to survive. Some institutions have differentiated subsystems with the primary task of working out innovative responses to the turbulent environment for the whole system. This task is inherently different in technology, territory, and time from the tasks of other subsystems. Adoption of the innovative products of the subsystem by other subsystems is essential to the effective response of the system to its environment; therefore, innovative and operational subsystems need to be appropriately linked.

On the assumptions that the decision makers of a system are in control of the system's response and that they act rationally, the degree of innovation and the mechanisms of differentiation and of integration depend on the decision makers' assessment of the changing environment. Downs' (1967) study explores some personality factors and develops a concept of "sunk costs" to deal with various patterns of disposing of resources in the system and the tendencies these patterns induce over time. Using assessment for this totality, even without examining its components, may at least indicate that important complexities may be worth studying in this connection and are not to be obscured by the more familiar label "perception."

Assessment, as used in this paper, is a function of the decision makers' joint preference for the goal of the institution after taking into account the costs of innovation. It is therefore both how the decision makers assess the environment and the terms in which they rationally explain the innovation and the mechanisms that they institute for it. In this model the decision makers' dominant assessment of environmental requirements for change is the independent variable, and differentiation and linkage devices are dependent variables.

Four major assessments are possible, each of which can lead to a hypothesis. They are ordered here in their historical sequence which also corresponds to ascending degrees of system change. Relating these situations and hypotheses to the two models shown in Figure 1, Hypotheses 2 and 3 restate the first model, and Hypothesis 4 is the second.

1. The decision makers see "nothing," "no need to change," therefore any need for change that actually arises will be met by *ad hoc* and informal arrangements, e.g., by individuals adjusting their roles.

Hypothesis 1. If, in a turbulent environment, decision makers assess the need for change as negligible, system differentiation and linkage will then be minimal and unorganized, and the effects on system structure will be negligible.

2. The decision makers assess the need to change as temporary, and will organize to meet it by temporary measures, such as meetings and temporary allocations of resources. In short, they will treat it as a project.

Hypothesis 2. If, in a turbulent environment, decision makers assess the need for change as temporary, they will tend to expect the ratio of benefits and costs to be most favorable if they institute temporary differentiation and linkages; the effects of innovation on system structure will then be small and unpredictable.

3. The decision makers assess the need for change as frequent, but specific, and differentiate a permanent subsystem to deal with these specific changes, while the institution as a whole will proceed as before.

Hypothesis 3. If, in a turbulent environment, decision makers assess the need for change as frequent and specific: (*a*) they will formally differentiate an innovative subsystem, but limit its linkages to specific purposes and subject them to direct hierarchical control; (*b*) when this kind of device and procedure is found to be ineffective, additional system changes will tend to take the form of further differentiation of innovative subsystems, not of changing the linkage devices and procedures.

4. The decision makers assess turbulence in the environment as a continuous state, which calls for sensitive, multifaceted, often unpredictable, and general innovative responses from the system as a whole. Differentiation and integration are then major foci of attention—for system design and preventive maintenance. Subsystems become sharply differentiated and deliberately linked on a permanent basis, and innovation affects the whole institution to a major extent.

Hypothesis 4. If, in a turbulent environment, decision makers assess the need for change as continuous and major, they will tend to differentiate innovative subsystems clearly and integrate them closely into the rest of the system; the effects on the system as a whole will then be major.

The rest of the paper examines the linkage mechanisms characteristic for these four situations and hypotheses.

NEED FOR CHANGE IS ASSESSED AS NEGLIGIBLE

Situations in which responses within the normal range of adaptation of the system as currently constituted, and of the individuals in it, can cope with changes in the environment are outside the scope of this paper. Its proper scope begins only when the integrative needs exceed normal adaptation. This commonly occurs imperceptibly as a system grows. If such a discrepancy occurs and persists unattended, unplanned linkage devices find their way into the system's routine workings. In society at large, these unplanned mechanisms are most widely diffused in the form of new norms through literature, art, folklore, mythology, beliefs, mores, orientations, "small talk," etiquette, and through institutional practices of many kinds (Thayer, 1968:241). This list, certainly from folklore onwards, could be applied to any institution. In industry, "management's attempts toward integration may just happen over time, arise as temporary expedients, or emerge as a solution to a crisis situation" (Jasinsky, 1964:329).

At some stage there is an innovator, perhaps a scientist, or a workman. The initiative for informal attempts on linkage usually lies with this innovator. The point here is that, for application, the innovation requires the involvement of people who are different from the innovative person, and who will perceive him as a stranger or agitator (Inkeles, 1951:38–135; Katz and Lazarsfeld, 1955: 162–208). Such informal personal linkage makes it possible to appreciate the roots from which formal differentiation and linkage devices commonly grow. Figure 2 illustrates the position of the informal direct linker in an undifferentiated system.

Kahn and his colleagues (1964:125–36) found that innovative ac-

FIGURE 2

Informal Direct Linkers in
Undifferentiated System

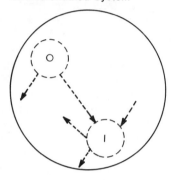

I = innovator, O = operator,
solid circle = formal boundary,
broken circle = informal boundary,
— → = informal contact.

tivity was associated "significantly and positively with both the degree of role conflict ($p < 0.01$) and the amount of tension that the role occupant experienced on the job ($p < 0.05$)." The innovator's interpersonal conflicts were "fought out around his proposals for innovation and vary in intensity according to the orientation of decision makers towards innovative functions." His intra-role conflicts arose from his characteristic impatience with uniform and routine procedures and from the lack of trust he felt "in (his) role senders, irrespective of the degree of conflict ($p < 0.06$)." He "had significantly less communication than others with their role senders ($p < 0.02$)," was also more self-confident in the face of role conflict than persons in uniform work ($p < 0.02$), was more highly involved than others in his job ($p < 0.05$), and attached more importance to his job relative to other areas of life than people in less innovative roles ($p < 0.05$).

The emergence and enthusiasm of such an innovator in the system certainly expands the system's capacity for adaptation, but the cost of informal linkage increases as the expanded limits in turn prove to be too narrow (Burns and Stalker, 1964). Where the main force of innovative activity is concentrated in an energetic scientist-enterpreneur, the linkage problems may overwhelm him.

Linkage may be attempted by an informal third party, most commonly a draughtsman, engineer, sales agent, or purchasing agent. Litwak (1961) calls their role a "transferral occupation." The studies of Burns and Stalker (1961:177) and Woodward (1965:138) are replete with cases in which managers either attach draughtsmen and engineers to development or to production, in the hope that they will "just get along" with their colleagues within the confines of normal bureaucratic rules, or leave them unattached to find their own place in the organization. Figure 3 illustrates the position of the informal third party in such an undifferentiated system.

Kahn and his colleagues conclude that "concentrating the functions of organizational liaison on a very few positions . . . risks much in a few hands, forces a search for champions to fill the crucial positions and is likely to create inter-organization struggle to insure that their interests will be well represented by the overworked representative" (1964:392–93). The concepts of marginality and tolerated deviance occur again and again in the literature concerning linkage and, along with them, the thought that these marginal men need bases from which to work effectively (Woodward, 1965:142).

FIGURE 3

Informal Third-Party Linkers in Undifferentiated System

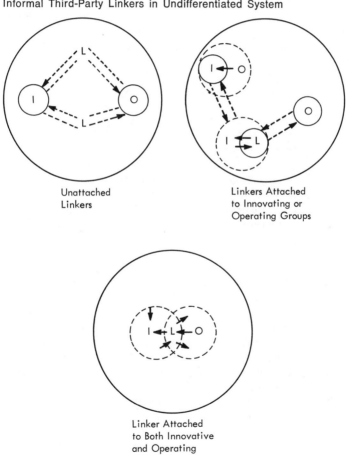

Unattached
Linkers

Linkers Attached
to Innovating or
Operating Groups

Linker Attached
to Both Innovative
and Operating
Groups

L = linker, I = innovator, O = operator, solid circle = formal boundary, broken circle = informal boundary, — → = informal contact.

NEED FOR CHANGE ASSESSED AS TEMPORARY
AND INFREQUENT

The logical response to needs assessed as temporary is to devise correspondingly temporary linkages, a particular type of what Matthew Miles (1964) terms "temporary systems." This paper deals only with situations where this assessment in fact is mistaken, since situations of turbulence are assumed to require permanent linkage devices of a pervasive kind. The situation here differs from that examined in the preceding section, in that (1) the decision makers do see the need for a response beyond the routine capacity of the system; and (2) linkage is made a formal occupation, albeit a temporary one. Formality means more security for the linker at the cost of more constraint. Whereas in the informal situation, the choice of attempting linkage activities, and the manner and timing, were largely the linker's, now the decision makers are likely to play a significant part in initiating, designing and controlling the linkage activities to conform to their assessment. The decision makers' direct interest makes the linker(s) *their* representative. This, in reverse, can give the linker(s) access to possible power and coercion over the people and groups involved in the problem. Figure 4 illustrates formal temporary linkers in a system with no innovative subsystem.

Linkage devices for this situation include temporarily freeing a staff member from his usual activities in order "to deal with the situation," *ad hoc* meetings of people involved in a change, calling in outside consultants, and sending some staff members(s) for training. The first three usually have overtones of "troubleshooting" and of latent coercion by decision makers and their representatives. The question here is whether these overtones promote effective linkage, deter it, or make no difference. The Lawrence and Lorsch (1967) study of within-system linkers found that access to coercive power was not associated with effective linkage. In two high-performing (innovative) companies the linkers were seen as very competent, and the subsystem managers ascribed their influence over decisions to this competence. The linkers in four low-performing (non-innovative) companies on the other hand, were accorded an important voice in decisions "because of the formal authority given them by top management and their close reporting relationship to top managers" (Lawrence and Lorsch, 1967:65–66).

Linkage activities in the lower-performing companies appeared to have many aspects characteristic of bargaining. Some linkers in this situation tried to avoid personal contact and preferred to resolve problems through formal memoranda. Similarly, Strauss (1962:182) found that linkers tended to expect that people in affected departments would not change their minds.

Resort to authority and coercion can be very tempting: forcing people to make the specific changes required seems quick and easily controllable, that is, inexpensive. Since the need is specific, it does not seem necessary that people should change their minds or appreciate the change. Many studies report *ad hoc* meetings that are so overshadowed

FIGURE 4

Formal Temporary Linkers in Systems with No Innovation Subsystem

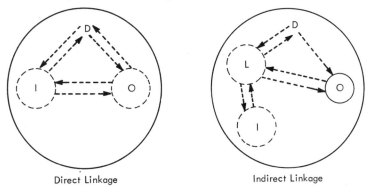

Direct Linkage   Indirect Linkage

L = linker, D = decision makers, I = innovator, O = operator, solid circle = formal boundary, broken circle = informal boundary, $- \rightarrow$ = informal contact.

by the latent power of absent decision makers, that joint decisions are avoided on the grounds that top management will make its own decision anyway (Lawrence and Lorsch, 1967:66; Burns and Stalker, 1961; Woodward, 1965). These meetings then serve either ritualistic purposes only or become rather elaborate and ill-constituted mechanisms for exchanging information. The ineffective use of time is itself costly, but the indirect costs of such coercion are likely to be higher still. For in a turbulent environment needs for change recur, and meeting these needs by coercion, open or implied, increases resistance to future changes and decreases initiative by people required to direct the changes.

Many decision makers who assess the need for change as temporary seek to reduce the costs of internal disturbance by contracting for temporary assistance from outside the system. Consultants are seen as offering the advantages, not so much of expertise beyond that of employees in the system, but of the outsider free of the normative and interpersonal constraints of the system, and free also of the need to carry on simultaneously their usual activities. Consultants clearly *are* temporary. In their turn, consultants who conceive of their task as consisting of diagnosis and prescription may thereby support the assessment of the decision makers that the need for change is temporary and specific, and calls only for a response equally temporary and specific.

Another means of dealing with linkage in this setting is to change some of the staff members who are important in the linkage process. This may be done by replacing personnel who have been unable to cope with the needs for change by others who are "better at getting on with people." The replacement may be temporary or permanent, depending on the frequency and importance of new linkages and alternative dispositions of staff resources. Changing the staff may also be attempted through training, particularly training in attitudes and interpersonal and organizational skills. The studies of Miles and others of innovation in educational systems indicate that changes of this kind call for *additional* linkage—that between the temporary training system and the parent system. The success of action decisions taken in the temporary system depends not only on the quality of these decisions, but also on the sophistication with which members have been able to anticipate the strategic problems they will encounter upon their return (Miles, 1964:484). Miles (1964:643) reported that in one three-day conference of a management team devoted to the improvement of operations in a chemical plant, the members agreed on a list of 47 action decisions, specified deadlines, and identified the persons responsible for ensuring action. When the consultant pointed to the need for some mechanism to guide implementation and cope with new problems, the top manager declared that implementation could easily be accomplished "through the regular line organization"; yet it was clear that the inability of the line organization to cope with these problems had necessitated the conference.

The probable usefulness of training, as of any other temporary devices for dealing with change, is enhanced if the decision makers of the system themselves take part in the temporary system, for studies show that recommendations are then more likely to be accepted (Miles, 1964). But this participation of decision makers also seems to ensure that any action recommendations will be moderate. This is merely another way of saying that meeting the realistic needs of the system is at variance with the constraints of normal system operations. That is the dilemma with which decision makers in these situations seem unable to cope.

NEED FOR CHANGE PERCEIVED AS FREQUENT, BUT SPECIFIC

When decision makers assess the need for change as frequent but still discrete, then the costs of setting up a formal subsystem with the primary task of designing innovations for the system as a whole are seen as justified. But the change is still conceived as an addition to the system, not as a reshaping of the system as a whole for innovation.

The input for the new subsystem may be proposed by the decision makers, by others in the system, or by members in the new subsystem. The output of the new subsystem is expected to be innovative ideas which the rest of the system will then decide to use or not to use. The task of the subsystem is conceived as terminating with communicating the ideas clearly enough for decision makers to understand and so make their decisions rationally. Linkage mechanisms are then conceived as bridges across the gap that is recognized as existing between the differentiated innovative subsystem and other subsystems. Figure 5 illustrates this concept of linkage as bridging the gap between subsystems.

In their study of Scottish engineering units, Burns and Stalker (1961:194–98) described three situations of this type in different companies. In each, a newcomer was brought in to head a new, innovative subsystem. For each system, successful innovation was "quite essential" for survival as an "independent business with its own capacity to compete with others and to expand." In each case, the innovative subsystem was unable to get innovative ideas accepted, and its manager resigned.

Case 1: A new chief development engineer was brought in on the explicit understanding "that he should take over an increasing share of the general management, i.e., of production as well as of design," and so he himself would be an important linkage of the development subsystem with other subsystems. In fact, he was confined by decision makers to acting as a consultant, supplying technical information at the

FIGURE 5

Linkages between Innovative Subsystem and
Other Subsystems

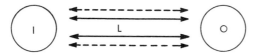

stage of negotiations for contracts with customers. At a critical point in his endeavors to extend his authority beyond the design department, he proposed a plan for reorganizing the system. He was immediately offered a rise in salary and a new title, but no significant organizational change. He resigned.

Case 2: To render it independent of obsolescent products and methods, an established company brought in an industrial scientist to take charge of the laboratory group. His understanding was that he would "take over the whole technical side and be responsible for technical development." After two years the scientist regarded this venture as "wholly unsuccessful. . . . Everything I've put up has been blocked . . . none accepted. It's been their view that whether or not such and such a development takes place is just not my affair; this is the manufacturer's concern . . . and similarly the question of how something is to be developed and produced." Prior to the scientist's resignation, after three years, interaction between him and the managing director in the adjoining office had ceased altogether.

Case 3: Following the Second World War, the directors of the third company decided to engage an industrial scientist as head of the research and development unit to do exploratory work among recent innovations and possible markets. The new research and technical manager was quickly effective in extending the firm's range of materials and processes along traditional lines, and in increasing technical sales. But these very successes, then, locked him into short-term activities with operating subsystems. When a major opportunity occurred for developing substitute plastic materials, he was not given the resources to explore it. Instead the scientist was appointed manager of a new department in addition to continuing as head of the research and development unit. (Development work then became "virtually impossible," while production in the new department was held back by traditional inappropriate practices.) So he too resigned.

Although the newcomer was placed formally on the same level as the departmental and functional managers immediately below the managing director, the role allowed him in practice added to the existing structure and was external to it: he was not permitted to infringe on the control exercised by established members of management. As a result, he was forced to act in a consultant, advisory role, while members of the established hierarchy retained the decision making and control functions. That is, the innovative subsystem had been appended, and no allowance had been made for altering the fields of control in the system structure. This happens, conclude Burns and Stalker (1961: 199), "when a special person or organized group of persons is designated as the agent of change and the existing organization is relegated to the role of spectator or patient."

The issue of significant structural change tends to be evaded in two ways. First, difficulties in effecting changes are ascribed to personal conflicts. Given the view of decision makers that an innovative subsystem is an addition, they then see any conflict as arising from the newcomers' personal interest in acquiring more control than they need. Factors of status and power develop and find expressions in increasingly devious and intractable ways, until system weaknesses are translated wholly into personal terms. Then, since experiences like those, being painful, are likely to be repressed rather than examined, disentangled and shared, this personal focus will tend to perpetuate the structure of the system.

The second way in which the significant system changes are evaded is by faulty system responses to difficulties that the subsystem has in effecting linkage. Although the difficulties really originate in the inadequate design and operation of linkage mechanism, decision makers respond by instituting further changes in the innovative subsystem itself. One change is to assign members of the innovative subsystem additional responsibilities in operating subsystems. Such a combination of roles does not prevent failure (Burns and Stalker, 1961; Havelock, 1964:25; Lawrence and Lorsch, 1967), although Pelz and Andrews (1966) consistently reported that scientists and engineers in industrial research and development departments who participated in management and knowledge dissemination were more effective and productive also as scientists, as judged by publications and by peer ratings of their scientific excellence and overall usefulness. There may be a cause-and-effect confusion here which only further empirical study can clarify: does such scientific productivity tend to lead to effective participation in the system? Or, does effective and acknowledged membership in the general system tend to promote scientific productivity? Another possibility is suggested by Sieber's (1966) study of the organization of educational research: that some functions may be combined effectively and profitably, as in the Pelz and Andrews (1966) study, while others, such as research and services, tend to combine badly. Kahn's study (1964:393) suggests that roles may also be effectively occupied in rotation.

A second change following linkage difficulties is to leave scientists alone to get on with their science. This is done by freeing them explicitly from the linking task in which they have failed and by establishing an additional group of a different kind of scientist specifically for the linking task. But without further structural change in the whole system, these different scientists are bound to fail at linkage in their turn; then the same process may be repeated, and yet another group may be entrusted with the linkage task. In time, several additional subsystems may get interposed in this manner between the original innovative subsystem and the operating system. Each of these subsystems has small-step linking tasks to the subsystem on either side of it in a lengthening chain of innovative subsystems. Burns and Stalker (1961:157–58) found that some organizations had separate departments for "pure" and "basic" research, and for "design" and "development," although the subsystems in each pair were "almost indistinguishable." Other studies (Kast and Rosenzweig, 1962; Quinn and Mueller, 1963) point to a similar proliferation of scientists and engineering specialists. This process fails to establish the required linkages to the operating system and at the same time makes the necessary system development more complicated and costly. The tendency to proliferate innovative subsystems in attempting to deal with linkage difficulties can be expected to be reinforced by the personal and professional inclinations of both researchers and operating men described in an earlier section of this paper.

The following hypothesis may be worth examining: In situations of stress between innovative and other subsystems, scientists will seek to safeguard their autonomy of work by proposing that the linkage function be carried out by new subsystems that are oriented more to application and less to science. Operating men support the same tendency when they overestimate the value, and underestimate the cost, of further differentiation in the innovative subsystem, following their erroneous impression that an up-to-date system consists of a multiplicity of subsystems.

All this elaboration of the system leaves the major difficulty untouched and is costly. The direct costs of coordinating go up with every differentiation of subsystems, and indirect costs may be even greater, including the creation of jobs, groups, and even departments of highly paid personnel whose survival in the system depends on the perpetuation of their subsystems. These costs are incurred in the attempt of decision makers to escape the problems of reshaping the system for continuous change.

NEED FOR CHANGE ASSESSED AS CONTINUOUS

The data for this section of the paper are drawn predominantly from four major field studies in industry to which frequent reference has already been made (Lorsch, 1965; Lawrence and Lorsch, 1967; Burns and Stalker, 1961; Woodward, 1965). All are published in the 1960's, three since 1965. Two present British data, two American. All contain comparison of linkage mechanisms in similar contexts of technology and uncertain environments. But linkage mechanisms effective in some systems seem to be ineffective in others.

The attempt in this section is to analyze the successes and failures in linkages, so that some common factors can be established. The overall issue is institutionalizing linkage mechanisms so that they function as permanent parts of a system that is continously and flexibly engaged in change; that is, a system that is functioning effectively in a turbulent environment.

The analysis will be in terms of four dimensions: differentiation of linkage mechanisms, the relation of such a mechanism to the subsystems it links, its norms of operation, and its congruence with system structure.

Differentiation

The studies showed effective linkage mechanisms to be formally organized on a permanent basis; for example, engineering department,

FIGURE 6

Differentiation of Linkage Mechanism from Subsystems and from Decision Makers

Effective Differentiation

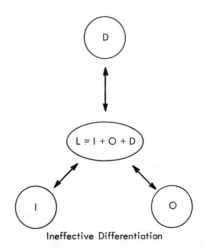

Ineffective Differentiation

D = decision makers, L = linker, I = innovator, O = operator, solid circle = formal boundary.

product management meeting, research and development committee, goal team. The effective linkage mechanisms consisted of people from the different subsystems to be linked; that is, the linkage mechanisms were cross-functional in composition and represented a particular type of differentiation.

The differentiation of effective mechanisms was clear both *vis à vis* the subsystems to be linked and *vis à vis* the decision makers for the whole system. Regarding the first, Lawrence and Lorsch found differentiation to be effective "to the degree to which managers in all the departments felt that they had influence over decisions" (1967:56). As to the second, they found the influence of effective linkage mechanisms to be "centered at the operating level, neither higher nor lower than where the decisions could be effectively reached and implemented" (1967:56).

Figure 6 illustrates effective and ineffective differentiation in these terms. Lorsch (1965:122) listed the differences between effective differentiation in one company and ineffective differentiation in another as follows:

Effective Mechanism	*Ineffective Mechanisms*
Limited attendance to four functional members from the subsystems	Large, included superiors and subordinates
Set out to solve problems and resolve conflicts	Set out primarily to exchange information
Did not involve decision makers	Dropped any difficult problems or asked decision makers to solve them

Linkage Mechanism in Relation to Innovative and Operating Subsystems

Effective linkage mechanisms were equidistant between the subsystems they were to link. Lawrence and Lorsch (1967:59) examined this distance in terms of four "orientations": to goals, time, interpersonal relations, and the structure of the linkage mechanism. Linkage mechanisms in high-performance company A were equidistant on all four dimensions, in low-performance company A on only one. The mechanisms of the other four companies they studied were equidistant on two dimensions.

A second kind of position factor relates the membership of the linkage mechanism with different hierarchical positions in the subsystems. According to Lawrence and Lorsch, a high-level production manager usually has sufficiently specialized knowledge of production processes to consider innovations proposed by scientists and can reach effectively decisions about them in a joint meeting. On the other hand scientific research and development is so specialized that the production manager's peer for linkage is a scientist actually working on the specific problem, or his immediate superior. And the corresponding peer in sales is likely to be a middle manager. The key linkers, whom Lawrence and Lorsch called "integrators," were at different hierarchical levels within their various subsystems and in the system as a whole. Where linkage mechanisms were effective, these hierarchical differences of members outside the linkage subsystem were relinquished upon entry into the linkage subsystem and neither blocked the decision making process nor reduced the acceptability of its results to other subsystems (1967:55–56). The differences in effectiveness related to "the degree to which these key integrators (were) seen by others in his organization as having the most important voice in decisions" (1967:62). In other words, while subsystems may have been formally represented at meetings by their managers, the major activities in and out of formal meetings were those of the various key linkers. Burns and Stalker (1961:88–89) found that the word "committee" captured the essential organizational equality of the members, irrespective of their status outside. In the companies they studied, they found linkage invariably associated with this equality. Effective linkage mechanisms "were indeed devices for abrogating for the length of each meeting the distribution of authority, information, and technical competence pictured in the hierarchic structure of the organization." They saw this abrogation as addressing appropriately the uncertainties in requirements for which the normal management structure was not appropriate (see also Lawrence and Lorsch, 1967:67, ff).

A third position factor is indicated by the reward system. Effective linkage seems to be associated with the extent to which linkers feel evaluated and rewarded according to the overall performance of the subsystems linked. According to Lawrence and Lorsch (1967:67, ff), it was important that linkers "had an incentive for effective integrative activities over and above the supervisory, planning and other functions" they performed within their own subsystems. Similarly, the goal team in Lorsch's (1965:122) study felt responsible for profits in their whole market area, so that each member "should be committed to the decisions we reach as a group."

Operating Norms of Linkage Mechanism

Operating norms relating to two aspects seem to be of particular importance: conflict resolution and flexibility. Lawrence and Lorsch (1967:73–77) list six factors in conflict resolution. According to them, linkage was effective to "the degree of openness with which conflicts were aired and hammered out in departmental meetings." Their Table III-6 sets out the partial determinants of effective conflict resolution in the six companies.

Flexibility appears to be the second essential norm for the operation of an effective linkage mechanism, with kind and degree of flexibility determined by the functional needs of the mechanisms and the particular situations with which it is dealing at any time. Burns and Stalker (1961:87–89) described the flexible operation of a new research and development committee. It began as "a large, comprehensive affair" but was reduced in size when it became unprofitable to conduct the meetings in a language that shop-floor supervisors could understand readily. Therefore the large meeting was changed. A small professional committee dealt with ongoing projects, and regular foremen's meetings were used to keep the supervisors routinely informed about the committee's work. In short, the hierarchical structure was used as a communication system for this as for other routine matters. However, the committee meetings were expanded to their earlier size whenever a major design change was developed for one of the two main products of the company. Smith mentions a similar flexibility around the question of inviting system decision makers to meetings concerned with linkage (in Bright, 1962: 49; see also Thompson, 1966:139–43).

Congruence with System Structure

Lorsch (1965:147) found the locus of formal authority congruent with the norms of conflict resolution: "The level with the authority within each subsystem about scientific transfer was also the level that the norms indicated should resolve differences between subsystems." This congruence existed with both ineffective and effective linkage mechanisms. The effective linkage system had fewer hierarchical levels than the ineffective, as would be expected from other research findings: in a turbulent environment a flatter system structure is associated with higher system performance.

In a turbulent environment, effective linkage mechanisms seem to have structures and operating norms that also characterize the whole system of which they are a part, and the system as a whole reflects more closely the structures and norms characteristic of innovative subsystems than those of traditional operating subsystems.

DISCUSSION AND CONCLUSION

This paper has been concerned with effective linkages in systems relating to turbulent environments. For these systems the problems of effective linkage are particularly difficult and particularly important (Lawrence and Lorsch, 1967:137).

Lawrence and Lorsch (1967) showed a direct relationship between degrees of turbulence, differentiation, and complexity of linkage mechanisms. The mechanisms were least differentiated and complex in the organization belonging to the relatively stable and undifferentiated container industry, more numerous and more complex in the moderately unstable and differentiated food organization, and most clearly differentiated and complex in the highly unstable plastics industry. Each of these high-performing organizations used a different combination of linkage devices (Lawrence and Lorsch, 1967:138, Table IV-1).

For systems in turbulent environments the design, development and operation of linkage mechanisms of appropriate complexity is costly. Decision makers will seek to avoid this investment. Even if they assess the environmental needs correctly, they are inclined to underestimate the response that would be effective and the costs involved in the response. Misjudgment of these costs is frequent among innovators who start their own organizations. As the organization grows, innovators often continue to assume that they can stretch linkage by personal contacts over ever greater numbers of people and finally to a new generation of decision makers and managers. Burns and Stalker (1961:92–94) described this personal pattern of attempted linkage. Characteristically, it results in people left "loose" to find their own way to get done whatever they see as necessary; in continuous definitions and redefinitions of tasks and groups "through a perpetual sequence of encounters" with laboratory chiefs, design engineers, and others "to see the job through"; and in increasing personal costs from "nervous preoccupation with the hazards of social navigation in the structure and with the relative validity of their own claims to authority, information and technical expertise." In one company these disruptive effects were counterbalanced by a general awareness of common purpose and of excitement at the tasks, and by the generally attractive atmosphere in the system. Joan Woodward's study (1965:196) compared the attempts of two other organizations, with similar technology and management structure, to establish a formal linkage mechanism. One organization, less than ten years old, with roles and functions not yet clearly established, persisted and designed an innovative system that fit the formal mechanism closely. The other organization was thirty years old and, the author concluded, too rigid to accept the innovation.

That the development of effective linkage mechanisms is associated with major changes in the system as a whole seems borne out by studies of organizations in a wide variety of settings. From a study of ten cases of organizational development in various settings, Buchanan (1967) established a list of 33 issues arising in each. The one issue that distinguished the seven most successful organizations conspicuously from the three unsuccessful ones was the linkage between the innovative subsystem and the rest of the system. "In two of the three unsuccessful cases, changes were initiated and progress was being made, only to come to a halt because of action by management. . . . Steps taken to accomplish linkage were not effective" (Buchanan 1967:62). In the successful cases, linkage between people at several levels of the system was established either as part of the change-induction plan or by steps taken early in the program. The linkage was in the form of working in small teams on operating problems or in the form of involving members in large numbers and at several levels during the early stages of identifying the linkage task (Buchanan, 1967:62). The successful organizations differed in the model they used for innovating, in the manner in which the model was introduced into the system, the location of the innovative subsystem in the system, and the time at which various levels of decision makers were involved. They were similar, on the other hand, in the following five aspects (Buchanan 1967:64):

1. They differentiated the innovative subsystem sharply and formally.
2. In the models they used they found a basis for establishing goals for improvement.

3. The models focused on problem-solving processes.
4. The models led to changes in the kind, distribution, and amount of power.
5. The models emphasized norms and skills that facilitated collaboration and problem solving rather than negotiation and bargaining.

It seems that as long as the model in its initial conception allows for these five dimensions, the system can develop the appropriate mechanisms for its singular circumstances and needs. In fact, working out its own distinctively appropriate pattern seems to be essential to effectiveness. The freedom to make changes, and the constraint in using this freedom, seem to stem from two sources. One is the shifting technological and informational needs of the system and the ways in which these can be met, given the resources of the system and the particular environment with which it interacts. The other source is the emotions of the people actually involved in the changes, which the linkage mechanisms are to mediate. Uncertainties from both sources have to be reflected in the openness of linkage mechanisms to further change; that is, their flexibility.

The emotional uncertainties seem the more complex. Sofer (1961) analyzed the emotional demands on linkage mechanisms that occurred in three very different organizations. He found that "the organization undertakes an emotional division of labor in attitudes towards innovators in just the same way it distributes any other organizational task" (Sofer, 1961:159). Some members will be for the new venture, others against it. "The relationship (of the innovative subsystem) with colleagues will be uneasy. There occur outbreaks of reciprocal paranoia as well as euphoria." When people in the innovative subsystem get discouraged and uncertain about whether they can perform the innovative task, they tend to withdraw and to create barriers rather than linkages. Sofer noted some of the mechanisms by which innovators tend to drain off internal tension. They called attention to the limitations under which decision makers expected them to succeed, their "rigidity," "shortsightedness," "intolerance" and "conservatism," their unwillingness to collaborate, their wish to see the innovation fail. "All parties will precipitate test cases. These test cases are variously used by all concerned to illustrate those aspects of the total situation that are most favorable to their rationale of the moment" (Sofer 1961:160). Sofer traced the system of defense by which people in the innovative subsystems sought to protect themselves and to explain their reluctance to devise, use, and strengthen linkage mechanisms (1961:160–62). The strains in the rest of the system and the ways people express them then tend in turn to isolate the new subsystem further (Burns and Stalker, 1961:171).

In studies of innovation in educational systems, Miles (1964) described how this progressive deterioration of linkage mechanisms made the innovators' fears come true. Support for the innovative subsystem withered, recruitment into it became more difficult, reports of innovative achievements came to lack credibility. Soon the subsystem faced problems of sheer survival (Miles, 1964:654). Under these circumstances intergroup conflict and hostility are very likely to occur, and opposing coalitions develop, with each group magnifying its virtues and the opponents' faults. Such conflict often produced high solidarity *within* the groups. The innovative subsystem, being usually the less powerful, tends to develop substitute satisfactions, like fantasies that some day the others will learn to appreciate them.

Two general conclusions follow. One is that linkage mechanisms need to be flexible and also strong to withstand such strains and such uncertainties. This usually means that they need to be differentiated clearly, be known to have the support of the system's decision makers, and remain highly functional over time. Chin developed the concept of "intersystem" to describe the linkage mechanisms between two subsystems. "The intersystem model exaggerates the virtues of autonomy and the limited nature of interdependence of the interactions between the two connected systems" (Bennis *et al.*, 1964:201–14). The intersystem allows linkage to be rooted in another system, and to offer those "marginal" people who do the linking a base from which to operate and in which to gain and regain perspective and strength.

The second conclusion is simply a forceful reminder that coordination is a heavy cost. This cost may be profitably incurred if, (1) a primary task is clearly differentiated that requires organizational differentiation and, (2) if the new subsystem is maximally autonomous, so that its connections with other subsystems are minimal. In short, the only justification for linkage mechanisms is strict functional interdependence (Kahn *et al.*, 1964:394; Quinn and Mueller, 1963:51). Even if these criteria are rigidly applied, the number of systems that need to design and operate linkage mechanisms will surely increase as more and more decision makers come to recognize the turbulence of the environment.

REFERENCES

Andrews, John H. M., and T. Barr Greenfield   1966. "Organizational themes relevant to change in schools." Ontario Journal of Educational Research, 8:81–99.

Barnes, Louis B.   1960. Organizational Systems and Engineering Groups: A Comparative Study of Two Technical Groups in Industry. Boston: Graduate School of Business Administration, Harvard University.

Bennis, Warren G., Kenneth D. Benne, and Robert Chin   1964. The Planning of Change: Readings in the Applied Behavioral Sciences. New York: Holt, Rinehart & Winston.

Bright, James R. (ed.)   1962. Technological Planning on the Corporate Level (Proceedings of a Conference sponsored by The Associates of the Harvard Business School. Boston: Graduate School of Business Administration, Harvard University.   1964. Research, Development and Technological Innovation: An Introduction. Homewood, Illinois: Irwin.

Buchanan, Paul C.   1967. "The concept of organization development, or self-renewal, as a form of social change." In Goodwin Watson (ed.), Concepts for Social Change: 1–9. Washington, D.C., National Training Laboratories.

Burns, Tom, and G. M. Stalker 1961. The Management of Innovation. London: Tavistock.

Carlson, R O. 1964. "School superintendents and the adoption of modern math: a social structure profile." In Matthew B. Miles (ed.), Innovation in Education: 329–342. New York: Teachers College, Columbia University.

Chandler, Alfred D., Jr. 1963. Strategy and Structure: Chapters in the History of the Industrial Enterprise. Cambridge: M.I.T. Press.

Churchman, C W., and A. H. Scheinblatt 1965. "The researcher and the manager: a dialectic of implementation." Management Science, II (February 1965): B-75.

Dalton, Melville 1950. "Conflicts between staff and line managerial officers." American Sociological Review, XV (June 1950): 342–51.

Dearborn, Dewitt C., and Herbert A. Simon 1958. "Selective perception: a note on the departmental identification of executives." Sociometry, XXI: 140–144.

Downs, Anthony 1967. Inside Bureaucracy. New York: Little, Brown.

Emery, F. E. 1967. "The next thirty years: concepts, methods and anticipations." Human Relations, XX: 199–238.

Emery, F. E., and E. L. Trist 1960. "Socio-Technical Systems." In Churchman and Verhulst (eds.), Management Science, Models and Techniques, Vol. II, Pergamon Press.

Gallagher, A. 1959. "Characteristics of socio-technical systems." Document 527. London: Tavistock Institute of Human Relations. 1963. The Role of the Advocate and Directed Change. Symposium paper. Lincoln, Nebraska.

Harvey, Edward 1968. "Technology and the structure of organizations." American Sociological Review, XXXIII (April 1968) 247–59.

Havelock, Ronald G. 1967. Dissemination and Translation Roles in Education and Other Fields. UCEA Career Development Seminar. Eugene, Oregon.

Havelock, Ronald G., and Kenneth Benne 1964. An Exploratory Study of Knowledge Utilization. Ann Arbor: Institute for Social Research, University of Michigan.

Herbst, P. G. 1959. "Task structure and work relations." Document 528. London: Tavistock Institute of Human Relations.

Hermann, Charles F. 1963. "Some consequences of crisis which limit the viability of organizations." Administrative Science Quarterly, 8 (June 1963): 61–82.

Hirschman, A. O., and C. E. Lindblom. 1962. "Economic development, research and development, policymaking: some convergent views." Behavioral Science, 7: 211–22.

Inkeles, Alex 1951. Public Opinion in Soviet Russia. Cambridge, Mass.: Harvard University Press.

Jasinski, Frank J. 1963. "Adapting organization to new technology." In Joseph A. Litterer (ed.), Organizations: Structure and Behavior. New York: Wiley.

Kahn, Robert L., and Donald M. Wolfe, Robert P. Quinn, and J. Diedrick Snock 1964. Organizational Stress: Studies in Role Conflict and Ambiguity. New York: Wiley.

Kast, Fremont E., and James E. Rosenzweig (eds.) 1962. Science, Technology, and Management. New York: McGraw-Hill.

Katz, Daniel, and Robert L. Kahn 1966. The Social Psychology of Organizations. New York: Wiley.

Katz, Elihu, and Paul F. Lazarsfeld 1955. Personal Influence. Glencoe, Ill.: Free Press.

Lawrence, Paul R., and Jay W. Lorsch 1967. Organization and Environment: Managing Differentiation and Integration. Boston: School of Business Administration, Harvard University.

Litterer, Joseph A. (ed.) 1963. Organizations: Structure and Behavior. New York: Wiley.

Litwak, Eugene 1961. "Models of bureaucracy which permit conflict." The American Journal of Sociology, LXVII (September 1961):177–84.

Lorsch, Jay W. 1965. Product Innovation and Organization. New York: Macmillan.

Lorsch, Jay W., and Paul R. Lawrence 1968. Environmental Factors and Organizational Integration. Paper prepared for 63rd Annual Meeting of the American Sociological Association, August 27, 1968, Boston, Mass.

Miles, B. Matthew (ed.) 1964. Innovation in Education. New York: Teachers College, Columbia University.

Miller, Eric J. 1959. "Part III: Internal differentiation and problems of management." Human Relations, XII: 261–70.

Miller, E. J., and A. K. Rice 1967. Systems of Organization. London: Tavistock.

Parsons, Talcott 1956. "Suggestions for a sociological approach to the theory of organizations—I." Administrative, Science Quarterly, 1:63–85.

Pelz, Donald C., and Frank M. Andrews 1966. Scientists in Organizations. New York: Wiley.

Perrow, Charles 1967. "A framework for the comparative analysis of organizations." American Sociological Review, XXXII (April 1967):194–208.

Pondy, Louis R. 1967. "Organizational conflict: concepts and models." Administrative Science Quarterly, 12 (September 1967):296–320.

Quinn, James Brian, and James A. Mueller 1963. "Transferring research results to operations." Harvard Business Review, XLI (January 1963): 49–66.

Rice, A. K. 1958. Productivity and Social Organization: The Ahmedabad Experiment. London: Tavistock.

Rogers, Everett M. 1962. Diffusion of Innovations. New York: Free Press.

Ross, D. H. (ed.) 1958. Administration for Adaptability. New York: Metropolitan School Study Council.

Selznick, P. 1957. Leadership in Administration. New York: Row Peterson.

Shepard, Herbert A. 1956. "Nine dilemmas in industrial research." Administrative Science Quarterly, 1 (December):293–309.

Sieber, Sam D. 1966. The Organization of Educational Research in the United States. New York: Columbia University.

Smith, Alfred G. 1966. Communication and Status: The Dynamics of a Research Center. Eugene, Oregon: Center for the Advanced Study of Educational Administration, University of Oregon.

Sofer, Cyril 1961. The Organization from Within: A Comparative Study of

Social Institutions Based on a Sociotherapeutic Approach. London: Tavistock.

Strauss, George 1962. "Tactics of lateral relationship: the purchasing agent." Administrative Science Quarterly, 7 (September): 161–86.

Thayer, Lee 1968. Communication and Communication Systems: In Organization, Management and Interpersonal Relations. Homewood, Ill.: Irwin.

Thompson, James D. 1967. Organizations in Action. New York: McGraw-Hill.

Thompson, Victor A. 1965. "Bureaucracy and innovation." Administrative Science Quarterly, 10 (June: 1–20).

Trist, E. L., and G. W. Higgin, H. Murray, A. B. Pollock 1963. Organizational Choice. London: Tavistock.

Watson, Goodwin (ed.) 1967. Change in School Systems. Washington, D.C.: NTL, NEA.

Woodward, Joan 1965. Industrial Organization: Theory and Practice. London: Oxford University Press.

Ned A. Rosen[1]

18. Open Systems in an Organizational Subsystem. A Field Experiment

RECENT LITERATURE in the field of organizational psychology has summarized a number of common characteristics of open systems which relate to the basic nature of organizations (see Katz & Kahn, 1966, for example). Systems theorists suggest that proper application of systems theory can enhance our understanding of complex organizations. Indeed, systems analysis appears valid on its face, corresponds with evidence from the physical and biological sciences, and seems compatible with many scholars' views of organizations. However, few, if any, experimental studies testing the predictions of systems theory in formal organizations appear to exist.[2]

The theoretical model used by Katz and Kahn (1966, p. 16) for analyzing organizations is that of an "energic input-output system in which the energic return from the output reactivates the system." Thus, work organizations as open systems transfer energy from their external environment, transform (through work activity) the energy, and return some product (output) back to the environment. The pattern of activities involved in the energy exchange has a cyclic character. That is, the exportation of final product produces new energy input that begins the cycle all over again.

[1] The data and some sections of this paper are based upon Rosen, N. A. *Leadership change and work group dynamics: An experiment*, Cornell University Press, 1969. The writer is indebted to the Cornell Press for permission to reproduce certain tables, graphs, and paragraphs. I am also indebted to Dr. Waino Suojanen, Chairman of the Management Department, School of Business Administration, University of Miami, for facilitating the preparation of this manuscript and for critically reviewing it while I was a visiting professor in his department in 1968–1969.

[2] It should be pointed out that while the experiment *enabled* a test on certain systems theory hypotheses, the experiment itself was not designed for that specific purpose alone. This paper, therefore, reflects to some extent a post-hoc construction of the data and deals only with selected aspects of the overall study.

From *Organizational Behavior and Human Performance*, 1970, vol. 5, pp. 245–65. Reprinted with permission of the author and the publisher.

Katz and Kahn point out (p. 20), "The energy reinforcing the cycle of activities can derive from some exchange of the product in the external world or from the activity itself. In the former instance, the industrial concern utilizes raw materials and human labor to turn out a product which is marketed and the monetary return is used to obtain more raw materials and labor to perpetuate the cycle of activities. In the latter instance, the voluntary organization (and probably work organizations, too) can provide expressive (and presumably other) satisfactions to its members so that the energy renewal comes directly from the organizational activity itself."

Two characteristics of open systems are of prime interest in this report. The first concerns information inputs as contrasted with energic inputs. "Inputs," according to Katz and Kahn (p. 22), "are also informative in character and furnish signals to the structure about the environment and about its own functioning in relation to the environment. Just as we recognize the distinction between cues and drives in individual psychology, so we must take account of information and energic inputs for all living systems." Feedback mechanisms operate in all systems to assist in regulating cycles of activities and adaptation to environmental or internal changes. The reactions of systems to environmental feedback will be important in the experiment to be described later.

The second major characteristic refers to the homeostatic tendencies of systems or their apparent trend toward a "steady state." Stagner (1951)[3] was probably the first industrially-oriented social psychologist to discuss the concept of homeostasis. However, he did so within the context of personality, rather than organization. Stagner, in common with Katz and Kahn, points out that the concept of homeostasis is not synonymous with stability, i.e., a steady state is *not* motionless or a true equilibrium. That is, the apparent, equilibrium-seeking process indigenous to open systems is dynamic and incorporates change. Stagner used the term "dynamic homeostasis" to describe the process by which organisms "reduce the variability and disturbing effects of external stimulation." In the process of adapting to such influences, the organism (system) changes in some respects. Systems theorists hypothesize, however, that there are internal and external factors which tend to *restore the system as closely as possible to its previous state.* Krech, Crutchfield, and Ballachey (1962) make the same point in connection with the reaction of individuals' cognitive structures to "disturbing" inputs.

The tendency of systems to retain their essence or restore themselves apparently is a function of their interrelated parts and internal variable networks which exert reciprocal, internal influences to produce a net effect—a change or "foreign" influence on one of these variables tends to activate energy transactions within the entire, interconnected system to facilitate return to "normal." The net effect (or equilibrium state) develops from the previously established pattern of energic influences within the system, although it will not necessarily be identical to the

---

[3] Kurt Lewin, of course, worked with the concept of equilibrium even earlier.

equilibrium that existed prior to the disturbing influence. Thus, a virus may cause damaging side effects to a vital organ in the body leading, in turn, to changes in the reciprocal relationships between that organ and others in the body. The basic character of the human system, however, is preserved as the body adapts unless the initial damage is too great or is in a completely critical organ. Two systems theorists summarize their discussion of homeostasis as follows:

In fine, living systems exhibit a growth or expansion dynamic in which they anticipate change through growth which assimilates the new energic inputs to the nature of their structure. In terms of Lewin's quasi-stationary equilibrium the ups and downs of the adjustive process do not always result in a return to the old level. Under certain circumstances a solidification or freezing occurs during one of the adjustive cycles. A new base line level is thus established and successive movements fluctuate around this plateau which may be either above or below the previous plateau of operation. (Katz & Kahn, 1966, p. 25).

The airline system of the U.S., despite its growth and dynamism, provides a useful model of systems-equilibrium theory dynamics. The total domestic airline passenger industry may be viewed as an organizational system. In the absence of such factors as bad weather, major strikes, and large-scale fleet changeovers, the system approaches a stable equilibrium, although minor departures from equilibrium in one airline may create adjustment problems for other carriers serving the same or related routes. Such minor changes or load factors tend to be absorbed with relatively little strain. However, a major snowstorm may simultaneously close several large metropolitan airports and cause a serious disequilibrium. Almost all component variables in the system will show changes, if measured, as equilibrium is reestablished. Energy will be mobilized throughout the system; key reservations and ticket clerks, maintenance and baggage handling crews will work harder and longer hours. Dispatchers, control tower personnel, and flight crews all will make various special efforts to return to normal schedules. This list could even be expanded to include airport restaurant personnel, hotel keepers, catering services, limousine and taxi drivers, plus others. Airlines not directly servicing the snow-stricken cities will feel the effects and also will initiate actions to help the overall industry return to equilibrium.

The industry will, of course, get back to its normal schedules. However, each such experience, including the storm and all of the direct and indirect concomitants of dealing with it, will leave the internal condition of the system at least a little different than it was before. Interpersonal relations among employees, group cohesion, supervisory behaviors, attitudes toward supervisors, customer behaviors, customer attitudes toward specific airlines, employee attitudes toward customers, and so on, all are likely to change in one way or another, as a function of the ways they influenced each other before and during the crisis. Thus, the new equilibrium state is likely to be at least a little different than the initial one prior to the storm.

Given an unusual opportunity to introduce a major leadership change under experimentally controlled conditions in a vital unit of a manufacturing organization, the writer found it possible to test empirically three explicit hypotheses extracted from the above discussion: (1) negative feedback to a stable system (or subsystem) from its environment coupled with a substantial manipulation of a central variable within the system (leadership in this case) will create disequilibrium and lead to corrective energy mobilization; (2) feedback mechanisms within the system will assist it to right itself and establish a new equilibrium; and (3) the essence of the system component relationships, after its temporary disruption, will restabilize in a form closely resembling its initial state.

Experimental Setting and Procedure[4]

The study was conducted in the upholstering department of a factory within a large furniture manufacturing organization. The upholstering department, for our purposes, is considered to be a subsystem with the same or very similar characteristics to those in the earlier example. The plant is a union shop, employs approximately 450 people (mostly male), and is located in a small, predominantly rural community. The upholstering department performs the final assembly operations on the furniture. To accomplish this end, the department receives wooden frames, cushions, precut and sewn covering materials, springs and other necessary components from several different departments in the factory and assembles chairs, couches, sleeper beds, and suites according to customer specifications.

The upholstering department is comprised of eight manually-paced work groups, each of which is organized as an assembly line to build furniture. The eight lines are parallel to each other at approximately 20-foot intervals and each is about 60 yards long from beginning to end. Each group is comprised of eight to ten men and has its own foreman.

The operations performed in the several groups are quite similar and are rationalized. Thus, each group has arm-makers, springers, back-makers, seat-makers, and final trimmers who manually move a piece of partially completed furniture forward to the next man in line after completing its specified operations. The movement of furniture is facilitated by four-wheeled "trucks" on which each piece of furniture in process is mounted.

One group was excluded from certain aspects of this study because its specialized product concentration is technologically different from the others. Although some of the groups are specialized according to product line, all seven groups make all kinds of furniture at one time or another and use very similar methods, tools, and organization.

Each group is characterized by task interdependence. As a result, the pace of each worker on the line is influenced by the man or men whose operations precede him, and also by the men whose operations

---

[4] The description of the research site and procedures here is an encapsulated version based on another work (Rosen, 1969). See the original for fuller details.

follow his. If a man works too slowly, the man preceding him in the line will fill up his limited temporary storage space on the floor and will be unable to continue work at his normal station. Thus as a group or *team* processes a stream of furniture, bottlenecks must be alleviated by workers leaving their normal work stations and moving to bottleneck areas. For example, a seat-maker might step back in the line to assist an arm-maker or he might move forward to assist a back-maker. (Fiedler, 1967, would call these teams "interacting groups.")

Employees are paid on an *individual* incentive basis (MTM). Workers prefer to work on their own specialties and tend to resist changing work stations. When a man does work out of his specialty, he receives full incentive credit for work performed on that station. Regular feedback is provided to all individuals concerning individual and group performance. Individual feedback is received weekly when the workers pick up their pay, which includes incentive earnings. Individual *and group* feedback is received quarterly when average earnings during the preceding calendar quarter for each team member and each overall team are routinely posted on bulletin boards alongside each of the eight lines. Thus, everyone is informed of his own performance, his group's performance, and the performance of all other individuals and groups in the department. Hypothesis 2, above, refers to feedback from these sources.

The groups are characterized by a remarkable degree of membership stability. Mean tenure at the time of the study was 10 years per man. Only two men in the entire department had been employed for less than two years. This is partly explained by the fact that the upholstering operations pay better than all others in the plant. Only five men left the department during the 68-week study; three of these men were members of the excluded group.

Supervisory and union personnel estimate that at least two years of experience are required for an upholsterer to become skilled. Given the long tenure of these men, however, their work might be called "stereotyped" in the same sense implied by Schachter and others (1961). Except for minor changes of style and raw materials, their functions are performed almost automatically. In other words, these men are skilled, "old pros" who can do most of their work habitually. The long tenure of the workers, the small size of the plant, department, and community, and the prevalence of car pools contribute to a high level of personal friendship and knowledge about fellow employees within the organization.

The eight foremen in the department are paid on a straight salary basis. All have "risen" from the ranks. The foremen's functions include all the leadership functions described by Krech *et al.* (1962). Thus, each foreman has some responsibility for making and executing policy, planning, providing expertise, representing his group to outsiders, controlling internal relations, administering rewards and punishments, and arbitrating and mediating conflicts. In Fiedler's terms (1967), the foremen have at least moderately high "position power." Much of this position power stems from long experience with the various products made at the factory and the attendant operating problems. Each foreman may be viewed as an integrated member of his group because of his task knowledge and the many functions he is expected to perform during the production process. It will be important that this point be kept in mind by the reader. The experiment to be described essentially involved the removal of more-or-less well-integrated group leaders (foremen) and their transfer to different groups where integration had to be achieved all over again.

The following measurements, relevant to this paper, were taken from each group[5] in the upholstering department by means of a questionnaire that was administered prior to the experiment and again on a later occasion after the experimental manipulation:

a) *Group attraction or cohesion,* as represented by the mean $(\bar{X})$ rank all group members assigned their own group when ranking all groups in the upholstering department. (This cohesion index is correlated with such factors as perceived financial rewards, friendliness of coworkers, nature of the work, and so on.)

b) *Foreman preference,* as represented by the mean $(\bar{X})$ rank all group members assigned their own foreman and each of the other foremen) when ranking all the foremen in the upholstering department. (Foreman preferences are correlated with a number of scales which load, factor analytically, on general personality, employee-centeredness, job-centeredness, and leader-member role differentiation.)

c) *Status consensus on foreman,* as represented by the standard deviation for each group of the ranks assigned by the members to their own foreman. This is an index of internal agreement showing how closely the group members agree on their leadership preference.[6]

d) *Group money motivation,* an indirect index of potential energy mobilization measured by the extent to which each group discriminated on a 7-point scale between its most and least preferred upholstering line in the department in terms of perceived money-making opportunities. A *t* test was calculated in each group for this purpose.

In addition to the above paper-and-pencil measurements, various others relevant to this paper were developed from the organization's records:

a) *Group productivity levels,* represented by means $(\bar{X})$ calculated for relevant time periods.

b) *Post-experimental percentage change in group productivity,* measured by comparing post-experimental productivity levels (following the foreman reassignments) with a pre-experimental 12-week base line period.

c) Mean $(\bar{X})$ length of longest style run for relevant time periods. This is a production scheduling-cycle variable reflecting the proportion

---

[5] All measures were tied to groups as the appropriate unit of analysis in a leadership study. While only seven groups of 8–10 members each were involved, the measures used appear to be reliable and the small number of groups enabled a relatively high degree of experimental control for a field experiment.

[6] See R. Heslin and D. Dunphy, 1964, for a brief review of several empirical studies in which this variable has been used productively.

of time a given group works uninterruptedly on the same kind of product, e.g., chair number ABC-12 or davenport number XYZ-15. Other things being equal, the longer the product runs a group customarily receives, the more momentum and practice effect it can generate.[7]

The Experimental Plan

Table 1 summarizes the experimental manipulation for each of the

TABLE 1

The Experimental Leadership Manipulation for Each Group

Work Group	Mean Worker Preference Rank for Initial Foreman[a]	Mean Worker Preference Rank for Experimental Foreman[a]
B	1.7	6.0
C	2.0	5.3
D	1.4	5.8
E	2.8	3.8
F	6.5	2.9
G	5.1	2.7
H	3.1	5.2

[a] Measured before experimental change, during the preliminary phase of the study. Rank of "1" is favorable.

seven groups. The manipulation was based on the various groups' foreman preferences. For example, Group B was given a relatively unpopular foreman it had ranked low ($\bar{X} = 6.0$) in place of its incumbent foreman that the members liked very well ($\bar{X} = 1.7$, a favorable mean). Since a major purpose of this experiment, beyond the systems hypotheses of this paper, was to test certain leadership influence hypotheses, the experimental plan called for a maximizing strategy. Two groups (F and G) received the most favorable foreman reassignment we could make and two others (B and D) received the least favorable or most unfavorable reassignments we could make. The others fell in between. As it happened, five of the seven groups received *negative* (sociometrically speaking) leadership changes because of the maximizing strategy and because we were working with a "closed system" in this case; once a foreman was "used" or assigned to a particular group, he couldn't be assigned to another as well.

The reassignments were made on two days' notice[8] by the plant manager one year after the sociometric data had been collected. (The sociometric items had been buried in a large questionnaire for screening

---

[7] Cycle length and related production scheduling factors play a significant part in the group life of his department. Extensive field notes on these matters are reported in Rosen, 1969.

[8] Some general information about the impending change was "leaked" by a middle-level manager ahead of schedule but apparently did not disrupt the experiment.

purposes.) The researcher, to avoid demand characteristics, played no role in introducing the changes at the factory. The plant manager explained the changes in a foremen's meeting as a personal decision of *his* to facilitate the general training of his foremen and "to shake the place up a little"; emphasis was placed on the training value that would result from broadened experience. No reference was made to the earlier study or to the researcher's influence over the reassignment plan.[9] The shift in assignments was based on an agreement between the plant manager and the principal investigator. The arbitrary methods used to introduce the change, the short notice given, and the fact that most of the changes were negative from the workers' viewpoint are presumed to have been negative information inputs for the workers and probably the foremen as well, and are basic to an understanding of the research design.

Figure 1 describes the overall timetable for the entire experiment.

FIGURE 1

Timetable for Experiment

April 1963	*Elapsed time 1 year +*	*April 1964*	*Elapsed time 10 weeks*	*June 1964*	*+ 6 More Weeks*
First measurement of group attraction, foreman preferences, status consensus on foremen and related items.	Assumption: Variable System remained stable during this time period.	Introduction of foreman reassignments on Monday A.M.	Collection and analysis of productivity data from records to test systems hypothesis.	Second measurement of group attraction, foreman preferences, status consensus on foremen and related items.	Further collection of productivity data to ascertain whether any experimental effects picked up in June were lasting.

The first questionnaire was administered in April 1963. These questionnaire data indicated a consistent set of relationships among foreman preferences, group attraction, status consensus on foreman and group productivity levels. Moreover, reliability analyses on productivity levels showed remarkable stability; correlations between group performance levels in several contiguous two-week time periods fell within a range of .88 to .94.

The leadership reassignments were made approximately one year later. We must assume that the interrelated variable system located the previous year remained relatively stable in the interim. (Certain aspects of the data will support this assumption.) During the 10 weeks following the foreman reassignments, productivity records were mailed secretly from the plant manager to the writer. Selected portions of the initial questionnaire were administered a second time in June 1964, on Monday of the 11th week following the experimental manipulation. The workers and foremen were advised of the intended questionnaire

---

[9] Post-experimental interview evidence indicates that the great majority of the men accepted the explanation given by the plant manager and did not attribute the change to the university.

activity at the last possible minute, through the mails on Friday-Saturday of the 10th week, to avoid demand characteristic problems through and including week 10. It is quite obvious both from our experiences in the plant that day and from our data that the entire department now had become aware of our research design. Finally, productivity data were collected secretly for another six weeks for the purposes of certain longitudinal analyses.

Findings

Initial Equilibrium and Variable System. Figure 2, which summarizes the baseline, pre-experimental data, illustrates the systems nature of the upholstering department. The circles in the figure represent variables measured one year prior to the experimental manipulation. The arrows connecting the variables indicate the strength of their relationships at that time. Many of the arrows have two points, indicating relationships that were found to be reciprocal either in the experiment or through observation at the site. Run-Length is an example of the latter. Figure 2 suggests a logically interconnected variable system encompassing the eight groups.[10] The correlations (rho) among group cohesion, foreman preference rankings, status consensus on leader and group performance, while in some cases not statistically significant, are consistent with a large body of survey research and small group laboratory evidence.[11]

The run-length variable in Fig. 2 requires some explanation which is based on several days of on-the-spot observations made during a study of production scheduling factors in the plant. That study revealed that several factors (listed under Run-Length in the Figure) contribute to run-length variance, i.e., space limitations for storing components, raw material and equipment availability, market factors, front office pressures, inter-departmental coordination, and traditional product specialization among the upholstering groups. Indeed, the pre-experimental rho of .79 between run-length and group performance levels suggests that structural, not psychological variables, control the entire system. This is not altogether the case, however. Our observational research (see Rosen, 1969) indicates that run-length is both a cause and an effect of

---

[10] Data from all eight groups are represented in this figure (except in the case of run length where N equals 7), because all eight groups were included in the pre-experimental perceptual system that existed at that time. Only the seven groups involved in the actual experiment are represented in Figs. 3–5, however. Thus, when the reader compares the systems diagrams in Figs. 3–5 with the one in Fig. 2, he will be comparing a post-experimental seven-group system with a pre-experimental eight-group system as a function of sample attrition.

[11] A note on statistical treatment is in order here. The writer has no idea to what population one may infer from this experiment. Moreover, the absence of control groups and the small number of groups involved does not encourage confidence. Consequently, only raw correlations are presented without regard for significance levels. All correlations are rhos, in view of the small N and the nature of the data. For those interested in statistical significance, a rho of .71 is required at the .05 level in one tail tests ($N = 7$). While many of the observed rhos reach or exceed this magnitude and others do not, the trends and patterns of the data appear to be more instructional than any specifically significant correlations.

FIGURE 2

System in Equilibrium before Experiment: Relationships among Interconnected Systems Variables (N = 8 groups)

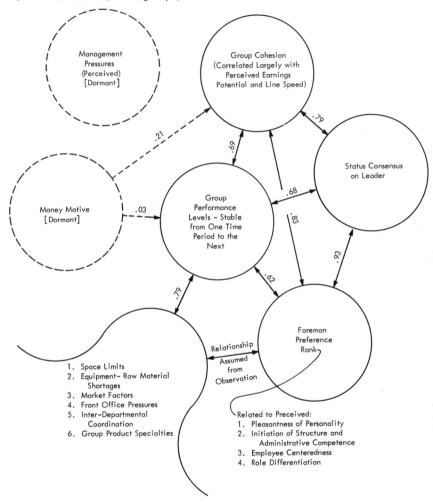

productivity levels. That is, the chief production dispatcher who allocates work to the groups is guided by structural considerations, but he also tends to give long runs to groups he knows to be viable.

The reader should also note that the Money Motivation index in Fig. 2 was uncorrelated with the system at the time it was measured prior to the experiment. We shall see shortly, however, what happened when the motives underlying this variable were aroused following the sudden leadership changes. In any event, the Money Motive variable is included in the system chart (Fig. 2) prior to the experiment because later developments indicate that it belongs here as a dormant influence. The same is true of the Management Pressures variable at the top of Fig. 2. This variable will be explained later.

In summary, Fig. 2 illustrates the variables that were identified in the upholstering department subsystem and shows how they were interconnected before the experiment began. Whether additional variables were present is unknown. The ones in the chart, however, represent commonly recognized group dimensions which seem to be the most relevant for the subsystem we are examining. The experimental data below should tell us something about the equilibrium restoration influence of these interconnected variables which, in combination define the preestablished pattern of energic influences within the upholstering department subsystem.

Given the experiment's conceptualization and hypotheses, it is essential to demonstrate that the subsystem described by Fig. 2 was in equilibrium when the leadership changes were introduced. We assume this to be the case for several reasons:

1. The groups involved were composed of "old pros" who had long tenure and experience and had been working "as usual" for a long time.
2. There were no unusual stresses in the environment or within the groups.
3. Working hours and work load were regular.
4. No major technological or administrative changes had occurred between the variable measurements in April 1963 and the experimental manipulation introduced a year later.[12]
5. Labor turnover, or group membership change, was virtually nil.
6. Group productivity levels over time remained quite stable.

Hypothesis 1.   Our first testable hypothesis states that negative feedback to this (stable) subsystem from its environment, coupled with a substantial manipulation of a central variable within the system, will create disequilibrium and lead to corrective energy mobilization. Since the foreman reassignments were introduced more than a year after the initial research and without the presence of researcher interventionists, the subjects involved should react as they normally would to a major administrative change imposed arbitrarily and on short notice by top management. Such leadership changes traditionally are made only when higher authorities are dissatisfied with performance levels of subordinates. Therefore, it was hypothesized that they would show a "pressure" reaction to perceived negative feedback. The following specific things should happen. (*a*) The groups should feel threatened in inverse relation to their customary performance levels. (*b*) Adaptation problems should occur early given the foreman's role as expeditor in emergencies and as the group's external representative. His relative lack of technological-product information appropriate on his new as-

---

[12] One foreman was reassigned by the plant manager about 7 months prior to the larger scale reassignment process used in the experiment. The effects of this change had ample time to settle down before the experiment began.

signment also should make a big initial difference. In short, the system should be knocked out of equilibrium due to a drastic change in leadership. (c) The money motive of the employees, originally measured before the experimental change, should become salient because the new foreman, in five out of seven groups, will be seen as inferior to the old one; he probably will not be seen by the workers as having immediately transferable and *necessary skills*, e.g., a line that usually assembles couches receiving a chair line foreman. Motivation will be aroused, therefore, to compensate for the disruptive effects of the leadership changes. *The increased saliency of the money motive should stimulate the mobilization of group energy to return to equilibrium.*[13]

Figure 3 provides data to support (a), above. The arrow labelled Management Pressure represents the finding that the initial, pre-experimental productivity levels (means) of the work groups did in fact correlate negatively ($-.61$) with *percentage* increases in productivity during the first two weeks following the foreman reassignment; the slowest groups, in other words, tended to show the biggest gains. Moreover, *all seven* experimental groups showed an increase in performance during the first week after the change. One group increased by 24 percent, five groups increased between 10 percent and 16 percent, and one increased by 6 percent. It is extremely rare for all groups in this department to move in the same direction and in such magnitudes in any given week, which suggests that the experimental manipulation created disequilibrium.[14]

Several additional bits of evidence also indicate that disequilibrium developed (b, above). First, Table 2 shows that the usually stable group performance level hierarchy broke down sharply during the first few weeks following the experimental manipulation. Moreover, the correlation between group performance levels summed over the 12 weeks preceding the experiment with performance levels over the entire 16 weeks following the change is only .46 (compared with .82 for the same periods in the previous calendar year and .62 the following calendar year). This suggests that the leadership changes had a major impact on the system. It also is clear from the table that group performance levels restabilized at about the 6th week, indicating that the disrupted system was beginning to settle down.

Further evidence of equilibrium disruption may be observed in Table 3 which shows, for each group, the group attraction, foreman preference, and status consensus measurements one year before the experimental manipulation and ten weeks after. A great deal of change is evident, especially in the group attraction variables where the pre- and post-experimental measurements are correlated *negatively* ($-.38$). In

---

[13] Considerable quantitative and qualitative evidence, gathered in the overall study from which this paper is adapted, clearly indicates strong interest in money among the upholsterers as a group. For example, both their line and job title preferences were correlated with perceived earnings prospects.

[14] These productivity gains are interesting for reasons beyond systems theory and are discussed, accordingly, in two other places. See Rosen, 1969 and 1969a.

FIGURE 3

Disequilibrium First Two Weeks after Change in Foreman Assignments ($N = 7$).

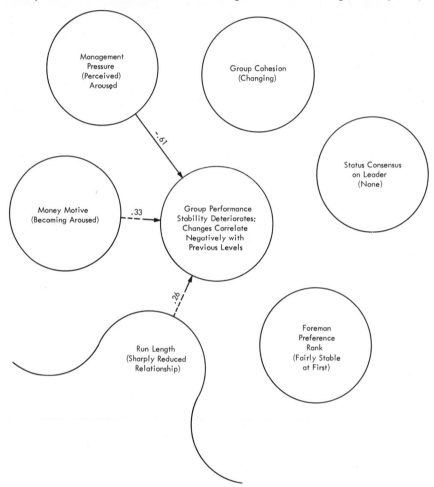

other words, the group cohesion measure showed no stability at all over time. On the contrary, groups that were cohesive before the experiment tended to lose their cohesion while less cohesive groups tended to become more cohesive. Considering the nature of the foreman reassignments, this is not surprising. Foreman preferences, on the other hand, were much more stable, according to Table 3. While all the pre-experimental mean rankings assigned by the groups to their new foremen a year prior to receiving them improved over time, their relative magnitudes changed only slightly (rho = .71). On the other hand, *status consensus* within the groups changed markedly in the period between the two sets of measurements. In this case the rho between the two sets of measurements is only .24. In other words, a great deal of within-

TABLE 2

The Correlations (Rho) of Group Mean Productivity Levels between the Pre-experimental Twelve-Week Base Period and the First Two Weeks Following the Change and between the Adjacent Two-Week Time Periods after the Experimental Change

12-Week base vs weeks 1 + 2	.26
Weeks 1 + 2 vs 3 + 4	.21
Weeks 3 + 4 vs 5 + 6	.43
Weeks 5 + 6 vs 7 + 8	.71
Weeks 7 + 8 vs 9 + 10	.75
Weeks 9 + 10 vs 11 + 12	.79
Weeks 11 + 12 vs 13 + 14	.75
Weeks 13 + 14 vs 15 + 16	.79

TABLE 3

Group Attraction, Foreman Preference and Status Consensus on New Foreman, by Group, before and after the Experimental Change

Group	Group Attraction (mean)[a]		Foreman Preference (mean)[a]		Status Consensus (S.D.)[b]	
	Before	After	Before	After	Before	After
F	3.6	1.4	2.9	1.4	2.0	0.5
G	3.4	1.1	2.7	1.0	1.3	0.0
E	2.6	2.7	3.8	2.5	2.7	1.8
H	1.5	2.4	5.2	5.8	1.8	1.7
C	1.0	1.6+	5.3	1.9	1.8	1.7
B	1.4	1.6	6.0	3.0	1.7	1.3
D	1.4	1.6+	5.8	2.6	1.5	1.9
	Rho = −.38		Rho = .71		Rho = .24	

[a] The smaller the mean, the more favorable the attitude.
[b] The smaller the standard deviation, the closer the consensus, before the new foreman was assigned.

group opinion shifting took place regarding the new foremen, and the groups differed in the extent to which they were able to reach internal agreement about their new leaders.[15]

In summary, then, disequilibrium did develop in the subsystem following the reassignments of foreman. The previously stable performance hierarchy of the groups broke down, atypical performance patterns emerged, group attraction or cohesion apparently changed markedly, and individual opinions about foremen underwent considerable change.

Figure 4, below, supports our hypothesis (c) above, that corrective energy mobilization would follow disequilibrium. Observe, first of all,

[15] Responses of the subjects to a large number of questionnaire items indicate that the perceived *personality* of the group leader plays an important part in the formation of within-group status consensus on him.

FIGURE 4

Energy Mobilization and Homeostatic Tendencies during First 10 Weeks Following
Experimental Change

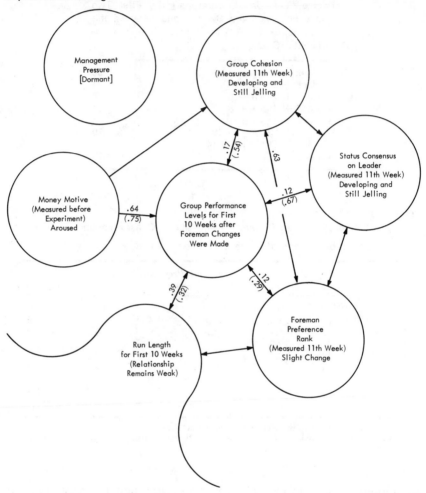

that the Money Motive index (measured *before* the experimental
change), which correlated only .03 with performance levels in Fig. 2
(before the experiment) and .33 in Fig. 3 (first 2 weeks after experiment began) is correlated .64 with productivity levels for weeks
1–10. *Meanwhile, the correlation between run-length and performance
dropped from .79 to .39.*[16] It appears that as the initial threat or shock

---

[16] The sharply reduced relationship between run-length and group performance
levels was *not* found to be a function of run-length or other product-mix variable
instability over time. Run-length remained stable over time. See Chapters 6–7 in
Rosen, 1969, for an extensive treatment of this matter.

from the simultaneous leadership changes wore off, the money motive was aroused, and the groups exerted *energy* or *effort* in direct relation to their interest in money. Performance changes accordingly occurred *in spite of* constraints imposed by structural factors influencing run-length. In other words, an indirect, paper and pencil measure of motivation was dormant when the system lay in a long-established equilibrium pattern (Fig. 2). Under threat induced by a major leadership change, coupled with perceived negative information inputs, this variable was activated and became an energy mobilizer in the subsystem (Figs. 3 and 4).

Hypothesis 2. This hypothesis states that feedback mechanisms within the system will assist it to adjust itself and establish a new equilibrium. In this connection, Fig. 4 shows the correlations involving group cohesion, status consensus on leader, and foreman preferences, *as measured 10 weeks after the leadership reassignments.* The relationships between these post-experimental measures, on the one hand, and performance levels (means) during the preceding 10 weeks are small. However, the parenthetical correlations in Fig. 4 reflect the use of a *percentage change* performance index rather than group performance means. These correlations are markedly greater. This suggests that when the subjects were asked after 10 weeks to rank their groups and foremen a second time, their responses clearly reflected performance *change feedback* that had affected their weekly paychecks during the preceding 10 weeks. That is, paycheck changes told the groups how well they were coping and influenced the development of new perceptions and attitudes. It seems, therefore, that Hypothesis 2 also is supported.

Hypothesis 3. Our third hypothesis states that the system-component relationships, after temporary disruption, will tend to restabilize in a form closely resembling the initial equilibrium state. Figure 5, below, can be compared with Fig. 2 to test this hypothesis. Figure 5 shows predictive correlations between the questionnaire measurements taken on group cohesion, status consensus, and foreman preferences on Monday of the 11th week, and performance criteria in the *subsequent* 6-week period. This Figure represents a time period that occurred after the system was beginning to stabilize and after the subjects found out the real reason behind the foreman changes. Figure 5 also shows the intercorrelations of these predictors and includes the Money Motive index measured more than a year earlier. (The pre- and post-experimental measurements on this variable correlate .79, indicating statistically significant reliability.) Comparison of Figs. 2 and 5, whose similar patterns of relationships certainly suggest the emergence of a new equilibrium quite similar to the old, tends to support Hypothesis 3, although the data indicate that the system may not yet have settled down completely.

Let us summarize all the evidence bearing on Hypothesis 3:

1. The set of interrelated variables, identified before the experiment, changed internally. That is, the measurements on these variables for

FIGURE 5

System Approaching New Equilibrium in Weeks 11–16, Using Group Productivity Means to Represent Performance. Correlations in parentheses reflect use of *percentage changes* in performance levels, rather than performance levels *per se.* (*N* = 7.)

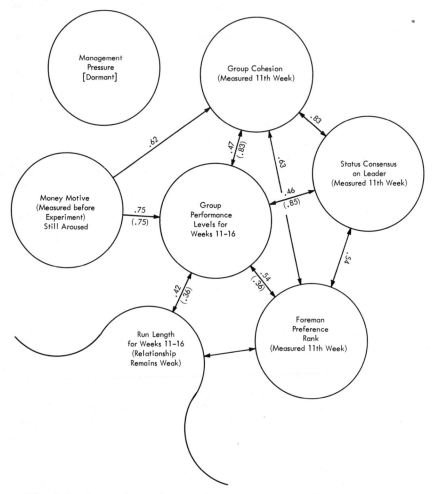

* Correlations in parentheses reflect use of *percentage changes* in performance levels, rather than performance levels *per se.* (N = 7)

the different groups moved around after the leadership manipulation; e.g., some groups became more cohesive while others became less cohesive.

2. Despite the changed measurements for the several groups, the variables continued to be correlated at a later time.

3. The magnitude and pattern of the latter, post-experimental correlations are similar to those found prior to the experiment. Given the

group changes that occurred, the similarity of the variable networks before and after the experiment supports the hypothesis of a self-correcting system restoring itself to equilibrium following a temporary disruption. The restoration process seems to be returning the system to a state very similar to the one existing before the experiment. There are two exceptions to the "similarity" argument made in paragraph 3, above. These are Money Motivation and Run-Length. Their relative influences within the pre-experimental system as shown in Fig. 2 seem to have changed (Fig. 5), and as of weeks 11–16 had not yet been restored to their pre-experimental levels. It seem likely that in time even these two correlations would tend to normalize. That is, many of the structural influences on run-length are beyond work group influence and eventually are likely to influence a return to equilibrium similar to the original equilibrium state. Moreover, such high levels of energy mobilization as exhibited by some of the groups probably can't be maintained indefinitely and eventually should taper off. In fact, one might argue that run-length is the crucial equilibrium restoration force in this subsystem. Run-length, however, is not the only restoration force in the system. One should remember that the pre-experimental and post-experimental foreman preference measurements correlate .71. In other words, previously formed perceptions of leader behavior were resistant to gross change and influenced the establishment of a new equilibrium that was similar to the original one.

Artifact versus Homeostasis. The evidence of equilibrium seeking would be more convincing, if we could show that the process began several weeks prior to the post-experimental measurements. Otherwise, one might be tempted to conclude that the researcher intervention on Monday of the 11th week *created* the new attitudes through demand characteristics generated by the questionnaire items and circumstances. One could also argue that the system variable network that developed in weeks 11–16 (see Fig. 5) similarly was "caused" by the intervention. Strictly speaking, a longitudinal design incorporating several repeated measurements on the important variables throughout the experimental period seems to be needed to get around this problem. "Panelitis" problems (Remmers, 1954), however, would be likely in such a setting. The writer firmly believed that the groups had to be left to their own devices if we were to learn anything valid about the effects of leadership change.

The question is, then, how can we demonstrate that the workers were assessing the new foremen in the light of performance levels *prior* to the second round of attitude measurements on Monday of the 11th week? Can we show any kind of legitimate trend data that will reveal the disintegration, over time of pre-experimental variables, and the emergence, over time of their post-experimental counterparts? Such data are reported and discussed in Rosen, 1969, and strongly suggest that the equilibrium-seeking process began well *before* the second round of measurements were taken in the 11th week. The findings tend to discredit a purely artifactual explanation of the new equilibrium and suggest the longitudinal adaptation of a multivariate system to a change in equilibrium state.

SUMMARY

This paper reports a field experiment whose design enabled a test on selected aspects of open systems theory. A manufacturing department characterized by equilibrium was disrupted by a simultaneous reassignment, on short notice, of all but one of its foremen. Seven intact work groups were involved. The seven foremen were reassigned among the groups by the plant manager without researcher intervention. The longitudinal relationships among several interconnected variables were studied. The findings lend support to the following hypotheses:

1. Negative feedback to a stable system (or subsystem) from its environment, coupled with a substantial manipulation of a central variable within the system (leadership in this case), will create disequilibrium and lead to corrective energy mobilization.
2. Feedback mechanisms (weekly paychecks reflecting incentive earnings) within the system will assist it to right itself and establish a new equilibrium.
3. The essence of the system component relationships, after its temporary disruption, will restabilize in a form closely resembling its initial state.

A combination of several structural variables, all of which affect work group behavior through the variable "run-length" or work-cycle length, were found to be major forces toward stabilization. Widely shared perceptions of leadership behavior also tended to act as a stabilizing influence.

REFERENCES

Fiedler, F. *A Theory of Leadership Effectiveness.* New York: McGraw-Hill, 1967.

Katz, D., & Kahn, R. L. *The Social Psychology of Organizations.* New York: John Wiley, 1966.

Krech, D., Crutchfield, R., & Ballachey, E. *Individual in Society.* New York: McGraw-Hill, 1962.

Remmers, H. H. *Introduction to Opinion and Attitude Measurement.* New York: Harper, 1954.

Rosen, N. A. Demand Characteristics in a Field Experiment. *Journal of Applied Psychology,* in press.

Rosen, N. A. *Leadership Change and Work Group Dynamics.* Ithaca, N.Y.: Cornell University Press, 1969.

Schachter, S., et al. Emotional Disruption and Productivity. *Journal of Applied Psychology,* 1961, 45, 201–213.

Stagner, R. Homeostasis as a Unifying Concept in Personality Theory. *Psychological Review,* 1951, 58, 5–17.

A. K. Rice

19. Individual, Group, and Intergroup Processes

THIS PAPER is an attempt to apply to individual and group behaviour a system theory of organisation, normally used for the analysis of enterprise processes. The use of such a theory will inevitably concentrate on the more mechanistic aspects of human relationships, but I hope that the approach will help to clarify some of the differences and similarities between individual, group and intergroup behavior and throw some light on the nature of authority.

To perform a task, activities must be carried out. But however automated their processes, enterprises have to employ human resources for some activities. Such activities seldom, if ever, use the total capacities of the individuals so employed and individuals can seldom, if ever, give to an enterprise only the capacities it requires. If an enterprise fails to provide outlets for the unused capacities, they are likely to interfere with task performance. But to provide them it has to use resources that could otherwise be used for task performance or could be dispensed with. The provision of such outlets inevitably, therefore, reduces the efficiency of task performance, measured as the difference between intakes and outputs. I hope that the application of a system theory of organisation to human behaviour will, therefore, not only help to clarify that behaviour, but may also enrich the theory itself and, in time, thus enable enterprises to make better use of their human resources.

I have another minor aim: To try to find concepts and language that can be applied both to institutional processes and to human behaviour. Psychologists and psychiatrists have enormously enlarged our understanding of individual and group behaviour, and anthropologists and sociologists of institutional behaviour, but when we try to use their insights to gain understanding of enterprise organisation and institutional processes we are frequently faced with the difficulty of marrying different theoretical frameworks and different languages. The

From *Human Relations*, 1969, vol. 22, pp. 565–84. Reprinted with permission of the Tavistock Institute of Human Relations, on behalf of the author, and the publisher.

problem is to raise the level of conceptual abstraction; the danger that in so doing we shall not only lose the simplicity and richness of existing descriptive concepts, but lose also ourselves in an arid complexity of irrelevant variables.

My basic propositions are that:

1.   The effectiveness of every intergroup relationship is determined, so far as its overt purposes are concerned, by the extent to which the groups involved have to defend themselves against uncertainty about the integrity of their boundaries, and
2.   Every relationship—between individuals, within small groups and within large groups as well as between groups—has the characteristics of an intergroup relationship.

A corollary to the first proposition is that the making of any intergroup relationship carries with it the possibility of a breakdown in authority, the threat of chaos, and the fear of disaster.

In addition to a system theory of organisation, I shall also use concepts derived from the object-relations theories of psychoanalysis, and I hope the findings will be consistent with the work of Bion on the nature of small group behaviour.[1]

The first part of the paper summarises briefly the relevant concepts of enterprise and institutional organisation. Subsequent parts then apply these concepts to individual behaviour, to group and intergroup relations, and to the role of leadership.

A SYSTEM THEORY OF ORGANISATION

The Enterprise as an Open System

The theory treats any enterprise or institution, or a part of any enterprise or institution, as an open system. Such a system must exchange materials with its environment in order to survive. The difference between what it imports and what it exports is a measure of the conversion activities of the system. Thus a manufacturing company imports raw materials, converts them, and exports finished products (and waste). For the outputs it receives a pay-off, from which it acquires more intakes. The intakes into a university, on the other hand, are students; and the outputs graduates (and failures).

Such intakes and outputs are the results of import-conversion-export processes that differentiate enterprises from each other. But every enterprise has many import conversion-export processes; a manufacturing company, for example, recruits employees, assigns them to jobs, and sooner or later exports them through retirement, resignation, dismissal or death. It imports and consumes power and stores; it collects data about markets, competitors and suppliers' performance and converts the data into plans, designs and decisions about products and prices.

[1] Bion, W. R. (1961), Experiences in Groups. London, Tavistock Publications.

The nature of the many processes and their intakes and outputs reveals the variety of relationships that an enterprise, or part of it, makes with different parts of its environment and within itself, between its different parts. The processes also reveal the variety of tasks that the enterprise performs as a whole and the contributions of its different parts to the whole.

Every enterprise, or part enterprise, has, however, at any given time a primary task—the task it must perform to survive.[2] The dominant import-conversion-export process is that process by which the primary task is performed. It is this dominant process that defines the essential relationship of an enterprise to its environment, and to which other tasks and other throughputs are subordinate.

Boundary Controls

A *system of activities* is that complex of activities which is required to complete the process of transforming an intake into an output. A *task system* is a system of activities plus the human and physical resources required to perform the activities. The term 'system,' as it is used here, implies that each component activity of the system is interdependent with at least some of the other activities of the same system, and that the system as a whole is identifiable as being in certain, if limited, respects independent of related systems. Thus a system has a boundary which separates it from its environment. Intakes cross this boundary and are subjected to conversion processes within it. The work done by the system is, therefore, at least potentially measurable by the difference between its intakes and outputs.

What distinguishes a system from an aggregate of activities is the existence of regulation. Regulation relates activities to throughput, ordering them in such a way as to ensure that the process is accomplished, that the different import-conversion-export processes of the system are related to each other and that the system as a whole is related to its environment.

The most important *management* control in any organisation is, therefore, the control of the boundaries of systems of activity, since it is only at boundaries that the difference between intake and output can be measured. Task management then is essentially:

a)   the definition of boundaries between task systems
b)   the control of transactions across boundaries.

The boundary of a system of activities, therefore, implies both a discontinuity of activity and the interpolation of a region of control. The location of the boundary control function is shown in Figure 1.

Those systems of activity that lie on the main stream of the dominant import-conversion-export process by which the primary task is performed are the operating systems. Where, in any enterprise, there

---

[2] For a fuller description of this concept and the system theory of organisation see Miller & Rice (1967), Rice (1958 and 1963).

is more than one operating system, a differentiated managing system is required to control, co-ordinate, and service the activities of the different operating systems. This will include the management of the total system, management of each discrete operating system, and also those nonoperating systems that do not perform directly any part of the primary task of the whole, but which provide controls over, and services to, the operating systems. An enterprise with three first-order

FIGURE 1

A Task System and Its Boundary Control Function

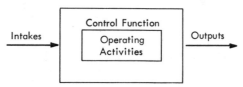

operating systems, three control and service functions in the first-order managing systems, two second-order operating systems, and two second-order control and service functions is shown in Figure 2. In it,

FIGURE 2

Organisational Model

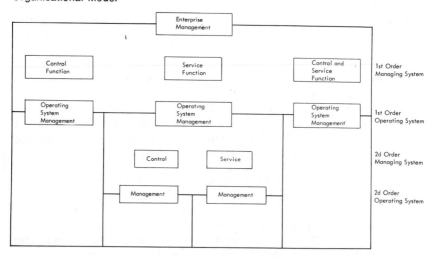

to avoid complexity the topological form of Figure 1 has been simplified by locating the boundary control region at one point on the boundary of the operating system viz:

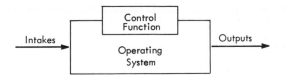

Members of an enterprise occupy roles in these various systems of activity. One member may take more than one role; and one role may be taken by more than one member. In assigning activities to roles and roles to people, the enterprise cannot always predetermine the role-sets that will emerge or the relative *sentience* of the various groups to which each individual will belong. These factors are, nevertheless, relevant to the effectiveness of task performance, supporting or opposing it.

Management of an enterprise requires therefore three kinds of boundary control:

1. regulation of task system boundaries (i.e., regulation of the whole enterprise as an import-conversion-export system, and regulation of constituent systems of activity);
2. regulation of sentient system boundaries (the boundaries of the group to which individuals belong, either directly through their roles in systems of activity or indirectly through their consequential role-sets and personal relationships); and
3. regulation of the relation between task and sentient systems.

Organisation Models

Organisation is the instrument through which an enterprise assigns activities to roles and roles to individuals and groups. Organisation is thus a means to an end and the most appropriate organisation is the one that best fits primary task performance. It follows that for every task an organisational model is required, which will define the boundaries of operating systems and the control and service functions that are required to co-ordinate, control, and service the operating systems. Such definitions of the boundaries of the systems will determine the roles and role-relationships that provide for effective performance.

In building an organisational model, the starting point is the process flow. The dominant process identifies the nature of the intakes, the activities required to convert these into, and dispose of, outputs, and the human and physical resources required to provide or to facilitate these activities. The next step is to discover the discontinuities in a process that mark the boundaries of systems of activity. These are the appropriate points at which to draw organisational boundaries and these in their turn define management commands.

Since the performance of any task is subject to complex constraints, the actual organisation of the enterprise will inevitably be a compromise between the model and the constraints. In the same way, since each part of any enterprise has its own primary task and thus requires an organisational model for itself, the organisation for the whole will be constrained by the need to integrate the organisations of the parts, and the organisations of the parts will be constrained by the need to fit into the whole. The model provides a basis against which to examine the reality of the constraints and the consequent compromises.

To the three boundary controls given above, therefore, must be added the regulation of organisational boundaries where these, because of unalterable constraints, do not coincide with boundaries of activity systems.

Task, organisational and sentient boundaries may coincide. Indeed they must coincide to some extent at the boundary of the enterprise if it is to continue to exist. The enterprise may also be differentiated into parts which are similarly defined by coinciding boundaries. There are dangers in such coincidence. One danger is that members of a group may so invest in their identity as a group that they will defend an obsolescent task-system from which they derive membership. One can add the possibility that the identification of change in task-system boundaries, and even the identification of the boundaries themselves, can be made difficult by the existence of group boundaries that are strongly defended.

In general, it can be said that without adequate boundary definitions for activity systems and sentient groups, organisational boundaries are difficult to define and frontier skirmishing is inevitable. It is perhaps a major paradox of modern complex enterprises that the more certainly boundaries can be located, the more easily formal communications systems can be established. Unless a boundary is adequately located, different people will draw it in different places and hence there will be confusion between inside and outside. In the individual this confusion leads to breakdown, in enterprises to inefficiency and failure.

Because an enterprise is an open system, the nature of the constraints within which it operates is constantly changing. Internally, a change in technology may remove old constraints and introduce new ones. Externally, changes may range from a minor statutory requirement to a major shift in definition of the primary task. Such changes, even if they do not demand a redefinition of the primary task of the whole, frequently redefine the primary tasks of parts and modify the strategies through which an enterprise relates its internal and external environments so as to achieve the most effective performance of its primary task. Changes in strategy may not always be explicit; they may be merely reflected in changes in the behaviour of the enterprise. Different forms of organisation differ in their capacity to respond and adapt to changes in strategy. Strategic changes, whether or not they are explicit, and even if they do not entail a redefinition of the primary task, may require changes in the form of organisation if this is to retain its effectiveness.

Multiple Task Systems

A simplified form of a multiple task system is shown in Figure 3. Theoretically, two tasks, and thus two systems of activity, can be identified, but those who perform the tasks—the human resources of the two systems—constitute a single and identical sentient group. Thus the strength of the sentient boundary of the group is affected by what happens in both activity systems, and, by way of the common sentient group, the activities of each task system are affected by those of the other. Involvement of a group in two activity systems may require the co-existence of two different arrays of roles and role relationships that may relate the individual members together in different ways.

FIGURE 3

One Group—Two Task Systems

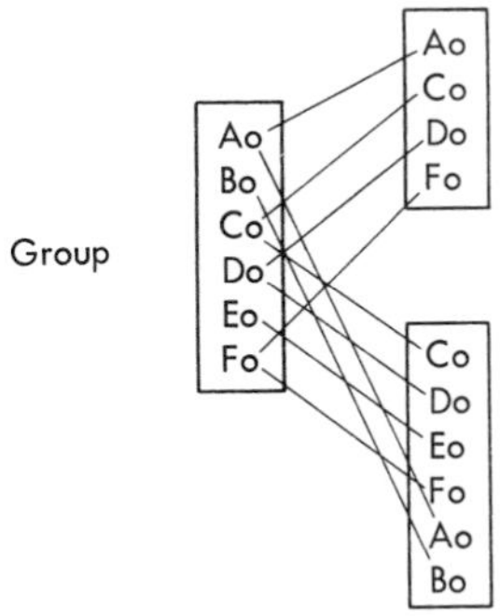

System conflict does not 'arise in conditions of stable equilibrium—in other words, where environmental forces are unchanging or do not impinge too differentially on the three systems. For example, primitive societies often seem to have had relatively closed system characteristics over a long time. A stable equilibrium was established between the different systems of activity in which the tribal group engaged. But in contemporary society with its increasingly rapid social and technological change, disequilibrium is common.

Family businesses provide many examples of disequilibrium because of the differential pressures on the different systems of activity: those pertinent to the family, as a family, and those pertinent to the business as a business. In the kind of business that requires increasing capital to maintain parity with competitors, it is difficult for any but the most wealthy family to provide enough to maintain control. If, in addition, the business requires increasing numbers of technicians, scientists, and managers, to handle the more sophisticated technologies of production, marketing and control, it is unusual for one family to be able to provide them all. As others outside the family are introduced into positions of power in the business task system, they tend to usurp the expected roles of family members, and thus distort role relationships in the family system.

In conditions of social and technical change, the attitudes and behaviour of members of a group to each other, and to the external environment may not only jeopardise the survival of the task systems but put such strain on internal group relationships that group survival is also jeopardised.

Temporary and Transitional Task Systems

By definition, temporary and transitional task systems require temporary and transitional organisations (for convenience called *project organisations*). The essential feature of a project type of organisation is that the group brought together to perform a particular task is disbanded as soon as the task is completed. The group as a group has no further *raison d'etre* in terms of task performance. But the theoretically finite life of a project team is frequently prolonged either as a result of a redefinition of its task or of the accretion of new tasks. A research team, for example, either because it has invented a new technique or because its members have become devoted to working together, generates further problems to which it can apply its technique, or which will keep the team intact—irrespective of whether the generated problems are relevant to the overall task of the research enterprise of which they are a part.

But project groups cannot by definition provide either permanent sentience or career patterns for their members. Or if they do, they become difficult to disband at the conclusion of task performance. A successful project type organisation requires, therefore, control and service functions in the managing system:

a)   to ensure that adequate resources, both human and physical, are available for every project undertaken;

b)   to provide sentient groups to which members can commit themselves and to which they can return for reallocation at the conclusion of each project; and

c)   to provide pools in which physical resources can be stored and maintained.

The general form of organisation is illustrated in Figure 4. In a small enterprise, the total enterprise can be the only sentient system required. But a large and complex enterprise can seldom provide sufficient personal identity, and separate differentiated systems are then essential.

Transactional Task Systems

By definition, a transaction between an enterprise and its environment must take place across the boundary of the enterprise. The activities of the transaction involve those parts of the enterprise and of its environment through which the transaction is made. The task system of the transaction, however temporary, therefore has a boundary which cuts across enterprise boundaries, a boundary that in any genuine two-way transaction cannot be fully controlled by the enterprise. Moreover if the enterprise is large, not all of its members can take part in the transac-

FIGURE 4

General Form of Project Organisation

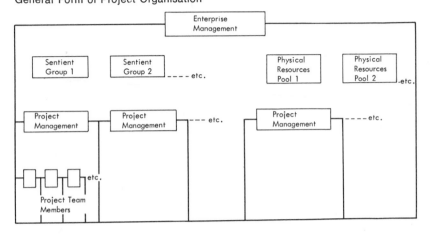

tion and one or more have to 'represent' the enterprise. This condition inevitably involves the control of boundaries between the enterprise and its representative or representatives.

A simple example is illustrated in Figure 5. It represents a transaction between two enterprises A & B; 'a' conducting the transaction on behalf of A, and 'b,' on behalf of B. For the duration of the transaction, the task system (ab) boundary cuts across the enterprise boundaries of both A & B. So far as there is any uncertainty in A, B, a, b, or ab about the relative strengths of the A, B, a, b, or ab boundaries so will there be doubts about the control of transactions across the boundaries between

FIGURE 5

Simple Transactional Task System

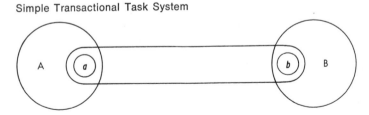

A and a, B and b and a and b. Control of the ab boundary must be sufficiently strong to perform the task of transaction, and yet, if it becomes too strong it jeopardises control of the Aa and the Bb boundaries, and hence the integrity of the A and B boundaries. Any uncertainty can be exacerbated when either a or b consists of more than one individual. Then not only may there be uncertainty about the Aa and *Bb* boundary but the intra-group relations of '*a*' and '*b*' may also cause anxiety.

Examples of difficulty are common. If '*a*' is a sales representative of a supplier *A* and '*b*' the buyer of customer *B*, the management of *A* will require its representative '*a*' to make good relationships with '*b*' and, at the same time, to remain loyal to *A*. A representative who is suspected of favouring '*b*' at the expense of *A* does not usually last long. Similarly a buyer who is believed to use the *ab* relationships for his own benefit rather than for the benefit of *B* gets short shrift as soon as he is caught. In the same way, negotiations between groups acting on behalf of institutions, or even nations, may not only create uncertainty about control of the boundaries of the institutions they represent, but different opinions between the delegates of the negotiating groups about the task they are engaged on can threaten the whole transaction.

Some transactions by their nature give power and privilege to one party to the transaction. Most such transactions are governed by social and even legal sanctions. If *A* in Figure 5 represents the medical profession and '*a*' a member of it, *B* the community and '*b*' a patient, very strict rules with legal backing govern the nature of the transactions and the role relationships within the task system boundary. Other professions have equally strict rules.

But even in those transactions in which no recognised professions are involved, the (*ab*) boundary (of the transactional task system) is usually 'controlled' by cultural conventions to which both *A* and *B* subscribe. In the absence of such conventions, or without acceptance of recognised conventions, much time and effort has to be spent in establishing rules of procedure before performance of the real transaction can start.

More generally, any transaction between an enterprise and its environment introduces some uncertainty into the relative strengths of the boundaries of the enterprise, the environment, and the transactional task system. If chaos is defined as uncertainty about boundary definition, or more colloquially, as not knowing who or what belongs where, then every transaction is potentially chaotic. If we go further and suggest that the major characteristic of disaster is the obliteration of known boundaries (of the guides and directories which govern existence), then every transaction can be said to have built into it the elements of incipient disaster. The doctor who has sexual intercourse with his patient, that is, who allows personal relationships in the task system boundary to obliterate the boundary of the medical profession to which he belongs is, in reality, courting disaster.

The transactional task system is temporary and transitional. When the task has been performed it should be discontinued. If it is prolonged beyond task completion, it uses resources unnecessarily. By so doing it must reduce the efficiency of task performance. It requires therefore a project type organisation which may or may not be renewed.[3] At the end

---

[3] In this conceptual framework the calls made by a sales representative are seen as separate 'projects,' between calls he 'returns' to his sentient group within his company.

of the transaction 'a' is once more enclosed in A; and 'b' in B. But whatever the outcome of the transaction, the relationships of 'a' to other parts
of A and to the whole A (and of 'b' to B) are likely to have changed.
And it is at this stage that any disagreements within 'a' (or 'b') are likely
to affect the formation of future 'project teams' for transactional tasks.

The general point is that, in terms of its transactions with the environment, an enterprise is a multi-task system, forming and disbanding
temporary project teams (of one or more) for task performance. For
the duration of the transaction a project team operates with the authority of the enterprise. Every transaction tests the integrity of the boundaries of the enterprise and the project team and the control that the
enterprise can exercise over its own project team; that is over the
extent to which the project team acts with less, equal, or more authority
than it has been given. The outcome of every transaction can thus
change the nature and strength of the controls.

The common defences against uncertainty of control are the precise
definition of terms of reference for the 'project team' and the prescription of rules and procedures for dealing with any unforeseen or unplanned activities in the transaction. But the defences, by adding
constraints to the transactional task system, must put limits on performance.

THE INDIVIDUAL

The theories of human behaviour and of human relationships are in
many ways analogous to those of system theory as applied to institutions. Like an institution, an individual may be seen as an open system.
He exists and can exist only through processes of exchange with his
environment. Individuals, however, have the capacity to mobilise themselves at different times and simultaneously into many different kinds of
activity systems, and only some of their activities are relevant to the
performance of any particular task.

The personality of the individual is made up of his biological in-
heritance, his learned skills and experiences through which he passes,
particularly those of early infancy and childhood. A baby is dependent
on one person—his mother. He gradually assimilates into his patterns
of relationships his father and any brothers and sisters. As he grows into
childhood he includes other members of his extended family and of the
family network. The first break with this pattern is usually made when
the child goes to school and encounters for the first time an institution
to which he has to contribute as a member of a wider society. It is his
preliminary experience of what, in later years, will be a working environment.

The hopes and fears that govern the individual's expectations of how
he will be treated by others, and the beliefs and attitudes on which he
bases his code of conduct derive from these relationships and are built
into the pattern that becomes his personality. They form part of his
internal world. Besides the skills and capabilities he develops, this con-

tains his primitive inborn impulses and the primitive controls over them that derive from his earliest relations with authority, together with the modifications and adaptations he incorporates as he grows up.

In the mature individual, the ego-function mediates the relationships between the external and the internal worlds and thus takes in relation to the individual a 'leadership' role and exercises a 'management' control function. The mature ego is one that can differentiate between what is real in the outside world and what is projected on to it from 'inside,' between what should be accepted and incorporated into experience and what should be rejected. In short, the mature ego is one that can define the boundary between what is inside and what is outside and can control the transactions between the one and the other. Diagramatically the individual can be represented at any one time, therefore, as a system of activity. The ego-function is located in the boundary control region, checking and measuring intakes, controlling conversion activities and inspecting outputs. It uses the senses as instruments of the import system; thinking, feeling and other processes to convert the intakes; then action, speech or other means of expression to export the outputs.

But the individual is not just a single activity system with an easily defined primary task. He is a multi-task system and capable of multiple activities. The activities become bounded and controlled task systems when they are directed to the performance of a specific task, to the fulfilling of some specific purpose. The difficulty then is the control of internal boundaries and dealing with activities that are not relevant to task performance. And these controls are the result of the built-in attitudes and beliefs, born of previous experience, which may or may not be relevant to the specific task or system of activities required for its performance.

To take a role requires the carrying out of specific activities and the export of particular outputs. To take a role an individual could be said to set up a task system; and the task system to require the formation of a 'project team' composed of the relevant skill, experience, feelings and attitudes. Different roles demand the exercise of different skills and different outputs. The task of the ego-function is then to ensure that adequate resources are available to form the 'project' team for role performance, to control transactions with the environment so that intakes and outputs are appropriate, and to suppress or otherwise control irrelevant activities. When the role changes, the 'project team' has to be disbanded and reformed.

The individual as a multiple task enterprise is shown in simplified form Figure 6. Task systems I (T_1) and II (T_2) require the individual to take roles 1 & 2 (R_1 & R_2). R_1 and R_2 overlap to the extent that they use some, but not all, of the capabilities of the individual. The task systems are related to different but neighbouring parts of the environment. The management controls required will also therefore be similar, but not necessarily the same. In contrast, task system III (T_3) requires the individual to take role 3 (R_3). This requires quite different capabilities, is related to a quite different part of the environment, and hence requires

FIGURE 6

The Role System of the Individual

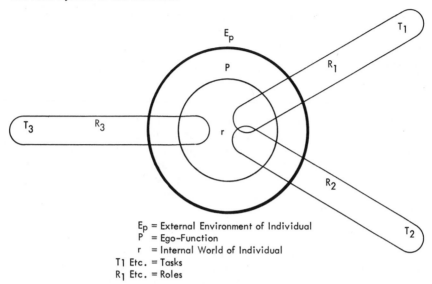

E_p = External Environment of Individual
P = Ego–Function
r = Internal World of Individual
T1 Etc. = Tasks
R_1 Etc. = Roles

a different kind of managerial control. In practice, such complete splits are not usual (except in the schizophrenic), but it is possible to recognise, on the one hand, those individuals who are always the same no matter what the situation is or with whom they are in contact; and, on the other, those who appear to be quite different people in different situations.

More generally we can say the ego-function has to exercise different kinds of authority and different kinds of leadership in different roles and in different situations. Dislike of the role and of the activities or behaviour required in it, and the demonstration of the dislike by attempts to change the role or modify the behaviour, or the intrusion of feelings or judgments that contradict role requirements, inevitably distort intakes, modify conversion processes, and can only result in inappropriate outputs. It is as though the management of a multiple task enterprise, with an organisation such as that outlined in Figure 4, were to set up a project team for the solution of a particular problem but not only could not be sure whether the team was working on the right problem, but could not even control membership of the team or the resources they used or squandered.

In effect, I wish to suggest that the general diagram of a project type organisation, as shown in Figure 4, can be used, however crudely, to represent the individual as a role-taking but sentient being. In the individual, the sentient groups and resource pools of the enterprise become the repositories of the capacities of the individual to fill different roles. The resource pools hold the intellectual power, cognitive and motor skills, experience and other capabilities; the sentient groups the attitudes, beliefs, and feelings—the world of objects and part objects—resulting from up-bringing. In effect, because a role demands specific skills and the exercise of specific authority in a particular context, it is unlikely to require every personal attribute of a given individual. Some attitudes and some skills will always be unused by any given role. Maintaining a role over a long time leads therefore either to the atrophy of unused attributes, or to the need to find other means of expressing them.

I recognise, of course, that for human beings, the many import-conversion-export processes cannot be so easily defined as the previous paragraphs might suggest, and that 'productivity' is seldom a simple measure of the difference between known intakes and known outputs. I hope, however, that this way of thinking about an individual will help to clarify some of the problems of role-taking when we have to consider group and intergroup processes.

Before moving on to consider the group, however, I would like to use the concept of a transactional task system as an example of a pair relationship. When an individual takes a role and makes a relationship with his environment, the parts of him that are 'used' are the equivalent of the representative team 'a' of Figure 5 acting on behalf of the whole A. If the other party to the transaction is another individual, then Figure 5 represents a transaction between project teams a and b on behalf of A and B. In the same way as for a transaction between two institutions, uncertainties about the relative strengths of A, B, and ab boundaries can interfere with the effectiveness of the transaction, as, of course, can uncertainties about the a and b boundaries. In an 'ideal' pair relationship of the kind frequently imagined by the more romantic novelists, all parts of the personality are invested in the relationship and the A and 'a' and the B and 'b' boundaries coincide, and the ab boundary is the AB boundary. No relationships ever achieve such a romantic ideal, and the effectiveness of the pair relationship is frequently diminished by the relationships of other parts of A with 'a,' with 'b' and with other parts of B, as well as by the many other roles taken by A and B in relation to others.[4]

The ego-function has therefore to control not only transactions across the individual/environment boundary but also between role and person. When the ego-function fails to locate boundaries precisely and fails to control transactions across those boundaries, confusion is inevitable, confusion in roles and in the authorities exercised in roles. Authority and responsibility appropriate in one role are used inappropriately in other roles. To be continuously confused about the role/person boundaries or completely unable to define and maintain boundaries is to be mentally sick.

---

[4] I am perhaps using 'role' in a rather limited sense in this context. I am ignoring for example the difference between overt and covert expectations, and between anticipated and realised interactions as elements of 'role' and role relationships.

THE GROUP

The internal world of the individual includes the objects and part-objects derived from the relationships he has made, particularly in his early life. His attitudes towards authority, both of his own and that of others, are conditioned by his earliest experiences of authority, usually that exercised by his parents. 'Individual' has, therefore, little meaning as a concept except in relationships with others. He uses them and they him, to express views, take action and play roles. The individual is a creature of the group, the group of the individual. Each, according to his capacity and experience, carries within him the groups of which he has been and is a member. His experiences as an infant, child, adolescent and adult within his family, at school and at work, and the cultural setting in which he has been brought up will thus affect, by the way in which they are moulded into his personality, the contemporary and future relationships he makes, in his family, his work and his social life.

A group always meets to do something. In this activity the members of the group co-operate with each other; and their co-operation calls on their knowledge, experience and skill. Because the task for which they have met is real, they have to relate themselves to reality to perform it. The members of the group have, therefore, to take roles and to make role relationships with each other. The work group is now a task system. It may or may not have very much sentience depending on the extent to which its members are committed to each other. Even as a sentient system it may, or may not, support task performance. Controls are then required:

a)   to regulate transactions of the whole, as a task system, with the environment and of the constituent systems with each other,

b)   to regulate sentient group boundaries,

c)   to regulate relationships between task and sentient groups.

But, in the discussion of the individual, I wrote that the role taken by each member of a group is also a 'task system' and the 'management' of each of these (the ego-function) has to control the relations between the 'task' and 'sentient' systems of the individual. So long as the role taken by each individual member is supported by that member's own individual 'sentient' system, so the task group and sentient group tend to coincide. But individual members may not be aware of all the elements either of their own individual or of total group sentience, even if such exists. To put this another way: task roles are unlikely to use all attributes of every member's personality; the unused portions may or may not support role—and hence group task performance—but neither individual member nor group may be aware of the discrepancies between individual and group sentience or of changes over time.

More importantly, the unused attributes of individuals may themselves have such powerful sentience attached to them that they have to be expressed in some way. That is, an individual, though a member of a task group, may be unable to control those personal attributes that are not relevant to task performance and may seek other outlets for the emotions and feelings that the unused attributes, and the inability to control them, give rise to. This represents a breakdown in the 'management control' of the individual so far as role performance is concerned. Group task leadership may still so be able to control group sentience, as not only to overcome individual discrepancies but also to harness group emotions and feelings in favour of group task performance. The charismatic leader, for example, can be said to attract to himself as a person the unused sentience of group members and, since he is concerned with task performance, can thus control any group opposition to that performance. If task leadership cannot either harness group feelings in favour of task performance or contain opposing feelings by personal leadership, then other 'groups' consisting of some or all of the task group members may be formed to express opposing sentience. Such groups may seek and 'appoint' other 'leaders.' If the other 'group' gets support from all other members of the task system, however unaware they may be of this support (since individual management control has broken down), then the other 'group' can become more powerful than the task group.

Bion postulated that a group always behaves on two levels: the sophisticated level and the basic assumption level.[5] He also postulated a protomental system in which the inoperative basic assumptions were held. He described three basic assumptions which determine group behaviour:

Dependency: to obtain security from one individual;
Fight/Flight: to attack or run away from somebody or something; and
Pairing: to reproduce itself.

He suggested there may be others. He could not at the time define them. He showed that when the basic assumption was appropriate to task performance, the group culture was a powerful reinforcement to that performance; when it did not there was conflict of such an order that it could lead to task distortion or redefinition.

In the terms used here, Bion describes the situation in which the sentience of the roles taken by the members of a group in the task system may or may not be stronger than other possible sentient systems. If the sentient systems of the individual members coalesce, that is, individual members find a common group sentience, then the group can be said to be behaving as if it had made a basic assumption. If the common group sentience is opposed to task performance, that is, the control is not maintained by task leadership, other leaders will be found.

I now feel that Bion's concepts describe special cases which are most easily observable in small groups, because they are large enough to give recognisable power to an alternative leadership, and yet are not so large as to provide support for more than one kind of powerful alternative leadership at any one time. As Bion points out, the capacity for co-operation among the members of a task group is considerable; that is, role

5 Bion, ibid.

sentience in a task group is always likely to be strong. Hence, while the group maintains task definition, the strength of the sentience supporting task performance at the reality level makes the life of leadership opposing task performance precarious.

A pair who have met to perform an agreed task can hardly provide alternative leadership and remain a task system. With three, an alternative leader is rapidly manifest and either immediately outnumbered or at once destroys co-operation in task performance, i.e. the three cannot easily remain a task group. (Two is company, three is none). A quartet can provide some support for alternative leadership by splitting into pairs, but cannot sustain the split for very long without destroying the quartet as a task system. In groups of five and six, the interpersonal transaction systems are still relatively few and task leadership can be quick to recognise alternative leadership, usually before it can manifest powerful opposition to task performance. Above six, the number of interpersonal transactions becomes progressively larger, and hence it may be more difficult to detect their patterning.

In general, the larger the number of members of a group there are, the more members there are to find an outlet for their non-task related sentience, and hence the more powerful can be its expression, and the more support can an alternative leader obtain. Equally, because of the large number, the more futile and useless can group behaviour appear when there is no sentient unanimity among the membership either in support of, or in opposition to, group task performance. In other words, the larger the group the more opportunities members have to divest themselves of their unwanted or irrelevant sentience, by projecting it into so many others.

But the individual is a multiple task enterprise, and his various sentient systems can be in conflict with each other. When he joins a group to perform a group task, he must by his very joining to some extent commit himself to take the role assigned to him, and hence to control irrelevant activities and sentience. Mature individuals thus find themselves distressed and guilty when in any attempt to reassert 'management control' over their own individual boundaries they recognise, however vaguely, the number of different hostages they have given to so many conflicting sentient groups.

The situation of the group can be roughly approximated symbolically:

Let the members of a group be: $I_1, I_2, I_3 \ldots I_n$
Each is capable of taking many roles: $R_1, R_2, R_3 \ldots R_n$.

Each role, in the way the term is used here, is a task system in itself. It comprises a number of specific activities together with the necessary resources for its performance. The resources should include not only the skills, but also the appropriate attitudes, beliefs, and feelings derived from the individual's 'sentient' groups. But not all individuals are capable of taking all roles, and role performances by different individuals in the same role also differ.

If the role performance is represented by IR then:

$$I_1R_1 \neq I_2R_1 \neq I_3R_1 \ldots \text{etc.}$$
$$\text{and } I_1R_1 \neq I_1R_2 \neq I_1R_3 \ldots \text{etc.}$$

Ideally a task system requires only activities and we could then write

$$T = f(R_1 + R_2 + R_3 + \ldots R_n) = f\Sigma(R). \ldots$$

But because roles are taken by individuals, we have to write

$$TP \ (task \ performance) = f(I_1R_1 + I_2R_2 + \ldots + I_nR_n)$$
$$= f\Sigma(IR). \ldots \ldots \tag{1}$$

assuming R_1 to be taken by I_1, R_2 by I_2 . . . etc. But when an individual takes a specific role not all his aptitudes are likely to be used, and his performance in any specific role is likely to be reduced by the amount of 'energy' he devotes to other aptitudes and to other sentience. If we represent these other irrelevant activities and their related sentience by R^0_1, R^0_2 . . . R^0_n then any given role performance R_1 by an individual l_1 will have to be written:

$$I_1R_1 - I_1(R^0_1 + R^0_2 \ldots R^0_n)$$

in which R^0_1, R^0_2 etc. can have zero, or positive values so far as they do not affect or oppose I_1R_1. (I assume that all tasks supporting sentience are included in R_1.) Equation (1) therefore has to be written:

$$TP = f[\Sigma(IR) - I_1(R^0_1 + R^0_2 + \ldots R^0_n) - I_2(R^0_1 + R^0_2 \ldots R^0_n)$$
$$\ldots -I_n(R^0_1 + R^0_2 + \ldots R^0_n)] \tag{2}$$
$$= f[\Sigma(IR) - \Sigma(IR^0)]$$

Even if $\Sigma (IR^0) \neq 0$ and has a positive value it can still be small enough to be controlled either because of the discrepancy between the many different roles taken by the different I's or because the combinations of different numbers are themselves small. Nevertheless, the sentience invested in the R^0's can still produce such disagreements between I's that a sense of futility can grow as I's spend more time and energy trying to find agreement between themselves in roles irrelevant to TP than in R_1, R_2 etc. that are relevant. If overtly or covertly they all agree on a role that is irrelevant to TP (say R^0_m) then equation (2) becomes:

$$TP = f[\Sigma(IR) - R^0_m\Sigma(I)] \tag{3}$$

If I write out equation (3) more fully:

$$TP = f[(I_1R_1 + I_2R_2 + I_3R_3 + \ldots + I_nR_n) - R^0_m(I_1 + I_2 + I_3 + \ldots + I_n)] \tag{4}$$

It can be seen that because R^0_m is taken by all group members it can become a considerable threat to TP which requires different members to take different roles. If R^0_m is large enough and is a consciously agreed role, there is revolt; if members are both unaware of their agreement and of the role they have agreed upon, they are then behaving 'as if' they have made a basic assumption opposed to task performance.

It can also be seen that the more I's there are the greater the threat of R^0_m ($I_1 + I_2$ . . . I_n) but, at the same time, the more difficulty there is likely to be in getting agreement on R^0_m. It can also be seen why, with small numbers, alternative leadership is difficult to sustain without immediate destruction of task performance. From equation (4) $TP = f$ $[(I_1 R_1 + I_2 R_2) - R^0_m (I_1 + I_2)]$ for a pair. If now R^0_m has a large value, and is reinforced by $I_1 + I_2$ it will almost certainly give TP a negative value.

INTERGROUP PROCESS

I have tried to show that all transactions, even the intra-psychic transactions of the individual, have the characteristics of an intergroup process. As such they involve multiple problems of boundary control of different task systems and different sentient systems and control of relations between task and sentient systems. Each transaction calls into question the integrity of boundaries across which it takes place and the extent to which control over transactions across them can be maintained. Every transaction requires the exercise of authority and calls into question the value of and sanction for that authority.

In the examination of a simple intergroup transaction between two groups in which individuals represent the two groups (as in Figure 5), account has to be taken, therefore, of a complex pattern of intergroup processes: within the individuals who represent their groups, within the transactional task system, between the groups and their representatives, within the groups, and within the environment that includes the two groups. Even a simple intergroup transaction is, therefore, affected by a complex pattern of authorities, many of which are either partially or completely covert. If I now extend the analysis to more than two groups and each with more than one representative, the pattern becomes still more complex. A meeting of pairs of representatives from four groups is illustrated in Figure 7. It will be seen that in the meeting of representa-

FIGURE 7

Meeting of Representatives' Group—One Pair from Each of Four Groups

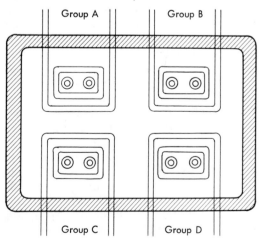

tives alone, transactions across seventeen different pairs of boundaries have to be controlled: four pairs for each pair of representatives, and one pair for the group of representatives as a group.

To understand the nature of the authority of a representative, or of a group of representatives, appointed to carry out a transaction on behalf of a group involves, therefore, the understanding of multiple and complex boundary controls. In other words, the appointment of a representative or representatives is never just a simple matter of representing a task system to carry out a task-directed transaction with the environment. To put the same thing more colloquially: representatives are invariably chosen not only to carry out the specific transaction, but also to convey the mood of the group about itself and about its representative, and its attitude, not only to the specific part of the environment with which the transaction is intended, but to the rest of it as well. And not all the 'messages' are explicit and overt; many, if not most of them, are implicit and covert.

But the representative has his own intra-psychic processes, and his own intra-psychic 'group' has had to make 'intergroup' relations with the group he represents. The same mixture of transactions, overt and covert, have, or should have, taken place before he starts the intergroup transaction for which he has been appointed. The results of these transactions can seldom endow the representative with personal attributes that he did not previously possess, at least latently. The choice of representative(s) therefore offers important data, about the group attitude not only towards its task, but also towards itself and its environment. Further important data can be gathered from the extent to which the representative is given the authority to commit his group, and by his status within the group.

One further dimension of complexity has to be mentioned: time. I have spoken about the problems of the control of the representative's own boundaries, of the boundaries between the representative and his group, and of the relative strengths of the individual, group and transactional task system boundaries. It is surely rare for them all to be perfectly controlled in the interests of task performance. Even if they are, a transaction takes time, and during the transaction the representative cannot be in continuous communication with the group he represents, not, that is, if he is anything more than a relay system. During the transaction the individual, group, and task-system sentiences may change. Indeed, in any critical negotiation they are almost bound to change, as hopes and fears of the outcome increase and decrease.

The past, during which decisions were made, attitudes formed and resources collected is always the past; a transaction is the present and if it is to have any meaning, must determine a future. Individuals and even groups with strongly defended boundaries can, by staying firmly within them, occasionally live in the past; intergroup relations never.

The number and complexity of the boundary controls required for even comparatively simple transactions between groups might make one wonder how any negotiation is ever successful, how any salesman ever got an order for anything. The reality is, of course, that the preponderance of intergroup transactions takes place in settings in which the conventions are already established and mutual pay-offs understood. Nevertheless, I suggest that it is this complex authority pattern, imperfectly understood and imperfectly comprehended, together with the need to defend each of the boundaries in the multiple transactional systems against uncertainty, chaos and incipient disaster, that give rise to the futility of so many negotiations and to the unexpected results that often emerge. The conventions and pay-offs for the majority of intergroup transactions are defences against chaos and disaster. In new kinds of negotiations without established defences, the feat of chaos and disaster often makes procedure more important than content.

There is perhaps small wonder that international negotiating institutions find it so difficult to satisfy the hopes of their creators. Indeed, unless the boundary of the negotiating group itself becomes stronger than the boundaries that join the representatives and those they represent, there seems little hope of successful negotiations. But this means that not only the group of representatives but the groups they represent have to invest the representative task system with more sentience than they invest in their own groups. The United Nations cannot, in other words, be fully effective until not only the members of its Council, but the nations they represent, invest more sentience in the United Nations than they do in their nationalisms.

THE ROLE OF LEADERSHIP

Finally, I turn to the role of leadership, which can be conceived of as a special case of representation: representation with plenipotentiary powers. Conceptually, it is irrelevant whether the role is taken by an individual or by a group. For convenience, I shall discuss it in terms of an individual leader.

As a member of a task group every individual has to take a role and through it control his task transactions with his colleagues individually and collectively; as a person he also has to control his own person/role transactions as well as his interpersonal relationships with his colleagues. In addition to these, a leader has to control transactions between the group and relevant agencies in the environment in the interests of task performance; without such control, task performance is impossible. In this sense, the role taken by the leader and boundary control function of the group must have much sentience in common. For the leader, at least, sentient group and task group *must* reinforce each other. So far as task performance is unsatisfactory, either by reason of inadequate resources or of opposing group sentience, transactions with the environment are likely to be difficult and the task sentience of the leader weakened if not destroyed.

If I return to the notation I used earlier and let R^L represent the role of leader taken by an individual I then leadership task performance can be written:

$$TP = IR^L - I(R^0_1 + R^0_2 + R^0_3 + \ldots \ldots R^0_n)$$
$$= I(R^L - \Sigma R^0)$$

For the leader at least ΣR must be close to zero. What he has to provide is an IR model that is task oriented. The model, however, must be a credible one. A leader who puts all his energy into IR^L (with $\Sigma R^0 = O$) is hardly credible and gives no reinforcement to group members in controlling their own ego boundaries; on the other hand, a leader who puts too much energy into ΣR^0 encourages his followers to do the same and their Σ (IR^0) may only temporarily take the same form as his with consequent detriment to task performance.

More generally, since transactions with the environment can only be based on adequate task performance, the leader's authority has to be based on sufficient group sentience that is supportive of such performance. It follows that the mobilisation of group sentience for any other reason than task performance, for example, personal loyalty, friendship or ideology, always leaves a task group vulnerable. It also follows that any change in the group task, by change in either in the environment or in the group, changes not only the internal transactions between the members but also those with the environment, and hence the role of leadership and the appropriate sentience that has to be mobilised.

In practice, groups use all kinds of feelings and attitudes to maintain co-operation in task performance: love, affection, friendship, hatred, dislike and enmity as well as commitment to the group task. So far as a group is committed to its task, contrary sentience, including leadership's own, can be contained and controlled within the group; so far as commitment is tenuous, so far will the group find it impossible to control the contrary sentience. Under such circumstances, task leadership is castrated, the task redefined, or irrelevant transactions with the environment have to be used to cope with the discordant feeling and attitudes.

Section V

INTERORGANIZATIONAL SYSTEMS RELATIONS

THE ENVIRONMENT of an organization is largely made up of other organizational systems and their related individuals and groups. The external relations of an organization include its interactions with those organizations which receive the benefits of the focal organization's activities, as well as those organizations which supply necessary inputs. The organizations in the environment play various roles and constitute sources of varied external influence on a given organization.

Levine and White explore the nature of interorganizational relationships among health and social welfare agencies. Proposing exchange as a conceptual framework, they see community health organizations as comprising a system in which interdependence of systems components is determined by: (1) the environmental distribution and accessibility of resources to various organizations; (2) organizational functions; and (3) the degree of domain consensus existing among various organizations.

Aldrich criticizes approaches to organizations which overemphasize the functional interdependence and stability of organizations. He notes that the open system perspective forces the organizational theorist to consider variation both in the organization and in the environment. Conceptualizing organizations as boundary-maintaining systems, he develops the concepts of authority, membership, and organizational conflict. Aldrich discusses strategies for securing the necessary member participation for use by the organization in its competition with other organizations. He also offers predictions about the conditions in which an organizational system will constrict or expand its boundaries in dealing with interorganizational conflict.

In the last paper in this section, Baker and O'Brien also recognize the problems inherent in assuming interdependence among organizational systems. They discuss an intersystem model which focuses on determining the degree and nature of interchanges between systems rather than assuming the operation and organization of a large suprasystem composed of well-integrated organizational systems. Examining intersystem relations in an interorganizational field of community health and welfare organizations, they offer some hypotheses concerning the interorganizational interaction of such human service systems.

Sol Levine
Paul E. White

20. Exchange as a Conceptual Framework for the Study of Interorganizational Relationships

SOCIOLOGISTS HAVE devoted considerable attention to the study of formal organizations, particularly in industry, government, and the trade union field. Their chief focus, however, has been on patterns within rather than between organizations. Studies of interrelationships have largely been confined to units within the same organizational structure or between a pair of complementary organizations such as management and labor. Dimock's study of jurisdictional conflict between two federal agencies is a notable exception.[1] Another is a study of a community reaction to disaster by Form and Nosow in which the authors produce revealing data on the interaction pattern of local health organizations. The authors observe that "organizational cooperation was faciliated among organizations with similar internal structures."[2] March and Simon suggest that interorganizational conflict is very similar to intergroup conflict within organizations but present no supporting data.[3] Blau has commented on the general problems involved in studying multiple organizations.[4] In pointing up the need to study the organization in relation to its environment, Etzioni specifies the area of interorganizational relationships as one of the three meriting further intensive empirical study.[5]

From *Administrative Science Quarterly*, 1961, vol. 5, pp. 583–601. Reprinted with permission of the authors and the publisher.

[1] Marshall E. Dimock, "Expanding Jurisdictions: A Case Study in Bureaucratic Conflict," in Robert K. Merton, Ailsa P. Gray, Barbara Hockey, Hanan C. Selvin, eds. *Reader in Bureaucracy* (Glencoe, 1952).

[2] William H. Form and Sigmund Nosow, *Community in Disaster* (New York, 1958), p. 236.

[3] James G. March and H. A. Simon, *Organizations* (New York, 1958).

[4] Peter M. Blau, Formal Organization: Dimensions of Analysis, *American Journal of Sociology*, 63 (1957), 58.

[5] Amitai Etzioni, New Directions in the Study of Organizations and Society, *Social Research*, 27 (1960), 223–28.

Health and social welfare agencies within a given community offer an excellent opportunity for exploring patterns of relationship among organizations. There are an appreciable number of such organizations in any fairly large urban American community. Most of them are small so that relatively few individuals have to be interviewed to obtain information on their interaction. Within any community setting, varying kinds of relations exist between official and voluntary organizations concerned with health and welfare. Thus welfare agencies may use public health nursing services, or information on the status of families may be shared by such voluntary organizations as the Red Cross and the Tuberculosis and Health Association.

Facilitating communication between local organizations has been a major objective of public health administrators and community organizers. Their writings contain many assertions about the desirability of improving relationships in order to reduce gaps and overlaps of medical services to the citizens, but as yet little effort has been made to appraise objectively the interrelationships that actually exist within the community.

In the following pages we should like to present our theoretical interpretation of interorganizational relationships together with a discussion of our research approach and a few preliminary findings, pointing up some of the substantive areas in organizational sociology for which our study has relevance. Our present thinking is largely based on the results of an exploratory study of twenty-two health organizations in a New England community with a population of 200,000 and initial impressions of data on a more intensive study, as yet unanalyzed, of some fifty-five health organizations in another New England community of comparable size.[6]

The site of our initial investigation was selected because we found it fairly accessible for study and relatively independent of a large metropolis; moreover, it contained a range of organizations which were of interest—a full-time health department, a welfare department, autonomous local agencies, local chapters or affiliates of major voluntary health and social welfare organizations, and major community hospitals. Of the twenty-two health organizations or agencies studied, fourteen were voluntary agencies, five were hospitals (three with outpatient clinics and two without) and three others were official agencies —health, welfare, and school. Intensive semistructured interviews were conducted with executive directors and supervisory personnel of each organization, and information was obtained from members of the boards through brief semistructured questionnaires. In addition, we used an adaptation of an instrument developed by Irwin T. Sanders to locate the most influential leaders in the community for the purpose of

[6] The project is sponsored by the Social Science Program at the Harvard School of Public Health and supported by Grant 8676–2 from the National Institutes of Health. Professor Sol Levine is the principal investigator of the project and Benjamin D. Paul, the director of the Social Science Program, is coinvestigator. We are grateful for the criticisms and suggestions given by Professors Paul, S. M. Miller, Irwin T. Sanders, and Howard E. Freeman.

determining their distribution on agency boards.[7] The prestige ratings that the influential leaders assigned to the organizations constituted one of the independent variables of our study.

EXCHANGE AS A CONCEPTUAL FRAMEWORK

The complex of community health organizations may be seen as a system with individual organizations or system parts varying in the kinds and frequency of their relationships with one another. This system is enmeshed in ever larger systems—the community, the state, and so on.

Prevention and cure of disease constitute the ideal orientation of the health agency system, and individual agencies derive their respective goals or objectives from this larger orientation. In order to achieve its specific objectives, however, an agency must possess or control certain elements. It must have clients to serve; it must have resources in the form of equipment, specialized knowledge, or the funds with which to procure them; and it must have the services of people who can direct these resources to the clients. Few, if any, organizations have enough access to all these elements to enable them to attain their objectives fully. Under realistic conditions of element scarcity, organizations must select, on the basis of expediency or efficiency, particular functions that permit them to achieve their ends as fully as possible. By function is meant a set of interrelated services or activities that are instrumental, or believed to be instrumental, for the realization of an organization's objectives.

Although, because of scarcity, an organization limits itself to particular functions, it can seldom carry them out without establishing relationships with other organizations of the health system. The reasons for this are clear. To fulfill its functions without relating to other parts of the health system, an organization must be able to procure the necessary elements—cases, labor services, and other resources—directly from the community or outside it. Certain classes of hospitals treating a specific disease and serving an area larger than the local community probably most nearly approximate this condition. But even in this case other organizations within the system usually control some elements that are necessary or, at least, helpful to the carrying out of its functions. These may be money, equipment, or special personnel, which are conditionally lent or given. Usually agencies are unable to obtain all the elements they need from the community or through their individual efforts and, accordingly, have to turn to other agencies to obtain additional elements. The need for a sufficient number of clients, for example, is often more efficiently met through exchanges with other organizations than through independent case-finding procedures.

Theoretically, then, were all the essential elements in infinite supply there would be little need for organizational interaction and for sub-

---

[7] Irwin T. Sanders, The Community Social Profile, *American Sociological Review*, 25 (1960), 75–77.

scription to cooperation as an ideal. Under actual conditions of scarcity, however, interorganizational exchanges are essential to goal attainment. In sum, organizational goals or objectives are derived from general health values. These goals or objectives may be viewed as defining the organization's ideal need for elements—consumers, labor services, and other resources. The scarcity of elements, however, impels the organization to restrict its activity to limited specific functions. The fulfillment of these limited functions, in turn, requires access to certain kinds of elements, which an organization seeks to obtain by entering into exchanges with other organizations.

Interaction among organizations can be viewed within the framework of an exchange model like that suggested by Homans.[8] However, the few available definitions of exchange are somewhat limited for our purposes because they tend to be bound by economics and because their referents are mainly individual or psychological phenomena and are not intended to encompass interaction between organizational entities or larger systems.[9]

We suggest the following definition of organizational exchange: *Organizational exchange is any voluntary activity between two organizations which has consequences, actual or anticipated, for the realization of their respective goals or objectives.* This definition has several advantages. First, it refers to activity in general and not exclusively to reciprocal activity. The action may be unidirectional and yet involve exchange. If an organization refers a patient to another organization which then treats him, an exchange has taken place if the respective objectives of the two organizations are furthered by the action. Pivoting the definition on goals or objectives provides for an obvious but crucial component of what constitutes an organization. The co-ordination of activities of a number of individuals toward some objective or goal has been designated as a distinguishing feature of organizations by students in the field.[10] Parsons, for example, has defined an organization as a "special type of social system organized about the primacy of interest in the attainment of a particular type of system goal."[11] That its goals or objectives may be transformed by a variety of factors and that,

---

[8] George C. Homans, Social Behavior as Exchange, *American Journal of Sociology*, 63 (1958), 597–606.

[9] Weber states that "by 'exchange' in the broadest sense will be meant every case of a formally voluntary agreement involving the offer of any sort of present, continuing, or future utility in exchange for utilities of any sort offered in return." Weber employs the term "utility" in the economic sense. It is the "utility" of the "object of exchange" to the parties concerned that produces exchange. See Max Weber, *The Theory of Social and Economic Organization* (New York, 1947), p. 170. Homans, on the other hand, in characterizing interaction between persons as an exchange of goods, material and nonmaterial, sees the impulse to "exchange" in the psychological make-up of the parties to the exchange. He states, "the paradigm of elementary social behavior, and the problem of the elementary sociologist is to state propositions relating the variations in the values and costs of each man to his frequency distribution of behavior among alternatives, where the values (in the mathematical sense) taken by these variables for one man determine in part their values for the other." See Homans, op. cit., p. 598.

[10] Talcott Parsons, Suggestions for a Sociological Approach to the Theory of Organizations—I, *Administrative Science Quarterly*, I (1956), 63–85.

[11] Ibid., p. 64.

under some circumstances, mere survival may become primary does not deny that goals or objectives are universal characteristics of organizations.

Second, the definition widens the concept of exchange beyond the transfer of material goods and beyond gratifications in the immediate present. This broad definition of exchange permits us to consider a number of dimensions of organizational interaction that would otherwise be overlooked.

Finally, while the organizations may not be bargaining or interacting on equal terms and may even employ sanctions or pressures (by granting or withholding these elements), it is important to exclude from our definition relationships involving physical coercion or domination; hence emphasis is on the word "voluntary" in our definition.

The elements that are exchanged by health organizations fall into three main categories: (1) referrals of cases, clients, or patients; (2) the giving or receiving of labor services, including the services of volunteer, clerical, and professional personnel; and (3) the sending or receiving of resources other than labor services, including funds, equipment, and information on cases and technical matters. Organizations have varying needs of these elements depending on their particular functions. Referrals, for example, may be seen as the delivery of the consumers of services to organizations, labor services as the human means by which the resources of the organization are made available to the consumers, and resources other than labor services as the necessary capital goods.

THE DETERMINANTS OF EXCHANGE

The interdependence of the parts of the exchange system is contingent upon three related factors: (1) the accessibility of each organization necessary elements from sources outside the health system, (2) the objectives of the organization and particular functions to which it allocates the elements it controls, and (3) the degree to which domain consensus exists among the various organizations. An ideal theory of organizational exchange would describe the interrelationship and relative contribution of each of our preliminary findings to suggest possible relationships among these factors and to indicate that each plays a part in affecting the exchange of elements among organizations.

Gouldner has emphasized the need to differentiate the various parts of a system in terms of their relative dependence upon other parts of the system.[12] In our terms, certain system parts are relatively dependent, not having access to elements outside the system, whereas others, which have access to such elements, possess a high degree of independence or functional autonomy. The voluntary organizations of our study (excluding hospitals) can be classified into what Sills calls either

---

[12] Alvin W. Gouldner, Reciprocity and Autonomy in Functional Theory, in Llewellyn Gross, ed., *Symposium on Sociological Therapy* (Evanston, Ill., 1959); also The Norm of Reciprocity: A Preliminary Statement, *American Sociological Review*, 25 (1960), 161–78.

corporate or federated organizations.[13] Corporate organizations are those which delegate authority downward from the national or state level to the local level. They contrast with organizations of the federated type which delegate authority upward—from the local to the state or national level.

It appears that local member units of corporate organizations, because they are less dependent on the local health system and can obtain the necessary elements from the community or their parent organizations, interact less with other local agencies than federated organizations. This is supported by preliminary data presented in Table 1. It is also suggested that by carrying out their activities without entering actively into exchange relationships with other organizations, corporate organizations apparently are able to maintain their essential structure and avoid consequences resulting in the displacement of state or national goals. It may be that corporate organizations deliberately choose functions that require minimal involvement with other organizations. An examination of the four corporate organizations in our preliminary study reveals that three of them give resources to other agencies to carry out their activities, and the fourth conducts broad educational programs. Such functions are less likely to involve relationships with other organizations than the more direct service organizations, those that render services to individual recipients.

An organization's relative independence from the rest of the local health agency system and greater dependence upon a system outside the community may, at times, produce specific types of disagreements with the other agencies within the local system. This is dramatically demonstrated in the criticisms expressed toward a local community branch of an official state rehabilitation organization. The state organization, to justify its existence, has to present a successful experience to the legislators—that a minimum number of persons have been successfully rehabilitated. This means that by virtue of the services the organization has offered, a certain percentage of its debilitated clients are again returned to self-supporting roles. The rehabilitative goal of the organization cannot be fulfilled unless it is selective in the persons it accepts as clients. Other community agencies dealing with seriously debilitated clients are unable to get the state to accept their clients for rehabilitation. In the eyes of these frustrated agencies the state organization is remiss in fulfilling its public goal. The state agency, on the other hand, cannot commit its limited personnel and resources to the time-consuming task of trying to rehabilitate what seems to be very poor risks. The state agency wants to be accepted and approved by the local community and its health agencies, but the state legislature and the governor, being the primary source of the agency's resources, constitute its significant reference group. Hence, given the existing definition of organizational goals and the state agency's relative independence

---

[13] David L. Sills, *The Volunteers: Means and Ends in a National Organization* (Glencoe, 1957).

TABLE 1

Weighted Rankings* of Organizations Classified by *Organizational Form* on Four Interactions Indices

Interaction Index	Sent by	N	Sent to					Total Interaction Sent
			Voluntary		Hospitals			
			Corporate	Federated	Without Clinics	With Clinics	Official	
Referrals	Vol. corporate	4	4.5	5	3.7	4.5	5	5
	Vol. federated	10	3	4	3.7	3	4	3
	Hosps. w/o clinics	2	4.5	3	3.7	4.5	3	4
	Hosps. w. clinics	3	1	1	1.5	2	1	1
	Official	3	2	2	1.5	1	2	2
Resources	Vol. corporate	4	5	2	1	4	5	3.5
	Vol. federated	10	4	3	3	4	4	3.5
	Hosps. w/o clinics	2	2	4.5	4.5	5	3	5
	Hosps. w. clinics	3	1	1	2	1	2	1
	Official	3	3	4.5	4.5	2	1	2
Written and verbal communication	Vol. corporate	4	5	3	2	4	5	4
	Hosps. w/o clinics	2	2	5	4.5	5	4	5
	Hosps. w. clinics	3	4	4	4.5	1	1.5	2.5
	Official	3	1	2	1	2	1.5	1
Joint activities	Vol. corporate	4	4.5	4	3	5	3.5	5
	Vol. federated	10	3	3	5	3	1	3
	Hosps. w/o clinics	2	2	5	1	2	3.5	4
	Hosps. w. clinics	3	4.5	2	2	1	5	1.5
	Official	3	1	1	4	4	2	1.5

* Note: 1 indicates highest interaction; 5 indicates lowest interaction.

of the local health system, its interaction with other community agencies is relatively low.

The marked difference in the interaction rank position of hospitals with out-patient clinics and those without suggests other differences between the two classes of hospitals. It may be that the two types of hospitals have different goals and that hospitals with clinics have a greater "community" orientation and are more committed to the concept of "comprehensive" care than are hospitals without clinics. However, whether or not the goals of the two types of hospitals do indeed differ, those with out-patient departments deal with population groups similar to those serviced by other agencies of the health system, that is, patients whom other organizations may also be seeking to serve. Moreover, hospitals with out-patient clinics have greater control over their clinic patients than over those in-patients who are the charges of private physicians, and are thereby freer to refer patients to other agencies.

The functions of an organization not only represent the means by which it allocates its elements but, in accordance with our exchange formulation, also determine the degree of dependence on other organizations for specific kinds of elements, as well as its capacity to make certain kinds of elements available to other organizations. The exchange model leads us to explain the flow of elements between organizations largely in terms of the respective functions performed by the participating agencies. Indeed, it is doubtful whether any analysis of exchange of elements among organizations which ignores differences in organizational needs would have much theoretical or practical value.

In analyzing the data from our pilot community we classified agencies on the basis of their primary health functions: resource, education, prevention, treatment, or rehabilitation. Resource organizations attempt to achieve their objectives by providing other agencies with the means to carry out their functions. The four other agency types may be conceived as representing respective steps in the control of disease. We have suggested that the primary function determines an organization's need for exchange elements. Our preliminary data reveal, as expected, that treatment organizations rate highest on number of referrals and amount of resources received and that educational organizations, whose efforts are directed toward the general public, rate low on the number of referrals (see Table 2). This finding holds even when the larger organizations—official agencies and hospitals—are excluded and the analysis is based on the remaining voluntary agencies of our sample. As a case in point, let us consider a health organization whose function is to educate the public about a specific disease but which renders no direct service to individual clients. If it carries on an active educational program, it is possible that some people may come to it directly to obtain information and, mistakenly, in the hope of receiving treatment. If this occurs, the organization will temporarily be in possession of potential clients whom it may route or refer to other more appropriate agencies. That such referrals will be frequent is unlikely however. It is even less likely that the organization will receive many referrals from

TABLE 2

Weighted Rankings* of Organizations, Classified by *Function* on Four Interaction Indices

Interaction Index	Received by	N	Received from					Total Interaction Received
			Education	Resource	Prevention	Treatment	Rehabilitation	
Referrals	Education	3	4.5	5	5	5	5	5
	Resource	5	3	4	2	4	1	3
	Prevention	5	2	1	3	2	2.5	2
	Treatment	7	1	2	1	1	2.5	1
	Rehabilitation	2	4.5	3	4	3	4	4
Resources	Education	3	4.5	5	4	5	4.5	5
	Resource	5	1.5	3	3	4	3	3.5
	Prevention	5	1.5	4	2	3	4.5	3.5
	Treatment	7	3	2	1	2	2	1
	Rehabilitation	2	4.5	1	5	1	1	2
Written and verbal communication	Education	3	4	5	4.5	5	5	5
	Resource	5	3	2	2	3	2	2.5
	Prevention	5	2	4	3	4	4	3
	Treatment	7	1	1	1	2	3	1
	Rehabilitation	2	5	3	4.5	1	1	2.5
Joint activities	Education	3	4	4	1	3	4.5	4
	Resource	5	2	1	3	4	1	3
	Prevention	5	1	2	2	2	3	1
	Treatment	7	3	3	4	1	2	2
	Rehabilitation	2	5	5	5	5	4.5	5

* Note: 1 indicates highest interaction; 5 indicates lowest interaction.

other organizations. If an organization renders a direct service to a client, however, such as giving X-ray examinations, or polio immunizations, there is greater likelihood that it will send or receive referrals.

An organization is less limited in its function in such interagency activities as discussing general community health problems, attending agency council meetings or cooperating on some aspect of fund raising. Also, with sufficient initiative even a small educational agency can maintain communication with a large treatment organization (for example, a general hospital) through exchanges of periodic reports and telephone calls to obtain various types of information. But precisely because it is an educational agency offering services to the general public and not to individuals, it will be limited in its capacity to maintain other kinds of interaction with the treatment organization. It probably will not be able to lend or give space or equipment, and it is even doubtful that it can offer the kind of instruction that the treatment organization would seek for its staff. That the organization's function establishes the range of possibilities for exchange and that other variables exert influence within the framework established by function is suggested by some other early findings presented in Table 3. Organizations were classified as direct or indirect on the basis of whether or not they provided a direct service to the public. They were also classified according to their relative prestige as rated by influential leaders in the community. Organizations high in prestige lead in the number of joint

TABLE 3

Weighted Rankings* of Organizations Classified by *Prestige of Organization* and by General *Type of Service Offered* on Four Interaction Indices

| Interaction Index | Received by | N | High Prestige | | Low Prestige | | Total Interaction Received |
			Direct Service	Indirect Service	Direct Service	Indirect Service	
Referrals	High direct	9	1	1	1	1	1
	High indirect	3	3	3.5	3	3.5	3
	Low direct	6	2	2	2	2	2
	Low indirect	4	4	3.5	4	3.5	4
Resources	High direct	9	2	2	2	2	2
	High indirect	3	3	3	3	3.5	3
	Low direct	6	1	1	1	1	1
	Low indirect	4	4	4	4	3.5	4
Written and verbal communication	High direct	9	2	2	3	1	2
	High indirect	3	3	3	1	3	3
	Low direct	6	1	1	2	2	1
	Low indirect	4	4	4	4	4	4
Joint activities	High direct	9	1	1.5	2	2	2
	High indirect	3	2	1.5	1	1	1
	Low direct	6	4	3	3	4	3
	Low indirect	4	3	4	4	3	4

Received from

* Note: 1 indicates highest interaction; 5 indicates lowest interaction.

activities, and prestige seems to exert some influence on the amount of verbal and written communication. Yet it is agencies offering direct services—regardless of prestige—which lead in the number of referrals and resources received. In other words, prestige, leadership, and other organizational variables seem to affect interaction patterns within limits established by the function variable.

An obvious question is whether organizations with shared or common boards interact more with one another than do agencies with separate boards. Our preliminary data show that the interaction rate is not affected by shared board membership. We have not been able to ascertain if there is any variation in organizational persons with high status or influence. In our pilot community, there was only one instance in which two organizations had the same top community leaders as board members. If boards play an active role in the activities of health organizations, they serve more to link the organization to the community and the elements it possesses than to link the organization to other health and welfare agencies. The board probably also exerts influence on internal organizational operations and on establishing or approving the primary objective of the organization. Once the objective and the implementing functions are established, these functions tend to exert their influence autonomously on organizational interaction.

ORGANIZATIONAL DOMAIN

As we have seen, the elements exchanged are cases, labor services, and other resources. All organizational relationships directly or indirectly involve the flow and control of these elements. Within the local health agency system, the flow of elements is not centrally co-ordinated, but rests upon voluntary agreements or understanding. Obviously, there will be no exchange of elements between two organizations that do not know of each other's existence or that are completely unaware of each other's functions. Even more, there can be no exchange of elements without some agreement or understanding, however implicit. These exchange agreements are contingent upon the organization's domain. The domain of an organization consists of the specific goals it wishes to pursue and the functions it undertakes in order to implement its goals. In operational terms, organizational domain in the health field refers to the claims that an organization stakes out for itself in terms of (1) disease covered, (2) population served, and (3) services rendered. The goals of the organization constitute in effect the organization's claim to future functions and to the elements requisite to these functions, whereas the present or actual functions carried out by the organization constitute *de facto* claims to these elements. Exchange agreements rest upon prior consensus regarding domain. Within the health agency system, consensus regarding an organization's domain must exist to the extent that parts of the system will provide each agency with the elements necessary to attain its ends.

Once an organization's goals are accepted, domain consensus continues as long as the organization fulfills the functions adjudged appropriate to its goals and adheres to certain standards of quality. Our data show that organizations find it more difficult to legitimate themselves before other organizations in the health system than before such outside systems as the community or state. An organization can sometimes obtain sufficient elements from outside the local health system, usually in the form of funds, to continue in operation long after other organizations within the system have challenged its domain. Conversely, if the goals of a specific organization are accepted within the local agency system, other organizations of the system may encourage it to expand its functions and to realize its goals more fully by offering it elements to implement them. Should an organization not respond to this encouragement, it may be forced to forfeit its claim to the unrealized aspect of its domain.

Within the system, delineation of organizational domains is highly desired.[14] For example, intense competition may occur occasionally between two agencies offering the same services, especially when other agencies have no specific criteria for referring patients to one rather than the other. If both services are operating near capacity, competition between the two tends to be less keen, the choice being governed by the availability of service. If the services are being operated at less than capacity, competition and conflict often occur. Personnel of referring agencies in this case frequently deplore the "duplication of services" in the community. In most cases the conflict situation is eventually resolved by agreement on the part of the competing agencies to specify the criteria for referring patients to them. The agreement may take the form of consecutive handling of the same patients. For example, age may be employed as a criterion. In one case three agencies were involved in giving rehabilitation services: one took preschool children, another school children, and the third adults. In another case, where preventive services were offered, one agency took preschool children and the other took children of school age. The relative accessibility of the agencies to the respective age groups was a partial basis for these divisions. Another criterion—disease stage—also permits consecutive treatment of patients. One agency provided physical therapy to bedridden patients; another handled them when they became ambulatory.

Several other considerations, such as priorities in allocation of elements, may impel an organization to delimit its functions even when no duplication of services exists. The phenomenon of delimiting one's role and consequently of restricting one's domain is well known. It can be seen, for instance, in the resistance of certain universities of high prestige to offer "practical" or vocational courses, or courses to meet the needs of any but high-status professionals, even to the extent of foregoing readily accessible federal grants. It is evidenced in the insistence of certain psychiatric clinics on handling only cases suitable for psychoanalytic treatment, of certain business organizations on selling only to wholesalers, of some retail stores on handling only expensive merchandise.

---

[14] In our research a large percentage of our respondents spontaneously referred to the undesirability of overlapping or duplicated services.

The flow of elements in the health system is contingent upon solving the problem of "who gets what for what purpose." The clarification of organizational domains and the development of greater domain consensus contributes to the solution of this problem. In short, domain consensus is a prerequisite to exchange. Achieving domain consensus may involve negotiation, orientation, or legitimation. When the functions of the interacting organizations are diffuse, achieving domain consensus becomes a matter of constant readjustment and compromise, a process which may be called negotiation or bargaining. The more specific the functions, however, the more domain consensus is attained merely by orientation (for example, an agency may call an X-ray unit to inquire about the specific procedures for implementing services). A third, less frequent but more formalized, means of attaining domain consensus is the empowering, licensing or "legitimating" of an organization. Negotiation as a means of attaining domain consensus seems to be related to diffuseness of function, whereas orientation, at the opposite extreme, relates to specificity of function.

These processes of achieving domain consensus constitute much of the interaction between organizations. While they may not involve the immediate flow of elements, they are often necessary preconditions for the exchange of elements, because without at least minimal domain consensus there can be no exchange among organizations. Moreover, to the extent that these processes involve proferring information about the availability of elements as well as about rights and obligations regarding the elements, they constitute a form of interorganizational exchange.

DIMENSIONS OF EXCHANGE

We have stated that all relationships among local health agencies may be conceptualized as involving exchange. There are four main dimensions to the actual exchange situation. They are:

1. *The parties to the exchange.* The characteristics we have thus far employed in classifying organizations or the parties to the exchange are: organizational form or affiliation, function, prestige, size, personnel characteristics, and numbers and types of clients served.

2. *The kinds and quantities exchanged.* These involve two main classes: the actual elements exchanged (consumers, labor services, and resources other than labor services), and information on the availability of these organizational elements and on rights and obligations regarding them.

3. *The agreement underlying the exchange.* Every exchange is contingent upon a prior agreement, which may be implicit and informal or fairly explicit and highly formalized. For example, a person may be informally routed or referred to another agency with the implicit awareness or expectation that the other organization will handle the case. On the other hand, the two agencies may enter into arrangements that stipulate the exact conditions and procedures by which patients are referred from one to another. Furthermore, both parties may be actively

involved in arriving at the terms of the agreement, or these terms may be explicitly defined by one for all who may wish to conform to them. An example of the latter case is the decision of a single organization to establish a policy of a standard fee for service.

4. *The direction of the exchange.* This refers to the direction of the flow of organizational elements. We have differentiated three types:

a) *Unilateral:* where elements flow from one organization to another and no elements are given in return.

b) *Reciprocal:* where elements flow from one organization to another in return for other elements.

c) *Joint:* where elements flow from two organizations acting in unison toward a third party. This type, although representing a high order of agreement and co-ordination of policy among agencies, does not involve the actual transfer of elements.

As we proceed with our study of relationships among health agencies, we will undoubtedly modify and expand our theoretical model. For example, we will attempt to describe how the larger systems are intertwined with the health agency system. Also, we will give more attention to the effect of interagency competition and conflict regarding the flow of elements among organizations. In this respect we will analyze differences among organizations with respect not only to domain but to fundamental goals as well. As part of this analysis we will examine the orientations of different categories of professionals (for example, nurses and social workers) as well as groups with varying experiences and training within categories of professionals (as nurses with or without graduate education).

In the meantime, we find the exchange framework useful in ordering our data, locating new areas for investigation, and developing designs for studying interorganizational relationships. We feel that the conceptual framework and findings of our study will be helpful in understanding not only health agency interaction but also relationships within other specific systems (such as military, industrial, governmental, educational, and other systems). As in our study of health agencies, organizations within any system may confidently be expected to have need for clients, labor, and other resources. We would also expect that the interaction pattern among organizations within each system will also be affected by (1) organizational function, (2) access to the necessary elements from outside the system, and (3) the degree of domain consensus existing among the organizations of the system. It appears that the framework also shows promise in explaining interaction among organizations belonging to different systems (for example, educational and business systems, educational and governmental, military and industrial, and so forth). Finally, we believe our framework has obvious value in explaining interaction among units or departments within a single large-scale organization.

Howard Aldrich

21. Organizational Boundaries and Interorganizational Conflict[1]

OPEN SYSTEMS theory has stimulated a number of theoretical discussions, yet many implications of the theory still remain to be explored (Katz & Kahn, 1966; Miller, 1955; Buckley, 1967). The purpose of this paper is to suggest two applications of the concept of organizations as boundary-maintaining systems to the study of organizational phenomena: (1) power and authority, and (2) interorganizational conflict and member compliance. It will be shown that conceptualizing organizations in terms of boundary-maintaining systems provides a theoretical link between several concepts that have previously not been treated together.

THE SYSTEMS PERSPECTIVE

Before discussing the concept of boundary maintenance, it is useful to consider some of the ways in which a systems, or organization-environment, perspective differs from traditional approaches to the study of organizations. An ever-present problem is the danger of reasoning from an analogy of social organizations to physical organisms. When we study individual members of a biological species the task of distinguishing the organism from its environment is rather easy; the parameters of the environment stop at the organism's skin. However, as Ashby has pointed out, this dichotomy is overly simplistic (Ashby, 1952). In order for the environment to have an effect on the organism, it must somehow interact with it. Obviously, the environment must penetrate the organism's boundaries to achieve any immediate effect. However, the point still holds that we are able to use the 'skin' or outer covering of the organism as a frame of reference in delimiting where the environment ends and the organism begins. No such short cut exists for the

From *Human Relations*, 1971, vol. 24, pp. 279–93. Reprinted with permission of the author and the publisher.

[1] I have benefited from the comments of Edward Laumann, Allan Levett, John Magney, James L. Norr, Ted Reed, and Albert J. Reiss, Jr. William Gamson provided the stimulus for the original draft.

study of social organization, as social theorists are still seeking the fundamental social units that define social organization.

A second difficulty involves using the individual organism instead of the species as the comparative reference point (Buckley, 1967, pp. 7–40), or, in the case of formal organizations, using single organizations instead of populations of organizations as the frame of reference. In studying physical organisms, it is clear that a definite steady state exists, e.g. body temperature. The preservation of the character of the system is a necessity for the organism if it is to continue to exist, since the viable range of its essential functions is rather narrow. However, from the point of view of the species, the survival of a particular organism is irrelevant, as long as enough of the species is kept alive to reproduce. Thus evolutionary theory in biology focuses on the species, not the organism. The potential for adaptation of any one organism is limited by its genotype, as manifested in its phenotype, but at the level of the species the possibility exists that mutation will make enough organisms adaptable so that survival to the reproductive state can provide for the preservation and continued existence of the species.

There is also an intraorganism environment affecting the survival potential of specific organisms, as Williams (1966) has pointed out. Williams stresses 'the unity of the genotype and the functional subordination of the individual genes to each other and to their surroundings' (Williams, 1966, p. 57). A particular phenotype is created by the entire genetic mix of the organism, and any individual gene may have one effect in a particular genotype and an entirely different one in another. Thus the total organism is the product, not the sum, of its constituent parts. This perspective has direct implications for the study of formal organizations, since it suggests the need to conceptualize organizations not in terms of global or idealized types, but in terms of total outcomes of the interaction of many specific organizational variables. Organizational change is a result of variation within the organization as well as in the environment.

Buckley makes this point in his listing of the elements of the adaptive process. The first element is the continuous introduction of 'variety' into the system which acts to rearrange the pool of shared information and pattern of intraorganizational interactions in such a way as to allow variation in the environment to be 'mapped' in some way by the organization (Buckley, 1967, p. 206). Ashby (1952), Sjoberg (1960), Gouldner (1959), and others have made use of this conception of organizations as loosely joined systems.

Concretely, adopting this perspective means viewing organizations as possessing varying degrees of interdependence among their constituent parts; that is, leaving open the question of the degree of interdependence between organizational subunits. Conversely, the possibility exists for the elements of the system to have a great deal of functional autonomy. Indeed, if we insist on viewing complex systems as richly joined, i.e. highly interdependent, then it becomes difficult to account for system adaptation, since a change in any variable in the system would have an immediate effect on every other variable, leaving the system in a perpetual state of change, counter-change, and chaos.

There are thus two major differences between an open-systems or organization-environment approach and the traditional approach to the study of organizations: (1) an emphasis on factors making for intra-organizational conflict and internal variation as opposed to a concern for internal stability and cooperation; and (2) an emphasis on turn-over and interorganizational conflict within a population of organizations as opposed to a concern with the characteristics of a single organization. These points will be expanded upon below.

Several organizational theorists have expressed concern over the preoccupation of organizational research with the internal stability and interdependence of organizations. Sjoberg (1960) points out that theorists have over-emphasized the functional interdependence and operating harmony of social systems. There are a number of sources of disintegrative or independence-generating forces at work in formal organizations. First, within the organization, there are often conflicts over priorities in maximizing values, as between the values of efficiency and effectiveness. Also within the organization, conflicts arise because of the authority structure, e.g. conflict between management's desire for control versus the employees' desires for participation. Second, conflicts arise from differences between internal and external sources. For example, in the realm of national security there is the government's attempt to maintain secrecy in its dealings with other nations versus the public's freedom to know what the government is doing. With reference to manufacturing firms, there is the customer's demand for individualized production versus the organization's need for operating certainty and standardized production. Third, conflicts arise because of multiple and conflicting demands from an organization's task environment (Dill, 1958). For example, there is the problem of multiple constituencies, faced by organizations attempting to sell to more than one market, e.g. European automobile manufacturers who ship their products to the United States face a conflict between the attempt to maintain a lower price versus the need to meet safety regulations.

Gouldner has also made several points with regard to the need to focus on internal variation in organizations. Closed-system models assume a high degree of connectedness among the various elements of the organization. However, it is evident that in many organizations each department and major power center has the potential for operating more or less independently of the other elements. In addition to being structurally loosely joined, other forces at work in organizations lead to the functional autonomy of its elements, mainly the presence of individuals, groups, or departments in the organizations with a vested interest in maintaining their autonomy.

Being aware of these sources of internal variability in organizations enhances our ability to understand organizational change. As Campbell (1969) has stated, the occurrence of variations is essential if change is to take place. Campbell's argument is couched in evolutionary terms and was drawn upon by Buckley in his construction of an 'abstract model of morphogenesis' (Buckley, 1967, p. 62). Buckley notes that sources of variety act as 'a potential pool of adaptive variability to meet the problem of mapping new or more detailed variety and constraints in a changeable environment' (Buckley, 1967, p. 63). There also must be a selection mechanism that selectively chooses or reinforces those aspects of organizational variability that more closely map the environment than those that do not. In the perspective discussed, the selection function is performed by the organization's environment.

The second emphasis referred to above is a concern for turnover and conflict in a *population* of organizations as opposed to the study of individual organizations. The problem with traditional approaches to the study of individual organizations has been that organizational properties ('structure') have been investigated without regard to their contributions to fitness in varying or diverse organizational environments. Studies have not investigated what particular combinations of internal organizational characteristics are most effective in permitting organizational survival.

This point is directly related, of course, to the point that we too often forget—that the organizations we study are a highly select (and therefore biased) group of organizations, compared to the total population of new starts, successes and failures. For example, between 1944 and 1954, over 5.4 million small firms were established, and 4.5 million were transferred to another owner. During this same period, 7.8 million were disposed of, with about 40 percent being sold to a new owner and the rest going out of business. The study of only the surviving and successful organizations may be quite misleading insofar as it blinds us to essential differences between them and the organizations no longer around to be studied (Aldrich, 1969; Mayer & Goldstein, 1961).

The open-systems or organization-environment perspective turns our attention to *turnover* in the population of organizations, and makes use of our knowledge of high failure rates and transiency of most organizations that are begun. Indeed, it makes more theoretical sense to deal with concepts and variables that closely resemble the actual organizational population rather than to theorize about organizations as if most were stable and lasted forever. The open-systems perspective also alerts us to the fact that the source of many organizational problems lies *outside* the organization, in the organization's environments. The organizational theorist is thus forced to concentrate his attention on variation both in the organization and in the environment, since the organization-environment perspective requires one to look at both sides of a relationship to understand or predict an outcome. This implies that a theorist cannot make accurate predictions about the affect of organizational variables without some knowledge of the organization's environments.

The above discussion of the open-systems perspective was only intended to highlight certain key differences between it and more traditional approaches to the study of organizations. For a more extended discussion of the perspective the reader is referred to Buckley (1967),

Thompson (1967), and Lawrence & Lorsch (1967). The remainder of the paper will attempt to demonstrate the utility of the perspective by applying it to the study of organizational power, authority, membership, interorganizational conflict, and member compliance. Our aim is to integrate these concepts by means of a common perspective.

BOUNDARIES, AUTHORITY, AND ORGANIZATIONAL MEMBERSHIP

Social theorists have traditionally grounded their conceptualization of power and authority in the coercion that is exercised by the powerful over the powerless. For example, the power of the state, as usually defined, rests on the state's control of the collective use of violence. Power and authority as grounded in force are clearly relevant concepts for nation-states and governments, but this conceptualization breaks down when we consider other units of social organization.

Weber defined power as 'the chance of a man or a number of men to realize their will in a communal action even against the resistance of others who are participating in the action' (Gerth & Mills, 1958, p. 180). Or consider André Beteille's statement: 'The power of the state is backed ultimately by the control and use of physical force' (Beteille, 1965, p. 143). As most theorists have been concerned with the power of the state, and the state has control over the means of violence, power has been viewed as resting ultimately on the control of violence. Lasswell & Kaplan, for example, write of 'severe sanctions' and of violence as the most severe sanction (Lasswell & Kaplan, 1950, pp. 74–102).

Instead of having different notions of power and authority for the various types of units of social organization, it would be theoretically useful to have concepts that are general enough to encompass all forms of social organization. The question to be asked, then, is what is it about the attributes of formal authority in any organization which make it the effective authority and allow it to wield power. The answer to this question should apply to both nation-states, business firms, and the administration of a public university.

Conceptualizing organizations as boundary-maintaining systems provides us with a scheme for dealing with this problem. The concept of boundaries is present in Weber's definition of a formal organization:

A [social] relationship will . . . be called "closed" against outsiders so far as, according to its subjective meaning and the binding rules of its order, participation of certain persons is excluded, limited or subjected to conditions . . . A social relationship which is either closed or limits the outsiders by rules, will be called a "corporate group" so far as its order is enforced by the action of specific individuals whose regular function this is, of a chief or "head" and usually also an administrative staff (Weber, 1947, pp. 139–46).

The crucial element that defines an organization is the fact that a distinction is made between members and nonmembers, with an organization existing to the extent that entry into and exit out of the organization are limited. Some persons are admitted, while others are excluded.

An organization is a formal organization, following Weber's definition, when there is a person in authority present (Weber, 1947, p. 146). *Authorities* are those persons who 'act in such a way as to tend to carry out the order governing the group' (Weber, 1947, p. 146). Rephrasing the role of authorities in formal organizations in terms of the boundary-maintaining perspective, authorities can be defined as those persons who are given the task of applying organizational rules in making of decisions about entry and expulsion of members.

In practice, the exercise of authority often takes the form of setting conditions for *entry* into the system. For example, students must pay tuition as a condition of their membership in the university. Business organizations, with the advent of formally free labor, use the acceptance of a proffered wage rate as a condition of entry. In a nation-state, denial of entrance or of continued membership takes the form of immigration laws or of loss of citizenship, imprisonment, or deportation. Control of *exit* also is a defining condition of authority. Authorities are invested with the power to sanction deviants in the system, and the ultimate sanction they wield is expulsion from the system. Note that no assumption is made about the attainment of collective goals—who benefits from the authorities' decisions is an empirical question. By keeping the conceptual scheme as parsimonious as possible, we can avoid the difficult problems posed by conceptual distinctions that have no relationship to an underlying theoretical dimension, as is the case with the *cui bono* typology of organizations (Blau & Scott, 1962).

LIMITS ON ORGANIZATIONAL AUTHORITY AND AUTONOMY

Organizations vary greatly in the degree to which they have the ultimate authority to control entry and exit. Actually there are two dimensions involved: *dependence*, or the degree to which the organization has control over persons admitted to the organization; and *member autonomy,* or the degree to which members or clients control their own participation in a particular organization. Carlson (1964) has developed a typology of organizations based on these two dimensions and the following discussion depends heavily on his typology, given in Table 1.

With regard to dependence, many organizations are dependent upon others for their membership: schools have to keep children until they

TABLE 1

Limits on Organizational Authority and Autonomy: Organizations Classified by Control over Entry and Member Control over Participation

	Member Control over Participation	
	Yes	*No*
Organizational control over admission:		
Yes	I ('wild' organizations)	III
No	II	IV ('domesticated' organizations)

reach a certain age because of state laws; prisons get their members from the legal system; army units are dependent upon the Selective Service System; and some state universities have to take all high school graduates in their state. In other words, the fact that an organization is highly dependent upon another organization for resources (in whatever form) may mean a loss in its authority to control entry and exit; that is, a loss of functional autonomy. This is especially true of organizations which are made up of other organizations, e.g. school systems, labor unions, and diversified but centralized business firms.

The second dimension is the degree of choice exercised by potential members in deciding whether to become affiliated with an organization. The implication, of course, is that the greater the degree of potential member autonomy, the more the organization must modify its selection and entry control mechanisms to make itself attractive to the potential member. Thus, the greater the member autonomy, the more the organization is dependent upon its members.

Table 1 is derived by dichotomizing and cross tabulating the dimensions to obtain a typology of four organizations. Carlson labels Type 1 organizations 'wild' organizations to emphasize their lack of dependence on other organizations and their need to compete in the marketplace for clients/members. Examples of Type I organizations are private hospitals, private medical clinics, and private voluntary associations. At the other extreme are Type IV organizations, labeled 'domesticated' organizations. Such organizations have neither the power to control entry nor the need to compete for members, e.g. public schools, reform schools, and prisons.

Type III organizations, according to Carlson, are very rare, at least among service type organizations. Such organizations control entry over members who have no choice in the matter, e.g. the Selective Service System. Among business firms, monopolies most closely approximate the organizational conditions of Type III organizations, since they are the only source of a desired good and, if privately owned, can be selective in distribution. Type II organizations have less autonomy than the other types, since they are dependent upon other organizations for their clients/members and must compete for those they are able to secure, e.g. state universities or adult education units (Clark, 1956).

These limits on organizational authority over boundary maintenance have several implications for organizational behavior. Organizations with no control over entry usually are assured of at least a minimally steady flow of members and thus their existence is guaranteed. Thus, survival is sacrificed for autonomy; the opposite is true for those organizations with a great deal of control over entry. In situations where members or clients have a great deal of choice of affiliation, organizations must develop methods for attracting and motivating members. Where members have no freedom of choice, as in prisons, the opposite is true. Another prediction has to do with the propensity of organizations to vary their activities. 'Domesticated' organizations, because of assured resources and ability to disregard the desires of potential members, will be less likely to vary their activities over time than 'wild' organizations, which face a more uncertain environment. As was pointed out earlier, survival in an uncertain and varying environment depends, in part, on an organization varying enough internally so as to be able to 'map' relevant characteristics of the environment into organizational routines. In the following sections we will discuss the implications of these dimensions for organizational strategies in interorganizational conflict.

VIOLENCE AND THE DEFINITION OF ORGANIZATIONAL MEMBERSHIP

Before discussing interorganizational conflict, it should be noted that two issues, the role of violence in defining authority and the definition of organizational membership, are partially clarified by the above discussion. First, with regard to violence, the use of force can be viewed as a special case of the ability of authorities to control entry and exit of the elements of the system through the use of sanctions. 'Force' is simply one method for removing deviants from the system and a means for carrying out the ultimate sanction, not the ultimate sanction itself. This result demonstrates the generality of conceptualizing authority in terms of boundary-maintaining systems.

The second issue concerns the definition of an organizational member. Such a definition is needed if we are to be able to mark off the organization from its environment. As mentioned in the first section of the paper, organizational theorists have a great deal of difficulty in agreeing on how to locate the boundaries of an organization. Following our earlier statement of intent, our goal is to define membership in the same terms as authority and on the same level of generality.

Weber again provides a helpful lead, as he defines a 'member' in the same terms as his definition of formal organization: 'A party to a closed social relationship will be called a "member"' (Weber, 1947, p. 140). This suggests that, in our terms, a member may be defined as follows: a person whose entry and exit into and from the organization is controlled by the authorities of the organization. The organization sets the conditions for entry, controlling such things as wages or salaries, hours of work, amount of work expected, and so forth. As for controlling exit, the organization can terminate at will (or with certain contractual obligations) the association of an organizational member with the organization.

While this definition seems to resemble the traditional definition of what a member is, it should be noted that it potentially includes much more. The definition emphasizes that membership is not an 'either-or' distinction but instead is a matter of degree. Considering only entry and exit, it is evident that suppliers, customers, inmates, and other types of persons in the organization's domain become potential members insofar as the organization gains some control over their actions at one time or another. This implication makes theoretical sense, since there is no inherent reason why organizational members should take precedence over other parts of the organization's environment, except insofar as the behavior of employees can be predicted with a higher degree of certainty by the organization.

There is precedence in previous organizational theorizing for the above definition of an organizational member. In his book *Organization and Management,* Chester Barnard stated:

Thus I rejected the concept of organization as comprising a rather definite group of people whose behavior is coordinated with reference to some explicit goal or goals. On the contrary, I included in organization the actions of investors, suppliers, and customers or clients. Thus the material of organization is personal services, i.e. actions contributing to its purposes (Barnard, 1948).

Barnard's insight has yet to find its way into actual research on formal organizations.

Another way to think of the problem of membership is in terms of roles or role behavior, instead of individuals, as the units subject to organizational authority. In fact, if we are to refine the concept of organizational boundaries, we will eventually have to grapple with this formulation of the problem, viz. organizations as the intersection of role behaviors contributed by a number of individuals and/or organizations.

ORGANIZATIONAL BOUNDARIES, INTERORGANIZATIONAL CONFLICT, AND MEMBER COMPLIANCE

The definition of an organizational member leads logically to the subject of problems of organizational strategy concerning member compliance in a condition of interorganizational conflict. While many theorists have treated membership compliance as problematic, they have not generally done so within the framework of systems theory and interorganizational conflict. For example, Etzioni's theory of member compliance focuses on the relationship between the goals of the organization and member compliance, instead of on relationships between organizations and their effect on member compliance. In this section of the paper two organizational strategies for dealing with member compliance are discussed, and the probability of using one or the other strategy is related to problems of boundary maintenance during conflict between organizations.

Member participation is important to an organization in normal circumstances because the maintenance of the organization is dependent upon the activities of the membership. In a conflict situation, the members of an organization constitute resources for the organization to use in its competition with other organizations whose goals are partly or wholly in conflict with its own. The more active the member participation, other things being equal, the more the chances of success in the conflict.

Organizations have available two strategies that may be used in interorganizational conflict to heighten the value of member participation. The push to use either strategy arises from the authorities' desires to overcome elements of uncertainty in the conflict situation. Such uncertainty is caused by the authorities' lack of knowledge as to how members will respond to the conflict and from the organization's lack of control over many external elements in the conflict. One method of girding the organization for conflict is to tighten the organizational boundaries while the other is to expand them. In propositional form, these strategies are:

1. Organizational strategy in a situation of conflict may take the form of constructing the boundaries of the organization by strengthening the requirements of participation, with more being asked of each member in the way of conformity to organizational rules and ideology.
2. Organizational strategy may also take the form of expanding the boundaries of the organization so as to take in persons from the competing organizations or groups in order to make them members of the focal organization.

These strategies will be described briefly and then will be related to the limits of organizational authority discussed in the previous sections.

Tightening and strengthening the organization's boundaries means either raising performance standards or appealing to the member's identification with the organization. Simmel described this process in his paper on conflict: 'This need for centralization, for the tight pulling together of all elements, which alone guarantees their use, without loss of energy and time, for whatever the requirements of the moment may be, is obvious in the case of conflict' (Simmel, 1955, p. 88). Simmel goes on to point out that there are two ways for an organization to achieve consensus in the fact of an external threat: 'either to forget internal counter-currents or to bring them to unadulterated expression by expelling certain members' (Simmel, 1955, p. 92). If an organization is able to paper over internal rifts then all members will be counted on to participate in the struggle. For example, after the nominating convention, political parties are usually expected to forget internal differences as they prepare for election, with members called to a 'higher loyalty.'

The second organizational strategy for dealing with member compliance in a conflict situation is an oblique one: the organization expands to take into its boundaries persons or organizations from the competing organizations or factions. This process can take the form of either absorption, co-optation, or amalgamation (Etzioni, 1961, pp. 103–4). Obviously, any of these methods is a secondary strategy, since they all necessitate the organization's taking on new, unsocialized members, thus adding an element of uncertainty to the organization's operations. Thus the problem of member compliance can arise because of an attempted solution to a conflict. For example, in the case of the TVA this mode of dealing with the conflict increased the problem of compliance since members came into the organization with values and goals in some ways opposed to those of the TVA (Selznick, 1953). However, without an expansion of the organization's boundaries it is doubtful whether it would have survived.

Utilization of these strategies can be expected to vary, depending on the limits on organizational autonomy discussed in previous sections. The key dimension is the type of control an organization has over its members, as affected by member autonomy or freedom of choice. At one extreme are organizations whose members have a wide latitude in choosing an organization, while at the other extreme are organizations with members exercising no freedom of choice, as dichotomized in Table 1. If we add a midpoint to this dimension, we can derive Etzioni's classification of organizational compliance structures: normative, utilitarian, and coercive (Etzioni, 1961). Potential members have the greatest freedom of choice in choosing a normative organization, since the initiative for membership is expected to come from congruence between the member's beliefs and values and those of the organization. Membership in a utilitarian organization is required in a society where one's economic rewards depend on being associated with a wage or salary-paying organization. Thus, while technically the 'right to work' exists, the right is constrained by necessity in most cases. Finally, coercive type organizations are usually not entered by member choice.

Type of control structure is associated with the conflict strategy chosen because of its relation to the organization's capacity for varying its boundaries and to the cost of each strategy in terms of organizational resources. After discussing these factors we will summarize the argument by making predictions about strategy choices in the context of particular examples.

With respect to the organization's capacity for varying its boundaries, there are two dimensions to consider: (1) the degree to which the organization's boundaries include the life-space of members; and (2) the potency or degree of the member's involvement in the organization (Katz & Kahn, 1966, pp. 119–21). The degree of inclusion of a member's personality or life-space is an indication of the member's dependence on the organization. In normative and coercive organizations the degree of inclusion is quite high, albeit for differing reasons. In these organizations we would expect additional claims upon the members to meet with less resistance than in utilitarian organizations, where the degree of inclusion, except for top management, is much less. At the same time, the extensive inclusion of members' life-spaces poses a difficult problem of new member socialization for normative and coercive organizations and makes the introduction of new members a risk-laden venture.

The second dimension of the organization's capacity for varying its boundaries is the potency of a member's involvement. The higher the member's involvement, the more likely it is that membership itself will be considered a reward. The possibility of increasing demands upon the present membership is directly related to potency of involvement. In normative organizations potency is high for many, if not most members. In utilitarian organizations the degree of involvement is directly proportional to the economic remuneration received by a member, other things being equal. Thus, added activity can probably be counted on only if remuneration is raised to match the demands; this has implications for the cost of each strategy, as will be pointed out below. Finally, while the degree of inclusion is undoubtedly high in coercive organizations, the degree of involvement is rather low for most members (inmates), except for those whom Goffman labels 'colonizers' (Goffman, 1961, pp. 171–320).

The cost of each organizational strategy is also related to the organization's control structure. In normative organizations, expansion of the organization's boundaries brings with it the potential cost of seriously compromising the organization's original mission by contaminating the 'purity' of the organization's message (Zald & Ash, 1966). Utilitarian organizations expand at the cost of added complexity and increased monetary outlays. The coercive organization's tenuous control over its members would be seriously threatened by expansion. Constriction of the boundaries seems to be a cheaper solution in most cases. The major cost in each type of organization is the loss of contributors to the organization's activity system, but this loss is certainly less harmful to coercive organizations. The structural complexity of most utilitarian organizations implies that a reduction in membership will require a structural reorganization which could come at a substantial cost.

Several predictions can be derived from the above discussion concerning the relationship between conflict strategies and the organization's degree of boundary control. These predictions will be considered by type of organization in the following section. While a thorough examination of this subject requires data on organizations in conflict, few studies have been carried out from the perspective described here and so examples have been chosen more as illustrations than evidence. As a result, the following discussion is highly speculative and primarily discursive.

Organizations with normative control structures are more likely to follow the tightening-up strategy than the expansion strategy, since admission to the good graces of the organization constitutes one of their major control mechanisms. Expansion to take in new members raises the prospect of hordes of unsocialized members who will be difficult to control. Moreover, performance standards for many members are ordinarily slight and therefore many of them are likely to desert when a true test of their convictions comes up. In addition, conflict is likely to increase the visibility of ideological divisions within the organization, with each side calling for a purge of the other. For example, the divisions of the Communist party in the Soviet Union, the split in the Social Democratic Party in Germany in the latter half of the nineteenth century, and the recent purges in the Black Panther party in the United States have followed this pattern.

One factor that might lead normative organizations to employ an expansion strategy is the absence of a rigid ideology to which all members must subscribe. Political parties in the United States are known for their ability to encompass within themselves a wide variety of positions (the 'new politics' movement in the Democratic Party may change this). Also, normative organizations with pure social goals sometimes face conflicts over member loyalty. In the absence of a highly specific belief system, these types of organizations are able to solve this problem by using the expansion strategy and adding new interest groups or auxiliaries. For example, the Los Angeles branch of MENSA International has over thirty special interest groups, all peacefully co-existing. Member compliance is thus retained by incorporating the potential conflicting relationship into the organization.

Organizations with utilitarian control structures seem to have a very wide range of tolerance of member behaviour, outside of purely functional activities. Indeed, conflict between utilitarian organizations is ordinarily so benign that only a minimal degree of member compliance at the lower levels of the organization is required. Given the prevailing business values of 'size' and 'growth,' it is highly unlikely that a utilitarian organization will use the constriction strategy except as a short-run tactic that is taken out of absolute economic necessity. It is more likely that such organizations will choose the expansion strategy when engaged in interorganizational competition and conflict.

Industries often attempt to combat employee militancy and union organizers by absorbing them into a company union or even into the company 'family' (Blauner, 1964, Chapter Four). Industrial espionage is common, as is the pirating away of a competitor's skilled workers (Wilensky, 1967). Such actions are possible because a utilitarian organization's control over its boundaries is highly uncertain, with entry and exit of members depending ultimately on the financial health of the organization. If another organization is able to penetrate the organization's boundaries with a lucrative offer to a high-level employee, the low-level of potency of involvement is easily broken.

Organizations with coercive control structures are a different case. They are continually oriented to minimal member participation; the primary type of conflict they face is inadequate control of the behaviour of their inmates. Carrying the argument to its logical extreme, one could predict that if such an organization were physically attacked by an external force attempting to free the controlled population, as in prisoner of war camps, the conditions for continued presence in the organization could be raised by liquidating most of the inmates.

Although the authorities of coercive organizations are not likely to use expansion as a strategy, expansion can be used by the opposition to co-opt or capture the lower-level functionaries of the coercive-type organization. Sykes has described such a process of expansion in an article on prisons, calling the result the 'corruption of authority' (Sykes, 1956). The case of organized crime and the urban police is instructive. Organized crime is in conflict with the police department, yet instead of conflict leading to a higher degree of compliance on the part of police officers as they battle an external foe (as would be predicted by traditional organization theory), the conflict seems in many cases to lead to decreased compliance with official duties. This is to be expected, since while the police department is quasi-coercive in its command structure, officers are recruited and held on a utilitarian basis and may transfer their loyalty to the higher bidder, be he inside or outside the organization. From another point of view, however, this process involves a *de facto* co-optation of the 'syndicate' in exchange for a certain degree of ordered criminality (Whyte, 1937). This is increased compliance, of a sort, if we follow our earlier definition of organizational member and include criminals in the police organization.

CONCLUSION

The intent of this paper was to bring to the attention of organizational theorists the potential that open systems theory has for integrating the study of topics heretofore considered theoretically and conceptually heterogeneous. Traditional approaches to the study of organizations have been hampered by their focus on the internally stable characteristics of organizations and on samples of successful organizations, ignoring the high rate of failure in the total population of organizations. Several reasons were given for concentrating our energies on the neglected topic of internal variation in organizations and on the characteristics that distinguish successful from unsuccessful organizations.

The concepts of authority, membership, and organizational autonomy were developed in the context of organizations conceptualized as boundary-maintaining systems. Authority as control over organizational boundaries was shown to be theoretically linked to member compliance. Finally, organizational strategies to secure member compliance under conditions of interorganizational conflict were discussed and several predictions made about the conditions under which boundary constriction or expansion will be used in a conflict situation.

Hopefully, the degree of theoretical integration of diverse topics that was achieved through the use of the boundary-maintaining perspective will serve as an invitation to other organizational theorists to undertake similar work.

REFERENCES

Aldrich, H. (1969). *Organizations in a Hostile Environment: A Panel Study of Small Businesses in Three Cities.* Ph.D. Dissertation, University of Michigan.

Ashby, W. R. (1952). *Design for a Brain.* New York: John Wiley & Sons.

Barnard, C. (1948). *Organization and Management.* Cambridge, Mass.: Harvard University Press.

Beteille, A. (1965). *Caste, Class and Power.* Berkeley: University of California Press.

Blau, P. & Scott, W. (1962). *Formal Organizations.* San Francisco: Chandler Publishing Co.

Blauner, R. (1964). *Alienation and Freedom.* Chicago: The University of Chicago Press.

Buckley, W. (1967) *Sociology and Modern Systems Theory.* Englewood Cliffs, New Jersey: Prentice-Hall.

Campbell, D. T. (1969) "Variation and Selective Retention in Socio-Cultural Evolution." *General Systems 14,* 69–85.

Carlson, R. (1964). "Environmental Constraints and Organizational Consequences: The Public School and its Clients." In *Behavioral science and educational administration*. Chicago: National Society for the Study of Education.

Clark, B. (1956). "Organizational Adaptation and Precarious Values." *Am. Soc. Rev. 21*, 323–36.

Dill, W. (1958). "Environment as an Influence on Managerial Autonomy." *Admin. Sci. Quart. 2*, 395–420.

Etzioni, A. (1961). *A Comparative Analysis of Complex Organizations*. Glencoe: The Free Press.

Gerth, H. & Mills, C. Wright (1958). *Essays in Sociology* (ed.) Max Weber. New York: Oxford University Press.

Goffman, E. (1961). *Asylums*. New York: Doubleday & Co., Inc.

Gouldner, A. (1959). "Reciprocity and Autonomy in Functional Theory." In *Symposium on Sociological Theory*. (ed.) Llewellyn Gross. Evanston, Illinois: Row, Peterson & Co.

Katz, D. & Kahn, R. (1966). *The Social Psychology of Organizations*. New York: John Wiley & Sons.

Lasswell, H. & Kaplan, A. (1950). *Power and Society: A Framework for Political Inquiry*. New Haven: Yale University Press.

Lawrence, P. & Lorsch, J. (1967). *Organization and Environment*. Boston: Graduate School of Business Administration, Harvard University.

Lipset, S. M. (1963). *Political Man*. Garden City: Doubleday & Co.

Litwak, E. & Hylton, L. (1962). "Interorganizational Analysis: An Hypothesis about Coordinating Agencies." *Admin. Sci. Quart. 6*, 395–420.

Mayer, K. & Goldstein, S. (1961). *The First Two Years: Problems of Small Firm Growth and Survival*. Washington, D.C.: Small Business Administration.

Miller, J. (1955). "Toward a General Theory for the Behavioral Sciences." *Am. Psych. 10*, 513–31.

Selznick, P. (1953). *TVA and The Grass Roots*. Berkeley: University of California Press.

Simmel, G. (1955). *Conflict*. New York: The Free Press.

Sjoberg, G. (1960). "Contradictory Functional Requirements and Social Systems." *J. Conflict Resolution 4*, 198–208.

Sykes, G. (1956). "The Corruption of Authority and Rehabilitation." *Social Forces 34*, 257–62.

Thompson, J. (1967). *Organizations in Action*. New York: McGraw-Hill.

Weber, M. (1947). *The Theory of Social and Economic Organization*. New York: Oxford University Press.

Whyte, W. F. (1937). *Street Corner Society*. Chicago: University of Chicago Press.

Wilensky, H. (1967). *Organizational Intelligence: Knowledge and Policy in Government and Industry*. New York: Basic Books.

Williams, G. (1966). *Adaptation and Natural Selection*. Princeton: Princeton University Press.

Zald, M. & Ash, R. (1966). "Social Movement Organizations: Growth, Decay and Change." *Social Forces 44*, 327–41.

Frank Baker
Gregory O'Brien

22. Intersystems Relations and Coordination of Human Service Organizations

GAPS IN coordination among health and welfare agencies have long been considered to be a major problem for community organizers, community planners, and other theorists and practitioners attempting to understand and improve human services delivery. Why has the relative absence of coordination in the activities and programs of human service organizations persisted over the years despite the best efforts of a variety of professional groups? We suggest that the answer lies partially in an inadequate knowledge of the essential factors involved in organizational coordination. Effective action in dealing with problems of community agency coordination requires conceptual and empirical knowledge about interorganizational systems relations. This paper attempts to introduce an initial set of concepts and hypotheses which are being developed in the authors' research on intersystems relations in the field of human services delivery.

The authors and their colleagues in the Program Research Unit of the Laboratory of Community Psychiatry, Harvard Medical School, have been developing an open-systems conceptual formulation of health and welfare systems for several years. Beginning with an attempt to study multiple interacting variables affecting the growth of a state mental hospital into a community mental health center (Baker, 1969; Baker, Schulberg and O'Brien, 1969; Schulberg and Baker, 1968; Schulberg and Baker, 1969), the Program Research Unit is now shifting focus from studying the organizational structures and processes of community mental health programs to examining the interorganizational environments of such organizational systems in depth (Baker and Schulberg, 1970; Chin and O'Brien, 1970).

Plans for comprehensive community health services fostered by both

From *American Journal of Public Health*, 1971, vol. 61, pp. 130–37. Reprinted with permission of the authors and the publisher.

the Community Mental Health Center Act of 1963 (P.L. 88–164) and the Partnership for Health legislation of 1966 (P.L. 89–749) have recently reemphasized the importance of interagency cooperation in the coordination of all parts of a community health care delivery system and the necessity of including all community human services, both public and private, in plans for comprehensive community-based delivery of human services. Considering this need for greater interagency cooperation and broadened participation in planning for human services delivery, it is not surprising that much of the previous research and development of theoretical approaches to interorganizational relations comes from examination of hospitals (Baker and Schulberg, 1970; Elling and Halebsky, 1961) and social agencies (Chin and O'Brien, 1970; Levine and White, 1961; Litwak and Hylton, 1962; Warren, 1967a, 1967b). The considerable attention paid to the relations between organizations in recent years represents not only an increased need by practitioners to gain understanding of interorganizational relations (Lawrence and Lorsch, 1967), but also a progression toward an open-systems formulation of organization theory.

An open-systems formulation of organizational structure and dynamics emphasizes the study of a bounded interacting set of components engaging in an input-output commerce with an external environment in the processing of material objects, information, and people. As a distinct systemic entity, an organization must maintain some discontinuity with its environment to continue to exist. This "boundary" of a system may be a territorial line; but, in a social system such as a human service organization, it exists more importantly as a boundary in "social space" representing discontinuity in patterns and clusterings of human interaction. The boundary of a system functions to separate it from its environment. The system takes in inputs across the boundary, engages in some conversion processes within the boundaries of the system, and then exports the products of the system throughout as outputs across the boundary. There is a discontinuity at the boundary of an organization constituted by a differentiation of technology, territory, or time—or some combination of these (Miller, E. J., 1959). Thus, applying the term "system" to an organization implies *interdependence* in the sense of necessary input and output linkages, but also *independence* in the sense of maintenance of the integrity of system elements through boundary control processes.

Baker and Schulberg (1970), in pointing to the need for, and difficulties of, interorganizational analysis, have emphasized the need for further development of an open-systems model to deal with the unique complexities of interorganizational relations among human service care-givers. There is strong appeal in shifting focus from the level of the single organization to that of a complex network of organizations and to speak of a community interorganizational system in which individual organizations constitute components or subsystems. From a general systems perspective it is appropriate to apply the neutral terms of open-systems theory in making cross-system-level generalizations, so long as it is recognized that a molar system may have different characteristics from a molecular one (Miller, J. G., 1965). However, this last qualification is a most important factor in the attempt to extend open-systems organizational theory to an interorganizational system level. As Gross (1966) has observed, larger and more differentiated interorganizational systems are likely to be marked by more complicated, divergent, and conflicting roles and relationships than those found within a single organizational system.

There is an inherent danger in the assumption that there is one coherent system of human service care-givers whose boundaries can be well defined both conceptually and empirically. Chin and O'Brien (1970) have shown the importance of developing a general intersystem model for the study of relations among care-givers. Such an intersystem model focuses on determining the degree and nature of interchanges between systems rather than assuming the operation and organization of a single large suprasystem. There are a number of independent and quasi-independent systems which are simultaneously engaged in the administration, planning, and provision of human services. Interchange and interdependence exist among some care-givers, but not necessarily all. A general intersystem model stresses the relative autonomy of interaction systems rather than assuming interdependencies which might or might not exist.

From the intersystem perspective, the "suprasystem hypothesis" must be tested empirically; that is, the conceptualization of a complex interorganizational system in which organizations exist as components or subsystems in such a larger interdependent macro-social system must be compared against the evidence of the empirical existence of such a system. Any system exists only to the extent that its parts or components are linked in a network of internal relations. Thus, the demonstration of *component interdependence* is a primary test of the empirical existence of a system, since any system exists only to the extent that its components or parts are linked in some network of internal relations. Component interdependence is a measure of the interconnectedness of a system, and may be defined as the extent to which a component depends upon other components for resource inputs, decision premises, or as receivers of the component's actions.

All organizational systems need not be assumed to be subsystems or components in some well-defined suprasystem. Some theorists have studiously avoided labeling a complex array of organizations as a system. Evan (1966), borrowing the idea from role theory, has developed the concept of organization-set to refer to a network of organizations in the environment of a particular focal organization. Chin and O'Brien (1970) also adopt the set concept as a label for the care-givers who deal with focal organizations of their research. Perhaps the term which most accurately communicates the nature of the relationships among care-givers is Warren's (1967b) concept of the interorganizational field of human service care-givers; when conceptualized as a field, the relevance of grouping the care-givers together is apparent. All organizations in the field are concerned or related to human services in some way, either directly or indirectly. At the same time, being a part of the interorganizational field does not imply or deny membership in the same suprasystem or systems.

A general intersystem model assumes that relations between systems in an interorganizational field may differ qualitatively as well as quantitatively as a function of the types of interaction which are being used to assess interdependence. Two systems may be highly interdependent in terms of information flow but may be mutually autonomous with regard to financial exchange. Baker (1969) and Chin and O'Brien (1970) point out that the relative amount of interchange between organizational systems is a function of the permeabilities of the boundaries between the two related, but partially autonomous, systems. Boundaries will be differentially permeable to different media of exchange; they may be permeable in one direction for one type of exchange and in the opposite direction for another type of exchange.

The method of examining intersystem relations in an interorganizational field as they occur for different types of interaction is analogous to examining the same sample of material under a microscope, using different types of cross-sectional slices. Depending upon the type of slice made, certain components or systems may appear very close, but may be quite distant when examined from another viewpoint. The intersystem model focuses on differences in the types of interaction between systems and may be employed in order to develop more accurate views of the totality of intersystem relations operating at different levels simultaneously.

Considering the fact that community organizers and planners have been bemoaning the lack of coordination in the community human service interorganizational field for many years, it seems highly useful to examine the conditions of intersystem relations that relate to problems of coordination and interdependence among human service organizations. It must be recognized, however, that greater understanding of intersystem relations may raise questions about the desirability of "coordination." Actually, coordination is sought because of an implicit assumption which is often left unverbalized and which probably has only rarely been empirically tested. This hidden assumption is that coordination is an intervening variable between human and material resource allocation and "effective" or "comprehensive" human services. It is possible that "coordinated" human service organizations might, in particular circumstances, produce a less optimal output mix of services than competing organizations attempting to further their own individual goals.

SOME GENERAL HYPOTHESES ABOUT INTERSYSTEM RELATIONS

Whether or not the intersystem conceptualization of problems of interorganizational relations will prove useful in understanding human service systems depends largely upon the development of related empirical research. In the interest of furthering inquiry in this area, the rest of this paper will be concerned with the presentation of several hypotheses on intersystem relations. These hypotheses assume a *ceteris paribus* condition.

RESOURCE EXCHANGE AND INTERSYSTEM RELATIONS

As open systems, organizations depend upon exchange with the environment in order to obtain resources necessary for organizational maintenance and growth. To ensure continued survival, organizations are forced to adapt to changing conditions in the environment and are motivated to adapt, where possible, the environment itself. In an inter-organizational field, control of resources necessary for the operations of other organizations offers the controlling organization a position of considerable influence in the field.

Hypothesis 1

Organizations which control the input resources for other organizational systems in the field will exert greater influence on the goal selection and decision-making of those other organizational systems. The extent of such influence will be directly proportional to the scarcity of resources and the availability of alternative sources of the resource to members of the interorganizational field.

Levine and White (1961) and Aiken and Hage (1968) have demonstrated the impact of the degree and form of interorganizational exchange of resources such as personnel, cases, and funds on the relative interdependence of care-givers. Such interchanges are probably more common between organizations which act as subsystem components of some larger suprasystem. Traditionally, financial exchange, for example, has occurred most often between members of the private sector such as the United Community Services Organization and its member care-giving social agencies, or between members of the public sector such as state budgeting for local public welfare agencies. While this is still the case, there is increasing financial exchange between private and public sectors as efforts in planning and provision of comprehensive services have been increased.

In a time of constant reexamination of national priorities, the supply of resources for various parts of the human services interorganizational field will be in continuous fluctuation. In developing viable structures and processes for broadened interagency cooperation and comprehensive planning, it will be imperative to understand how changes in the nature of input resources will affect intersystem relationships.

Hypothesis 2

The degree of cooperation between agencies which have similar functions will decrease as the extent to which they draw from a common and limited supply of input resources increases. Also, the more limited the supply of resources, the greater will be the degree of competition that will be perceived between the agencies.

INFORMATION FEEDBACK DEPENDENCE

Feedback refers to the situation in which a portion of the output of a system is returned from the environment to the system so as to modify subsequent outputs of the system. "Feedback dependence" is a condition in which a focal organizational system depends upon another system for sensing, detecting, and processing of information regarding the effects of the action of the focal system (Chin and O'Brien, 1970).

Hypothesis 3

To the degree an organization is dependent upon another system or other systems for information used in guiding its own system functioning, the other system, by its control of needed information, will be able to exert control on the actions of the local organization.

If feedback mechanisms are highly centralized, the source of feedback will be so influential that it might threaten the autonomy of the other agencies and organizations within the field. The threatened autonomy of organizations within the field will reduce interagency trust and cooperation as it fosters increased perception of competitive relations between organizations.

Hypothesis 4

The greater the degree of feedback dependence which organizations in a field have with any one organization, the lower the degree of trust among members of the interorganizational system.

Feedback dependency in the interorganizational field of human services delivery may prove to be one of the major factors in determining the climate of interagency cooperation. As there develops a broader base of agency and citizen participation and planning, new types of feedback relationships between agencies will also emerge. Comprehensive health planning agencies may have the potential to develop either highly centralized or widely based feedback loops. Areas where such feedback can be broadened should facilitate cooperation in intersystem relations.

DECISION-MAKING AND INFLUENCE

In every system, certain decision-making processes are essential. As indicated above, however, the interorganizational field does not necessarily meet the requirements of a unified system. Although a well-organized system may not exist, decisions within the organizational field are still made which affect the actions of numerous organizations. A decision made by one organization (for example, a community welfare council or a united community services organization) may affect the operation of other organizations, although no direct linkages between the two systems may exist. This flow of influence and the location of key points of decision-making for the various types of issues about which decisions must be made will provide a third viewpoint of the linkages between systems in the interorganizational field.

Hypothesis 5

In a large organizational field a number of large suprasystems will exist, each with considerable decision-making and influencing capabilities, which will affect intersystem relations throughout the interorganizational field.

Evan (1966) has pointed out that the larger the size of an organization-set, the higher will be the centralization of authority within that set. A consequence of such centralization will be the use of formal rules as a means of influence. Extending Evan's concept to an interorganizational field, there are a number of very large multiorganizational suprasystems, both governmental and voluntary, involved. Each of such suprasystems has its own organization-set. There evolve a number of centralized community-decision organizations which share responsibility and power in the field. To the extent to which the organizational suprasystems differ, the decisions made and the direction of influence asserted by them will be independent. Thus, the greater the independence, the greater the possibility of conflicting influences from these multiple sources.

As responding to the need for broadened participation and comprehensive planning continues, it must be recognized that the organizational sets of key community-decision organizations will overlap in different ways. The dual problem of maintaining autonomy and coordination will become increasingly complex as greater interdependencies evolve among community-decision organizations.

ORGANIZATIONAL ROLE COMPLEMENTARITY

In addition to the exchanges and influences discussed above, which represent the interaction between organizations in the interorganizational field, linkages exist between systems in areas of interaction concerning goals and role expectations which members of various organizations share. Influence or material exchange are not needed for two systems to be linked through similarity of goals or complementary role expectations. It is probable that systems linked in role, or related by similar goals, will often have interaction by other means; but, here again, these variables offer an additional viewpoint from which to examine the relations within the interorganizational field of human services delivery. Agencies with similar goals may see themselves as more directly related to other agencies which have either competing or complementary roles. Complementarity of roles will foster personnel exchange or joint planning efforts.

Hypothesis 6

Organizations with highly similar goals will tend to see each other as competitive, while organizations with complementary goals will tend to see each other as more cooperative.

Interagency cooperation, a cornerstone of comprehensive planning, is a function of developing complementary goal structures and complementary role expectations among organizations. Comprehensive planning will require that agencies obtain clear understandings of their own goals and primary tasks and that these understandings be shared by the entire interorganizational field.

Hypothesis 7

Cooperation will be greatest among organizations with complementary organizational role expectations. When interorganizational role ambiguity exists, conflicts and gaps in intersystem relationships will occur.

It is expected that the broader the degree of participation in planning is, the greater will be the clarity about organizational roles which will exist among organizations in the field. The greater the centralization of the planning process and the more impermeable the boundaries of the community-decision organizational system to informational output, the greater will be the organizational role ambiguity and thus the greater will be the felt competition among members of the interorganizational field.

Loomis (1959) developed the concept of *systemic linkage* which he defines as "the process whereby the elements of two social systems come to be anticipated so that in some ways they function as a unitary system" (p. 55). Hanson (1962) reformulated the systemic linkage concept into a more general proposition for describing individual role expectations, suggesting that individuals in related positions in a given social system develop different expectations for other position incumbents depending upon the nature of their relations to particular social structures. Moving from this social system level to the intersystem level, the following hypothesis can be stated.

Hypothesis 8

Organizations linked to other organizations by frequent interaction of a particular type will express different expectations for these other organizations, determined by the character of the particular enduring interaction form.

ENVIRONMENTAL CHANGE

Given that the interorganizational field of human services constitutes what Emery and Trist (1965) have described as a turbulent environment with conditions of increasingly rapid change, there should be an increased openness to change on the part of human service organizations. The movement in the field toward comprehensiveness of planning should find greater receptivity in organizations sensitive to the changes in the environment. Changes in key systems within the field may be expected to further alter the balance in intersystem relations and this will bring about a period of reverberating change in all systems within the interorganizational field. Evan (1966) predicted that there would be greater openness to change on the part of an organization when rapid change was going on within its organizational set. Such periods of rapid change are followed by periods of reorganization and new organizational liaisons.

Hypothesis 9

Rapid change in relationships and operation of systems in an interorganizational field will encourage the openness of organizational systems within that field toward innovation in operations and new intersystem cooperation.

CONCLUSION

Examining the complex array of physical health, mental health, and social welfare organizations as an intersystem field may provide new guidelines for research on the problems of delivering services. The foregoing discussion and suggested hypotheses are offered as a stimulus to such empirical work with the assumption that such effort is necessary to remove us from our present blind alleys in talking about, but not being able to do much about, coordinated comprehensive community service delivery systems.

REFERENCES

Aiken, M., and Hage, J. "Organizational Interdependence and Intraorganizational Structure." *Am. Sociol. Rev.* 33:912–30, 1968.

Baker, F. "An Open Systems Approach to the Study of Mental Hospitals in Transition." *Community Ment. Health J.* 5:403–12, 1969.

Baker, F., and Schulberg, H. "Community Health Caregiving Systems: Integration of Interorganizational Networks." In: A. Sheldon, F. Baker, and C. McLaughlin (eds.). *Systems and Medical Care.* Cambridge, Mass.: MIT Press, 1970.

Baker, F.; Schulberg, H. C.; and O'Brien, G. "The Changing Mental Hospital: Its Perceived Image and Contact with the Community." *Ment. Hyg.* 53:237–44, 1969.

Chin, R., and O'Brien, G. "General Intersystem Theory: The Model and a Case of Practitioner Application." In: A. Sheldon, F. Baker, and C. McLaughlin (eds.). *Systems and Medical Care.* Cambridge, Mass.: MIT Press, 1970.

Elling, R. H., and Halebsky, S. "Organizational Differentiation and Support: A Conceptual Framework." *Administrative Sc. Quart.* 6:185–209, 1961.

Emery, S. E., and Trist, E. L. "The Causal Texture of Organizational Environments." *Human Relations* 18:21–32, 1965.

Evan, W. M. "The Organization Set: Toward a Theory of Interorganizational Relations." In: J. D. Thompson (ed.). *Approaches to Organizational Design.* Pittsburgh: University of Pittsburgh Press, 1966.

Gross, B. M. "The State of the Nation: Social Systems Accounting." In: R. A. Bauer (ed.). *Social Indicators.* Cambridge, Mass.: MIT Press, 1966.

Hanson, R. C. "The Systemic Linkage Hypothesis and Role Consensus Patterns in Hospital-Community Relations." *Am. Sociol. Rev.* 27:304–13, 1962.

Lawrence, P. R., and Lorsch, J. L. *Organization and Environment: Managing Differentiation and Integration.* Boston: Division of Research, Graduate School of Business Administration, Harvard Business School, 1967.

Levine, S., and White, P. E. "Exchange as a Conceptual Framework for the Study of Inter-organizational Relationships." *Administrative Sc. Quart.* 5:583–601, 1961.

Litwak, E., and Hylton, L. "Interorganizational Analysis: A Hypothesis on Co-ordinating Agencies." *Administrative Sc. Quart.* 6:395–420, 1962.

Loomis, C. P. "Tentative Types of Directed Social Change Involving Systemic Linkage." *Rural Sociol.* 24:383–90, 1959.

Miller, E. J. "Technology, Territory, and Time." *Human Relations* 8:292–316, 1959.

Miller, J. G. "Living Systems: Structure and Process." *Behavioral Sc.* 10:337–79, 1965.

Schulberg, H., and Baker, F. "The Changing Mental Hospital: A Progress Report." *Hospital & Community Psychiat.* 20:159–65, 1969.

———. "Program Evaluation Models and the Implementation of Research Findings." *A.J.P.H.* 58:1248–55, 1968.

Warren, R. "The Interaction of Community Decision Organizations: Some Basic Concepts and Needed Research." *Social Service Rev.* 41:261–70 (a), 1967.

———. "The Interorganizational Field as a Focus for Investigation." *Administrative Sc. Quart.* 12:396–419 (b), 1967.

Section *VI*

SYSTEMS THEORY AND ORGANIZATIONAL MANAGEMENT

CONCEPTUAL frameworks can play an important cognitive role in organizing the manager's view of his organization and of his role within it. Johnson, Kast, and Rosenzweig provide a useful framework with which to describe the internal and external relationships of a business organization and also to act as a sophisticated guide in the recognition and solution of managerial problems. They discuss the strengths and limitations of systems theory as a framework for integrated decision making and provide an illustrative model of the systems concept which can be generalized for use in a variety of modern business organizations.

Wendell French offers a clarification of the relationship between "system" and "process." In clarifying the definition and characteristics of processes as these relate to systems, French develops a model of the organization and of the manager's job.

Both planning and goal setting are major functions of management and it is these organizational tasks that Swinth discusses as he combines ideas about organizational systems phenomena and concepts of individual problem solving. He provides a limited experimental test of his model using different conditions for allocating goals related to conditions of subsystem autonomy and interdependence. This study demonstrates the way in which systems concepts may be integrated with small group experimentation in developing increased understanding of effective models of organizational design.

Richard A. Johnson, Fremont E. Kast,
James E. Rosenzweig

23. Systems Theory
and Management

INTRODUCTION

THE SYSTEMS concept can be a useful way of thinking about the job of managing. It provides a framework for visualizing internal and external environmental factors as an integrated whole. It allows recognition of the proper place and function of subsystems. The systems within which businessmen must operate are necessarily complex. However, management via systems concepts fosters a way of thinking which, on the one hand, helps to dissolve some of the complexity and, on the other hand, helps the manager recognize the nature of the complex problems and thereby operate within the perceived environment. It is important to recognize the integrated nature of specific systems, including the fact that each system has both inputs and outputs and can be viewed as a self-contained unit. But it is also important to recognize that business systems are a part of larger systems—possibly industry-wide, or including several, maybe many, companies and/or industries, or even society as a whole. Further, business systems are in a constant state of change—they are created, operated, revised, and often eliminated.

What does the concept of systems offer to students of management and/or to practicing executives? Is it a panacea for business problems which will replace scientific management, human relations, management by objective, operations research, and many other approaches to, or techniques of, management? Perhaps a word of caution is applicable initially. Anyone looking for "cookbook" techniques will be disappointed. In this article we do not evolve "ten easy steps" to success in management. Such approaches, while seemingly applicable and easy to grasp, usually are shortsighted and superficial. Fundamental ideas, such as the systems concept, are more difficult to comprehend, and yet they present a greater opportunity for a large-scale payoff.

From *Management Science*, January 1964, vol. 10, No. 2, pp. 367–84. Reprinted with permission of the authors and the publisher.

SYSTEMS DEFINED[1]

A system is "an organized or complex whole; an assemblage or combination of things or parts forming a complex or unitary whole." The term system covers an extremely broad spectrum of concepts. For example, we have mountain systems, river systems, and the solar system as part of our physical surroundings. The body itself is a complex organism including the skeletal system, the circulatory system, and the nervous system. We come into daily contact with such phenomena as transportation systems, communication systems (telephone, telegraph, etc.), and economic systems.

A science often is described as a systematic body of knowledge; a complete array of essential principles or facts, arranged in a rational dependence or connection; a complex of ideas, principles, laws, forming a coherent whole. Scientists endeavor to develop, organize, and classify material into interconnected disciplines. Sir Isaac Newton set forth what he called the "system of the world." Two relatively well known works which represent attempts to integrate a large amount of material are Darwin's *Origin of Species* and Keynes's *General Theory of Employment, Interest, and Money*. Darwin, in his theory of evolution, integrated all life into a "system of nature" and indicated how the myriad of living subsystems were interrelated. Keynes, in his general theory of employment, interest, and money, connected many complicated natural and man-made forces which make up an entire economy. Both men had a major impact on man's thinking because they were able to conceptualize interrelationships among complex phenomena and integrate them into a systematic whole. The word system connotes plan, method, order, and arrangement. Hence it is no wonder that scientists and researchers have made the term so pervasive.

The antonym of systematic is chaotic. A chaotic situation might be described as one where "everything depends on everything else." Since two major goals of science and research in any subject area are explanation and prediction, such a condition cannot be tolerated. Therefore there is considerable incentive to develop bodies of knowledge that can be organized into a complex whole, within which subparts or subsystems can be interrelated.

While much research has been focused on the analysis of minute segments of knowledge, there has been increasing interest in developing larger frames of reference for synthesizing the results of such research. Thus attention has been focused more and more on over-all systems as frames of reference for analytical work in various areas. It is our contention that a similar process can be useful for managers. Whereas managers often have been focusing attention on particular functions in specialized areas, they may lose sight of the over-all objectives of the business and the role of their particular business in even larger systems. These individuals can do a better job of carrying out their own responsibilities if they are aware of the "big picture." It is the familiar problem

[1] For a more complete discussion see: Johnson, Kast, and Rosenzweig [3], pp. 4–6, 91, 92.

of not being able to see the forest for the trees. The focus of systems management is on providing a better picture of the network of subsystems and interrelated parts which go together to form a complex whole.

Before proceeding to a discussion of systems theory for business, it will be beneficial to explore recent attempts to establish a general systems theory covering all disciplines or scientific areas.

GENERAL SYSTEMS THEORY

General systems theory is concerned with developing a systematic, theoretical framework for describing general relationships of the empirical world. A broad spectrum of potential achievements for such a framework is evident. Existing similarities in the theoretical construction of various disciplines can be pointed out. Models can be developed which have applicability to many fields of study. An ultimate but distant goal will be a framework (or system of systems) which could tie all disciplines together in a meaningful relationship.

There has been some development of interdisciplinary studies. Areas such as social psychology, biochemistry, astrophysics, social anthropology, economic psychology, and economic sociology have been developed in order to emphasize the interrelationships of previously isolated disciplines. More recently, areas of study and research have been developed which call on numerous subfields. For example, cybernetics, the science of communication and control, calls on electrical engineering, neurophysiology, physics, biology, and other fields. Operations research is often pointed to as a multidisciplinary approach to problem solving. Information theory is another discipline which calls on numerous subfields. Organization theory embraces economics, sociology, engineering, psychology, physiology, and anthropology. Problem solving and decision making are becoming focal points for study and research, drawing on numerous disciplines.

With these examples of interdisciplinary approaches, it is easy to recognize a surge of interest in larger-scale, systematic bodies of knowledge. However, this trend calls for the development of an over-all framework within which the various subparts can be integrated. In order that the *interdisciplinary* movement does not degenerate into *undisciplined* approaches, it is important that some structure be developed to integrate the various separate disciplines while retaining the type of discipline which distinguishes them. One approach to providing an over-all framework (general systems theory) would be to pick out phenomena common to many different disciplines and to develop general models which would include such phenomena. A second approach would include the structuring of a hierarchy of levels of complexity for the basic units of behavior in the various empirical fields. It would also involve development of a level of abstraction to represent each stage.

We shall explore the second approach, a hierarchy of levels, in more detail since it can lead toward a system of systems which has application in most businesses and other organizations. The reader can undoubtedly call to mind examples of familiar systems at each level of Boulding's classification model.

1. The first level is that of static structure. It might be called the level of *frameworks;* for example, the anatomy of the universe.
2. The next level is that of the simple dynamic system with predetermined, necessary motions. This might be called the level of *clockworks*.
3. The control mechanism or cybernetic system, which might be nicknamed the level of the *thermostat*. The system is self regulating in maintaining equilibrium.
4. The fourth level is that of the "open system," or self-maintaining structure. This is the level at which life begins to differentiate from not-life: it might be called the level of the *cell*.
5. The next level might be called the genetic-societal level; it is typifield by the *plant*, and it dominates the empirical world of the botanist.
6. The animal system level is characterized by increased mobility, teleological behavior, and self-awareness.
7. The next level is the "human" level, that is, of the individual human being considered as a system with self-awareness and the ability to utilize language and symbolism.
8. The social system or systems of human organization constitute the next level, with the consideration of the content and meaning of messages, the nature and dimensions of value systems, the transcription of images into historical record, the subtle symbolizations of art, music and poetry, and the complex gamut of human emotion.
9. Transcendental systems complete the classification of levels. These are the ultimates and absolutes and the inescapables and unknowables, and they also exhibit systematic structure and relationship.[2]

Obviously, the first level is most pervasive. Descriptions of static structures are widespread. However, this descriptive cataloguing is helpful in providing a framework for additional analysis and synthesis. Dynamic "clockwork" systems, where prediction is a strong element, are evident in the classical natural sciences such as physics and astronomy; yet even here there are important gaps. Adequate theoretical models are not apparent at higher levels. However, in recent years closed-loop cybernetic, or "thermostat," systems have received increasing attention. At the same time, work is progressing on open-loop systems with self-maintaining structures and reproduction facilities. Beyond the fourth level we hardly have a beginning of theory, and yet even here system description via computer models may foster progress at these levels in the complex of general systems theory.

Regardless of the progress at any particular level in the above

_____
[2] Boulding [2], pp. 202–5.

scheme, the important point is the concept of a general systems theory. Clearly, the spectrum, or hierarchy, of systems varies over a considerable range. However, since the systems concept is primarily a point of view and a desirable goal, rather than a particular method or content area, progress can be made as research proceeds in various specialized areas but within a total system context.

With the general theory and its objectives as background, we direct our attention to a more specific theory for business, a systems theory which can serve as a guide for management scientists and ultimately provide the framework for integrated decision making on the part of practicing managers.

SYSTEMS THEORY FOR BUSINESS

The biologist Ludwig von Bertalanffy has emphasized the part of general systems theory which he calls open systems [1]. The basis of his concept is that a living organism is not a conglomeration of separate elements but a definite system, possessing organization and wholeness. An organism is an open system which maintains a constant state while matter and energy which enter it keep changing (so-called dynamic equilibrium). The organism is influenced by, and influences, its environment and reaches a state of dynamic equilibrium in this environment. Such a description of a system adequately fits the typical business organization. The business organization is a man-made system which has a dynamic interplay with its environment—customers, competitors, labor organizations, suppliers, government, and many other agencies. Furthermore, the business organization is a system of interrelated parts working in conjunction with each other in order to accomplish a number of goals, both those of the organization and those of individual participants.

A common analogy is the comparison of the organization to the human body, with the skeletal and muscle systems representing the operating line elements and the circulatory system as a necessary staff function. The nervous system is the communication system. The brain symbolizes top-level management, or the executive committee. In this sense an organization is represented as a self-maintaining structure, one which can reproduce. Such an analysis hints at the type of framework which would be useful as a systems theory for business—one which is developed as a system of systems and that can focus attention at the proper points in the organization for rational decision making, both from the standpoint of the individual and the organization.

The scientific-management movement utilized the concept of a man-machine system but concentrated primarily at the shop level. The so-called "efficiency experts" attempted to establish procedures covering the work situation and providing an opportunity for all those involved to benefit—employees, managers, and owners. The human relationists, the movement stemming from the Hawthorne–Western Electric studies, shifted some of the focus away from the man-machine system per se to interrelationships among individuals in the organization. Recognition of the effect of interpersonal relationships, human behavior, and small groups resulted in a relatively widespread reevaluation of managerial approaches and techniques.

The concept of the business enterprise as a social system also has received considerable attention in recent years. The social-system school looks upon management as a system of cultural interrelationships. The concept of a social system draws heavily on sociology and involves recognition of such elements as formal and informal organization within a total integrated system. Moreover, the organization or enterprise is recognized as subject to external pressure from the cultural environment. In effect, the enterprise system is recognized as a part of a larger environmental system.

Since World War II, operations research techniques have been applied to large, complex systems of variables. They have been helpful in shop scheduling, in freightyard operations, cargo handling, airline scheduling, and other similar problems. Queuing models have been developed for a wide variety of traffic- and service-type situations where it is necessary to program the optimum number of "servers" for the expected "customer" flow. Management-science techniques have undertaken the solution of many complex problems involving a large number of variables. However, by their very nature, these techniques must structure the system for analysis by quantifying system elements. This process of abstraction often simplifies the problem and takes it out of the real world. Hence the solution of the problem may not be applicable in the actual situation.

Simple models of maximizing behavior no longer suffice in analyzing business organizations. The relatively mechanical models apparent in the "scientific management" era gave way to theories represented by the "human relations" movement. Current emphasis is developing around "decision making" as a primary focus of attention, relating communication systems, organization structure, questions of growth (entropy and/or homeostasis), and questions of uncertainty. This approach recognizes the more complex models of administrative behavior and should lead to more encompassing systems that provide the framework within which to fit the results of specialized investigations of management scientists.

The aim of systems theory for business is to develop an objective, understandable environment for decision making; that is, if the system within which managers make the decisions can be provided as an explicit framework, then such decision making should be easier to handle. But what are the elements of this systems theory which can be used as a framework for integrated decision making? Will it require wholesale change on the part of organization structure and administrative behavior? Or can it be woven into existing situations? In general, the new concepts can be applied to existing situations. Organizations will remain recognizable. Simon makes this point when he says:

1. Organizations will still be constructed in three layers; an underlying *system* of physical production and distribution processes, a layer of programmed (and probably largely automated) decision processes

for governing the routine day-to-day operation of the physical *system,* and a layer of nonprogrammed decision processes (carried on in a man-machine system) for monitoring the first-level processes, re-designing them, and changing parameter values.

2. Organizations will still be hierarchical in form. The organization will be divided into major subparts, each of these into parts, and so on, in familiar forms of departmentalization. The exact basis for drawing departmental lines may change somewhat. Product divisions may be-come even more important than they are today, while the sharp lines of demarcation among purchasing, manufacturing, engineering, and sales are likely to fade.[3]

We agree essentially with this picture of the future. However, we want to emphasize the notion of systems as set forth in several layers. Thus the systems that are likely to be emphasized in the future will develop from projects or programs, and authority will be vested in managers whose influence will cut across traditional departmental lines. This concept will be developed in more detail throughout this article.

There are certain key subsystems and/or functions essential in every business organization which make up the total information-decision system, and which operate in a dynamic environmental system subject to rapid change. The subsystems include:

1. A *sensor subsystem* designed to measure changes within the system and with the environment.
2. An *information-processing subsystem* such as an accounting, or data processing system.
3. A *decision-making subsystem* which receives information inputs and outputs planning messages.
4. A *processing subsystem* which utilizes information, energy, and materials to accomplish certain tasks.
5. A *control component* which ensures that processing is in accord-ance with planning. Typically this provides feedback control.
6. A *memory or information storage subsystem* which may take the form of records, manuals, procedures, computer programs, etc.

A goal-setting unit will establish the long-range objectives of the organization, and the performance will be measured in terms of sales, profits, employment, etc., relative to the total environmental system.

This is a general model of the systems concept in a business firm. In the following section a more specific model illustrating the applica-tion of the systems concept is established.

AN ILLUSTRATIVE MODEL OF THE SYSTEMS CONCEPT

Traditionally, business firms have not been structured to utilize the systems concept. In adjusting the typical business structure to fit within the framework of management by system, certain organizational changes may be required. It is quite obvious that no one organizational

---

[3] Simon [4], pp. 49–50. (Italics by authors.)

structure can meet operational requirements for every company. Each organization must be designed as a unique system. However, the illustrative model set forth would be generally operable for medium- to large-size companies which have a number of major products and a variety of management functions. The primary purpose of this model is to illustrate the application of systems concepts to business organizations and the possible impact upon the various management functions of planning, organizing, communication, and control. The relationships which would exist among the top management positions are shown in Figure 1.

The master planning council would relate the business to its environmental system, and it would make decisions relative to the products of services the company produced. Further, this council would establish the limits of an operating program, decide on general policy matters relative to the design of operating systems, and select the director for each new project. New project decisions would be made with the assistance and advice of the project research and development, market

FIGURE 1

The Systems Model: Top Management

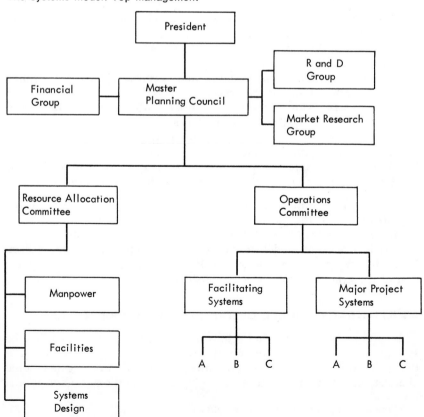

research, and financial groups. Once the decision was made, the resource allocation committee would provide the facilities and manpower
for the new system, and supply technical assistance for systems design.
After the system had been designed, its management would report to
the operations committee as a major project system, or as a facilitating
system.

Facilitating systems would include those organized to produce a service rather than a finished product. Each project system would be
designed toward being self-sufficient. However, in many cases this
objective may not be feasible or economical. For example, it may not
be feasible to include a large automated mill as a component of a major
project system, but the organization as a whole, including all of the
projects, might support this kind of a facility. A facilitating system
would be designed, therefore, to produce this kind of operating service
for the major project systems. The output of the facilitating system
would be material input for the project system and a fee should be
charged for this input, just as if the input had been purchased from
an outside source.

A soap manufacturer could have, for example, major project systems
in hand soap, laundry soap, kitchen soap, and toothpaste. A facilitating
system might be designed to produce and *sell* containers to the four
project systems.

Operating Systems

All operating systems would have one thing in common—they would
use a common language for communicating among themselves, and
with higher levels of management. In addition, of course, each system
designed would be structured in consideration of company-wide policies. Other than these limits, each operating system would be created
to meet the specific requirements of its own product or service. A model
of an operating system is shown in Fig. 2.

Figure 2 illustrates the relationship of the functions to be performed
and the flow of operating information. The operating system is structured to (1) direct its own inputs, (2) control its own operation, and
(3) review and revise the design of the system as required. Input is
furnished by three different groups: technical information is generated
as input into the processing system, and in addition, technical information is the basis for originating processing information. Both technical
and processing information are used by the material input system to
determine and supply materials for processing. However, corrective
action, when necessary, would be activated by input allocation.

This model can be related to most business situations. For example, if
this represented a system to produce television sets, the technical information would refer to the design of the product, processing information would include the plan of manufacture and schedule, and the
material input would pertain to the raw materials and purchased parts
used in the processing. These inputs of information and material would
be processes and become output. Process control would measure the

FIGURE 2

An Operating System Model

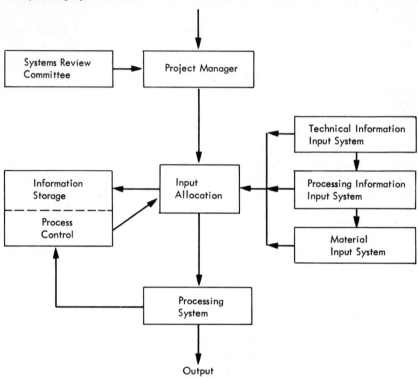

Output

output in comparison to the standard (Information Storage) obtained from input allocation, and issue corrective information whenever the system failed to function according to plan. The design of the system would be reviewed continually and the components rearranged or replaced when these changes would improve operating efficiency.

Basically, the operating systems would be self sustaining with a high degree of autonomy. Therefore, they could be integrated into the over-all organizational structure (Figure 1) with a minimum of difficulty.

SYSTEMS CONCEPTS AND MANAGEMENT

Managers are needed to convert the disorganized resources of men, machines, and money into a useful and effective enterprise. Essentially, management is the process whereby these unrelated resources are integrated into a total *system for objective accomplishment*. A manager gets things done by working with people and physical resources in order to accomplish the objectives of the system. He coordinates and integrates the activities and work of others rather than performing operations himself.

Structuring a business according to the systems concept does not eliminate the need for the basic functions of planning, organization, control, and communication. However, there is a definite change of emphasis, for the functions are performed in conjunction with operation of the system and not as separate entities. In other words, everything revolves around the system and its objective, and the function is carried out only as a service to this end. This point can be clarified by reviewing each of the functions in terms of their relation to the model of the systems concept illustrated previously.

Planning

Planning occurs at three different levels in the illustrative model. These levels are shown in Figure 1. First, there is top-level planning by the master planning council. Second, the project and facilitating systems must be planned and resources allocated to them. Finally, the operation of each project and facilitating system must be planned.

The master planning council establishes broad policies and goals and makes decisions relative to the products or services the company produces. It decides upon general policy matters concerning the design of the operating systems and selects the director for each new program. It is the planning council which receives informational inputs from the environmental and competitive systems. It combines these inputs with feedback information from the internal organizational system and serves as the key decision-making center within the company. Much of the decision making at this level is non-programmed, unstructured, novel, and consequential. While some of the new techniques of management science may be helpful, major reliance must be placed upon mature assessment of the entire situation by experienced, innovative top executives.

Once these broad decisions have been made, the planning function is transferred to the resource allocation and operating committees. They plan and allocate facilities and manpower for each new system and supply technical assistance for individual systems design. At this planning level it is possible to utilize programmed decision making— operations research and computer techniques.

The third level, planning the operations of each project or facilitation system, is concerned primarily with the optimum allocation of resources to meet the requirements established by the planning council. This planning can most easily be programmed to automatic decision systems. However, the project director would still have to feed important non-quantifiable inputs into the system.

Under the systems concept of planning there is a direct relationship between the planning performed at each of the three levels. The first planning level receives informational inputs from the environment and competitive system and feedback information from within the organization. It translates this into inputs for the next planning level which in turn moves to a more detailed level of planning and provides inputs for the third or project level. One of the major advantages of this systems model is to provide a clear-cut delineation of the responsibility for various types of planning.

This concept facilitates integrated planning on a systems basis at the project level within the organization. Given the inputs (premises, goals, and limitations) from the higher levels the project managers are delegated the function of integrated planning for their project.

Organization

Traditional organization theory emphasized parts and segments of the structure and is concerned with the separation of activities into tasks or operational units. It does not give sufficient emphasis to the interrelationships and integration of activities. Adapting the business organization to the systems concept places emphasis upon the integration of all activities toward the accomplishment of over-all objectives but also recognizes the importance of efficient subsystem performance.

The systems basis of organization differs significantly from traditional organization structures such as line and staff or line, staff, and functional relationships. As shown in Figure 1, there are three major organizational levels, each with clearly delineated functions. The master planning council has broad planning, control, and integrative functions; the resource allocation committee has the primary function of allocating manpower and facilities, and aids in systems design for the facilitating or project systems. One of the major purposes of this type organization is to provide an integration of activities at the most important level—that is, the individual project or program.

Staff specialization of skills is provided for the master planning council through such groups as financial, research and development, and market research. Their activities, however, are integrated and coordinated by the planning council. There are specialists at the operating level who are completely integrated into each project system. Thus, the activities of these specialists are geared to the effective and efficient performance of the individual project system. This type organization minimizes a major problem associated with staff and functional personnel—their tendency to associate their activities with specialized areas rather than with the optimum performance of the over-all operation. Yet, under the model the importance of initiative and innovation are recognized. In fact, the major function of the master planning council is planning and innovation. Specific provision for receiving information inputs from product and market research are provided in the model.

There are other advantages of the systems concept. Business activity is dynamic, yet the typical organization is structured to perpetuate itself rather than change as required. There is generally resistance by the various specialized functions to change in order to optimize organization performance. For example, Parkinson's Law states that there is an ever-increasing trend toward hierarchies of staff and functional personnel who are self-perpetuating and often do not contribute significantly to organizational effectiveness, or in extreme cases may be dysfunctional. In contrast, a system is designed to do a particular task. When the task is completed, the system is disbanded.

Systems are created from a central pool of resources. Facilities, machines, and manpower are assigned to specific projects or programs. The approach is to create and equip the project system with a complete arrangement of components to accomplish the job at hand. This may result in the duplication of certain activities in more than one operating system; however, this disadvantage is not as serious as it may seem. For example, it may be more efficient to have several typewriters assigned to each system rather than a center pool of typewriters. In the first instance, the typewriters may be utilized less than 100 per cent of the time, but the problems of scheduling the work at the central pool, delays, accountability, measurement of contribution, etc., would soon offset the advantages of centralizing equipment. Too much effort may be spent in creating processing information which accomplishes no objective other than keeping the machines utilized. A reasonable amount of redundancy or extra capacity will provide more flexibility, protect against breakdowns, reduce flow time, require less planning, eliminate many problems associated with interdepartmental communication, and reduce the amount of material handling.

Obviously, there are situations when it is impractical to decentralize a particular facility, because individual systems cannot utilize it sufficiently to warrant its incorporation into each separate operation. In these instances, a facilitating system would be created which would sell its services to any or all of the major project systems. These service systems would have to justify their existence and compete with outside vendors as suppliers to the major project system.

One of the great advantages of the systems concept for organizing pertains to the decentralization of decision making and the more effective utilization of the allocated resources to the individual project system. This has the merit of achieving accountability for performance through the measurability of individual systems of operation.

Control

The systems concept features control as a means of gaining greater flexibility in operation, and, in addition, as a way of avoiding planning operations when variables are unknown. It is designed to serve the operating system as a subsystem of the larger operation. Its efficiency will be measured by how accurately it can identify variations in systems operation from standard or plan, and how quickly it can report the need for correction to the activating group.

We must conclude that error is inevitable in a system which is subject to variations in input. When the lag in time between input and output is great, more instability is introduced. Feedback can reduce the time lag; however, corrective action which is out of phase will magnify rather than overcome the error. Every system should be designed to make its own corrections when necessary. That is, a means should be provided to reallocate resources as conditions change. In our model the Systems Review Committee (see Figure 2) should be aware of any change in operating conditions which might throw the system "out of control." Replanning or redesign may be required.

In controlling a system it is important to measure inputs of information, energy, and materials; and outputs of products and/or services. This will determine operating efficiency. In addition it may be important to establish points of measurement during critical or significant stages of processing. Such measurements would be used principally to help management analyze and evaluate the operation and design of individual components. The best approach is to spotlight exceptions and significant changes. Management can focus their attention on these areas. One important thing to remember is that the control group is not a part of the processing system—it is a subsystem, serving the operating system. Cost control can be used as an example to illustrate this point. The cost accountant must understand that his primary objective is to furnish managers with information to control costs. His task is to inform, appraise, and support; never to limit, censure, or veto. The same principle applies to every control group serving the operating system.

Communication

Communication plays a vital role in the implementation of the systems concept. It is the connecting and integrating link among the systems network. The flow of information, energy, and material—the elements of any processing system—are coordinated via communication systems. As shown in the model (Figure 2) the operating system requires information transmission to ensure control. Communication systems should be established to feed back information on the various flows—information, energy, and material. Information on the effectiveness of the planning and scheduling activities (as an example of information flow) would be helpful in adjusting the nature of this activity for the future. Similarly, reports on absenteeism are examples of communication concerning the energy flow (the people in the system) to the processing activity. Information on acceptance inspection is an example of information stemming from the material flow aspect of an operating system. All of these feedback communication systems provide for information flow to a sensor and a control group. Comparison between the information received and the information stored (the master plan for this particular operating system) would result in decisions concerning the transmission of corrective information to the appropriate points.

Relationships within and among various project systems and between the levels of the system as a whole are maintained by means of information flow which also can be visualized as a control device. Moreover, any operating system maintains contact with its environment through some sensory element. Referring to Figure 1, the sensory elements in this case are the groups reporting to the master planning council. The master planning council makes decisions, concerning the product or service the organization will produce, based on information gained from market research, research and development, and financial activities. In a sense, these activities function as the antenna of the organization, maintaining communication with the external environment. The master planning council melds the information received through these activities with other premises covering the internal aspects in order to make decisions about future courses of action. Here again, communication or information flow can be visualized as a necessary element in controlling the course of action for the enterprise as a whole. Based on the feedback of information concerning the environment in general, the nature of competition, and the performance of the enterprise itself, the master planning council can continue its current courses of activity or adjust in light of developing circumstances. Thus, communication or information flow facilitates the accomplishment of the primary managerial functions of planning, organizing, and controlling.

Communication by definition is a system involving a sender and a receiver, with implications of feedback control. This concept is embodied in the lowest level projects or subsystems, in all larger systems, and in the system as a whole. Information-decision systems, regardless of formal charts or manuals, often flow across departmental boundaries and are often geared to specific projects or programs. The systems concept focuses on this approach and makes explicit the information-decision system which might be implicit in many of today's organizations.

The systems concept does not eliminate the functions of management, i.e., planning, organizing, control, and communication. Instead, it integrates these functions within a framework designed to emphasize their importance in creating more effective systems. Because of the great diversity of operations and environments, particular missions of organizations differ and each system must be unique or at least have some unique elements. Nevertheless, the illustrative model and its application to the management functions of planning, organizing, controlling, and communication can serve as a point of departure in systems design.

PERVASIVENESS OF SYSTEM CONCEPTS

Many of the most recent developments in the environment of businessmen and managers have involved systems concepts. For example, the trend toward automation involves implementation of these ideas. Automation suggests a self-contained system with inputs, outputs, and a mechanism of control. Yet the concept also recognizes the need to consider the environment within which the automatic system must perform. Thus the automated system is recognized as a subpart of a larger system.

The kinds of automation prevalent today range in a spectrum from sophisticated mechanization to completely automatic, large-scale production processes. Individual machines can be programmed to operate automatically. Large groups of machines also can be programmed to perform a series of operations, with automatic materials-handling devices providing connecting links among components of the system. In such a system, each individual operation could be described as a system and could be related to a larger system covering an entire processing operation. That particular processing operation could also be part of the total enterprise system, which in turn can be visualized as a part of an environmental system.

Completely automated processing systems such as oil refineries are also commonplace today. In such cases the entire process from input of raw material to output of finished products is automated with preprogrammed controls used to adjust the process as necessary, according to information feedback from the operation itself.

The systems concept is also apparent in other aspects of automation. The above examples deal with physical processing; another phase which has been automated is information flow. With the introduction of large-scale, electronic-data-processing equipment, data-processing systems have been developed for many applications. Systems concepts are prevalent, with most applications built around the model of input-processor-output and with feedback control established within the instructions developed to guide the processing of data. Here again, there is an entire spectrum of sophistication leading from simple, straightforward data-reduction problems to elaborate, real-time data-processing systems.

Physical distribution systems have received increasing attention on the part of manufacturers and shippers. The concept of logistics, or materials management, have been used to emphasize the flow of materials through distribution channels. The term *rhochrematics* has been coined to connote the flow process from raw-material sources to final consumer.[4] In essence, these ideas embrace systems concepts because emphasis is placed on the total system of material flow rather than on functions, departments, or institutions which may be involved in the processing.

In recent years increasing attention has been focused upon massive engineering projects. In particular, military and space programs are becoming increasingly complex, thus indicating the need for integrating various elements of the total system. Manufacturing the product itself (a vehicle or other hardware) is quite complex, often involving problems of producibility with requirements of extremely high reliability. This is difficult to ensure for individual components or subsystems. In addition, each subsystem also must be reliable in its interrelationship with all other subsystems. Successful integration of subcomponents, and hence successful performance of a particular product, must also be integrated with other elements of the total system. For example, the functioning of the Nike-Zeus antimissile missile must be coordinated with the early warning system, ground facilities, and operating personnel. All elements must function as an operating, integrated whole.

---

[4] Rhochrematics comes from two Greek roots; rhoe, which means a flow (as a river or stream), and chrema, which stands for products, materials, or things (including information). The abstract ending -ics has been added, as for any of the sciences.

The previous discussion has emphasized the mechanistic and structural aspects of the systems concept. Yet, we cannot forget that business organizations are social systems; we are dealing with man-made systems. Obviously, a great deal could be said about the possible consequences of applying systems concepts to human relationships, but such a task is beyond the scope of this article. However, in discussing the impact of the systems concept it should not be assumed that people basically resist systems. Much of man's conscious activities since the dawn of history has been geared to creating system out of chaos. Man does not resist systematization of his behavioral patterns per se. Rather, the normal human being seeks satisfactory systems of interpersonal relationships which guide his activities. Without systematization, behavior would be random, nongoal-oriented, and unpredictable. Certainly, our complex, modern, industrial society demands more systemized human behavior than older, less-structured societies. A common characteristic in a rapidly advancing society is to make systems of interpersonal relationship more formal. While many of these systems have been implicit in the past, they are becoming more explicit. This remains one of the basic precepts of our systems model; systematic interpersonal relationships are necessary for accomplishing group objectives and an effective organizational system should be designed to meet this need.

SUMMARY

General systems theory is concerned with developing a systematic, theoretical framework for describing general relationships of the empirical world. While a spectrum or hierarchy of systems can be established over a considerable range, the systems concept is also a point of view and a desirable goal, rather than a particular method or content area. Progress can be made as research proceeds in various specialized areas but within a total system context.

The business organization is a man-made system which has a dynamic interplay with its environment—customers, competitors, labor organizations, suppliers, government, and many other agencies. In addition, the business organization is a system of interrelated parts working in conjunction with each other in order to accomplish a number of goals, both those of the organization and those of individual participants. This description parallels that of open systems in general, which maintain a constant state while matter and energy which enter them keep changing; that is, the organisms are influenced by and influence their environment, and reach a state of dynamic equilibrium within it. This concept of the organization can be used by practicing managers in order to integrate the various ongoing activities into a meaningful total system. Regardless of specific adjustments or organizational arrangements, there are certain subsystems or essential functions which make up a total information-decision system. However, the exact form utilized by a particular organization may depend upon the task orientation. We have presented a generalized illustrative model which indicates an approach that may be appropriate for a large segment of modern business organizations.

Managers are needed to convert disorganized resources of men, machines, and money into a useful, effective enterprise. Essentially, management is the process whereby these unrelated resources are integrated into a total *system for objective accomplishment*. The systems concept provides no cookbook technique, guaranteed to provide managerial success. The basic functions are still planning, organization, control, and communication. Each of these activities can be carried out with or without emphasis on systems concepts. Our contention is that the activities themselves can be better accomplished in light of systems concepts. Furthermore, there can be a definite change in emphasis for the entire managerial process if the functions are performed in light of the system as a whole and not as separate entities.

The business organization as a system can be considered as a sub-system of a larger environmental system. Even industry or inter-industry systems can be recognized as sub-elements of the economic system, and the economic system can be regarded as a part of society in general. One of the major changes within business organizations of the future may be the breakdown of traditional functional specialization geared to optimizing performance of particular departments. There may be growing use of organizational structures designed around projects and information-decision systems. The systems concept calls for integration, into a separate organizational system, of activities related to particular projects or programs. This approach currently is being implemented in some of the more advanced-technology industries.

The breakdown of business organizations into separate functional areas has been an artificial organizational device, necessary in light of existing conditions. Management-science techniques, computer simulation approaches, and information-decision systems are just a few of the tools which will make it possible for management to visualize the firm as a total system. This would not have been possible two decades ago; it is currently becoming feasible for some companies; and it will become a primary basis for organizing in the future.

REFERENCES

1. Bertalanffy, L. von, "General System Theory: A New Approach to Unity of Science," *Human Biology*, December 1951, pp. 303–61.
2. Boulding, K., "General Systems Theory: The Skeleton of Science," *Management Science*, April 1956, pp. 197–208.
3. Johnson, R. A., Kast, F. E. and Rosenzweig, J. E., *The Theory and Management of Systems*, McGraw-Hill Book Company, Inc., New York, 1960.
4. Simon, H. A., *The New Science of Management Decision*, Harper & Brothers, New York, 1960.

Wendell French

24. Processes vis-à-vis Systems: Toward a Model of the Enterprise and Administration

IT IS THE central thesis of this paper that the enterprise can usefully be described as a man-directed, dynamic network of processes and facilitating systems, and that a useful model of the administrator's job logically follows from this description. We shall first contrast the concepts of "process" and "system" and then integrate these concepts with certain other constructs toward building (*a*) a model of the enterprise and (*b*) a model of the administrator's job.[1,2]

THE CONSTRUCT OF "PROCESS"

A Definition of "Process." The word "process" is used extensively in the literature of many disciplines. For example, authors in management theory write about "the decision-making process," "the planning process," "the collective bargaining process," "the management process," and "the administrative process." Authors in the physical sciences talk about chemical, biological, and physical processes. In the social

From *Academy of Management Journal*, 1963, vol. 6, pp. 46–57. Reprinted with permission of the author and the publisher.

[1] Our use of the word "enterprise" will be consistent with Barnard's use of the term. See Chester I. Barnard, *The Functions of the Executive* (Cambridge, Massachusetts: Harvard University Press, 1958), pp. 65–67. "The organization" will be used synonymously with "enterprise." In addition, we shall use the terms "administration" and "management" synonymously.

[2] In such model building, perhaps we can avoid the "jungle warfare" which Koontz perceives as being prevalent among management theorists. There is no intention of denying the usefulness of other models, nor of claiming any ubiquitous utility in what is proposed. Different models have different uses. See Harold Koontz, "The Management Theory Jungle," *Journal of the Academy of Management*, 4:174–88 (December 1961).

sciences there are references to social, psychological, and economic processes.

Processes are also seen as impinging upon and interacting with one another. In ecology, for example, there are descriptions of the interaction between living organisms and the environment,[3] and in psychology there is reference to biological and social processes interacting through the human personality.[4] In the field of management, processes are seen as being in a state of interdependence.[5]

Typically, authors use the word "process" without definition, which suggests that there may be a commonly held meaning attached to the word which is generally consistent with dictionary definitions. Occasionally, however, authors desire more precision in the use of the word. For example, Newman and Summer define process as ". . . a series of actions that lead to the accomplishment of objectives."[6] Argyris defines process as ". . . any course or sequence of behavior accomplishing a necessary purpose."[7] According to Guest: "process" denotes "progressive action or a series of acts performed by persons in the course of moving the organization from one state to another."[8] Lundberg defines process as ". . . a sequence of events which collectively yield a specified product or state and which together exhibit a particular function."[9]

The following definition of "process" is consistent with the above definitions, but lends itself more directly to the purposes of this paper: *a process is a flow of interrelated events moving toward some goal, purpose, or end.* In this definition, "flow" implies movement through time in the direction of a consequence. "Interrelated" denotes interaction within the process, and events which are highly relevant one to another. "Events" are changes or happenings that occur at one point or period of time, and may be any of an infinite number of phenomena. "Goal" suggests a human objective, while "purpose" suggests either human objectives or objectives in a metaphysical sense. "End" implies some conclusion or consequence which may not necessarily be sought or

---

[3] See John H. Storer, *The Web of Life* (New York: The Devin-Adair Company, 1954).

[4] See Gardner Murphy, *Personality: A Bio-Social Approach to Origins and Structure* (New York: Harper and Brothers, 1947), pp. 1–37. For other references on the interactions of processes, See Kurt Lewin, *Field Theory in Social Science* (New York: Harper and Brothers, 1951), p. 57; Harold J. Leavitt, "Management According to Task: Organizational Differentiation," *Management International* (January–February 1962), p. 22; and Chester I. Barnard, *op. cit.,* p. 6.

[5] See E. Wight Bakke and Chris Argyris, *Organizational Structure and Dynamics* (New Haven: Yale University, Labor and Management Center, 1954), pp. 10, 26; and M. E. Salveson, "An Analysis of Decisions," *Management Science,* 4:212 (April 1958).

[6] William H. Newman and Charles E. Summer, Jr., *The Process of Management* (Englewood Cliffs, N.J.: Prentice-Hall, 1961). p. 9.

[7] Chris Argyris, *Organization of a Bank* (New Haven: Yale University Labor and Management Center, 1954), p. 9.

[8] Robert H. Guest, *Organizational Change: The Effect of Successful Leadership* (Homewood, Illinois, The Dorsey Press and Richard D. Irwin, Inc., 1962), pp. 139–49.

[9] Craig C. Lundberg, "Administrative Decisions: A Scheme for Analysis," *Journal of the Academy of Management,* 5:168 (August 1962).

planned by man. Thus, a process may or may not have man-intended (or human-oriented) consequences.[10]

Examples of Characteristics of "Processes." A familiar process in nature is erosion. Here is seen a flow of events including the occurrence of rain, the flow of water, and fluctuations in temperature, resulting in the washing away of tillable soil. An example of a process from management is the "staffing process," a flow of events which results in the continuous filling of positions within the organization. This flow is comprised of such elements as advertising, interviews, decisions to hire, transfers, and promotions. (It should be pointed out at the outset that "process" does not preclude "system." Man-directed processes always involve systematization, as will be elaborated upon later.)

Processes have certain characteristics which are obvious and others that are not so apparent. Unless the most minute processes have been identified (which is doubtful), all processes will contain a number of sub-processes. Similarly, a given process will constitute a segment of a broader process unless the most universal process has been identified (which is also questionable).

An example of a process which is at once a sub-process and a process containing sub-processes is tree growth. This process can be sub-divided into the interrelated processes of plant food assimilation and distribution, photosynthesis, pollination, etc. But at a higher level of abstraction, the tree can be perceived as a part of the life-process of a forest in which there are interacting processes of tree growth, moisture distribution, soil-building, animal growth, food growth, erosion, aging, decay, combustion, and others.

Using our example from management, the staffing process may be seen as a network of sub-processes, including the recruiting process, the interviewing process, etc. In turn, the staffing process is, itself, one of several interacting sub-processes within a broader management process. Thus, a process may be narrow or broad, simple or abstract, depending upon the particular frame of reference or focus of the perceiver.

The "length" of process is defined in terms of some goal, purpose, or end. For example, the employment of college students can be considered the culmination of the college placement process. But these events may also be considered early events in the staffing process of each employing organization. Thus, any given process will be a continuation of one or more other processes, and its dimensions are defined in terms of the relevance of events to a particular consequence.

Another important characteristic of a process, which is probably not so obvious, is that a process will usually, if not always, contain elements which are simultaneously components of one or more other processes. For example, the process of the decay of a plant will also be a component of a soil-building process. The staffing process, while an intra-enterprise process, has at the same time components pertaining to the

---

[10] It is not implied that the word "process" automatically denotes these phenomena. The word "process" simply seems to be the best label to summarize these conditions. Hopefully, for the purposes of this paper, the word "process" will conjure the intended picture.

broader life or career processes of the individuals involved. Similarly, the staffing process overlaps other processes within the organization. For example, a decision to offer a particular initial salary is a component of the financial management process as well as a component of the staffing process.[11]

Along with other characteristics of processes, this phenomenon of mutuality of process components is a particularly useful concept in building a model of the administrator's job, as we shall see later. Examined next, however, will be the concept of "system" and certain other constructs.

"SYSTEM" DEFINED AND CONTRASTED WITH "PROCESS"

A Definition of "System." The concept of "system" has been defined and utilized by many authors. The construct has been utilized, for example, in cybernetics,[12] biology and physics,[13] psychology,[14] sociology,[15] management,[16] and in a variety of other basic and applied fields.[17]

The definition of system which follows will be consistent with the construct as it has evolved in the literature, but traditional usage will be enlarged to include the concept of "process," and the relationship between "system" and "process." Thus, *a system is a particular linking of elements which has a facilitating effect, or an intended facilitating effect, on the carrying out of a process.*

In this definition, "particular" suggests that System₁ is different in some respects from System₂, although both may be contributing to the carrying out of similar processes. The "elements" of a system may be components such as devices, raw materials, techniques, procedures,

---

[11] For references on the interrelatedness of events, see especially George A. Kelly, *The Psychology of Personal Constructs,* vol. 1 (New York: W. W. Norton and Company, 1955), p. 6; John H. Storer, op. cit.; and Chester I. Barnard, op. cit., pp. 78–79.

[12] See, for example, Stafford Beer, *Cybernetics and Management* (New York: John Wiley and Sons, 1959).

[13] See Ludwig van Bertalanfly, "The Theory of Open Systems in Physics and Biology," *Science,* 111:23 ff. (January 13, 1950).

[14] See Floyd Allport, *Theories of Perception and the Concept of Structure* (New York: John Wiley and Sons, 1955), pp. 469 ff.; and George A. Kelly, *op. cit.,* p. 57.

[15] See Talcott Parsons, Robert F. Bales, and Edward A. Shils, *Working Papers in the Theory of Action* (Glencoe, Illinois: The Free Press, 1953). These authors also use the word "process" extensively.

[16] See Richard A. Johnson, Fremont E. Kast, and James E. Rosenzweig, *The Theory and Management of Systems* (New York: McGraw-Hill Book Company, 1963); Stanford L. Optner, *Systems Analysis for Business Management* (Englewood Cliffs, N.J.: Prentice-Hall, 1960); Chester I. Barnard, *op. cit.,* p. 77; Robert V. Presthus, "Toward A Theory of Organizational Behavior," *Administrative Science Quarterly,* 3:50 (June, 1958); and William G. Scott, "Organization Theory: An Overview and An Appraisal," *Journal of the Academy of Management,* 4:15–20 (April 1961).

[17] Robert M. Gagne, et al., *Psychological Principles in System Development* (New York: Holt, Rinehart, and Winston, 1962); and Roy R. Grinker, ed., *Toward a Unified Theory of Human Behavior* (New York: Basic Books, 1956). The contributors to both books also use the word "process" a good deal. See also Allen Newell and Herbert A. Simon, "The Logic Theory Machine," *IRE Transactions On Information Theory* (New York: Institute of Radio Engineers, 1956), p. 61. Newell and Simon subsume "process" within their complex information-processing system.

plans, and people. "Facilitating effect" suggests that systems are ordinarily designed (or they evolve) as a means of carrying out or regularizing processes. "Intended" suggests (*a*) that mistakes can be made in the design of a system so that it works at cross-purposes with intended consequences or in opposition to the survival of an organism or organization, and (*b*) that a system can be in a static condition. The reader will also note that this definition, since it incorporates the concept of process, includes the important idea that a system is goal, consequence, or end oriented. And finally, in the context of the above definition, "carrying out" means unfolding or progressive development.

Some Examples and Characteristics of Systems. Stafford Beer provides useful examples of systems:

. . . It is legitimate to call a pair of scissors a system. But the expanded system of a woman cutting with a pair of scissors is also itself a genuine system. In turn, however, the woman-with-scissors system is part of a larger manufacturing system—and so on.[18]

Thus, the definition of a system is always an arbitrary matter, as is the definition of a process. It should be noted, however, that a system can be static (the scissors), whereas, by definition, a process is always dynamic (the cloth-cutting process).

The current weapon-system concept relative to missiles development, as described by Optner, provides further example of systems and related sub-systems:

. . . the nose cone is conceived as a subsystem; propulsion and guidance are subsystems; and the missile, its ground support equipment, material and personnel are the system.[19]

In this example can be seen a particular linking of devices, materials, and people held in readiness to facilitate a process—namely, the process of delivering a projectile on a target. Other systems might also be used to facilitate the same process. For example, the atomic submarine with its Polaris missiles, electronic devices, men, procedures, equipment, and supplies form another integrated system to accomplish the same purpose. Similarly, a supersonic bomber with its ground support, crew, weaponry, and other components constitute yet another system designed to facilitate this same process.

This same concept of systems may also be seen in the familiar examples of forests and trees. A given forest is a particular linking of plant and animal life, non-living organic material, atmosphere, moisture, streams, and so forth, all of which facilitate the processes described earlier. The tree itself is a system—a particular, integrated linking of

---

[18] Stafford Beer, op. cit., p. 9.

[19] Stanford L. Optner, op. cit., p. 4, To Optner, a system consists of inputs (energy, materials, information), a processor (the machine or human mechanism doing the work), and outputs (the product). The system may also include a feedback mechanism. Of course, these are the characteristics of a system in operation—in its dynamic state. In a static state the system might include all of these elements with the exception of the outputs.

elements including roots, sap, trunk, leaves and cells which carry out the process of tree growth. The leaf itself can be considered another system (or subsystem) employed to facilitate the process of photosynthesis—i.e., the utilization of sunlight to convert carbohydrates to sugar from carbon dioxide and water.

Using an example from management, the employment process in a given firm is facilitated by a particular linkage of people and employment devices and procedures, such as application blanks, a battery of tests and a series of interviews. In all likelihood, the employment systems of different firms will differ, although some of the major components may be identical—e.g., the test battery.

Optner provides some additional important characteristics of systems by differentiating between "structured" and "incompletely structured" systems. A structured system has highly regulated inputs, is essentially free from disturbance, and has outputs within highly predictable limits. Examples are hydraulic and electrical systems. In contrast, an incompletely structured system has inputs which are variable in terms of quality and quantity; it is subject to considerable disturbance, and has relatively unpredictable outputs. An example would be industrial enterprise, or any system involving people.[20]

It would appear that the less structured a system, the more that unsystematized processes must be inferred. Or to say it another way, the non-structured part of a system implies process.

The above definition and characteristics of "systems" vis-à-vis "process" will be employed below in developing a model of the organization and of the manager's job. First, however, Fayol's classification scheme and the concept of "organizational resources" will be examined as to their mutual compatability and as to their relevance to the concept of "process."

FAYOL'S CONSTRUCTS

Dating from the early 1900s and the writings of Fayol, authors in the field of management theory have stressed certain managerial activities common to the manager's job, i.e.; ". . . planning, organization, command, co-ordination, control. . . ."[21] Koontz and O'Donnell call these activities "functions."[22] Newman labels these activities "processes," as does Catheryn Seckler-Hudson.[23]

These authors and others since Fayol have slightly modified the above

---

[20] Ibid., pp. 3–9. Beer labels a company an "exceedingly complex" and "probabilistic" system. Stafford Beer, op. cit., pp. 9–18. In addition, any human organization would be a bio-social-physical system since all organizations utilize physical devices, machines, or artifacts of some kind.

[21] Henri Fayol, *General and Industrial Management*, Constance Storrs, trans. (London: Sir Isaac Pitman and Sons, 1961), p. 3.

[22] Harold Koontz and Cyril O'Donnell, *Principles of Management*, 2d ed. (New York: McGraw-Hill Book Company, 1959), p. 35.

[23] William H. Newman, *Administrative Action* (New York: Prentice-Hall, 1950), pp. 5–6; and Catheryn Seckler-Hudson, "Major Processes of Organization and Management," in Harold Koontz and Cyril O'Donnell, *Readings in Management* (New York: McGraw-Hill Book Company, 1959), p. 22.

categories to suit their purposes.[24] In general, however, management theorists of the "Fayol School" have retained Fayol's classification scheme with only minor modifications, and this scheme will be a major ingredient in the model of the manager's job to be discussed below.[25] Within the scope of this paper, reference will be made to Fayol's categories as "administrative processes."

THE CONCEPT OF ORGANIZATIONAL RESOURCES

Another classification scheme relative to the manager's job has had wide acceptance, particularly among economists. This classification scheme differentiates between productive factors—e.g., labor, money, materials—which the manager utilizes in the proper proportions in attempting to maximize profits.[26]

Although there has been criticism of this classification scheme by those who prefer to stress that management has to do with "getting things done through people,"[27] a number of authors of the "Fayol school" find the resource classification scheme compatible with the constructs of administrative processes. For example, according to Terry, "Management is the activity which plans, organizes, and controls the operations of the basic elements of men, materials, machines, methods, money, and markets . . ."[28] Newman makes a similar statement when he suggests: "The work of executives may be divided into subject fields, such as sales, production, or finance. It may also be divided into administrative processes; for example, planning, directing, and controlling."[29]

Such an integration was implicit even in Fayol's writings. After discussing the six activities (or functions) of industrial undertakings—technical, commercial, financial, security, accounting, and managerial—he stated: "To organize means building up the dual structure, material and human, of the undertaking . . . To . . . co-ordinate means building together, unifying and harmonizing all activity and effort."[30] Further:

. . . To govern is to conduct the undertaking towards its objective by seeking to derive optimum advantage from all available resources and to assure the smooth working of the six essential functions.[31]

---

[24] For example, Terry refers to planning, organizing, directing, leading, coordinating, and controlling. Newman refers to planning, organizing, assembling resources, directing, and controlling. Le Breton and Henning refer to planning, organizing, staffing, directing, coordinating, and controlling. See George R. Terry, *Principles of Management* (Homewood, Illinois: Richard D. Irwin, Inc., 1953), p. 5; William H. Newman, op. cit., p. 6; and Preston P. Le Breton and Dale A. Henning, *Planning Theory* (Englewood Cliffs, New Jersey: Prentice-Hall, Inc., 1961), pp. 3–4.

[25] The "Fayol school" has also been labeled "the management process school," the "traditional school," and the "universalist school." See Harold Koontz, op. cit., pp. 175–76.

[26] See Paul A. Samuelson, *Economics,* 5th ed. (New York: McGraw-Hill Book Company, 1961), p. 5.

[27] Koontz and O'Donnell, op. cit., p. 3, 45–46.

[28] George R. Terry, op. cit., p. 8.

[29] William H. Newman, op. cit., pp. 5–6.

[30] Henri Fayol, op. cit., p. 6.

[31] Ibid., p. 6.

Thus, Fayol seems to suggest an integration of resources and functions.

Bakke, in particular, has made considerable use of the "resource" concept in recent years. Writing in the fields of industrial relations and management theory, he refers to ". . . human, material, capital, ideational, and natural . . ." resources and to the necessity for their integration.[32]

Using a similar classification scheme, it appears useful to categorize the resources of the enterprise in the following way: human resources, financial resources, material (including capital equipment), markets[33] and all of the accumulated ideas and knowledge pertaining to these resources, including the methods and techniques of science. These resources may be equated with the following sub-processes, which, for want of a better term, we shall call "resource procurement and utilization processes," or as a way of abbreviating further, "operational processes": (a) personnel management, (b) financial management, (c) materials and production management, (d) marketing management, and (e) research, development, and engineering management.

TOWARD A MODEL OF THE ENTERPRISE

Within the framework of the concepts described above, the following definition of the organization can now be suggested: *An enterprise is an essentially-man-directed and multiple-goal-oriented network of interacting administrative and operational processes and corresponding facilitating systems, and is immersed in a broader network of processes and systems with which it interacts.*

This definition recognizes several important characteristics of organizations which, although set forth as theoretical propositions, in the author's opinion reflect the facts of enterprise life:

1. The "management" people in the enterprise are simultaneously designers, managers, and components of systems. The so-called "rank-and-file" members of the enterprise, while perhaps not participating in the design of systems nor in the direction of human components of systems, nevertheless, may be acting as "managers" of segments of systems. (The above definition does not deny the authority network of the enterprise, nor the concept of the enterprise as a socio-technical system.)

2. An organization is "essentially-man-directed" in that human planning and direction, whether efficient or not, are major characteristics of organization. The organization is only *essentially* man-directed, however,

---

[32] E. Wight Bakke, "The Concept of Social Organization," in Mason Haire, ed., *Modern Organization Theory* (New York: John Wiley and Sons, 1959), p. 37; and E. Wight Bakke, *The Human Resources Function* (Champaign, Ill.: Institute of Labor and Industrial Relations, University of Illinois, Lecture Series No. 21), pp. 3–5. Bakke also utilizes the term "process" extensively. See his "The Concept of Social Organization," op. cit., pp. 59–73; and *The Fusion Process* (New Haven: Yale University, Labor and Management Center, 1953).

[33] Markets have characteristics different from the other resources, of course. For example, markets are necessarily located in the external environment. See Talcott Parsons, "Suggestions for a Sociological Approach to the Theory of Organization—I," *Administrative Science Quarterly*, 1:62 (June 1956).

since the nature of the organization is also a function of a network of external environmental processes over which managers may have little or no control. In addition, since an organization is an "incompletely structured" system, by definition it is implied that there are some unsystematized processes in existence, some of which may not be man-directed, at least at a conscious level. It follows that some processes may be flowing in a direction contrary to enterprise goals, i.e., that some may be anti-goal oriented.

3. As is well recognized, the organization is probably always directed toward attaining multiple goals. Multiple goals increase the complexity of the network of organizational processes and systems.

4. There can be no organization without a network of systems. Systems are essential for channeling processes in the direction of, and toward the fulfillment of, enterprise goals.

5. The organization is a complex network of interdependent flows of events having to do with human planning, organizing, directing, co-ordinating, and controlling of the procurement and utilization of various human and non-human resources. Thus, we can conceptualize two major types of processes: (a) *administrative processes*, i.e., planning, organizing, directing, coordinating, and controlling; and (b) *operational processes*, i.e., personnel management, financial management, materials and production management, marketing management, and research, development, and engineering management.[34] The first category is essentially a category including activities common to all management jobs. The second category, while roughly the rationale for the departmentalization of many industrial firms, includes processes which by definition cannot be completely departmentalized.[35]

6. Not only do these administrative and operational processes interact and impinge upon each other—e.g., the flows of events pertaining to financial resources impinge on and interact with flows of events pertaining to human resources—but the systems designed to facilitate these various processes impinge upon and interact with each other.

7. While individual executives and departments may be assigned to focus on, or specialize in the administration of specific processes, no executive or department, if the assumptions described below are true, can possibly have exclusive authority for the management of one of these processes on an organization-wide basis. The chief executive (or executive committee) will typically manage all of these processes within an enterprise, however, in the sense that the chief executive will usually have authority over all subordinate executives and employees.

8. Since a given organizational process will have some elements in

---

[34] It would not be accurate to visualize these processes in two-dimensional space with the administrative processes described by horizontal lines and the operational processes by vertical lines, if such pictorial representation is equated with organization charts and departmentalization. These categories are not mutually exclusive ways of sub-dividing enterprise activities.

[35] There are, of course, many other ways to cut up the total "pie." Bakke and Argyris, for example, identify "work flow," "authority," "reward and penalty," "perpetuation," "identification," "communication," and "evaluation," as "essential organizational processes." See Bakke and Argyris, op. cit., pp. 10–11.

common with other processes, *systems designed to facilitate major processes are likely to have components which are also components of other systems.* This is not so likely to be true in the case of subsystems.

9. Any given organization will possess a unique fabric of processes and systems. It is highly unlikely that the flows of events described above and the quality and quantity of the available resources external to and within any two organizations will be identical.

10. This fabric of interacting processes and facilitating systems which is the organization cannot adequately be visualized in three-dimensional space since it is an "n-dimensional" matter. However, simple illustrations showing the mutual components of a few inter-dependent systems can be drawn.

TOWARD A MODEL OF THE MANAGER'S JOB

Assuming these characteristics of the enterprise, what, then, is the nature of the job of the manager? A number of observations seem to follow logically. It would seem that it is the task of management (not necessarily in this order) to:

1. Define the goals of the organization (and redefine them as the situation requires).
2. Assemble the human and non-human resources necessary to obtain these goals.
3. Visualize the internal processes essential to attaining enterprise goals.
4. Understand the network of external processes and systems which impinge upon the organization and its attainment of goals.
5. Design, direct, coordinate, and control internal systems in order to facilitate and/or regularize the internal processes which have been visualized. (It would seem that processes must be visualized before systems can be designed to facilitate or control them.)
6. Recognize the overlap of broad systems, identify overlapping components, and plan the network of systems so that such overlapping will produce a minimum of undesirable disturbance.
7. Recognize that a given manager will simultaneously be a component in a wide array of systems, and design systems to minimize intra- and inter-person conflict and inefficiency.
8. Define (*a*) the segments of processes and systems over which different individuals are given authority, and (*b*) the components which are managed mutually with other managers.
9. Install a new system only after careful planning of how it will be linked into the existing network of interdependent and impinging systems.
10. Identify non-planned processes to ascertain to what degree systematization would be useful, or, if these processes conflict with goal-attainment, the degree to which they can be contained or eliminated. If uncontrollable, identify their possible impact on risk-taking and goal-attainment.

11. Ascertain the degree of structuring, particularly in the area of interpersonal processes and sub-processes, which will optimize the attainment of enterprise goals. It is likely that a number of enterprise subprocesses such as interviewing, participation, management development, etc., require a minimum of systematizing and that over-systematizing in these and other areas defeats organizational purposes.[36] Similarly, the manager needs to anticipate to what degree systematization of various processes will increase or decrease flexibility in meeting internal and external demands for change.

In general, then, it would seem that *the task of the administrator is to define enterprise goals, assemble resources, visualize the processes essential to the attainment of these goals, and to design the network of facilitating systems.* It is implied, of course, that the administrator will be an active component in certain systems in this network.

Perhaps the following illustration about forest management will serve to point up the above definition of the administrator's job (this illustration applies "process" and "systems" terminology to descriptions of forest management provided by colleagues):

The management of forest typically involves many goals—for example, the production of tree and plant products, the furnishing of recreational areas, water supply and conservation, soil conservation, wildlife production and protection, grass for the grazing of domesticated animals, and streams for the production and conservation of fish life. Thus, the management of a forest requires systematic manipulation and control of a highly complex network of interacting processes, including tree, plant, animal, fish, bird, insect, fungus, and virus growth, and other processes involving fire and erosion. All must be maintained or controlled in some kind of equilibrium which will optimize the attainment of multiple goals.

The interdependence of trees, low vegetation, and deer presents a typical forest-management problem. If the tree growth in the forest is too dense, shrubs and grasses will be choked out, and the deer, dependent upon these for food, will be unable to survive. On the other hand, if trees are selectively cut and conditions become more favorable for the growth of shrubs and other low vegetation, the deer population will rapidly multiply. Eventually, unless checked in some way, the deer may exceed the food supply provided by grasses and shrubs, and before starving, will eat the young tree growth necessary for replenishing the timber crop. Thus, it becomes the task of forest management to devise systems to control various impinging processes and to maintain a dynamic equilibrium appropriate to the attainment of the pre-established multiple goals. Not only does the forest manager design systems, but through decision making and other behavior he becomes simultaneously an active component in a wide variety of systems in this interdependent and dynamic network.

This, then, is the manager's job. When subordinates are added to the picture—with the concomitant complexities—the model still applies.

---

[36] The accumulated knowledge from such fields as non-directive counseling and psychotherapy suggests that personality maturation and growth of initiative and capability can be aided by an over-structuring of certain flows of events.

THE UTILITY OF THE PROCESS-SYSTEM APPROACH

The "process-system" approach toward a model of the organization and of the manager's job seems to have utility in explaining, clarifying, or analyzing certain enterprise phenomena. For example, the approach may be used to clarify or analyze concepts about authority. Since the broad processes (and systems) which comprise the organization are interdependent and have mutual components, no one manager, unless he is the chief executive, can completely preempt the management of any general administrative process (i.e., planning, controlling, etc.,) or any operational process (i.e., financial management, personnel management, etc.) on an enterprise-wide basis. This does not mean that an executive may not be assigned general responsibility over the management of one of these processes—indeed, this may be the wisest course of action—but it does mean that he is dependent upon the cooperation of other managers who will also have authority over certain mutually-held components within this assigned process. Furthermore, he will be dependent upon the effective management of the processes which precede and follow his arbitrary segment of responsibility.

The "process-system" approach also serves to focus attention on the question of efficiency of flows of events across traditional departmental lines and between organizations, thus permitting improvements in the design of facilitating systems. The many advantages of the flow concept have been discussed by Forrester,[37] Brewer,[38] Drucker,[39] Chapple and Sayles,[40] and others.

In addition, this approach sheds some light on the conditions which seem to differentiate successful vs. unsuccessful personnel systems. For example, an analysis of the literature of incentive systems reveals that successful systems feature considerable management attention to the prevailing fabric or enterprise processes and systems, while a conspicuous ingredient in many unsuccessful incentive systems is lack of attention to existing impinging and overlapping systems and processes.[41]

---

[37] Jay W. Forrester, "Industrial Dynamics," *Harvard Business Review*, 36:37–52 (July–August 1958). See also Jay W. Forrester, *Industrial Dynamics* (New York: John Wiley and Sons, 1961).

[38] Stanley H. Brewer, *Rhocrematics: A Scientific Appraoch to the Management of Material Flows* (Seattle: University of Washington, Management Series No. 2, 1960). See also Stanley H. Brewer and James Rosenzweig, "Rhocrematics and Organizational Adjustments," *California Management Review*, 3:52–71 (Spring 1961).

[39] Peter F. Ducker, "The Economy's Dark Continent," *Fortune*, April 1962, pp. 103–265.

[40] Eliot D. Chapple and Leonard R. Sayles, *The Measure of Management* (New York: The Macmillan Company, 1961).

[41] For examples of case descriptions of successful and unsuccessful applications of incentive systems see Thomas Q. Gilson and Myron J. Lefcowitz, "A Plant-Wide Productivity Bonus in a Small Factory: Study of an Unsuccessful Case," *Industrial and Labor Relations Review*, 10:284–96 (January 1959); and Joseph N. Scanlon, "Profit Sharing Under Collective Bargaining: Three Case Studies," *Industrial and Labor Relations Review*, 2:58–74 (October 1948). Katzell also emphasizes the importance of the "surrounding circumstances or parameters." See Raymond A. Katzell, "Contrasting Systems of Work Organization," *American Psychologist*, 17:105 (February 1962).

In short, the installation of systems without planning for possible adjustments in prevailing systems and processes is likely to result in undesirable disturbances within the enterprise.

The "process-system" approach may also provide a framework for determining the extent to which event-flows relative to the utilization of human and other resources can be productively systematized, and to what degree. Finally, by focusing on *both* processes and systems, a wider range of enterprise characteristics and phenomena may be examined.

CONCLUSION

The concept of "process" has been discussed and contrasted with "system." Both have been integrated with concepts relating to general administrative processes and enterprise resources toward the development of models of the enterprise and of the administrator's job. These models seem to have utility in explaining certain intra-enterprise phenomena, and may be useful in further analysis of the enterprise and the role of the administrator.

Robert L. Swinth

25. Organizational Planning: Goal Setting in Interdependent Systems

THE PLANNER who turns to the literature on organizational structure for assistance in designing his organization will find a great number of books and studies dealing with the power issue—who should or shouldn't control whom. Recently expressed ideas on the nature of systems and individual problem-solving appear to permit the opening of a new front in organizational planning. In this paper an initial discussion of complex systems and human problem-solving behavior is translated into an organizational planning context and is related to the organizational task.

This analysis of the task permits one to draw inferences about the behavioral consequences of various goal-setting techniques and corresponding authority structures. Two main goal-setting techniques are considered: (1) assigning subgoals paralleling the subsystem task divisions and imposing a corresponding authority hierarchy to give the main goal precedence; and (2) assigning goals which bridge the subsystem interdependencies with no authority hierarchy necessary. The results of a limited experiment prove consistent with the propositions made here.

SYSTEM GOALS AND LOCAL GOALS

Herbert Simon has argued that most complex systems commonly encountered in the world are hierarchic.[1] By hierarchic he means, "a system that is composed of interrelated subsystems, each of the latter

From *Industrial Management Review*, 1966, vol. 7, pp. 57–70. Reprinted with permission of the author and the publisher. © 1966 by the Industrial Management Review Association; all rights reserved.

This research was supported in part by a Ford Foundation grant to Carnegie Institute of Technology for research on the theory of organization.

[1] Simon, [7].

being, in turn, hierarchic in structure until we reach some lowest level of elementary subsystem." He also points out that in hierarchic systems, we can distinguish between the interactions *among* subsystems, on the one hand, and the interactions *within* subsystems—i.e., among the parts of those subsystems—on the other.[2] Furthermore, the *state* of a system or subsystem is denoted by its attributes.

For example, the automobile is a hierarchic complex system. Such elementary parts as pistons, crankshaft and carburetor fit together into the subsystem, engine. The subsystems, engine, transmission, and body, combine into the system, automobile. There are interactions within subsystems—the size of the pistons affects the size required for the carburetor. And, of course, any automobile will always be in a particular state. That is, will be a Volkswagen, a Ford, or a Cadillac, etc., with a fixed set of attributes.

The state of any system is affected by the interaction among its subsystems. If one desires that the system and the subsystems be in particular states, i.e., one has a goal and subgoals, this interaction takes on a critical importance. For example, if the interaction between two subsystems (say the body and engine of an automobile), occurs in the space in the body taken up by the engine, providing sufficient space for a powerful engine may make it difficult to design a compact body. A compact body may not permit enough room for a powerful engine. The opportunity to attain the desired subgoal (or state) for any subsystem may be affected by the interaction between that subsystem and other subsystems which one has subgoals for as well.

The preceding comments illustrate an important technique in problem-solving which Newell, Shaw, and Simon have discussed in other contexts.[3] In problem-solving or task performance, the performer divides the system into subsystems and attempts to master each subsystem. That is, he steps through the problem from subsystem to subsystem because the system in beyond his capacity to attain in one step. What state, i.e., what subgoal, does he aim for in each subsystem? "In problem solving, a partial result that represents recognizable progress toward the goal plays the role of a stable sub-assembly."[4]

One might call this subgoal a local goal. It is a goal that brings the subsystem to a locally optimum state which at the same time contributes to the system goal. After he has achieved several subgoals and he turns to combining them into his main goal, or after attaining a subgoal and he desires to build on this to the next higher subgoal, the problem solver will find—if there are interdependencies between subsystems—that to achieve his main goal or higher subgoal, he must modify the states attained while optimizing his subsystems. Since he is interested only in the subgoals as he steps to higher goals, he will not hesitate to modify the state of a subsystem if it enables him to come closer to his goal, even if the subsystem is no longer at its locally optimum state.

---

[2] Ibid, p. 473.

[3] Newell, Shaw, and Simon, [6].

[4] Simon, op. cit., p. 472.

GOAL-SETTING TECHNIQUES

The insights gained from this discussion of complex hierarchic systems and human problem-solving behavior can be translated into a planning context and applied to the setting of certain organizational goals. Many tasks found in organizations require the performance of a series of operations or steps to achieve some end goal. What goals should the organizational planner assign to those who must perform these operations?

I. The Autonomous Goal

The organizational planner faced with the problem of mastering a complex system will, of course, take advantage of the inherent hierarchy—much as the problem solver does—by breaking his system down into subsystems and putting one man or group in charge of one subsystem, another in charge of a second subsystem, and so on. He breaks down the series of operations to be performed into sets and assigns these sets to the various subsystems. The result of the operations performed in one subsystem is passed to the next subsystem where further operations in the series are performed. This result is passed to the next subsystem—much like the flow down an assembly line—until the last operations have been performed and the over-all goal reached.

If the planner is still following the approach used by individuals in problem-solving, he will give each subsystem a local or autonomous goal which, if achieved, will optimize the state of the subsystem. For example, products are usually made in the production department and then passed on to the sales department to be sold. The first department may have the subgoal of minimizing the cost of production and the other the subgoal of maximizing revenue from sales. Optimizing these two subgoals should lead to the optimization of the system goal: profit.

Yet, if the resultant subsystems are interdependent because of some variable or operation common to both, a conflict may arise when the problem is solved within an organization which did not arise when the whole problem was solved by one individual. When the problem solver turned to his main goal, he could usually drop his former subgoal and, if necessary, modify the subsystem. However, the individual or group in an organization perceives this subgoal as an end in itself. Having been given only a subgoal, his limited perception of the system does not let him see that his effort is only a temporary step to a larger solution.[5] As far as he knows, the organization wants to achieve the goal he is seeking, and he will not understand and perhaps will even oppose any efforts to modify the state he has attained for his subsystem.[6] To ensure

---

[5] March and Simon make a similar point in their discussion of problem-solving in organizations. They argue that dividing up the task and assigning subgoals lead to subgoal identification and result in conflict between departments [5].

[6] It should be pointed out that this conflict is often compounded by basing an individual's rewards on how close he comes to optimizing the subgoal rather than on how well the organization does as a whole. This reward basis makes it even more difficult for him to accept modifications.

that the main goal takes precedence over any subgoals, organizations usually find it necessary to superimpose an authority hierarchy over the goal-subgoal structure. In fact, one could argue that this concern for "proper" goal precedence is one of the main reasons why organizations have authority hierarchies. Without this, one goal is just as worthwhile as any other.

To give the main goal precedence, the central authority must change the perceptions of his subordinates. That is, he must resolve the conflict between his goal and theirs. To do this, he must first be aware of their perceptions; he must then show, persuade, or order them to permit the system goal to take precedence over their subgoals; and finally, he must check up on them to make sure they have changed their perceptions and complied with his wishes. The superior will not always be able or willing or even realize that it is necessary to do this. Yet, when the conflict goes unresolved because the leader did not meet the challenge or he did but was unable to resolve the conflict, performance will be less than optimal.[7] On the other hand, attempts by the central authority to control and persuade his subordinates may lessen their motivation to perform the task. Rather than doing the work themselves, they will let the central authority do it for them.

The central authority may try to circumvent the conflicting subsystems by requiring that the results obtained by them be forwarded to him, and he then makes the final decision himself. This has its limitations because it may: (*a*) overload the center, or (*b*) lead to information transmission losses.[8]

II. The Joint Goal

When two subsystems are interdependent because an operation performed in one affects the other, the organizational planner may try the following approach instead: Even while continuing to put individuals in charge of subsystems as in I., the planner can assign goals which bridge subsystem interdependencies rather than subgoals which do not. To illustrate, if we have one man per subsystem, wherever we have an interdependency stemming from a common operation, two or more men will be given the same goal. While each is in charge of a different subsystem, all are interested in achieving the optimum state for the system as a whole.

In seeking the main goal, each is led toward perceiving the relation

---

[7] Such difficulties may be circumvented in some situations by imposing various side constraints which force one into line with the system goal. The designer of the car body may be told that a certain number of cubic inches must be left for the engine, and the engine designer may be told that the engine must not exceed a certain number of cubic inches. But even this technique may be barely tolerable. Ideally all the relevant constraints must be specified in advance, and to conceive of all of them may be very difficult except by trial and error. Furthermore, appropriate weightings must be given constraints—i.e., if one can get a large increase in engine horsepower by just barely exceeding the cubic inch restraint, it may be in the best interests of the over-all design to do so. Inflexible constraints may severely hamper creativity.

[8] Bavelas [1].

of his subsystem to the over-all system; and just as the individual problem-solver is able to modify the subgoals because his ultimate interest is in the main goal, so the members of the organization are willing to modify operations performed in their subsystems because they too are ultimately seeking the main goal rather than any subgoal. All the interdependent departments, those controlling the common variable and those affected by it, will be led—by the goal—toward the same conclusion. This goal promotes rather than interferes with perception and makes compromise or command behaviors unnecessary. The centralizing of decision making, with the consequence of lowered motivation, does not take place. In fact, participants are led toward an over-all concern with the performance of the system and an interest in the effect of their own choices on others.

The above model presumes a situation where there is interaction between subsystems. If there is no interaction or if the optimums for the various subsystems all coincide and result in the optimum for the system, then the planner's choice of goals—autonomous or joint—will not have an effect across subsystems and the model we have laid out here does not apply. Furthermore, we have presumed that sorting out the interdependencies is a manageable task and worth the effort. It may not always be; for example, steel mills which pollute the air and the general public which desires clean air may be interdependent, but rather than trying to change the goals of either group, we set up air pollution ordinances and live with the conflict.

TESTING THE MODEL

Having used various notions about complex hierarchic systems and human problem-solving behavior to build up a model of goal-setting techniques for interdependent systems, we are now ready for a limited test to see if the predicted consequences do in fact follow.

To test the model we need to observe performance and behavior in a system where the problem has been broken down into subproblems, subgoals assigned to each subproblem, an individual or subgroup put in charge of each subproblem, and an authority hierarchy imposed over the system. We also need to observe performance and behavior in a system similar to the above, except for one difference. The difference is that the individuals or subgroups in charge of each subproblem seek the system goal (the goal that bridges the interdependency between subgroups stemming from their common operation) rather than their own subgoal and are not subject to an authority hierarchy. In both environments, information should be distributed throughout the system rather than centralized, and in the first case the person at the center should have high leadership potential (he should at least be willing to resolve conflict).

The laboratory is a good place to check the model since subjects can be put in closely controlled problem-solving situations similar to those described above, and their performance and behavior can be measured and observed.

A key to the experiment is a task which can be appropriately broken down into subproblems and goals or subgoals assigned. The following example illustrates such a task.

Suppose one of the operations of a company is to cast and then machine steel shafts. Suppose further that the cost of casting and machining the steel depends mainly on its metallic properties or characteristics. Yet, the characteristics adopted for casting cannot be changed once the steel has been cast, and thus the characteristics chosen for casting determine the characteristics the steel will have for machining. The casting department can determine what the casting cost would be for each of the possible characteristics; the machining department can do likewise. Furthermore, the organizational planners are aware of all this, and their goal is to minimize the cost of casting plus machining steel.

The Experimental Environment

A means of observing behavior in the kind of situation illustrated by this example was achieved by setting up the following experiment:

Men in groups of five were brought into the laboratory, put in separate, private booths, and told to imagine that each was a department head in a company. They were given titles such as: Alpha Department, Beta Department, Gamma Department, Delta Department, and Eta Department to avoid having them assume role behavior beyond that which was specified. They were informed that their duties consisted of achieving goals and making decisions, and that they were to be evaluated on how well the company performs. They were then given a series of documents:

1. Each was given an organization chart. In the autonomous goal condition they were given Chart A in Figure 1, and were told that Alpha Department heads all other Departments, Beta and Gamma report to Alpha, and Delta and Eta report to Beta; in the joint goal condition they were given Chart B in Figure 1, and were told that it is a list of depart-

FIGURE 1

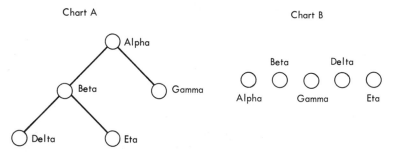

Chart A: Organization chart in autonomous goal condition.

Chart B: List of departments in joint goal condition.

ments, there is no special hierarchy, and they may communicate with anyone at any time.

    2. Each was next given a sheet which told them the goal(s) they were trying to achieve, and the decisions they had to make for each of four decision periods (each period is unrelated to the other three in any way). To illustrate a decision period, the goals and decisions of each Department in Decision Period 2 for the autonomous goal condition are shown in Figure 2. By way of comparison, the same decision period in

FIGURE 2

Autonomous Goal Condition		*Joint Goal Condition*	
Goals	*Decisions*	*Goals*	*Decisions*
Alpha Maximize the unit quality of Rads*	Quantity of Rads Quality of Rads	*Alpha* Maximize the unit quality of Rads	Quantity of Rads Quality of Rads
Beta Minimize the cost per ton of Coa *Minimize the unit cost of casting plus machining Shasts***	Tons of Coa Cost of Coa per ton	*Beta* Minimize the cost per ton of Coa	Tons of Coa Cost of Coa per ton
Gamma Minimize the cost per *unit* of Tranors	Characteristic of Tranors Cost of Tranors	*Gamma* Minimize the cost per *unit* of Tranors	Characteristic of Tranors Cost of Tranors
Delta *Minimize the unit casting cost of Shasts*	Characteristic of Shasts Casting cost of Shasts	*Delta* *Minimize the unit cost of casting plus machining Shasts*	Characteristic of Shasts Casting cost of Shasts
Eta *Minimize the unit machining cost of Shasts*	Characteristic of Shasts Machining cost of Shasts	*Eta* *Minimize the unit cost of casting plus machining Shasts*	Characteristic of Shasts Machining cost of Shasts

    * Fictitious words were used to avoid giving subjects any expectations about what properties an item was *supposed* to have in contrast to what we actually told them.
    ** The goals which differ between the two conditions are italicized.
    The goals and decisions given to each department in period two. On the left side is shown what each Department in the autonomous goal condition receives; on the right side is shown the joint goal condition.

the condition where goals bridge interdependencies is also shown in Figure 2 (joint goal condition).

    3. Each was given a sheet with information relevant to the goals and decisions of his group. The information relevant to Decision Period 2 is illustrated in Figure 3.

    Each Department was given a chart. (These charts were the same in

FIGURE 3

Information

Alpha:
Rads are used by the parent corporation.

Beta:
Coa is used by Beta Department for smelting aluminum.*
Shasts are cast in Delta Department and are then sent to Eta Department for machining.**
Shasts must be machined with the characteristics given to them in the casting process.**

Gamma:
Tranors may vary only between characteristics d and h.

Delta:
Shasts are cast in Delta Department and are then sent to Eta Department for machining.
Shasts must be machined with the characteristic given to them in the casting process.

Eta:
Shasts are cast in Delta Department and are then sent to Eta Department for machining.
Shasts must be machined with the characteristic given to them in the casting process.

---

* Given to Beta in both conditions.
** Given to Beta in the autonomous goal condition only.
Subjects in both conditions receive the same information, except where noted for Beta.

both conditions.) The Chart given in Decision Period 2 to each Department is shown in Figure 4.

The parallel to the machining and casting of steel discussed in the earlier illustration can be seen in the Machining and Casting of Shasts by Delta and Eta Departments in Period 2 in the experiment.

FIGURE 4

Charts Issued to Each Department in Decision Period 2

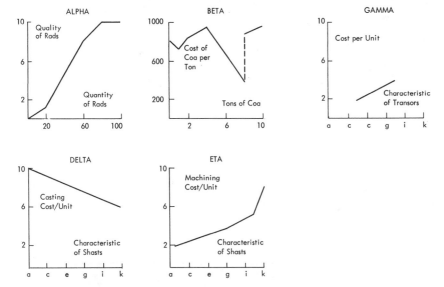

In the autonomous goal condition, Beta (the man with authority over Delta and Eta) is given the general goal of minimizing the cost of casting plus machining Shasts; Delta seeks to minimize the cost of casting, and Eta seeks to minimize the machining. Beta, Delta, and Eta all have the same information that Shasts are cast in Delta Department for machining, and that they must be machined with the characteristic given to them in the casting process. Delta has a chart showing the relationship between casting cost and characteristics of Shasts. Eta has a chart showing the relationship between machining cost and characteristics of Shasts. Beta has neither of these charts. Delta must decide what characteristic to give Shasts in the casting process and the resultant cost, while Eta must decide what characteristic to give Shasts in the machining process and the resultant cost.

In the joint goal condition, Beta has no goals or information related to Shasts. Delta and Eta each has the same goal of minimizing the cost of casting plus machining Shasts. Delta and Eta have the same information, charts, and decisions to make as Delta and Eta in the above autonomous-goal condition.

In this same decision period, Alpha, Beta, and Gamma have separate problems to solve in which their decisions are independent of any other department. They can achieve their goals by merely using their own charts and information, and they need not consult with each other.

Subjects are given two hours to go through four decision periods similar to the one just described. The decision periods are completely unrelated to each other, and in each period a different pair of departments have the interdependent problem. New tasks are used each period in all departments. That is, in Period 2 Delta and Eta might be interdependent over the cost of casting and machining Shasts as a function of its characteristics, while in Period 3 Gamma and Eta might be interdependent over the cost of producing and the revenue from selling Swishing Machines as a function of quantity. Of course, the other departments have new independent problems each time. Figure 5 shows the interdependent and independent decisions for each period. (In the auton-

FIGURE 5

Interdependent and Independent Decisions for Each Period

Departments	*Decision Periods*			
	1	2	3	4
ALPHA	*	*	*	*
BETA	Beta-Delta	*	Beta-Gamma	*
GAMMA	*	*	Beta-Gamma	Gamma-Eta
DELTA	Beta-Delta	Delta-Eta	*	*
ETA	*	Delta-Eta	*	Gamma-Eta

* = independent decision.

omous goal condition, the authority level above the highest level involved in the conflict has the system goal.)

In both conditions, the subjects are paid and told that their reward depends on how well the organization performs as a whole. That is, their earnings depend on how close their decisions are to optimum for the *over-all goal*, regardless of whether they have been assigned a subsystem goal or a system goal. Thus, anyone behaving with a limited perspective is doing so because of the goal effect, not the reward effect.

At the end of each decision period, a subject is paid one dollar for a perfect decision and correspondingly less for poorer performance (i.e., 62 percent of optimum earns 62 cents). With one decision per decision period and four decision periods, each subject could earn up to four dollars.

Subjects were masters and Ph.D. candidates in the Graduate School of Industrial Administration at Carnegie Institute of Technology. The school population was ranked from highest to lowest according to its scores on the Admission Test for Graduate Study in Business and then divided into five equal-sized groups. Scores ranged between the 99th and 46th percentiles. The highest fifth was made Betas, the second highest—Alphas, the third highest—Gammas, fourth—Deltas, fifth or lowest—Etas. This technique was used to reduce variance due to differential problem-solving ability; the Alpha Beta reversal was made because some researchers claim that the top man in an organization is smart but not the smartest.[9] Subjects were told that the experiment would be stopped at the end of two hours, regardless of whether or not they had worked through all four decision periods.

RESULTS

We now turn to an analysis of the experiment to see if behavior corresponded to that predicted by the model. The organizational goal in the interdependent problem in each decision period is the same in both conditions. For example, in Period 2 Beta in the autonomous goal condition and Delta and Eta in the joint goal condition have the same goal of minimizing the unit cost of casting plus machining Shasts. In other words, the organization is seeking the same end in both conditions, but, of course, starting from different places. Several routes are available in each condition to reach this over-all goal; some of them produce incorrect solutions. In order to understand the decision-making processes and to assess the consequences of various approaches, the author analyzed in detail the message flow of each group in each period and identified the steps taken by the participants to reach their goals.

An information-processing approach was used to represent this flow. From a single starting state in each condition the participants could move through the problem tree to the goal state along any one of several paths. One advanced his state by applying operators or means which

---

[9] Harrell [2].

moved him ahead. A general illustration of a problem tree: the paths, the states, and the means or steps are laid out in Figure 6.

The initial states and the various routes used by subjects in each condition to reach the goal are described below.

FIGURE 6

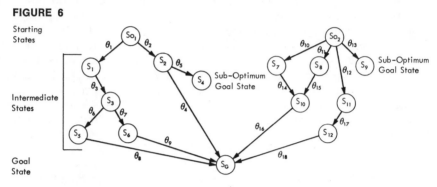

Illustration of routes through a problem tree. $S_1 =$ state i; $\theta_1 =$ the taking of step i or the application of means i to move from one state S_1 to state S_k.

Initial State in Each Decision Period in the Two Conditions

In the autonomous goal condition the central authority (CA) is seeking the general goal. One subordinate, person B, is seeking local goal B, and the other subordinate, person C, is seeking local goal C. Person B has chart B pertaining to his area, person C has chart C pertaining to his area, and all three have the same general information (See Figure 3). In the joint goal condition, persons B and C are seeking the general goal. Person B has chart B, person C has chart C, and both have the same general information.

Routes to Goal in Autonomous Condition

Route 1-A: The central authority (CA) instructs or gets person B to adopt or seek the general goal. In this intermediate state, B realizes he needs person C's chart to make a decision and sends a message to C requesting it. C complies with the request and sends his chart to B. B merges his chart with C's and finds the optimum decisions for the general goal. He then makes the part of this decision pertaining to his department, and sends a message to C telling him what decision to make in his (C's) department. C then makes his decision, as instructed. See Figure 7 for a flow diagram of the steps involved. This diagram shows the means or steps taken and omits the intermediate states. If they were shown, a state would appear between each step.

Route 2-A: The CA instructs *both* B and C to seek the general goal. B then asks C for his chart and C asks B for his. They exchange charts and independently find the over-all optimum. They then check with each

FIGURE 7

Flow Diagram of the Decision Routes through the Problem Tree of the Autonomous Goal and Joint Goal Conditions

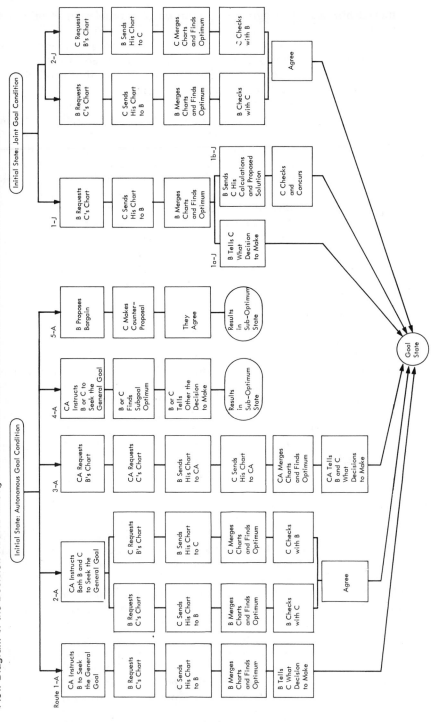

other to see that they agree, and then each makes his part of the decision.

Route 3-A: Whereas in routes 1 and 2 the decision-making process is passed down to B and/or C, in route 3 it is centralized in the CA. Here the CA asks B for his chart and C for his. He merges the two to find the optimum and tells each what decision to make.

The two routes described below (4-A and 5-A) will not lead to the goal state. In Route 4-A, neither the CA nor B nor C realizes that their decisions are interdependent. They all presume that, by maximizing the subgoals, the general goal will be maximized. Thus in period 2, B would attempt to minimize the casting cost of Shasts and C would attempt to minimize the machining costs of Shasts. In Route 5-A the CA says nothing and B and C, unaware of the general goal, attempt to reach a bargained, or compromise, decision.

Route 4-A: The CA instructs B and/or C to seek the general goal (but omits mention of their interdependence). B or C optimizes his subgoal, and C or B is idle. B tells C what decision to make, or C tells B.

Route 5-A: B makes a compromise proposal. C makes a counterproposal. They repeat the above cycle until they finally agree on a choice.

Routes to Goal in Joint Condition

Route 1-J: B realizes he needs C's chart to make a decision and sends a message to C requesting it. C sends his chart to B and B merges the two and finds the optimum decisions for the general goal. At this point the decision process branches in one of two directions: (Route 1a-J)— B then tells C what decision to make; (Route 1b-J)—B sends his calculations and proposed solution to C to check, C checks and concurs— or points out errors, if any.

Route 2-J: B asks C for his chart and C asks B for his. They exchange charts and independently find the over-all optimum. They then check with each other to see that they agree.

Eight groups were run in each condition, and each group could go through as many as four decision periods. In Table 1 we have indicated the performance of each group in each decision period. For example, group one in the autonomous goal condition took problem-solving Route 4-A in decision period 1 and reached a sub-optimum state. In period 2 their performance was 100 percent of optimum using Route 2-A. In period 3 they again suboptimized on 4-A. Time then ran out and they did not get into period 4. Group one in the joint goal condition used Route 1a-J in period 1 and performed less than optimally because of a calculation error by the decision maker. In periods 2 and 3 they hit 100 percent each time on Route 1b-J. They too had no time for period 4.

As indicated by the summary of results at the end of the table, all 16 groups finished at least 3 decision periods, and while only 2 of the autonomous goal groups finished the fourth period, 5 of the joint goal groups finished their last period.[10] Nineteen percent of the time the

---

[10] The difference in the number of groups in the two conditions finishing the fourth period is not significant (by the Fisher exact probability test).

decision-maker in the autonomous goal groups (5 of 26 decisions) did not see the interdependence between departments and chose Route 4-A. None of the joint goal groups suboptimized around one of the departments on any of their 29 decisions.[11]

A detailed look at the decision routes used in the two conditions explains the 52 percent errors (11 of 21) on the 21 decisions in the autonomous goal groups where the general goal was sought compared to the comparable 10 percent errors in the joint goal groups (3 of 29). Routes 1-A and 1-J are analogous. Ten of the 11 decisions in the autonomous goal condition using 1-A did not include the check step (see footnote 3 in Table 1) and 5 of these contained errors. Similarly 5 of the 11 joint goal condition decisions using 1-J did not include the check step and 3 of these contained errors. The one decision in 1-A in which there was a check and the 6 decisions in 1-J in which there was a check hit 100 percent. Not checking on Route 1 had about the same effect in both conditions, but far more of the autonomous decisions omitted this step.

Routes 2-A and 2-J are also comparable. Sixty-two per cent of the joint group decision-makers followed the procedure of exchanging charts, separately merging them, and then comparing their results with each other, and they made no errors. But only 27 percent of the time was this procedure followed in the autonomous groups and here in 2 of the 7 decisions, one of the parties to the decision stopped part way through the route (see footnote 4 in Table 1). Of course, this made it impossible to compare results and in both of these cases, decision errors were made. Finally, in the autonomous groups, 3 of the decisions were centralized (Route 3-A) and in two of these the party taking over made calculation errors.

Much more can be said about how the basic conflict induced by the autonomous goals was treated. As was pointed out in the statement of the model, the central authority may not fathom that the departments are interdependent and therefore let them attempt to optimize their subgoals. This happened five times (decision route 4-A). Or the central authority may perceive the existence of a conflict, yet he may do nothing. Ten of the decisions in the autonomous condition began with Route 5-A (bargaining). Here the subordinates were initially left to themselves and then shifted to another route, either after discovering the general goal on their own and found it a good way to resolve their conflict, or after the central authority initiated some control procedures.

A number of control procedures were used. Some groups used more than others but none were effectual. Three decisions were directly centralized (Route 3-A), or Alpha instructed Beta to make the decision rather than a lower level subordinate such as Delta or Eta (5 of the 11 Route 1-A decisions). Here the decision-maker (Alpha or Beta) often asked for either wrong or insufficient information, or their subordinates sent them incorrect information which they did not check or in which they had made calculation errors.

---

[11] While one would not always expect this to be so, performances in conditions I and II in this experiment were significantly different (by the chi square test at the 0.05 level).

TABLE 1

Performance of Each Group in Each Decision Period

Group	Decision Period	Autonomous Goal Condition (Route) 1-A	2-A	3-A	4-A	Joint Goal Condition (Route) 1a-J	1b-J	2-J
1	1		100%*		sub-optimum[1]	calc. error	100%[2]	100%
	2			—No decision—	sub-optimum		100%	100%
	3						—No decision—	100%
	4	100%*		100%				100%
2	1							100%
	2		100%*				100%	100%
	3	trans error					100%	100%
	4	100%*			sub-optimum			
3	1		100%*			100%		100%
	2							100%
	3	100%[3]		—No decision—				100%
	4							
4	1	trans error*				calc. error		100%
	2	100%						100%
	3			calc. error		—No decision—		100%
	4			—No decision—				100%
5	1					100%[5]		100%
	2							100%
	3		trans error*	calc. error*	sub-optimum			100%[5]
	4		trans & calc. errors[4] 100%*					100%[5]
6	1	calc. error						100%
	2							100%
	3			—No decision—	sub-optimum			100%
	4	100%*						100%
7	1		calc. error*					100%
	2							100%
	3	trans error		—No decision—				100%
	4	100%*						100%
8	1		trans error					100%
	2							100%
	3			—No decision—			100%	100%
	4	calc. error						100%
Number of decisions		11	7	3	5	5	6	18
Error or suboptimum		5	4	2	5	3	0	0
% of total decisions		42%	27%	12%	19%	17%	21%	62%
% of errors in the route		45%	57%	67%	100%	60%	0%	0%

Autonomous Goal Condition: All 8 groups completed 3 periods; 2 of these groups completed all 4.

Joint Goal Condition: All 8 groups completed 3 periods; 5 of these groups completed all 4.

Another control procedure used was to intercept messages. In seven of the autonomous goal groups some of the messages sent from one member to another were intercepted by Alpha or Beta and not allowed to reach their destination. This occurred from two to nine times in each of these groups. In about one-half of these groups, Alpha or Beta made the decisions for their subordinates on their independent problems. That is, they took the decisions out of their hands in situations where there was no conflict and where they had no unique information or special reason for making the decision. In all cases the subordinates acquiesced without complaint.

Finally, in all but six of the autonomous goal condition decisions, one or more of the following control steps were invoked by the central authority: He required that the subordinate making the decision keep him posted as to his progress. He required that the subordinate tell him what his proposed decision was before making it final. He took his subordinate's proposed decision and made it final himself. He temporarily held up information sent between subordinates. He repeated the general goal or he applied pressure to work correctly or faster, making such comments as, "What in Hell's name are you doing" "Come on, come on," "Please do not delay," "How's it coming?" "Make decision immediately."

These control procedures appeared to be ineffectual, since the proportion of errors or suboptimal decisions was about the same in the 20 decisions where they were applied as in the six where they were not (about 50 percent in each case).

In none of the joint goal condition groups did the members start off bargaining with each other. No messages were intercepted; all were allowed to reach their destinations. Nor were any of the decisions for the independent problems taken out of the hands of the assigned decision-maker.

CONCLUSIONS

The limited experimental test proved consistent with the model:
1. Using the goal-setting technique of allocating subgoals to the

Table 1 (continued)

References for Table 1
  ° The participants started out on Route 5-A (bargaining) but then shifted to this route.
  1 Explanation of code:
    100% = optimum decision, system goal achieved.
    calc. error = System goal sought but a calculation error by one of the participants caused the decision to be less than optimal.
    trans error = System goal sought but a transmission error by one of the participants caused the person making the decision to make a less than optimal choice.
    sub-optimum = the participant(s) optimize one of the subgoals resulting in a sub-optimum state being reached.
  2 One of the two participants involved in this decision initially perceived the two departments as independent, and intended to achieve the general goal by optimizing his subgoal. He was quickly set straight by the other participant who recognized their interdependence.
  3 In making this decision, person B sent his calculations to C to be checked rather than merely telling him what decision to make.
  4 Only one of the two participants completed all the steps. The other stopped part way through and therefore no checking was done.
  5 All decisions made by this group were also sent to a third party to be checked.

subsystems and imposing an authority hierarchy to give precedence to the main goals induced certain actions by men in the central positions. Either they attempted to impose control procedures and centralize the decision-making around themselves, or they avoided the conflict altogether. The former approach led to a high percentage of errors on their part; the latter resulted in subordinates making suboptimal decisions. The central members attempted both to manipulate and control their subordinates by such tactics as intercepting and stopping messages which passed over their desks, and by taking local independent decisions out of subordinates' hands. They frequently pressured them to work faster as well. The subordinates responded by withdrawing into their own portion of the system—spending little time checking to see what effect their choices had on the decisions others would reach. When asked to send information into the central men, they often sent wrong or insufficient data.

2. Using the goal-setting technique of allocating goals which bridge the subsystem interdependencies while not imposing an authority hierarchy resulted in both parties to any decision working closely together in making their choice without the necessity of an overseer. No member made any attempts to control, pressure, or manipulate other members, yet there existed a general desire to contribute to the success of the system as a whole. All the men continually checked to see what effect their choices had on others, and did not go ahead with their decisions until they were clear on the nature of the interactions. The lower percentage of errors indicates that the members provided each other with good information.

An important final question to ask is this—do the imposed differences in the two conditions produce any differences in the way the participants define their roles in relation to each other and to the task? The data analysis of routes through the problem tree indicates that they do. In the autonomous goal condition with its superimposed authority hierarchy, the participants adopted a dependent role relationship with a hierarchy of responsibility. Either responsibility was perceived as less the lower one's authority or task level, or it was compartmentalized by department such that the success of a decision was wholly one department's obligation.

On the other hand, in the joint goal condition, roles were not structured; here each felt responsible for the success of the whole system. People felt equally responsible for the task as a whole and did not adopt subordinate and superordinate roles in which one need consider only part of the task, or in which one person would take all the responsibility for a decision and the other would take none of it.

For example, the data indicate that in the autonomous goal condition almost 20% of the time decision-makers, in seeking the system goal, fell into the trap of assuming that their subsystem results were additive and that by maximizing each subsystem the system would be maximized. It is important to note that, in the joint goal condition, the participants have no additional clues to key them to the fact that the subsystems were actually interdependent rather than additive, yet none of the decisions made in these groups reflected an additive interpretation of the system goal. That is, in one case each saw his role as limited to part of the task, whereas in the other, one's responsibility was for the whole task.

The contrast in the pattern of checking and simultaneous decision-making in the two conditions (Routes 1b and 2) reveals the other part of the difference in role dependency. In the autonomous condition little cross checking was carried through (29 percent of the decisions in Routes 1, 2, or 3), indicating the responsibility was largely compartmentalized, with one person being wholly responsible. In the joint goal condition the participants generally maintained equal responsibility for the task (83% of the decisions were cross-checked).

Most studies in this field focus on organizational design in terms of the power aspect—who should or shouldn't control whom.[12] Nevertheless, the results of this experiment support the contention that models of organizational design can be at least partially based on the task to be performed, using the notions about problem-solving and task structure applied to organization contexts. Our analysis revealed a clear relationship between the task, the goal given to the participants, and the behavioral consequences.

REFERENCES

1. Bavelas, Alex, "Communication and Organization," in *Management Organization and the Computer*. George P. Schultz and Thomas L. Whisler (eds.), Free Press, Glencoe, Illinois, 1960, 119–30.

2. Harrell, T., *Managers' Performance and Personality*. South-Western Publishing Co., Chicago, Ill., 1961.

3. Koontz, H. and O'Donnell, C., *Principles of Management*. McGraw-Hill, New York, 1955.

4. Likert, R., *New Patterns of Management*, McGraw-Hill, 1961.

5. March, J. G. and Simon, H. A., *Organizations*. Wiley and Sons, New York, 1958.

6. Newell, A., Shaw, J. C., and Simon, H. A., "Elements of a Theory of Human Problem Solving." *Psychological Review*, 1958, 65, 151–66.

7. Simon, H. A., "The Architecture of Complexity." *Proceedings of the American Philosophical Society*, 1962, 106, 467–82.

---

[12] Koontz and O'Donnell [3], and Likert [4] are illustrative.

Section VII

EVALUATION OF ORGANIZATIONAL SYSTEMS

ALL ORGANIZATIONS are increasingly being held accountable by their members, management, and environmental constituents. As organizations move toward increasing differentiation in response to highly turbulent environments, evaluators of organizational performance are forced to recognize the multilevel, multigoal nature of such social systems. Traditional models for measuring organizational effectiveness are being challenged and replaced by systems evaluation models.

In the first paper of this last section, Etzioni criticizes the traditional goal model for measuring effectiveness and suggests that a system model has unique advantages over such a goal model. Among other advantages, Etzioni sees a system model as having more theoretical power and being less subject to the value judgments of the observer.

Building on Etzioni's distinction between the goal model and system model for organizational evaluation, Baker and Schulberg discuss the open systems evaluation model which they developed in the course of a long-term study of a changing mental hospital. They suggest that two methodological strategies be employed in organizational systems evaluative research: multiple operationism and the supplementation of quantitative data with qualitative data.

Yuchtman and Seashore also agree with Etzioni's criticism of goal attainment models of evaluation and propose a model for assessing organizational effectiveness based on a model of an organization as an open system. These authors emphasize the idea of treating formal organizations as distinctive, identifiable social structures which can be analyzed at their own level rather than as phenomenally incidental to the behavior of individuals. They also emphasize the transactional interdependence of the organization with its environment. Yuchtman and Seashore argue that the best expression of an organization's overall effectiveness is its performance in the exchange of scarce resources under competitive conditions. They argue for an "open-ended, multidimensional set of criteria" for the assessment of organizational effectiveness, based on a variety of resources and types of competitive relationships.

Bennis also finds fault with traditional techniques of organization evaluation and challenges the use of performance and satisfaction as criterion variables, since static output measures can be misleading as well as inadequate. He argues for the addition of information about organizational processes to supplement output measurements. Unlike Yuchtman and Seashore, he does not go so far as to substitute input measures for output criteria, but instead argues for the importance of evaluation of dynamic processes of organizational problem solving as they relate to the development of, and adaptation to, organizational goals.

Amitai Etzioni

26. Two Approaches to Organizational Analysis: A Critique and a Suggestion

ORGANIZATIONAL GOALS serve many functions. They give organizational activity its orientation by depicting the state of affairs which the organization attempts to realize. They serve as sources of legitimation which justify the organization's activities and its very existence, at least in the eyes of some participants and in those of the general public or subpublics. They serve as a source for standards by which actors assess the success of their organization. Finally, they serve as an important starting point for students of organizations who, like some of the actors they observe, use the organizational goals as a yardstick with which to measure the organization's performance. This paper[1] is devoted to a critique of this widespread practice and to a suggestion of an alternative approach.

GOAL MODEL

The literature on organizations is rich in studies in which the criterion for the assessment of effectiveness is derived from organizational goals. We shall refer to this approach as the goal model. The model is considered an objective and reliable analytical tool because it omits the values of the explorer and applies the values of the subject under study as the criteria of judgment. We suggest, however, that this model has some methodological shortcomings and is not as objective as it seems to be.

One of the major shortcomings of the goal model is that it frequently makes the studies' findings stereotyped as well as dependent on the model's assumptions. Many of these studies show (*a*) that the organization does not realize its goals effectively, and/or (*b*) that the organiza-

From *Administrative Science Quarterly*, 1960, vol. 5, pp. 257–78. Reprinted with permission of the author and the publisher.

[1] I am indebted to William Delany, William J. Goode, Terence K. Hopkins, and Renate Mayntz for criticisms of an earlier version of this paper.

tion has different goals from those it claims to have. Both points have been made for political parties,[2] trade unions,[3] voluntary associations,[4] schools,[5] mental hospitals,[6] and other organizations. It is not suggested that these statements are not valid, but it seems they have little empirical value if they can be deduced from the way the study is approached.[7]

Goals, as norms, as sets of meanings depicting target states, are cultural entities. Organizations, as systems of co-ordinated activities of more than one actor, are social systems.

There is a general tendency for cultural systems to be more consistent than social systems.[8] There are mainly two reasons for this. First of all, cultural images, to be realized, require investment of means. Since the means needed are always larger than the means available, social units are always less perfect than their cultural anticipations. A comparison of actual Utopian settlements with descriptions of such settlements in the books by the leaders of Utopian movements is a clear, although a somewhat disheartening, illustration of this point.[9]

The second reason for the invariant discrepancy between goals and social units, which is of special relevance to our discussion, is that all social units, including organizations, are multifunctional units. Therefore, while devoting part of their means directly to goal activities, social units have to devote another part to other functions, such as the creation or recruitment of further means to the goal and the maintenance of units performing goal activities and service activities.

Looking at the same problem from a somewhat different viewpoint, one sees that the mistake committed is comparing objects that are not on the same level of analysis as, for example, when the present state of an organization (a real state) is compared with a goal (an ideal state) as if the goal were also a real state. Some studies of informal organizations commit a similar mistake when they compare the blueprint of an organization with actual organizational practice and suggest that an organizational *change* has taken place. The organization has "developed" an informal structure. Actually, the blueprint organization never existed as a social fact. What is actually compared is a set of symbols on paper with a functioning social unit.[10]

---

[2] Robert Michels, *Political Parties* (Glencoe, Ill., 1949); Moise Ostrogorski, *Democracy and the Organization of Political Parties* (New York, 1902).

[3] Michels, op. cit.; William Z. Foster, *Misleaders of Labor* (Chicago, 1927); Sylvia Kopald, *Rebellion in Labor Unions* (New York, 1924).

[4] John R. Seeley et al., *Community Chest* (Toronto, 1957).

[5] A nonscientific discussion of this issue in a vocational school is included in E. Hunter's novel, *Blackboard Jungle* (New York, 1956).

[6] Ivan Belknap, *Human Problems of a State Mental Hospital* (New York, 1956), esp. p. 67.

[7] While such studies have little empirical value, they may have some practical value. Many of the evaluation studies have such a focus.

[8] Talcott Parsons, *The Social System* (Glencoe, Ill., 1951).

[9] See Martin Buber, *Paths in Utopia* (Boston, 1958).

[10] Actually, of course, in order for a blueprint to exist, a group of actors—often a future elite of the organization—had to draw up the blueprint. This future elite presumably itself had "informal relations," and the nature of these relations undoubtedly affected the content of the blueprint as well as the way the organization was staffed and so forth.

Measured against the Olympic heights of the goal, most organizations score the same—very low effectiveness. The differences in effectiveness among organizations are of little significance. One who expects a light bulb to turn most of its electrical power into light would not be very interested in the differences between a bulb that utilizes 4.5 percent of the power as compared with one that utilizes 5.5 percent. Both are extremely ineffective. A more realistic observer would compare the two bulbs with each other and find one of them relatively good. The same holds for organizational studies that compare actual states of organization to each other, as when the organizational output is measured at different points in time. Some organizations are found gradually to increase their effectiveness by improving their structure and their relations with the environment. In other organizations effectiveness is slowly or rapidly declining. Still others are highly effective at the initial period, when commitments to goals are strong, and less effective when the commitment level declines to what is "normal" for this organization. These few examples suffice to show that the goal model may not supply the best possible frame of reference for effectiveness. It compares the ideal with the real, as a result of which most levels of performance look alike—quite low.[11] Michels, for example, who applied a goal model, did not see any significant differences among the trade unions and parties he examined. All were falling considerably short of their goals.

When a goal model is applied, the same basic mistake is committed, whether the goals an organization claims to pursue (public goals) or the goals it actually follows (private goals) are chosen as a yardstick for evaluation of performance. In both cases cultural entities are compared with social systems as if they were two social systems. Thus the basic methodological error is the same. Still, when the public goals are chosen, as is often done, the bias introduced into the study is even greater.[12] Public goals fail to be realized not because of poor planning, unanticipated consequences, or hostile environment. *They are not meant to be realized.* If an organization were to invest means in public goals to such an extent that it served them effectively, their function, that is, improving the input-output balance, would be greatly diminished, and the oganization would discard them.[13] In short, public goals, as criteria, are even more misleading than private ones.

---

[11] Paul M. Harrison, *Authority and Power in the Free-Church Tradition* (Princeton, 1959), p. 6. Harrison avoids this pitfall by comparing the policy of the church he studied (The American Baptist Convention) at different periods, taking into account, but not using as a measuring rod, its belief system and goals.

[12] Some researchers take the public goals to be the real goals of the organization. Others choose them because they are easier to determine.

[13] Public goals improve the input balance by recruiting support (inputs) to the organization from groups which would not support the private goals. This improves the balance as long as this increase in input does not require more than limited changes in output (some front activities). An extreme but revealing example is supplied in Philip Selznick's analysis of the goals of the Communist party. He shows that while the private goal is to gain power and control, there are various layers of public goals presented to the rank and file, sympathizers, and the "masses" (*The Organizational Weapon* [New York, 1952], pp. 83–84).

SYSTEM MODEL

An alternative model that can be employed for organizational analysis is the system model.[14] The starting point for this approach is not the goal itself but a *working model of a social unit which is capable of achieving a goal*. Unlike a goal, or a set of goal activities, it is a model of a multifunctional unit.[15] It is assumed a priori that some means have to be devoted to such nongoal functions as service and custodial activities, including means employed for the maintenance of the unit itself. From the viewpoint of the system model, such activities are functional and increase the organizational effectiveness. It follows that a social unit that devotes all its efforts to fulfilling one functional requirement, even if it is that of performing goal activities, will undermine the fulfillment of this very functional requirement, because recruitment of means,[16] maintenance of tools, and the social integration of the unit will be neglected.[17]

A measure of effectiveness establishes the degree to which an organization realizes its goals under a given set of conditions. But the central question in the study of effectiveness is not, "How devoted is the organization to its goal?" but rather, "Under the given conditions, *how close does the organizational allocation of resources*[18] *approach an optimum distribution?*" "Optimum" is the key word: what counts is a balanced distribution of resources among the various organizational needs, not maximal satisfaction of any one activity, even of goal activities. We shall illustrate this point by examining two cases; each is rather typical for a group of organizational studies.

---

[14] Compare with a discussion of the relations between a model approach and a system approach in Paul Meadows, Models, Systems and Science, *American Sociological Review*, 22 (1957), 3–9.

[15] For an outline of a system model for the analysis of organizations see Talcott Parsons, A Sociological Approach to the Theory of Organizations, *Administrative Science Quarterly*, 1 (1956), 63–85, 225–39.

[16] The use of concepts such as goals, means, and conditions does not imply the use of a goal model as defined in the text. These concepts are used as defined on the more abstract level of the action scheme. On this scheme see Talcott Parsons, *The Structure of Social Action* (Glencoe, 1937).

[17] Gouldner distinguished between a rational model and a natural-system model of organizational analysis. The rational model (Weber's bureaucracy) is a partial model since it does not cover all the basic functional requirements of the organization as a social system—a major shortcoming, which was pointed out by Robert K. Merton in his "Bureaucratic Structure and Personality," *Social Theory and Social Structure* (rev. ed.; Glencoe, Ill., 1957), pp. 195–206. It differs from the goal model by the type of functions that are included as against those that are neglected. The rational model is concerned almost solely with means activities, while the goal model focuses attention on goal activities. The natural system model has some similarities to our system model, since it studies the organization as a whole and sees in goal realization just one organizational function. It differs from ours in two ways. First, the natural system is an observable, hence "natural" entity, while ours is a functional model, hence a construct. Second, the natural system model makes several assumptions that ours avoids, as, for example, viewing organizational structure as "spontaneously and homeostatically maintained," etc. See Alvin W. Gouldner, "Organizational Analysis," in Robert K. Merton, Leonard Broom, and Leonard S. Cottrell, Jr., eds., *Sociology Today* (New York, 1959), pp. 401 ff.

[18] "Resources" is used here in the wildest sense of the term including, for example, time and administration as well as the more ordinary resources.

Case 1: Function of Custodial Activities

One function each social unit must fulfill is adjusting to its environment. Parsons refers to this as the "adaptive phase" and Homans calls the activities oriented to the fulfillment of this function "the external system." This should not be confused with the environment itself. An organization often attempts to change some limited parts of its environment, but this does not mean that adjustment to the environment in general becomes unnecessary. The changes an organization attempts to introduce are usually specific and limited.[19] This means that, with the exception of the elements to be changed, the organization accepts the environment as it is and orients its activities accordingly. Moreover, the organization's orientation to the elements it tries to change is also highly influenced by their existing nature. In short, a study of effectiveness has to include an analysis of the environmental conditions and of the organization's orientation to them.

With this point in mind let us examine the basic assumptions of a large number of studies of mental hospitals and prisons conducted in recent years on the subject "from custodial to therapeutic care" (or, from coercion to rehabilitation). Two points are repeated in many of these studies: (1) The *goals* of mental hospitals, correctional institutions, and prisons are therapeutic. "The basic function of the hospital for the mentally ill is the same as the basic function of general hospitals . . . that function is the utilization of every form of treatment available for restoring the patients to health."[20] (2) Despite large efforts to transform these organizations from custodial to therapeutic institutions, little change has taken place. Custodial patterns of behavior still dominate policy decisions and actions in most of these organizations. "In the very act of trying to operate these institutions their *raison d'être* has often been neglected or forgotten."[21] Robert Vinter and Morris Janowitz stated explicitly:

Custody and care of delinquent youth continue to be the goals of correctional agencies, but there are growing aspirations for remedial treatment. The public expects juvenile correctional institutions to serve a strategic role in changing the behavior of delinquents. Contrary to expectations, persistent problems have been encountered in attempting to move correctional institutions beyond mere custodialism. . . . Despite strenuous efforts and real innovations, significant advances beyond custody have not been achieved.[22]

The first question the studies raise is: What are the actual organizational goals? The public may change its expectations without necessarily

---

[19] One way in which organizations can change their environment, which is often overlooked, is by ecological mobility, e. g., their textile industry moving to the less unionized South. But this avenue is open to relatively few organizations.

[20] Quoted from M. Greenblatt, R. H. York, and E. L. Brown, *From Custodial to Therapeutic Patient Care in Mental Hospitals* (New York, 1955), p. 3; see also H. L. Smith and D. J. Levinson, "The Major Aims and Organizational Characteristics of Mental Hospitals," in M. Greenblatt, D. J. Levinson, and R. H. Williams, eds., *The Patient and the Mental Hospital* (Glencoe, Ill., 1957), pp. 3–8.

[21] Greenblatt, York, and Brown, op. cit., p. 3.

[22] Effective Institutions for Juvenile Delinquents: A Research Statement, *Social Service Review*, 33 (1959), 118.

imposing a change on the organization's goals, or it may affect only the public goals. As Vinter and Janowitz suggest, much of the analysis of these organizations actually shows that they are oriented mainly to custodial goals, and with respect to these goals they are effective.[23]

But let us assume that through the introduction of mental health perspectives and personnel—psychiatrists, psychologists, social workers—the organization's goal, as an ideal self-image, changed and became oriented to therapy. We still would expect Vinter's and Janowitz's observation to be valid. Most prisons, correctional institutions, and mental hospitals would not be very effective in serving therapy goals. Two sets of reasons support this statement. The first set consists of internal factors, such as the small number of professionals available as compared to the large number of inmates, the low effectiveness of the present techniques of therapy, the limitations of knowledge, and so on. These internal factors will not be discussed here, since the purpose of this section is to focus on the second set, that of external factors, which also hinder if not block organizational change.

Organizations have to adapt to the environment in which they function. When the relative power of the various elements in the environment is carefully examined, it becomes clear that, in general, the subpublics (e.g., professionals, universities, well-educated people, some health authorities) which support therapeutic goals are less powerful than those which support the custodial or segregating activities of these organizations. Under such conditions, most mental hospitals and prisons must be more or less custodial. There is evidence to show, for example, that a local community, which is both an important segment of the organizational environment and which in most cases is custodial-minded, can make an organization maintain its bars, fences, and guards or be closed.

The [prison] camp has overlooked relations with the community. For the sake of the whole program you've got to be custodially minded. . . . The community feeling is a problem. There's been a lot of antagonism. . . . Newspapers will come out and advocate that we close the camp and put a fence around it.[24]

An attempt to change the attitudes of a community to mental illness is reported by Elaine and John Cumming. The degree to which it succeeded is discussed by J. A. Clausen in his foreword to the study. "The Cummings chose a relatively proximate goal: to ascertain whether the community educational program would diminish people's feelings

---

[23] R. H. McCleery, who studied a prison's change from a custodial to a partially "therapeutic" institution, pointed to the high degree of order and the low rate of escapes and riots in the custodial stage. See his *Policy Change in Prison Management* (East Lansing, 1957). See also Donald R. Cressey, Contradictory Directives in Complex Organizations: The Case of the Prison, *Administrative Science Quarterly*, 4 (1959), 1–19; and Achievement of an Unstated Organizational Goal: An Observation on Prisons, *Pacific Sociological Review*, 1 (Fall 1958), 43–49.

[24] Oscar Grusky, Role Conflict in Organization: A Study of Prison Camp Officials, *Administrative Science Quarterly*, 3 (1959), 452–72, quoted from p. 457. McCleery shows that changes in a prison he analyzed were possible since the community, through its representatives, was willing to support them, op. cit., pp. 30–31.

of distance and estrangement from former mental patients and would increase their feelings of social responsibility for problems of mental illness." They found that their program did not achieve these goals.[25] It should be noted that the program attempted by education to change relatively abstract attitudes toward *former* mental patients and to mental illness in general. When the rumor spread that the study was an attempt to prepare the grounds for the opening of a mental hospital in the town, hostility increased sharply. In short, it is quite difficult to change the environment even when the change sought is relatively small and there are special activities oriented toward achieving it.[26]

D. R. Cressey, addressing himself to the same problems, states: "In spite of the many ingenious programs to bring about modification of attitudes or reform, the unseen environment in the prisoner's world, with few exceptions, continues to be charged with ideational content inimical to reform."[27]

This is not to suggest that community orientation cannot be changed. But when the effectiveness of an organization is assessed at a certain point in time, and the organization studied is not one whose goal is to change the environment, the environment has to be treated as given. In contemporary society, this often means that the organization must allocate considerable resources to custodial activities in order to be able to operate at all.[28] Such activities at least limit the means available for therapy. In addition they tend to undermine the therapeutic process, since therapy (or rehabilitation) and security are often at least partially incompatible goals.[29] Under such circumstances low effectiveness in the service of therapeutic goals is to be expected.

This means that, to begin with, one may expect a highly developed custodial subsystem. Hence it seems justifiable to suggest that the focus of research should shift from the problem that, despite some public expectations, institutions fail to become primarily therapeutic to the following problems: To what degree are external and internal[30] organizational conditions responsible for the emphasis on security? Or are these conditions used by those in power largely to justify the elaboration of security measures, while the real cause for that elabora-

---

[25] See Elaine and John Cumming, *Closed Ranks* (Cambridge, Mass., 1957), p. xiv.

[26] Ibid. It is of interest to note that the Cummings started their study with a goal model (how effective is the educational program?). In their analysis they shifted to a system model (p. 8). They asked what functions, manifest and latent, did the traditional attitudes toward mental health play for the community as a social system (ch. vii). This explained both the lack of change and suggested possible avenues to future change (pp. 152–58).

[27] "Preface to the 1958 Reissue," in D. Clemmer, *The Prison Community* (New York, 1958), p. xiii.

[28] Grusky, op. cit., see also Cressey, "Foreword," to D. Clemmer, op. cit.

[29] See Cressey, Contrary Directives.

[30] It seems that some security measures fulfill internal functions as well. They include control of inmates till the staff has a chance to build up voluntary compliance and safety of other inmates, of the staff itself, of the inmate in treatment, of the institutional property, as well as others. These internal functions are another illustration of the nongoal activities that a goal approach tends to overlook and that a system approach would call attention to.

tion is to be found in the personal needs or interests (which can be relatively more easily changed and for which the organization is responsible) of part of the personnel, such as guards and administrators? To what degree and in what ways can therapy be developed under the conditions given? Do external conditions allow, and internal conditions encourage, a goal cleavage, i.e., making security the public goal and therapy the private goal of the organization or the other way around?

We have discussed the effect of the two models the researcher uses to study the interaction between the organization and its environment. We shall turn now to examine the impact each model has on the approach to the study of internal structure of the organization.

Case 2: Functions of Oligarchy

The study of authority structure in voluntary associations and political organizations is gradually shifting from a goal model to a system model. Michels' well-known study of socialist parties and trade unions in Europe before World War I was conducted according to a goal model.[31] These parties and unions were found to have two sets of goals: socialism and democracy. Both tended to be undermined: socialism by the weakening of commitments to revolutionary objectives and over-devotion to means activities (developing the organization) and maintenance activities (preserving the organization and its assets); democracy by the development of an oligarchic structure. A number of studies have followed Michels' line and supplied evidence that supports his generalizations.[32]

With regard to socialism, Michels claims that a goal displacement took place in the organizations he studied. This point has been extensively analyzed and need not be discussed here.[33] Of more interest to the present discussion is his argument on democracy. Michels holds that an organization that has *external* democracy as its goal should have an *internal* democratic structure; otherwise, it is not only diverting some of the means from goal to nongoal activities, but is also introducing a state of affairs which is directly opposed to the goal state of the organization. In other words, an internal oligarchy is seen as a dysfunction from the viewpoint of the organizational goals. "Now it is manifest that the concept *dictatorship* is the direct antithesis of the concept *democracy*. The attempt to make dictatorship serve the ends of democracy is tantamount to the endeavour to utilize war as the most efficient means for the defence of peace, or to employ alcohol in the struggle against alcoholism."[34] Michels goes on to spell out the conditions which make for this phenomenon. Some are regarded as unavoidable, some as

---

[31] Michels, op. cit.

[32] See Oliver Garceau, *The Political Life of the American Medical Association* (Cambridge, Mass., 1941); R. T. McKenzie, *British Political Parties* (London, 1955). See also note 3.

[33] Robert K. Merton, *Social Theory and Social Structure* (rev. ed.; Glencoe, Ill., 1957), pp. 199–201; Peter M. Blau, *The Dynamics of Bureaucracy* (Chicago, 1955), see index; David L. Sills, *The Volunteers* (Glencoe, Ill., 1957), pp. 62–69.

[34] Michels, op. cit., p. 401.

optional, but all are depicted as distortions undermining the effectiveness of the organization.[35]

Since then it has been suggested that internal oligarchy might be a *functional* requirement for the effective operation of these organizations.[36] It has been suggested both with regard to trade unions and political parties that conflict organizations cannot tolerate internal conflicts. If they do, they become less effective.[37] Political parties that allow internal factions to become organized are setting the scene for splits which often turn powerful political units into weak splinter parties. This may be dysfunctional not only for the political organization but also for the political system. It his also been pointed out that organizations, unlike communities and societies, are segmental associations, which require and recruit only limited commitments of actors and in which, therefore, internal democracy is neither possible nor called for. Developing an internal political structure of democratic nature would necessitate spending more means on recruitment of members' interests than segmental associations can afford. Moreover, a higher involvement on the part of members may well be dysfunctional to the achievement of the organization's goals. It would make compromises with other political parties or of labor unions with management rather difficult. This means that some of the factors Michels saw as dysfunctional are actually functional; some of the factors he regarded as distorting the organizational goals were actually the mechanisms through which the functions were fulfilled, or the conditions which enabled these mechanisms to develop and to operate.

S. M. Lipset, M. A. Trow, and J. S. Coleman's study of democracy in a trade union reflects the change in approach since Michels' day.[38] This study is clearly structured according to the patterns of a system model. It does not confront a social unit with an ideal and then grade it according to its degree of conformity to the ideal. The study sees democracy as a process (mainly as an institutionalized change of the parties in office) and proceeds to determine the external and internal conditions that enable it to function. It views democracy as a characteristic of a given system, sustained by the interrelations among the system's parts.

---

[35] The argument over the compatibility of democracy and effectiveness in "private government" is far from settled. The argument draws from value commitments but is also reinforced by the lack of evidence. The dearth of evidence can be explained in part by the fact that almost all voluntary organizations, effective and ineffective ones, are oligarchic. For the most recent and penetrating discussion of the various factors involved, see Seymour M. Lipset, "The Politics of Private Government," in his *The Political Man* (Garden City, 1960), esp. pp. 360 ff. See also Lloyd H. Fisher and Grant McConnell, "Internal Conflict and Labor Union Solidarity," in A. Kornhauser, R. Dublin, and A. M. Ross, eds., *Industrial Conflict* (New York, 1954), pp. 132–43.

[36] For a summary statement of the various viewpoints on the effect of democratic procedures on trade unions, see Clark Kerr, *Unions and Union Leaders of Their Own Choosing* (New York, 1957).

[37] Ibid.

[38] *Union Democracy* (Glencoe, Ill., 1956). See also S. M. Lipset, Democracy in Private Government, *British Journal of Sociology*, 3 (1952), 47–63; "The Political Process in Trade Unions: A Theoretical Statement," in Morroe Berger *et al.*, *Freedom and Social Control in Modern Society* (New York, 1954), pp. 82–124.

From this, a multifunctional theory of democracy in voluntary organizations emerges. The study describes the various functional requirements necessary for democracy to exist in an organization devoted to economic and social improvement of its members and specifies the conditions that have allowed these requirements to be met in this particular case.[39]

Paradox of Ineffectiveness.   An advantage of the system model is that it enables us to conceive of a basic form of ineffectiveness which is hard to imagine and impossible to explain from the viewpoint of the goal model. The goal approach sees assignment of means to goal activities as functional. The more means assigned to the goal activities, the more effective the organization is expected to be. In terms of the goal model, the fact that an organization can become more effective by allocating less means to goal activities is a paradox. The system model, on the other hand, leads one to conclude that, just as there may be a dysfunction of underrecruitment, so there may be a dysfunction of overrecruitment to goal activities, which is bound to lead to under-recruitment to other activities and to lack of co-ordination between the inflated goal activities and the depressed means activities or other nongoal activities.

Cost of System Models.   Up to this point we have tried to point out some of the advantages of the system model. We would now like to point out one drawback of this model. It is more demanding and expensive for the researcher. The goal model requires that the researcher determine the goals the organization is pursuing and no more. If public goals are chosen, they are usually readily available. Private goals are more difficult to establish. In order to find out how the organization is really oriented, it is sometimes necessary not only to gain the confidence of its elite but also to analyze much of the actual organizational structure and processes.

Research conducted on the basis of the system model requires more effort than a study following the goal model, even when private goals are chosen. The system model requires that the analyst determine what he considers a highly effective allocation of means. This often requires considerable knowledge of the way in which an organization of the type studied functions. Acquiring such knowledge is often very demanding, but it should be pointed out that (*a*) the efforts invested in obtaining the information required for the system model are not wasted since the information collected in the process of developing the system model will be of much value for the study of most organizational problems; and that (*b*) theoretical considerations may often serve as the bases for constructing a system model. This point requires some elaboration.

A well-developed organizational theory will include statements on

---

[39] Limitations of space do not allow us to discuss a third case of improved understanding with the shift from one model to another. Although apathy among members of voluntary associations as reflecting members' betrayal of their organizational goals and as undermining the functioning of the organization has long been deplored, it is now being realized that partial apathy is a functional requirement for the effective operation of many voluntary associations in the service of their goals as well as a condition of domestic government. See W. H. Morris Jones, In Defense of Apathy, *Political Studies*, 2 (1934), 25–37.

the functional requirements various organizational types have to meet. These will guide the researcher who is constructing a system model for the study of a specific organization. In cases where the pressure to economize is great, the theoretical system model of the particular organizational type may be used directly as a standard and a guide for the analysis of a specific organization. But it should be pointed out that in the present state of organizational theory, such a model is often not available. At present, organizational theory is dealing mainly with general propositions which apply equally well but also equally badly to all organizations.[40] The differences among various organizational types are great; therefore any theory of organizations in general must be highly abstract. It can serve as an important frame for specification, that is, for the development of special theories for the various organizational types, but it cannot substitute for such theories by serving in itself as a system model, to be applied directly to the analysis of concrete organizations.

Maybe the best support for the thesis that a system model can be formulated and fruitfully applied is found in a study of organizational effectiveness by B. S. Georgopoulos and A. S. Tannenbaum, one of the few studies that distinguishes explicitly between the goal and system approaches to the study of effectiveness.[41] Instead of using the goals of the delivery service organization, they constructed three indexes, each measuring one basic element of the system. These were: (a) station productivity, (b) intraorganizational strain as indicated by the incidence of tension and conflict among organizational subgroups, and (c) organizational flexibility, defined as the ability to adjust to external or internal change. The total score of effectiveness thus produced was significantly correlated to the ratings on "effectiveness" which various experts and "insiders" gave the thirty-two delivery stations.[42]

Further development of such system-effectiveness indexes will require elaboration of organizational theory along the lines discussed above, because it will be necessary to supply a rationale for measuring certain aspects of the system and not others.[43]

Survival and Effectiveness Models. A system model constitutes a statement about relationships which, if actually existing, would allow a

---

[40] The point has been elaborated and illustrated in Amitai Etzioni, Authority Structure and Organizational Effectiveness, *Administrative Science Quarterly,* 4 (1959), 43–67.

[41] A Study of Organizational Effectiveness, *American Sociological Review,* 22 (1957), 534–40.

[42] For a brief report of another effort to "dimensionalize" organizational effectiveness, see Robert L. Kahn, Floyd C. Mann, and Stanley Seashore, "Introduction" to a special issue on Human Relations Research in Large Organizations: II, *Journal of Social Issues,* 12 (1956), 2.

[43] What is needed from a methodological viewpoint is an accounting scheme for social systems like the one Lazarsfeld and Rosenberg outlined for the study of action. See Paul F. Lazarsfeld and Morris Rosenberg, eds., *The Language of Social Research* (Glencoe, Ill., 1955), pp. 387–491. For an outstanding sample of a formal model for the study of organizations as social systems, see Allen H. Barton and Bo Anderson, "Change in an Organizational System: Formalization of a Qualitative Study," in Amitai Etzioni, ed., *Complex Organizations: A Sociological Reader* (New York, forthcoming).

given unit to maintain itself and to operate. There are two major sub-types of system models. One depicts a *survival model,* i.e., a set of requirements which, if fulfilled, allows the system to exist. In such a model each relationship specified is a prerequisite for the functioning of the system, i.e., a necessary condition; remove any one of them and the system ceases to operate. The second is an *effectiveness model.* It defines a pattern of interrelations among the elements of the system which would make it most effective in the service of a given goal.[44]

The difference between the two submodels is considerable. Sets of functional alternatives which are equally satisfactory from the view-point of the survival model have a different value from the viewpoint of the effectiveness model. The survival model gives a "yes" or "no" score when answering the question: Is a specific relationship functional? The effectiveness model tells us that, of several functional alternatives, some are more functional than others in terms of effectiveness. There are first, second, third, and n choices. Only rarely are two patterns full-fledged alternatives in this sense, i.e., only rarely do they have the same effectiveness value. Merton discussed this point briefly, using the con-cepts functional alternatives and functional equivalents.[45]

The majority of the functionalists have worked with survival models.[46] This has left them open to the criticism that although society or a social unit might change considerably, they would still see it as the same system. Only very rarely, for instance, does a society lose its ability to fulfill the basic functional requirements. This is one of the reasons why it has been claimed that the functional model does not sensitize the researcher to the dynamics of social units.[47]

James G. March and Herbert A. Simon pointed out explicitly in their outstanding analysis of organizational theories that the Barnard-Simon analysis of organization was based on a survival model:

---

[44] For many purposes, in particular for the study of ascriptive social units, two submodels are required: one that specifies the conditions under which a certain *structure* (pattern or form of a system) is maintained, another which specifies the conditions under which a certain level of activities or *processes* is maintained. A model of effectiveness of organizations has to specify both.

[45] Robert K. Merton, *Social Theory and Social Structure* (rev. ed.; Glencoe, Ill., 1957), p. 52; see last part of E. Nagel's essay, "A Formalization of Functionalism," in *Logic without Metaphysics* (Glencoe, Ill., 1957).

[46] One of the few areas in which sociologists have worked with both models is the study of stratification. Some are concerned with the question: is stratification a necessary condition for the existence of society? This is obviously a question of the survival model of societies. Others have asked: which form of stratification will make for the best allocation of talents among the various social positions, will maximize training, and minimize social strains? Those are typical questions of the effectiveness model. Both models have been applied in enlightening debate over the function of stratification; see Kingsley Davis, A Conceptual Analysis of Stratification, *American Sociological Review,* 7 (1952), 309–21; Kingsley Davis and Wilbert E. Moore, Some Principles of Stratification, ibid., 10 (1954), 242–49; Melvin W. Tumin, Some Principles of Stratification: A Critical Analysis, ibid., 18 (1953), 387–94; Kingsley Davis, Reply, ibid., 394–97; W. E. Moore, Comment, ibid., 397. See also Richard D. Schwartz, Functional Alternatives to Inequality, ibid., 20 (1955), 424–30.

[47] For a theorem of dynamic functional analysis, see Amitai Etzioni and Paul F. Lazarsfeld, "The Tendency toward Functional Generalization," in *Historical Materials on Innovations in Higher Education,* collected and interpreted by Bernhard J. Stern (forthcoming).

The Barnard-Simon theory of organizational equilibrium is essentially a theory of motivation—a statement of the conditions under which an organization can induce its members to continue participation, and hence assure organizational *survival*. . . . Hence, an organization is "solvent"—and will continue in *existence*—only so long as the contributions are sufficient to provide inducements in large enough measure to draw forth these conditions.[48] [All italics supplied.]

If, on the other hand, one accepts the definition that organizations are social units oriented toward the realization of specific goals, the application of the effectiveness model is especially warranted for this type of study.

MODELS AND NORMATIVE BIASES

The goal model is often considered as an objective way to deal with normative problems. The observer controls his normative preferences by using the normative commitments of the actors to construct a standard for the assessment of effectiveness. We would like to suggest that the goal model is less objective than it appears to be. The system model not only seems to be a better model but also seems to supply a safety measure against a common bias, the Utopian approach to social change.

Value Projection

In some cases the transfer from the values of the observer to those of the observed is performed by a simple projection. The observer decides a priori that the organization, group, or public under study is striving to achieve goals and to realize values he favors. These values are then referred to as the "organizational goals," "public expectations," or "socitety's values." Actually they are the observer's values projected onto the unit studied. Often no evidence is supplied that would demonstrate that the goals are really those of the organization. C. S. Hyneman pointed to the same problem in political science:

A like concern about means and ends is apparent in much of the literature that subordinates description of what occurs to a development of the author's ideas and beliefs; the author's ideas and beliefs come out in statements that contemporary institutions and ways of doing things do not yield the results that society of a particular public anticipated.[49]

---

[48] *Organizations* (New York, 1958), p. 84. See also Gouldner, op. cit., p. 405, for a discussion of "organization strain toward survival." Theodore Caplow developed an objective model to determine the survival potential of a social unit. He states: "Whatever may be said of the logical origins of these criteria, it is a reasonable assertion that no organization can continue to exist unless it reaches a minimal level in the performance of its objective functions, reduces spontaneous conflict below the level which is distributive, and provides sufficient satisfaction to individual members so that membership will be continued" (The Criteria of Organizational Success, *Social Forces*, 32 [1953], 4).

[49] Charles S. Hyneman, Means/Ends Analysis in Policy Science, *PROD*, 2 (March 1959), 19–22.

Renate Mayntz makes this point in her discussion of a study of political parties in Berlin. She points out that the functional requirements which she uses to measure the effectiveness of the party organization are derived from *her* commitments to democratic values. She adds: "It is an empirical question how far a specific political party accepts the functions attributed to it by the committed observer as its proper and maybe noblest goals. From the point of view of the party, the primary organizational goal is to achieve power."[50]

There are two situations where this projection is likely to take place: one, when the organization is publicly, but not otherwise, committed to the same goals to which the observer is committed; the other, when a functional statement is turned from a hypothesis into a postulate.[51] When a functionalist states that mental hospitals have been established in order to cure the mentally ill, he often does not mean this as a statement either about the history of mental hospitals or about the real, observable, organizational goals. He is just suggesting that *if* the mental hospitals pursued the above goal, this *would* be functional for society. The researcher who converts from this "if-then" statement to a factual assertion, "the goal is . . . ," commits a major methodological error.

But let us now assume that the observer has determined, with the ordinary techniques of research, that the organization he is observing is indeed committed to the goals which he too supports; for instance, culture, health, or democracy. Still, the fact that the observer enters the study of the organization through its goals makes it likely that he will assume the position of a critic or a social reformer, rather than that of a social observer and detached analyst. Thus those who use the goal model often combine "understanding" with "criticizing," an approach which was recommended and used by Marx but strongly criticized and rejected by Weber. The critique is built into the study by the fact that the goal is used as the yardstick, a technique which, as was pointed out above, makes organizations in general score low on effectiveness scales.[52]

Effects of Liberalism

The reasons why the goal model is often used and often is accompanied by a critical perspective can be explained partially by the positions of those who apply it. Like many social scientists, students of organizations are often committed to ideas of progress, social reform,

---

[50] "Party Activity in Postwar Berlin," in Dwaine Marvick, ed., *Political Decision Makers* (forthcoming).

[51] On this fallacy, see Hans L. Zetterberg, *On Theory and Verification in Sociology* (New York, 1954), esp. pp. 26 ff.

[52] One of the reasons that this fallacy does not stand out in organizational studies is that many are case studies. Thus each researcher discovers that his organization is ineffective. This is not a finding which leads one to doubt the assumption one made when initiating the study. Only when a large number of goal-model studies are examined does the repeated finding of low effectiveness, goal dilution, and so on lead one to the kind of examination which has been attempted here.

humanism, and liberalism.[53] This normative perspective can express itself more readily when a goal model is applied than when a system model is used. In some cases the goal model gives the researcher an opportunity to assume even the indignant style of a social reformer.

Some writers suggested that those who use the system models are conservative by nature. This is not the place to demonstrate that this contention is not true. It suffices to state here that the system model is a prerequisite for understanding and bringing about social change. The goal model leads to unrealistic, Utopian expectations, and hence to disappointments, which are well reflected in the literature of this type. The system model, on the other hand, depicts more realistically the difficulties encountered in introducing change into established systems, which function in a given environment. It leaves less room for the frustrations which must follow Utopian hopes. It is hard to improve on the sharp concluding remark of Gresham M. Sykes on this subject:

> Plans to increase the therapeutic effectiveness of the custodial institution must be evaluated in terms of the difference between what is done and what might be done—and the difference may be dishearteningly small. . . . But expecting less and demanding less may achieve more, for a chronically disillusioned public is apt to drift into indifference.[54]

Intellectual Pitfall

Weber pointed out in his discussion of responsibility that actors, especially those responsible for a system, such as politicians and managers, have to compromise. They cannot follow a goal or a value consistently, because the various subsystems, which they have to keep functioning as well as integrated, have different and partially incompatible requirements. The unit's activity can be assured only by concessions, including such concessions as might reduce the effectiveness and scope of goal activities (but not necessarily the effectiveness of the whole unit). Barnard made basically the same point in his theory of opportunism.

Although the structural position of politicians and managers leads them to realize the need to compromise, the holders of other positions are less likely to do so. On the contrary, since these others are often responsible for one subsystem in the organization, they tend to identify with the interests and values of their subsystem. From the viewpoint of the system, those constitute merely segmental perspectives. This phenomenon, which is sometimes referred to as the development of departmental loyalties, is especially widespread among those who represent goal activities. Since their interests and subsystem values come closest

---

[53] A recent study of social scientists by P. F. Lazarsfeld and W. Thielens, Jr., demonstrates this point, *The Academic Mind* (Glencoe, Ill., 1958). Some additional evidence in support of this point is presented in S. M. Lipset and J. Linz, "The Social Bases of Political Diversity in Western Democracies" (in preparation), ch. xi, pp. 70–72.

[54] "A Postscript for Reformers," in his *The Society of Captives* (Princeton, 1958), pp. 133–34.

to those of the organization as a whole, they find it easiest to justify their bias.

In systems in which the managers are the group most committed to goal activities (e.g., in profit-making organizations), this tendency is at least partially balanced by the managers' other commitments (e.g., to system integration). But in organizations in which another personnel group is the major carrier of goal activities, the ordinary intergroup difference of interests and structural perspectives becomes intensified. In some cases it develops into a conflict between the idealists and the compromisers (if not traitors). In professional organizations such as mental hospitals and universities, the major carriers of goal activities are professionals rather than administrators. The conflict between the supporters of therapeutic values and those of custodial values is one case of this general phenomenon.[55]

So far the effect of various structural positions on the actors' organizational perspectives has been discussed. What view is the observer likely to take? One factor which might affect his perspective is *his* structural position. Frequently, this resembles closely that of the professional in professional organizations. The researcher's background is similar to that of the professionals he studies in terms of education, income, social prestige, age, language, manners, and other characteristics. With regard to these factors he tends to differ from managers and administrators. Often the researcher who studies an organization and the professionals studied have shared years of training at the same or at a similar university and have or had friends in common. Moreover, his position in his organization, whether it is a university or a research organization, is similar to the position of the physician or psychologist in the hospital or prison under study.[56] Like other professionals, the researcher is primarily devoted to the goal activities of his organization and has little experience with, understanding of, or commitment to, nongoal functions. The usual consequence of all this is that the researcher has a natural sympathy for the professional orientation in professional organizations.[57] This holds also, although to a lesser degree, for professionals in other organizations, such as business corporations and governmental agencies.

Since the professional orientation in these organizations is identical with the goal orientation, the goal model not only fails to help in checking the bias introduced by these factors but tends to enhance it. The system model, on the other hand, serves to remind one (*a*) that social units cannot be as consistent as cultural systems (*b*) that goals are

---

[55] Another important case is the conflict between intellectuals and politicians in many Western societies. For a bibliography and a recent study, see H. L. Wilensky, *Intellectuals in Labor Unions* (Glencoe, Ill., 1956).

[56] These similarities in background make communication and contact of the researcher with the professionals studied easier than with other organizational personnel. This is one of the reasons why the middle level of organizations is often much more vividly described than lower ranking personnel or top management.

[57] Arthur L. Stinchcombe pointed out to the author that organizations whose personnel includes a high ratio of professionals are more frequently studied than those which do not.

serviced by multifunctional units, and hence intersubsystem concessions are a necessary prerequisite for action, (c) that such concessions include concessions to the adaptive subsystem which in particular represents environmental pressures and constraints, and (d) that each group has its structural perspectives, which means that the observer must be constantly aware of the danger of taking over the viewpoint of any single personnel group, including that of a group which carries the bulk of the goal activities. He cannot consider the perspective of any group or elite as a satisfactory view of the organization as a whole, of its effectiveness, needs, and potentialities. In short, it is suggested that the system model supplies not only a more adequate model but also a less biased point of view.

Frank Baker
Herbert C. Schulberg

27. A Systems Model for Evaluating the Changing Mental Hospital

THIS PAPER presents an evaluation model developed by the authors in the course of conducting a five-year in-depth study of the processes occurring in a state mental hospital as it underwent changes which were to result in a more community-oriented facility. As we initiated our research in 1965, a review of the literature revealed two major approaches for studying the nature and extent of change within the mental hospital: a sociological process model and a goal-attainment evaluation model.

Sociological Process Model

This conceptual approach was largely favored by medical sociologists who considered the mental hospital to be a peculiar kind of social grouping which had emerged to meet specific societal needs. Sociologists were first encouraged to study the mental hospital by the availability of research funds in the 1950s, and by that organization's widespread accessibility as a handy field laboratory for the study of general social processes (Friedson, 1963). The mental hospital was of interest to sociologists for additional reasons: (*a*) hospitals with a treatment orientation welcomed studies whose findings could lead to improved services; (*b*) large captive groups of subjects were available for a variety of research purposes; (*c*) greater status for sociologists was achieved in this setting than in business and industrial settings; and (*d*) humanitarian social values and a concern for the alleviation of social ills could be expressed within this environment (Perrow, 1965). The varying lures of the mental hospital motivated sociologists to do

Preparation of this material was partially supported by NIMH Grants MH-09214 and MH-18382. It was prepared especially for this volume.

considerable amounts of research in that social setting. On the whole, however, most sociological research in the mental hospital had been focused upon a limited series of organizational characteristics, e.g., the nature of hierarchical "nondemocratic" structures and patterns of interpersonal relations among patients and staff. Thus a clear understanding of certain key aspects of the mental hospital's functioning had been achieved but these constituted only a fragment of the organization's complex operations.

The model for sociological research on psychiatric institutions was established by Stanton and Schwartz's two-year intensive study of Chestnut Lodge, a small, private psychiatric hospital. Their research focused on the interaction patterns of staff with staff, and staff with patients, rather than the organizational goals of the institution. The findings were published in 1954 in a book generically entitled, "The Mental Hospital," which placed a great emphasis on the communication problems created by an archaic formal structure.

Influenced by the work of Stanton and Schwartz, Caudill (1958) undertook a similar study at the Yale Psychiatric Institute. He described his work as an "attempt to take seriously the idea that the hospital is a small society, and that the ongoing function of such a society affects the behavior of the people who make it up in many ways in which they are unaware" (p. 3). The hospital reorganized its treatment procedures and increasingly emphasized psychodynamic principles during the ten-month period of Caudill's anthropological observation and interviewing. Caudill described the mental hospital as an encapsulated social system and emphasized that events occurring in one part of the institution have ramifications throughout the system. Like Stanton and Schwartz, Caudill restricted his view of the total organization to its internal problems of human relations, and he highlighted the stresses associated with a hierarchically organized social structure.

Even though the studies conducted by Stanton and Schwartz and by Caudill were of small, private, elite hospitals, their findings about the key problems of psychiatric institutions could easily be generalized to large public facilities. Belknap's (1956) research at a state mental hospital in Texas uncovered similar communication difficulties, incompatible hierarchical structures, and sharp differences in orientation and motivation among staff groups.

An additional major sociological study is *Asylums* by Erving Goffman, who conducted a year's field work at St. Elizabeth's Hospital in Washington, D.C., in an effort to "learn about the social world of the hospital inmate, as this world is subjectively experienced by him" (1962, p. ix). From his ethnographic study of patients' social life, Goffman developed dramatic insights into the nature of mental hospitals in the mid 1950s. Goffman saw these facilities as having a predominantly closed "organizational character," similar to prisons, concentration camps, army barracks, and monasteries because they had little or no contact with the external community. In order to dramatize the importance of this characteristic, Goffman labeled the mental hospital as a "total institution," since it is a "place of residence and work where a large number of like-situated individuals cut off from the wider society for an appreciable period of time, together lead an enclosed, formally administered round of life" (1962, p. xiii).

Goffman's description of the mental hospital in the 1950s as a self-contained total institution was misleading, despite its value in high-lighting the organization's relative isolation from the community and the patient's segregation from society. Goffman failed to note the variety of interactions between hospital and community and the many ways in which they are interdependent. Furthermore, although patients lead a circumscribed existence and iatrogenic maladjustment occurs, the majority of staff members live outside of the mental hospital and continue to be influenced by the social and professional pressures of the larger community.

In summary, the predominant emphasis which the early sociological studies of mental hospitals placed on human relations may be viewed as an extension of the sociologist's approach to study of the difficult plight of workers in industrial organizations (Etzioni, 1960b; Friedson, 1963). This past mental hospital research emphasized intraorganizational processes, assuming that the key to major problems lay in an analysis of internal structure and functions. Quantitative measures of the program's operations were given minimal consideration and the influence of the environment was largely ignored, as were such other significant variables as the availability of people, technology, and money. Thus, this purely sociological approach contained profound limitations for our study of a mental hospital which was striving to become a community mental health center.

Goal-Attainment Evaluation Model

The primary objective of the research just described was to enhance sociological knowledge. Although such organizational studies may produce direct implications for the operation of mental health programs, they differ from evaluative research, which is specifically directed toward the improved planning and operation of psychiatric services. This latter approach to the study of the mental hospital assumes that organizations are designed and operated to achieve specific predetermined objectives. By extending research techniques developed in the experimental laboratory into natural social settings, it is possible to determine whether objectives and goals are attained (Schulberg & Baker, 1968; Schulberg, Sheldon & Baker, 1969).

Assuming that a mental hospital's goals could be specified, goal-attainment evaluation could then be performed at one of a variety of levels which James (1970) classified in the following way:

1. Evaluation of effort: How do the practices of the hospital under study compare with local or national standards? Such yardsticks as patient-staff ratios and numbers of patients treated may be utilized on the assumption that the effort produces positive results. The more

untenable this assumption is, the more limited this approach to program assessment appears to be.

2. Evaluation of performance: What effects have the hospital program's efforts produced? Did the actual outcome attain the anticipated level specified in the program's objectives?
3. Adequacy of performance: To what extent has the hospital's total problem been solved by this program? From this perspective, services affecting only a minority of patients are less adequate than those focused on the total population.
4. Evaluation of efficiency: Can the same result be achieved at lower cost? Differing therapies may be evaluated in this manner by determining whether the same objectives can be attained more economically.

The goal-attainment model of evaluation has been very commonly used and many studies of psychiatric facilities have been based upon this approach, e.g., Ullman's (1967) analysis of the Veterans Administration Hospital's effectiveness in discharging newly admitted patients. This model presented an easily conceived relationship between a specific service and an ultimate effect while the methodology used in assessing outcome gradually was drawn from the well-established designs of experimental researchers.

Despite these advantages, the goal-attainment approach also contains key disadvantages. A major difficulty lies in identifying the independent variable or "cause" of an effect. As we proceeded with our study of a changing mental hospital, it soon became evident that this organization's effectiveness in achieving program objectives was not dependent upon the impact of any one variable. Rather, the hospital's programs were affected by the interaction of multiple, interdependent variables, and the same outcome could be attained by manipulating differing initial conditions in varying ways. For example, a reduced inpatient census could be achieved by limiting new admissions and/or discharging chronic patients, depending upon staff and community receptivity to these strategies. Thus, it seemed essential also to study the sociological processes associated with the successful, as well as unsuccessful, pursuit of specific objectives so that at a later time we could understand our findings and determine the extent to which they were generalizable to other changing mental hospitals.

An early example of mental health program evaluation which profited by combining process analysis and goal-attainment measurement is found in the work of Cumming and Cumming (1957). They studied an extensive six-month campaign by a team of psychiatrists, sociologists, and social workers which was intended to alter the attitudes of a Canadian community toward the mentally ill. At the end of the six-month educational program, virtually no change was found in the attitudes of the people in either the experimental or control towns. The Cummings found that in investigating the nature of the educational process they became aware of both the manifest and latent functions served by traditional mental health attitudes for the community as a social system. With this understanding, the researchers were able to explain the failure of their effort and to suggest avenues for producing attitudinal change in the future.

In summary, then, the sociological process model and the goal-attainment model were the evaluation precedents found in the mental health literature as we became immersed in our study of a state mental hospital. Even during the early phases of this study, however, the need to develop sophisticated indicators of outcome criteria, as well as descriptions of the processes by which the hospital searched for, adapted to, and achieved its changing goals, was clear to us. Therefore, any conceptual model we employed in studying the changing mental hospital had to relate the major variables of the organization's internal functioning, as well as account for the interrelations of the hospital and its community. Thus, the study of the hospital's role as a community mental health center required simultaneous analysis of intraorganizational, extraorganizational, and interorganizational processes.

OPEN SYSTEMS THEORY

In reviewing the literature of the late fifties and early sixties on organizational theory, one finds the expression of growing dissatisfaction with the then prevailing viewpoint that complex institutions can be studied primarily as closed social structures, that is, from an intraorganizational perspective. This approach was deemed fragmentary and overly restricted, and by the early 1960s organizational researchers were turning more and more to open systems theory as the more meaningful framework for the study of institutions with multiple functions (Emery & Trist, 1965; Katz & Kahn, 1966; Rice, 1958, 1963; Scott, 1964).

It is possible within a system to analyze the structure and processes of smaller units which singly, or in combination with other units, play specialized roles in the operation of the larger system (Miller, 1965b). In an open system, at least three identifiable subsystems are associated with its cyclic character: an input subsystem which admits resources, e.g., the admitting unit in a mental hospital; a conversion subsystem which modifies inputs, e.g., a therapeutic ward milieu for patient treatment; and an output subsystem which returns the system's products to the environment, e.g., a discharge and aftercare unit. The operating flow through the subsystems of a simple organization, i.e., the intraorganizational process, is presented in Figure 1.

Just as a living system may be analyzed in terms of its component subsystems, so it may also be viewed as part of a higher level system in which it then plays a subsystem role (Miller, 1965a). Thus, if a mental hospital is considered as a component of the broader caregiving network, its actions must then be understood within the context of this suprasystem of which it is a part. Again, this opens systems construct is relevant to the study of the changing mental hospital, since it permitted us to focus on the interorganizational processes of the caregiving network in which Boston State Hospital participated.

FIGURE 1

Operating Process Flow of a Simple Formal Organizational System*

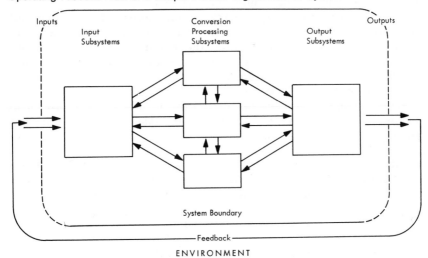

ENVIRONMENT

* This is a simplified and highly stylized diagram of the input-throughput-output-feedback flow through a formal organizational system. It is not intended to represent the full complexity of most organizations, and for the sake of simplicity does not show managerial subsystems or informal communication structures.

Although shown as a single line, the feedback loop represents numerous feedback channels. Internal feedback is represented by reversed arrows ($\rightleftarrows$).

As Etzioni (1960a) observed, the relationship of the systems model and the goal-attainment model of evaluation are neither divorced from nor in opposition to each other. Both models emphasize that any organization must assemble its human and material resources in such a way that goals can be achieved; the systems model, however, clarifies a basic confusion that is difficult to explain within the perspective of the goal-attainment model. Applying the latter framework, we would have predicted increased organizational effectiveness as additional resources were committed to specific goals. The systems model recognizes, however, that organizational resources must be allocated to activities other than those directly involved in goal achievement. For example, the mental hospital must divert some of its scarce personnel from direct patient treatment to maintain the physical plant, operate a medical records room, etc., if it is to function effectively. It is entirely conceivable that resources may be overallocated to goal activities, resulting in impairment of the organization's ability to maintain an integrated unity. Etzioni cautioned that "What counts is a balanced distribution of resources among the various organizational needs, not maximal satisfaction of one activity" (p. 262).

The open systems model is concerned with establishing a working design of a mental hospital which is capable of achieving a variety of goals, such as patient care, research, and training. Unlike the goal-attainment model of evaluation, which limits its assessment to the measurement of specified outcome criteria, the open systems model goes further by seeking to determine the degree to which an organization realizes its goals under a given set of conditions. What counts from this perspective is a balanced distribution of resources among all organizational objectives rather than the maximal attainment of any single goal. Thus a mental hospital would not be considered maximally effective if reorganization were possible that would improve a major part of the organization without damaging any other part.

Four dynamic characteristics of the mental hospital are crucial as they interact in determining the state of the system at any given time. These basic dimensions are:

1. Input: patients and their demographic characteristics and clinical histories; staff and their goals, ideologies, reference groups, abilities, and habits; financial resources; legitimation; and technology.
2. Throughput structures and processes: the internal operating and management subsystems of a hospital, including units, departments, and roles.
3. Output: the numbers and types of patients who leave the hospital, the students trained, research reports, and other distributed information.
4. Feedback: the patients, other people, and information which are returned to the hospital from other environmental systems.

These four groups of variables, as detailed in Table 1, are interdependent and interrelated and so must be considered together in evaluating the change which occurs in an organizational system when modifications are made in the pattern of its component parts. If, for example, the treatment units of a mental hospital and the professional roles of its staff are altered, the open systems model leads us to evaluate the impact of these changes on the system's output mix.

Change may also result from the fact that the environments of the mental hospital, like those of other organizational systems in modern society, are ever more turbulent. Consequently, the hospital possesses a decreasing degree of autonomy, and it is affected by change which tends, at an increasing rate, to be externally induced. The hospital's success in obtaining needed supplies, providing relevant services, and surviving in a complicated environment depends upon its ability to develop internal subsystems which allow the organization to adapt more favorably to environmental changes. If, however, reactive rather than proactive changes are the primary modality for handling these environmental pressures, they are unlikely to be adaptive in the long run, and the organization ultimately will be at a disadvantage in meeting the requirements of its input and output constituencies.

The study of a changing mental hospital which uses an open systems model as its conceptual guide is neither simple nor inexpensive. On the contrary, the researcher's task of considering the complex variety of interdependent variables which make up an organizational system is unusually complicated and extended. Additional problems also occur and are related to developing the procedural methods and data-gathering techniques to fulfill the requirements of our conceptual framework.

Methods of Procedure

Thus far we have described the many foci of concern in an open systems analysis of a changing mental hospital. Implicit in this ap-

TABLE 1

Major Groups of Variables in Model of the Mental Hospital as an Open Organizational System

Input	*Throughput Structures*	*Output*
I. Patients	I. Patient care subsystems	I. Discharged patients
A. Demographic characteristics	A. Patient admission	II. Trained or educated personnel
B. Personal histories	B. Therapy provision	III. Research
II. Staff and students	1. Verbal	A. Information distributed
A. Professionals, students, and other personnel	2. Somatic	IV. Change in level of health of the population
1. Professional skills, personality, attitudes, interests, goals, ideologies, habits, values, reference groups	3. Activity	V. Patient maintenance programs
	C. Custodial care provision	VI. Community education
	D. Community placement	
III. Legitimation	II. Training subsystems	
A. Political	A. Education of professionals	
B. Legal	1. Residents	
C. Financial	2. Nurses	
IV. Technology	3. Social workers	
V. Material	4. Psychologists	
A. Money	B. Training of nonprofessionals	
B. Supplies	1. Aides	
C. Facilities	2. Other nonprofessional personnel	
	III. Research subsystems	
	IV. Boundary-spanning subsystems	
	A. Aftercare	
	B. Primary prevention programs	
	C. Community relations	
	V. Managerial & administrative Subsystems	
	A. Receiving & processing information	
	B. Planning	
	C. Decision making	
	D. Integrating	
	E. Controlling & directing	
	F. Personnel functions	
	G. Physical plant maintenance	

Feedback: Patients, other people and information returned to the hospital from other systems in the environment.

proach is the realization that to understand the functioning of a total system, and to allow for the emergence of a complete *gestalt,* it is necessary to integrate numerous observational perspectives and data-collecting modalities. Such a study's research design, therefore, should incorporate two key methodological principles: (1) multiple operationism, and (2) the qualitative supplementation of quantitative data.

The strategy of multiple operationism recognizes that each methodological procedure employed to operationalize a variable adds particular types of error variance, i.e., each discrete research procedure adds its own "noise to the system" (Webb, *et al.,* 1966). This "noise" arises from the method of observation employed rather than from the phenomena being observed. By employing several maximally different methods for measuring a particular variable, the researcher obtains increased assurance that he is triangulating on the variable of concern. Since each method differs in the type of "noise" that it adds, the overlap or common variance across different methods (be they field observations, focused interviews, structured questionnaire surveys, or epidemiological analyses of patient cohorts) in combination, produces a more complete and accurate view of the organizational "elephant."[1]

A second methodological principle, the supplementing of quantitative data with qualitative material, emerges from several considerations pertinent to our open systems conceptual model which direct one's attention to an unusually broad array of systemic variables and levels of organizational analysis. Although statistically "harder" data generally are more highly valued by behavioral scientists because of their greater objectivity and reliability, rarely can such information reflect more than a skeletal impression of a complex state of social reality. Furthermore, in contrast to laboratory research, most organizational research tends to encounter difficulties in assembling all of the needed quantitative indices. Therefore, qualitative materials become invaluable in "fleshing out the skeleton" of the organization under study and in providing a keener understanding of the social processes affecting ultimate statistical measurements. For example, quantitative data are likely to be available about the input and output characteristics of the mental hospital's patients, but field observations provide a fuller picture of how the institution's throughput, or clinical structures, affect these variables. Admittedly, this type of qualitative data is more expensive to collect and requires more of the researcher's time than other types. It does,

---

[1] An Indian fable recorded by the Englishman John Godfrey Saxe (Cooper, F. T., Ed., *An Argosy of Fables.* New York: F. A. Stokes, 1921, pp. 306–8) recounts the experience of six blind men of Indostan who, upon encountering an elephant for the first time, came to six quite different conclusions about the nature of the beast. Each of these conclusions were dependent upon the limitations inherent in the different approaches of the six men. One, encountering the side, concluded that it was a wall; another, feeling the tusk, said it was a spear; the third thought the trunk was like a snake; the fourth, bumping into the leg, concluded it was a tree; the fifth, who chanced to touch an ear, said it was a fan; and the sixth thought the tail was a rope. They disputed loud and long, and, failing to put together the data from their different approaches, did not realize that each was partly right, and all were wrong. The moral is that, with organizations as with elephants, different methodologies yield only a partial knowledge which must be combined to yield a more complete picture of the whole system being observed.

however, have the quality of demanding less of the respondent's personal involvement. Since continued access to the mental hospital and the community under study are key issues in such a project's design, every effort was therefore made to conserve what we considered a scarce resource—respondents' cooperation in providing research data. Whereever feasible, nonreactive, indirect measures should be substituted for direct data collection procedures.

Blau and Scott (1962) have suggested that research designs be adapted to the level of the organization to be explained. The demand for complete, dense information often leads the investigator to choose the case study or the small sample as the preferred research design (Weiss, 1968). Consistent with these considerations, a study of a changing mental hospital might use field research designs to gain an overall picture of the organization as well as to obtain information about the interdependence of its constituent parts. For example, sample survey designs might be employed in studies concerned with characteristics of organizational members; epidemiological analyses might be conducted with prospective and retrospective patient cohorts to determine patient flow into and out of the hospital; and various types of organizational documents might be analyzed to supplement data obtained through observation, interviews, and questionnaires. In general, then, the following four types of data might be gathered, each offering a unique and meaningful vantage point for studying the many facets of a changing mental hospital:

1. *Field Observations.* A major source of data on organizational processes could consist of field observations made of formal and informal interaction among organization members. The physical presence of an observer may lead people to become self-conscious. This may create some problems in small groups where the researcher's non-participant behavior is highlighted. An attempt to blend into the background may tend at first to make staff group members uncomfortable and cause them to question the uses to which field notes would be put. Over time, however, the staff's initial suspicion and resistance will probably diminish and not create a significant obstacle to data collection.

2. *Interviews.* Senior staff and department heads might constitute a panel of informants who could be interviewed periodically over a long-term period of the study. Additional interview data could be obtained from other organizational members, as well as from key individuals in other components of the community's caregiving network. The interviews might be openended, as opposed to structured, oral questionnaires, and focused on particular substantive areas. They should be sufficiently flexible to permit the researcher to follow interesting leads. Openended interviews may often become the basis for developing more structured survey instruments.

3. *Surveys.* Questionnaires and structured interviews might be administered periodically to large groups of hospital staff in order to provide comparable data on a variety of variables across organizational subpopulations. The survey approach might be used to conduct studies

such as staff perceptions of organizational goals, degree of staff adherence to community mental health ideology, and staff attitudes toward unitization.

4. *Institutional Records.* Patient medical records provide a major source of archival data about a hospital's inputs and outputs. In addition, other institutional records such as memoranda, minutes of meetings, staff documents, annual reports, and grant applications might be subjected to content analysis. A clipping file from local newspapers could be maintained throughout the study and provide a source of valuable information about hospital-community relations as well as a barometer of general environmental conditions during the study's history.

Statistical Analysis Problems

Since a systems model emphasizes the reciprocal interdependencies of variables, usual bivariate statistical analyses must be supplemented by more appropriate techniques, including multivariate analysis and path analysis. Randell (1966) has addressed himself to the problem of applying appropriate multivariate techniques to the study of systems and suggested that canonical analysis is of value in apportioning the causal contributions of inputs, constraints, and treatments to outcomes. Canonical methods give the maximum correlation between two sets of variables and thus indicate the degree of association that could be obtained between systems variables and outcome measures if the set of criteria were optimally weighted.

In developing an understanding of the statistical relationships between variables in a systems model, recent discussions of linear causal models offer further intriguing possibilities. Blalock (1964) was one of the first to advocate the use of path analysis in clarifying complex causal arguments of the type encountered in the systems model. Diagrammatic representations of the systems are constructed and the numerical values of path coefficients are inserted to indicate the nature of the relationship between each determining variable and the variable dependent on it. Werts (1968) has described the advantages of a path diagram in bringing to light previously overlooked relationships that might well be brought into the theoretical analysis. As a technique for developing heuristic and evaluation models of complicated systems, path analysis offers great promise for systems analysis of data since it allows the researcher to deal with the reciprocal causality which is basic to the systems model.

Multivariate analysis of variance, discriminant analysis, and computer simulation are all useful procedures in systems studies. However, much work still remains to be done in developing and adopting adequate methodological procedures for examination of the interdependencies of variables examined in a systems model of program evaluation. Lacking knowledge of adequate procedures, the researcher must often fall back to bivariate analysis of variables taken two at a time.

In summary, a systems model is complex, expensive, and difficult to implement. However, on the basis of their experience, the authors believe that it provides a total perspective which is necessary to the objectives of some studies and that other researchers will find the effort worthwhile if they truly wish to try to understand the systemic complexity of modern organizations.

REFERENCES

Belknap, I. *Human Problems of a State Mental Hospital.* New York: McGraw-Hill, 1956.

Blalock, H. *Causal Inferences in Non-Experimental Research.* Chapel Hill: University of North Carolina, 1964.

Blau, P. M. & Scott, W. R. *Formal Organizations: A Comparative Approach.* San Francisco: Chandler, 1962.

Caudill, W. *The Psychiatric Hospital as a Small Society.* Cambridge: Harvard University Press & Commonwealth Fund, 1958.

Cumming, E. & Cumming, J. *Closed Ranks: An Experiment in Mental Health.* Cambridge: Harvard University Press, 1957.

Emery, F. E. & Trist, E. L. The Causal Texture of Organizational Environments. *Human Relations*, 1965, *18*, 21–32.

Etzioni, A. Two Approaches to Organizational Analysis: A Critique and a Suggestion. *Administrative Science Quarterly*, 1960a, 5, 257–78.

———— Interpersonal and Structural Factors in the Study of Mental Hospitals. *Psychiatry*, 1960b, 23, 13–22.

Friedson, E. Medical Sociology: A Trend Report. *Current Sociology*, 1963 (10 & 11).

Goffman, E. *Asylums: Essays on the Social Situation of Mental Patients and Other Inmates.* Garden City, N.Y.: Doubleday Anchor, 1962.

James, G. Evaluation in Public Health Practice. *American Journal of Public Health*, 1962, 52, 1145–54.

Katz, D. & Kahn, R. L. *The Social Psychology of Organizations.* New York: Wiley, 1966.

Miller, J. G. Living Systems: Basic Concepts. *Behavioral Science*, 1965a, *10*, 193–237.

———— Living Systems: Structure and Process. *Behavioral Science*, 1965b, *10*, 337–79.

Perrow, C. Hospitals, Technology, Structure, and Goals. In J. G. March (Ed.), *Handbook of Organizations*, 1965, pp. 910–71.

Randell, G. A Systems Approach to Industrial Behavior. *Occupational Psychology*, 1966, *40*, 115–27.

Rice, A. K. *The Enterprise and Its Environment.* London: Tavistock, 1963.

Rice, A. K. *Productivity and Social Organization: The Ahmedabad Experiment.* London: Tavistock, 1958.

Schulberg, H. C. & Baker, F. Program Evaluation Models and the Implementation of Research Findings. *American Journal of Public Health*, 1968, *58*, 1248–55.

Schulberg, H. C., Sheldon, A. S. & Baker, F. (Eds.) *Program Evaluation in the Health Fields.* New York: Behavioral Publications, 1969.

Scott, W. S.   Organization Theory: An Overview and an Appraisal.   *Academy of Management Journal,* 1964, 4, 7–26.

Stanton, A. H. & Schwartz, M. S.   *The Mental Hospital.*   New York: Basic Books, 1954.

Ullman, L.   *Institution and Outcome: A Comparative Study of Psychiatric Hospitals.*   New York: Pergamon Press, 1967.

Webb, E. J., Campbell, D. T., Schwartz, R. D. & Sechrest, L.   *Unobtrusive Measures: Nonreactive Research in Social Sciences.*   Chicago: Rand McNally, 1966.

Weiss, R.   Issues in Holistic Research.   In H. Becker, B. Geer, D. Riesman & R. Weiss (Eds.) *Institutions and the Person,* Chicago: Aldine, 1968.

Werts, C.   Path Analysis: Testimonial of a Proselyte.   *American Journal of Sociology,* 1968, 73, 509–12.

Ephraim Yuchtman
Stanley E. Seashore

28. A System Resource
Appı ch to Organizational
Effectiveness*

WE ARE badly in need of an improved conceptual framework for the description and assessment of organizational effectiveness. Nearly all studies of formal organizations make some reference to effectiveness; the growing field of comparative organizational study depends in part upon having some conceptual scheme that allows comparability among organizations with respect to effectiveness and that guides the empirical steps of operationalization and quantification.

Aside from these needs of social scientists, consideration should also be given to the esthetic and applied requirements of organization managers. They experience high emotional involvement, pleasurable or otherwise, in the assessment of the relative success of their organizations; they are, of course, intensively and professionally engaged, informally, in the formulation and testing of hypotheses concerning the nature of decisions and actions that alter organizational effectiveness. They need a workable conception of "effectiveness" to sustain their egos and their work.

The social scientist designing or interpreting an organizational study is presently in a quandary. Most of the research concerned with the problem has been devoted to the study of the *conditions* under which organizations are more or less effective. The classic paradigm consists of some measurement of effectiveness—productivity or profit, for example—as the dependent variable, and of various sociological and social-psychological measures as the independent variables. The independent variables are usually treated in a relatively sophisticated manner; little attention, however, has been given to the concept of effectiveness itself. The latter remains conceptually a vague construct;

* The preparation of this paper was financially supported by the National Science Foundation under Grant GS–70.

From *American Sociological Review,* 1967, vol. 32, pp. 891–903. Reprinted with permission of the authors and the American Sociological Association.

in consequence there is available a large amount of empirical data with little understanding of these data. As stated recently by Katz and Kahn:

There is no lack of material on criteria of organizational success. The literature is studded with references to efficiency, productivity, absence, turnover, and profitability—all of these offered implicitly or explicitly, separately or in combination, as definitions of organizational effectiveness. Most of what has been written on the meaning of these criteria and on their interrelatedness, however, is judgmental and open to question. What is worse, it is filled with advice that seems sagacious but is tautological and contradictory.[1]

Similar conclusions, on the same or on different grounds, have been reached by other students of organizations.[2] While emphasizing different aspects of the problem, all agree that results from studies of organizational effectiveness show numerous inconsistencies, and are difficult to evaluate and interpret, let alone compare. The inconsistencies arise, often, from discrepant conceptions of "organizational effectiveness." In the present paper an attempt is made, first, to show some of the limitations inherent in traditional approaches to organizational effectiveness and, second, to provide an improved conceptual framework for dealing with that problem.

TRADITIONAL APPROACHES TO ORGANIZATIONAL EFFECTIVENESS

In spite of the variety of terms, concepts, and operational definitions that have been employed with regard to organizational effectiveness, it is hardly difficult to arrive at the generalization that this concept has been traditionally defined in terms of goal attainment. More specifically, most investigators tend implicitly or explicitly to make the following two assumptions: (1) that complex organizations have an ultimate goal ("mission," "function") toward which they are striving and (2) that the ultimate goal can be identified empirically and progress toward it measured. In fact, the orientation to a specific goal is taken by many as the defining characteristic of complex organizations. A few organizational theorists[3] avoid making these assumptions, but they represent the exception rather than the rule.

Beyond these two common assumptions, however, one may discern

---

[1] Daniel Katz and Robert L. Kahn, *The Social Psychology of Organizations,* New York: Wiley, 1966, p. 149.

[2] Basil S. Georgopoulos and Arnold S. Tannenbaum, "A Study of Organizational Effectiveness," *American Sociological Review,* 22 (October 1957), pp. 534–40; Mason Haire, "Biological Models and Empirical Histories of the Growth of Organizations," in Mason Haire, ed., *Modern Organization Theory,* New York: Wiley, 1959, pp. 272–306; Amitai W. Etzioni, "Two Approaches to Organizational Analysis: A Critique and a Suggestion," *Administrative Science Quarterly,* 5 (September 1960), pp. 257–78; Robert M. Guion, "Criterion Measurement and Personnel Judgments," *Personnel Psychology,* 14 (Summer 1961), pp. 141–49; Charles Perrow, "Organizational Goals," in *International Encyclopedia of Social Sciences,* 1964 edition, pp. 854–66; Stanley E. Seashore, "Criteria of Organizational Effectiveness," *Michigan Business Review,* XVII (July 1965), pp. 26–30.

[3] James G. March and Herbert A. Simon, *Organizations,* New York: Wiley, 1958; Etzioni, op. cit.; Perrow, op. cit.; Seashore, op. cit.; Katz and Kahn, op. cit.

different treatments of the matter, especially with regard to the rationale and operations for identifying the goals of organizations. It is useful to distinguish between two major doctrines in this respect. The first may be called the "prescribed goal approach." It is characterized by a focus on the formal charter of the organization, or on some category of its personnel (usually its top management) as the most valid source of information concerning organizational goals. The second may be referred to as the "derived goal approach." In it the investigator derives the ultimate goal of the organization from his (functional) theory, thus arriving at goals which may be independent of the intentions and awareness of the members. The prescribed and derived doctrines will be referred to as the *goal approach* and the *functional approach,* respectively.

THE GOAL APPROACH TO ORGANIZATIONAL EFFECTIVENESS

The goal approach, which itself has taken many forms, is the most widely used by students of organizations. Some have adopted it only as part of a broader perspective on organizations.[4] Others have employed it as a major tool in their study of organizations.[5] The goal approach has been attacked recently on various grounds. Katz and Kahn, while noting that ". . . the primary mission of an organization as perceived by its leaders furnishes a highly informative set of clues," go on to point out that:

Nevertheless, the stated purpose of an organization as given by its by-laws or in the reports of its leaders can be misleading. Such statements of objectives may idealize, rationalize, distort, omit, or even conceal some essential aspects of the functioning of the organization.[6]

The goal approach is often adopted by researchers because it seems to safeguard them against their own subjective biases. But Etzioni attacks precisely this assumption:

The (goal) model is considered an objective and reliable analytical tool because it omits the values of the explorer and applies the values of the subject under study as the criteria of judgment. We suggest, however, that this model has some methodological shortcomings, and it is not as objective as it seems to be.[7]

---

[4] Chester I. Barnard, *The Functions of the Executive,* Cambridge: Harvard University Press, 1938; Peter F. Drucker, *The Practice of Management,* New York: Harper, 1954.

[5] Robert Michels, *Political Parties,* Glencoe, Ill.: The Free Press, 1949; William J. Baumol, *Business Behavior, Value and Growth,* New York: Macmillan, 1959; James K. Dent, "Organizational Correlates of the Goals of Business Management," *Personnel Psychology,* 12 (Autumn 1959), pp. 365–93; Carl M. White, "Multiple Goals in the Theory of the Firm," in Kenneth E. Boulding and W. Allen Spivey, editors, *Linear Programming and the Theory of the Firm,* New York: Macmillan, 1960, pp. 181–201; Bertram M. Gross, "What Are Your Organization's Objectives? A General-Systems Approach to Planning," *Human Relations,* 18 (August 1965), pp. 195–216.

[6] Katz and Kahn, op. cit., p. 15.

[7] Etzioni, op. cit., p. 258.

Furthermore, argues Etzioni, the assessment of organizational effective-ness in terms of goal attainment should be rejected on theoretical con-siderations as well:

> Goals, as norms, as sets of meanings depicting target states, are cultural entities. Organizations, as systems of coordinated activities of more than one actor, are social systems.[8]

We understand this statement as rejecting the application of the goal approach in the study of organizational effectiveness for two reasons: first, goals as ideal states do not offer the possibility of realistic assess-ment; second, goals as cultural entities arise outside of the organization as a social system and cannot arbitrarily be attributed as properties of the organization itself. A similar criticism is offered by Starbuck, who calls attention to a hazard in the inferring of organizational goals from the behavior of organizational members:

> To distinguish goal from effect is all but impossible. The relation between goals and results is polluted by environmental effects, and people learn to pursue realistic goals. If growth is difficult, the organization will tend to pursue goals which are not growth oriented; if growth is easy, the organiza-tion will learn to pursue goals which are growth oriented. What one observes are the learned goals. Do these goals produce growth, or does growth produce these goals?[9]

It should be noted that the authors cited above tend to treat the prob-lem as a methodological one even though, as we will show, theoretical differences and uncertainties are present as well. In order to escape some of these methodological shortcomings, several investigators have attempted to reply upon inferential or impressionistic methods of goal identification. Haberstroh, for example, makes the distinction between the formal objectives and the "common purpose" of the organization, the latter serving as the "unifying factor in human organizations."[10] But how, one may wonder, can that factor be empirically identified? Haberstroh maintains that it can be discovered through a systematic inquiry into the communication processes of the organization and by knowledge of the interests of its leadership, especially those in key positions. An empirical investigation conducted in accordance with that advice resulted in a list of operational (task) goals that, according to the investigator's own acknowledgment, do not adequately represent his notion of the "common purpose" of the organization. The latter remains therefore a rather vague concept and, it may be added, not surprisingly so. If one assumes that Haberstroh's "common purpose" stands for those objectives that are shared by the organization's members, he is reminded

---

[8] Etzioni, op. cit., p. 258.

[9] William H. Starbuck, "Organizational Growth and Development," in James G. March, ed., *Handbook of Organizations*, Chicago: Rand McNally, 1965, p. 465.

[10] Chadwick J. Haberstroh, "Organization Design and Systems Analysis," in James G. March, ed., *Handbook of Organizations*, Chicago: Rand McNally, 1965, pp. 1171–1211.

by several students of organizations[11] that such objectives are generally highly ambiguous, if not controversial, and therefore difficult to identify and measure.

The same kind of criticism can be applied to those who rely on the organization's charter, whether formal or informal, as containing the main identifying features of the organization, including its goals. Such an approach is represented by Bakke; he refers to the organization's charter, in the broad sense of the term, as expressing ". . . the image of the organization's unique wholeness." Such an image is created by ". . . selecting, highlighting, and combining those elements which represent the *unique* whole character of the organization and to which uniqueness and wholeness all features of the organization and its operations tend to be oriented."[12] The reader is left puzzled about how to discover the goals of the organization even after knowing that they are contained somewhere in the "image of the organization's unique wholeness."

The difficulty of identifying the ultimate goal of an organization is illustrated by some of the research on mental hospitals and other "total" institutions, as discussed by Vinter and Janowitz and, particularly, by Perrow and Etzioni.[13] Many of these institutions have been judged to be ineffective since they fail to achieve their presumed therapeutic goals. Vinter and Janowitz demonstrate, however, that the goal of therapy is held only by a limited segment of the public, and that the institutions themselves are oriented mainly to custody, not therapy.

Etzioni elaborates upon this issue as follows:

When the relative power of the various elements in the environment are carefully examined, it becomes clear that, in general, the sub-publics (e.g., professionals, universities, well-educated people, some health institutions) which support therapeutic goals are less powerful than those which support the custodial or segregating activities of these organizations. Under such conditions, most mental hospitals and prisons must be more or less custodial.[14]

This observation, like Starbuck's argument quoted above, amounts to saying that organizational goals are essentially nothing more than courses of action imposed on the organization by various forces in its environment, rather than preferred end-states toward which the organization is "striving." Such a perspective on the nature of organizational goals seems to undermine the rationale behind the use of goals as a yardstick for assessing organizational effectiveness. How, we may ask,

---

[11] Abraham D. H. Kaplan, Joel B. Dirlam, and Robert F. Lanzillotti, *Pricing in Big Business,* Washington: Brookings Institution, 1958; Richard M. Cyert and James G. March, "A Behavioral Theory of Organizational Objectives," in Mason Haire, ed., *Modern Organization Theory,* New York: Wiley, 1959, pp. 76–90.

[12] E. Wight Bakke, "Concept of the Social Organization," in Mason Haire, ed., *Modern Organization Theory,* New York: Wiley, 1959, pp. 16–75.

[13] Robert Vinter and Morris Janowitz. "Effective Institutions for Juvenile Delinquents: A Research Statement," *Social Service Review,* 33 (June 1959), pp. 118–30; Charles Perrow, "The Analysis of Goals in Complex Organizations," *American Sociological Review,* 26 (December 1961), pp. 854–66.

[14] Etzioni, op. cit., p. 264.

can a given social unit be regarded as "effective" if it cannot even determine its goals for itself, i.e., if the reference is wholly to the needs of entities other than itself? It would seem that the capacity of an organization to attain its own goals is a consideration of higher priority than that of success in attainment of imposed goals. An adequate conceptualization of organizational effectiveness cannot therefore be formulated unless factors of organization-environment relationships are incorporated into its framework.

Finally, it is not only in its eternal environment that the organization is faced with a variety of forces exerting influence on its behavior. The organization itself is composed of a large variety of individuals and groups, each having its own conceptions about any claims on the organization. The managers of an organization do not wholly agree among themselves about the organizational goals; in addition it is not certain that these goals, even if agreed upon, would prevail. This complicated reality is highlighted by the analysis of Cyert and March. They warn against the confusion in understanding organizational behavior whenever any one individual or group, such as the top management, is selected to represent the organization as a whole:

The confusion arises because ultimately it makes only slightly more sense to say that the goal of the business organization is to maximize profit than it does to say that its goal is to maximize the salary of Sam Smith, Assistant to the Janitor.[15]

These considerations, taken together, seem to cast a serious doubt on the fruitfulness of the goal approach to organizational effectiveness. This is not to suggest that the concept of organizational goals should be rejected *in toto*. For certain analytical purposes it is useful to abstract some goal as an organizational property. In the study of persons in organizational settings, the concept of goal is useful and perhaps essential.[16] In the study of organizational effectiveness, however, the goal approach has appeared as a hindrance rather than as a help.

THE FUNCTIONAL APPROACH TO ORGANIZATIONAL EFFECTIVENESS

The functional approach to organizational effectiveness can be characterized as "normative" in the sense that the investigator reports what the goals of an organization are, or should be, as dictated by the logical consistency of his theory about the relationship among parts of larger social systems. From this point of view, the functional, or derived goal, approach has an important advantage over the prescribed goal doctrine since it appears to solve the problem of identifying the ultimate goals of complex organizations: Given the postulates and premises of the functional model about the nature of organizations and their inter-

---

[15] Cyert and March, op. cit., p. 80.

[16] Alvin F. Zander and Herman M. Medow, "Individual and Group Levels of Aspiration," *Human Relations*, 16 (Winter 1963), pp., 89–105; Alvin F. Zander and Herman M. Medow, "Strength of Group and Desire for Attainable Group Aspirations," *Journal of Personality*, 33 (January 1965), pp. 122–39.

connectedness with the total social structure, one can derive from it the specific goals of an organization, or of a class of organizations. This is evident mainly in the work of Parsons, one of the outspoken advocates of functional analysis, in his suggestions for a theory of organizations.[17] The Parsonian scheme also illustrates, however, a major weakness inherent in the functional approach. This weakness can be usefully discussed in terms of "frames of reference."

Organizations, or other social units, can be evaluated and compared from the perspectives of different groups or individuals. We may judge the effectiveness of an organization in relation to its own welfare, or we may assess how successful the organization is in contributing to the well-being of some other entities. While the selection of a given frame of reference is a question of one's values and interests, the distinction among them must be clearly made and consistently adhered to. Vital as this requirement appears to be, one encounters various treatments of effectiveness that implicitly or explicitly refer to different frames of reference interchangeably, as if effectiveness from the point of view of the organization itself is identical with, or corresponds to, effectiveness viewed from the vantage point of some other entity, such as a member, or owner, or the community, or the total society.

The point of departure for Parsons' analysis of complex organizations is the "cultural-institutional" level of analysis. Accordingly, "The main point of reference for analyzing the structure of any social system is its value pattern. This defines the basic orientation of the system (in the present case, the organization) to the situation in which it operates; hence, it guides the activities of participant individuals."[18] The impact of the value pattern, furthermore, is felt through institutional processes which ". . . spell out these values in the more concrete functional contexts of goal-attainment itself, adaptation to the situation, and integration of the system."[19] These functional prerequisites, including the value pattern, are universally present in every social system. Their specific manifestations and their relative importance, however, vary according to the defining characteristic of the system and its place in the superordinate system. In the case of complex organizations, their defining characteristic is the primacy of orientation to the attainment of a specific goal. This goal, like all other organizational phenomena, must be legitimated by the value pattern of the organization. The nature of this legitimation is a crucial element in Parsons' analysis; the following quotation shows its relevance for the present discussion as well:

Since it has been assumed that an organization is defined by the primacy of a type of a goal, the focus of its value-system must be the legitimation of this goal in terms of the functional significance of its attainment for the superordinate system, and secondly, the legitimation of the primacy of this

---

[17] Talcott Parsons, "Suggestions for a Sociological Approach to a Theory of Organizations—I," *Administrative Science Quarterly*, 1 (June 1956), pp. 63–85; Talcott Parsons, *Structure and Processes in Modern Societies*. New York: The Free Press, 1960, pp. 16–96.

[18] Parsons, op. cit., 1956, p. 67.

[19] Parsons, op. cit., 1956, p. 67.

goal over other possible interests and values of the organization and its members.[20]

In terms of our analysis, this states explicitly that the focal frame of reference for the assessment of organizational effectiveness is not the organization itself but, rather, the superordinate system. Not only must the ultimate goal of the organization be functionally significant in general for that system but, in the case of a conflict of interests between it and the organization, the conflict is always resolved in favor of the superordinate system—since the value pattern of the organization legitimates only those goals that serve that system. In other words, the *raison d'être* of complex organizations, according to this analysis, is mainly to benefit the society to which they belong, and that society is, therefore, the appropriate frame of reference for the evaluation of organizational effectiveness. In order to avoid misunderstanding in this respect the following illustration is provided by Parsons:

> For the business firm, money return is a primary measure and symbol of success and is thus part of the goal structure of the organization. But it cannot be the primary organizational goal because profit-making is not by itself a function on behalf of the society as a system.[21]

Now there is no argument that the organization, as a system, must produce some important output for the total system in order to receive in return some vital input. However, taking the organization itself as the frame of reference, its contribution to the larger system must be regarded as an unavoidable and costly requirement rather than as a sign of success. While for Parsons the crucial question is "How well is the organization doing for the superordinate system?", from the organizational point of view the question must be "How well is the organization doing for itself?"

It was suggested earlier that a major weakness of the goal approach has been its failure to treat the issue of organizational autonomy in relation to organizational effectiveness. This seems to be the Achilles heel of the functional approach as well. In Parsons' conception of organizations, and of social systems in general, there exists the tendency to overemphasize the interdependence among the parts of a system and thus, as argued by Gouldner, fail ". . . to explore systematically the significance of variations in the degree of interdependence," ignoring the possibility that ". . . some parts may vary in their dependence upon one another, and that their interdependence is not necessarily symmetrical."[22]

Gouldner's proposition of "functional autonomy" may be examined on several different levels. For example, one may regard the organization itself as the total system, looking for variations in the degree of autonomy among its own parts; this has been the focus of Gouldner's analysis. But the same line of analysis can be attempted at a different

---

[20] Parsons, op. cit., 1956, p. 68.

[21] Parsons, op. cit., 1956, p. 68.

[22] Alvin W. Gouldner, "Organizational Dynamics," in Robert K. Merton et al., eds., *Sociology Today*, New York: Basic Books, 1959, p. 419.

level, where society is taken as the total system. Here the investigator may be exploring variations in the degree of autonomy of various parts and sub-systems, an instance of which are complex organizations. Such an analysis underlies the typology offered by Thompson and McEwen, in which the relations between organizations and their environments are conceived in terms of the relative autonomy, or dominance, of the organization vis-à-vis its environment.[23]

The proposition of functional autonomy implies that organizations are capable of gearing their activities into relatively independent courses of action, rather than orienting themselves necessarily toward the needs of society as the superordinate system. Under such assumptions it is difficult to accept as a working model of organizations the proposition that the ultimate goal of organizations must always be of functional significance for the larger system.

Comparing the goal and the functional approaches, it can be concluded that both contain serious methodological and theoretical shortcomings. The goal approach, while theoretically adhering to an organizational frame of reference, has failed to provide a rationale for the empirical identification of goals as an organizational property. The functional approach, on the other hand, has no difficulty in identifying the ultimate goal of the organization, since the latter is implied by the internal logic of the model, but the functional model does not take the organization as the frame of reference. Furthermore, neither of the two approaches gives adequate consideration to the conceptual problem of the relations between the organization and its environment.

A SYSTEM RESOURCE APPROACH TO ORGANIZATIONAL EFFECTIVENESS

The present need, to which we address our attention, is for a conception of organizational effectiveness that: (1) takes the organization itself as the focal frame of reference, rather than some external entity or some particular set of people; (2) explicitly treats the relations between the organization and its environment as a central ingredient in the definition of effectiveness; (3) provides a theoretically general framework capable of encompassing different kinds of complex organizations; (4) provides some latitude for uniqueness, variability, and change, with respect to the specific operations for assessing effectiveness applicable to any one organization, while at the same time maintaining the unity of the underlying framework for comparative evaluation; (5) provides some guide to the identification of performance and action variables relevant to organizational effectiveness and to the choice of variables for empirical use.

A promising theoretical solution to the foregoing problems can be derived from the open system model as it is applied to formal social organizations. This model emphasizes the distinctiveness of the or-

---

[23] James D. Thompson and William J. McEwen, "Organizational Goals and Environment: Goal-Setting as an Interaction Process," *American Sociological Review*, 23 (February 1958), pp. 23–31.

ganization as an identifiable social structure or entity, and it emphasizes the interdependency processes that relate the organization to its environment. The first theme supports the idea of treating formal organizations not as phenomena incidental to individual behavior or societal functioning but as entities appropriate for analysis at their own level. The second theme points to the nature of interrelatedness between the organization and its environment as the key source of information concerning organizational effectiveness. In fact, most existing definitions of organizational effectiveness have been formulated, implicitly or explicitly, in terms of a *relation* between the organization and its environment, since the attainment of a goal or the fulfillment of a social function imply always some change in the state of the organization vis-à-vis its environment. The crucial task, then, is the conceptualization of that relation. The system model, with its view of the nature of the interaction processes between the organization and its environment, provides a useful basis for such a conceptualization.

According to that model, especially as applied to the study of organizations by Katz and Kahn,[24] the interdependence between the organization and its environment takes the form of input-output transactions of various kinds relating to various things; furthermore, much of the stuff that is the object of these transactions falls into the category of *scarce and valued resources.* We shall have more to say about "resources" below. For the moment it will suffice to indicate that the value of such resources is to be derived from their utility as (more or less) generalized means for organizational activity rather than from their attachment to some specific goal. This value may or may not correspond to the personal values of the members of the organization, including their conception of its goals. It should be noted also that scarce and valued resources are, for the most part, the focus of competition between organizations. This competition, which may occur under different social settings and which may take different forms, is a continuous process underlying the emergence of a universal hierarchical differentiation among social organizations. Such a hierarchy is an excellent yardstick against which to assess organizational effectiveness. It reflects what may be referred to as the "bargaining position" of the organization in relation to resources and in relation to competing social entities that share all or part of the organization's environment.[25]

---

[24] Katz and Kahn, op. cit.

[25] The differential amounts of success of organizations with regard to their bargaining positions implies the possibility of exploitation of one organization by another, a possibility which may endanger the stability of social order. This asymmetry in interorganizational transactions and its consequences for the problem of social order underlie the sociological interest in exchange processes and their normative regulation. As pointed out recently by Blau:

"Without social norms prohibiting force and fraud, the trust required for social exchange could not serve as a self-regulating mechanism within the limits of these norms. Moreover, superior power and resources, which often are the results of competitive advantages gained in exchange transactions, make it possible to exploit others." (*Exchange and Power in Social Life,* New York: Wiley, 1964, p. 255)

Blau's discussion is concerned mainly with the more limited case of exchange between individuals as social actors. Nevertheless, it points to the potential asym-

We propose, accordingly, to define the effectiveness of an organization in terms of its bargaining position, as reflected in the ability of the organization, in either absolute or relative terms, to exploit its environment in the acquisition of scarce and valued resources.

The concept of "bargaining position" implies the exclusion of any specific goal (or function) as the ultimate criterion of organizational effectiveness. Instead it points to the more general capability of the organization as a resource-getting system. Specific "goals" however can be incorporated in this conceptualization in two ways: (1) as a specification of the means or strategies employed by members toward enhancing the bargaining position of the organization; and (2) as a specification of the personal goals of certain members or classes of member within the organizational system. The better the bargaining position of an organization, the more capable it is of attaining its varied and often transient goals, and the more capable it is of allowing the attainment of the personal goals of members. Processes of "goal formation" and "goal displacement" in organizations are thus seen not as defining ultimate criteria of effectiveness, but as strategies adopted by members for enhancing the bargaining position of their organizations.

The emphasis upon the resource-getting capability of the organization is not intended to obscure other vital aspects of organizational performance. The input of resources is only one of three major cyclic phases in the system model of organizational behavior, the other two being the throughput and the output. From this viewpoint the mobilization of resources is a necessary but not a sufficient condition for organizational effectiveness. Our definition, however, points not to the availability of scarce and valued resources as such, but rather to the bargaining position with regard to the acquisition of such resources as the criterion of organizational effectiveness. Such a position at a given point of time is, so far as the organization's own behavior is concerned, a function of all the three phases of organizational behavior—the importation of resources, their use (including allocation and processing), and their exportation in some output form that aids further input.

By focusing on the ability of the organization to exploit its environment in the acquisition of resources, we are directed by the basic yet often neglected fact that it is only in the arena of competition over scarce and valued resources that the performance of both like and unlike organizations can be assessed and evaluated comparatively. To put it somewhat differently, any change in the relation between the organization and its environment is affected by and results in a better or worse bargaining position vis-à-vis that environment or parts thereof.

It should be noticed that the proposed definition of effectiveness does not imply any specific goal toward which an organization is striving, nor does it impute some societal function as a property of the organiza-

metry involved in exchange processes in general and the consequences of such asymmetry, namely, the emergence of a hierarchical differentiation among the interacting units with regard to their exploitative ability. For the purposes of the present discussion it is important to note that such an advantageous bargaining position, which may be dysfunctional for the system as a whole, is from the organization's point of view a sign of its success.

tion itself. Our definition focuses attention on *behavior*, conceived as continuous and never-ending processes of exchange and competition over scarce and valued resources.[26] We shall now discuss some of the concepts central to our definition of organizational effectiveness.

COMPETITION AND EXCHANGE

Our emphasis upon the competitive aspects of interorganizational relations implies that an assessment of organizational effectiveness is possible only where some form of competition takes place. This raises the question of how general or limited is the scope of applicability of our definition, since interorganizational transactions take forms other than competition. An old and useful distinction in this respect has recently been formulated by Blau:

A basic distinction can be made between two major types of processes that characterize the transactions of organized collectivities—as well as those of individuals, for that matter—competitive processes reflecting endeavors to maximize scarce resources and exchange processes reflecting some form of interdependence. Competition occurs only among like social units that have the same objectives and not among unlike units . . . Competition promotes hierarchical differentiation between more or less successful organizations, and exchange promotes horizontal differentiation between specialized organizations of diverse sort.[27]

Blau's assessment that ". . . competition promotes hierarchical differentiation between more or less *successful* organizations" is, of course, in line with our definition of organizational effectiveness; furthermore, there is no question about the mainly competitive character of relations among "like" social units.

However, Blau's contention that competition occurs *only* among like organizations is an oversimplification. Indeed, it is difficult to point to any interrelated organizations that are not in competition with respect to some kinds of resources, and it is easy to point to organizations that are dominantly competitive, yet have some complementarity and interdependence in their relations. A university and a business firm, for example, may be involved in an exchange of knowhow and money, and still compete with respect to such resources as manpower and prestige. The type of pure complementarity of exchange is very limited indeed. We suggest, accordingly, that exchange and competition are the extremes of a continuum along which interorganizational transactions can be described. The proposed definition of effectiveness allows then for the comparative evaluation of any two or more organizations that have some elements of rivalry in their relations. Such a comparative evalua-

---

[26] One reader of an early draft of this paper, Dr. Martin Patchen, inquired about the sources of directive energy in goal-less organizations. The answer is that persons who are members of the organization, and acting both within their role prescriptions and in idiosyncratic deviations from role prescriptions, import personal values and goals which may modify the system in a directed way.

[27] Peter Blau, *Exchange and Power in Social Life*, New York: Wiley, 1964, p. 255.

tion becomes more meaningful—in the sense of encompassing the crucial dimensions of organizational behavior—as the variety and number of competitive elements in these relations increases. The clearest and most meaningful comparison obtains when the evaluated organizations compete directly for the same resources. This condition implies that the compared organizations are engaged in like activities and share to a large degree the same temporal and physical life space. In such cases the comparison is facilitated by the fact that the competition refers to the same kinds of resources and that the assessment variables —both of input and output—are measured in like units. Comparisons are also possible, however, in the case of organizations that do not compete directly, but that compete in environments that are judged to be similar in some relevant respects.

As the characteristic transactions between organizations come closer to the exchange pole of the continuum the problem of comparison becomes more complex: first, the elements of competition may be very few in number and peripheral in importance, thus making the comparison trivial; second, the more unlike the organizations, the more difficult it is to measure their performance units on common scales. In any case, the identification of the competitive dimensions in interorganizational transactions is the key problem in the assessment of organizational effectiveness. Some clarification and possible ways of solution for this problem can be achieved through an examination of the concept of "resources."

RESOURCES

A key element in this definition is the term "resources." Broadly defined, "resources" are (more or less) generalized means, or facilities, that are potentially controllable by social organizations and that are potentially usable—however indirectly—in relationships between the organization and its environment. This definition, it should be noted, does not attribute directionality as an inherent quality of a resource, nor does it limit the concept of resources to physical or economic objects or states even though a physical base must lie behind any named resource. A similar approach to "resources" is taken, for example, by Gamson. He argues that the "reputation" of individuals or groups as "influentials" in their community political affairs is itself a resource rather than simply ". . . the manifestation of the possession of large amounts of resources. . . ."[28]

One important kind of resource that is universally required by organizations, that is scarce and valued, and that is the focus for sharp competition, is energy in the form of human activity. The effectiveness of many organizations cannot be realistically assessed without some accounting for the organization's bargaining position with respect to the engagement of people in the service of the organization. One thinks, of course, of competition in the industrial or managerial labor market, but

---

[28] William A. Gamson, "Reputation and Resources in Community Politics," *American Journal of Sociology*, 72 (September 1966), pp. 121–31.

the idea is equally applicable to the competition, say, between the local church and the local political party, for the evening time of persons who are potentially active in both organizations.

Since human activity is such a crucial class of organizational resource, we elaborate on the meaning that is intended and one of the implications. We view members of an organization as an integral part of the organization with respect to their organizational role-defining and role-carrying activities, but as part of the environment of the organization with respect to their abilities, motives, other memberships, and other characteristics that are potentially useful but not utilized by the organization in role performance. An "effective" organization competes successfully for a relatively large share of the member's personality, engaging more of the personality in organizationally relevant ways, thus acquiring additional resources from its environment.

A number of other distinctions may usefully be made with respect to the resources that are involved in the effectiveness of organizations:

1. *Liquidity.* Some resources are relatively "liquid" in the traditional economic sense of that term and are readily exchangeable by an organization for resources of other kinds. Money and credit are highly liquid, being exchangeable for many other (but not all) kinds of resources. By contrast, the resource represented by high morale (among members) is relatively low in liquidity; under some conditions it is not directly exchangeable at full value in transactions with other organizations but must be internally transformed, e.g., into products or services, before exchange. Some organizations are characterized by having a large proportion of their resources in relatively nonliquid forms.

2. *Stability.* Some resources are transient in the sense that they must be acquired and utilized continuously by an organization, while other resources have the property of being stored or accumulated without significant depreciation. An organization that acquires a rapidly depreciating resource and fails to utilize this resource within an acceptable period will suffer loss of part of the value. The current high turnover among technical staff in some industrial firms is an example of loss of effectiveness through failure to utilize transient resources. By contrast, money is a highly stable resource that can be stored indefinitely at small loss and can be accumulated against future exchange requirements. Political influence is a resource of notorious instability.

3. *Relevance.* In principle, all resources are relevant to all organizations to the extent that they are capable of transformation and exchange. The degree of relevance, however, is of considerable interest, since identification of resources of high relevance offers a guide to a useful classification of organizations and serves to direct priority in comparative analyses to those kinds of resources that most clearly reflect the relative bargaining power of organizations. Degree of relevance also has a bearing upon the analysis of symbiotic relationships among organizations (high rates of exchange with relatively little bargaining and high mutual benefit) and upon the analysis of monopolistic forces (dominance of a given resource "market" and consequent enhancement of bargaining power). The degree of relevance of a given resource can be estimated on an *a priori* basis from a knowledge of the typical out-

puts of an organization and a knowledge of its characteristic through-put activities. Critical resources might be discerned from an analysis of changes in the pattern of internal organizational activity, for such changes can be interpreted to be a response to an enhanced requirement or a threatened deficit with respect to a given type of resource. Organizations are frequently observed to mobilize activities in a way that enhances their power to acquire certain resources. A judgment of future organizational effectiveness might accordingly be improved by information concerning the organization's ease of adaptation to shifts among classes of resources in their degree of relevance.

4. *Universality.* Some resources are of universal relevance in the sense that all organizations must be capable of acquiring such resources. The universally required classes include: (1) personnel; (2) physical facilities for the organization's activities; (3) a technology for these activities; and (4) some relatively liquid resource, such as money, that can be exchanged for other resources. The amount required of each class may in some cases be very modest, but all organizations must have, and must be able to replenish, resources of these kinds. The non-universal resources are, in general, those for which competition is limited, either because of irrelevance to many organizations or because the particular resource is ordinarily obtained amply through symbiotic exchange.

5. *Substitution.* Organizations with similar typical outputs competing in a common environment do not necessarily share the same roster of relevant and critical resources. One reason for this is that the internal processes of organizational life may be adapted to exploit certain readily available resources rather than to acquire alternative scarce resources in hard competition. An example of this is seen in the case of a small, ill-equipped guerilla army facing a force of superior size and equipment. While exploiting rather different resources, they may compete equally for the acquisition of territorial and political control.

A crucial problem in this context is the determination of the relevant and critical resources to be used as a basis for absolute or comparative assessment of organizational effectiveness. In stable, freely competitive environments with respect to relatively liquid resources, this determination may be rather easy to make, but under other conditions the determination may be problematic indeed. The difficulties arise primarily in cases in which the competing organizations have differential access to relatively rich or relatively poor environments, where symbiotic exchange relationships may develop, where the resources are not universal, and where the possibilities of substitution are great. In such situations, the analytic approach must employ not a static conception of the relationships between an organization and its environment but rather, a conception that emphasizes adaptation and change in the organizational patterns of resource-getting.

OPTIMIZATION VERSUS MAXIMIZATION

In their recent analysis of complex organizations, Katz and Kahn proposed defining organizational effectiveness as "the maximization of return to the organization by all means."[29] This definition shares with the one we propose an emphasis on resource procurement as the sign of organizational success; it differs, however, in invoking the notion of maximization, a concept we have avoided. The position taken here is that maximization of return, even if possible, is destructive from the viewpoint of the organization. To understand this statement it should be remembered that the bargaining position of the organization is equated here with the ability to exploit the organization's environment —not with the maximum use of this ability. An organization that fully actualizes its exploitative potential may risk its own survival, since the exploited environment may become so depleted as to be unable to produce further resources. Furthermore, an organization that ruthlessly exploits its environment is more likely to incite a strong organized opposition that may weaken or even destroy the organization's bargaining position. Thus, the short-run gains associated with overexploitation are likely to be outweighted by greater long-run losses. Also, the resource itself may lose value if overexploited; for example, an effective voluntary community organization may enjoy extraordinary bargaining power in the engagement of prestigeful people, but this power may not safely be used to the maximum, because excessive recruitment risks the diminishing of the value of membership when membership ceases to be exclusive.

These considerations lead to the proposition that the highest level of organizational effectiveness is reached when the organization maximizes its bargaining position and optimizes its resource procurement. "Optimum" is the point beyond which the organization endangers itself, because of a depletion of its resource-producing environment or the devaluation of the resource, or because of the stimulation of countervailing forces within that environment. As stated by Thompson and McEwen:

> It is possible to conceive of a continuum of organizational power in environmental relations, ranging from the organization that dominates its environmental relations to one completely dominated by its environment. Few organizations approach either extreme. Certain gigantic enterprises, such as the Zaibatsu in Japan or the old Standard Oil Trust in America, have approached the dominance-over-environment position at one time; most complex organizations, falling somewhere between the extremes of the power continuum, must adopt strategies for coming to terms with their environment.[30]

We may add, however, that the need "for coming to terms with their environment" applies to organizations that approximate the dominance-over-environment extreme as well. A powerful enterprise like General Motors must exercise its potential power with much restraint in order to avoid the crystallization of an opposition which may weaken its bargaining power considerably, through legislation or some other means.

[29] Katz and Kahn, op. cit., p. 170.
[30] Thompson and McEwen, op. cit., p. 25.

It is of course very difficult, if possible at all, to determine in absolute terms the organization's maximum bargaining position and the optimal point of resource procurement that is associated with that position. Since most organizations, however, fall short of maximizing their bargaining position, the optimization problem, though theoretically important, is only of limited empirical relevance. In practice, organizational effectiveness must be assessed in relative terms, by comparing organizations with one another. The above discussion on the nature of "resources" provides at best a general outline for carrying out such a task. A more detailed discussion and a preliminary effort to apply empirically the conceptual scheme presented here is reported elsewhere.[31] Briefly, the following steps seem necessary for a meaningful comparative assessment of organizational effectiveness: (1) to provide an inclusive taxonomy of resources; (2) to identify the different types of resources that are mutually relevant for the organizations under study; and (3) to determine the relative positions of the compared organizations on the basis of information concerning the amount and kinds of resources that are available for the organization and its efficiency in using these resources to get further resources.

SOME IMPLICATIONS

We end this discussion with a few speculations about the impacts that might arise from a general acceptance and use of the conception of organizational effectiveness that we have proposed. These may affect theorists, empirical researchers, and managers in various ways:

1. The rejection of the concept of an ultimate goal, and the replacement of this singular concept with one emphasizing an open-ended multidimensional set of criteria, will encourage a broadening of the scope of search for relevant criterion variables. Past studies have tended to focus too narrowly upon variables derived from traditional accounting practice or from functional social theory, or on narrowly partisan "goals" attributed to organizations. A conception of organizational effectiveness based upon organizational characteristics and upon resource-acquisition in the most general sense will encourage the treatment as criteria of many variables previously regarded as by-products or incidental phenomena in organizational functioning.

2. Past comparative studies of organizations have, in general, been of two kinds: (1) Comparison of organizations differing markedly in their characteristics, e.g., prisons and factories, so that issues of relative effectiveness were deemed irrelevant and uninteresting as well as impractical; and (2) comparisons among organizations of a similiar type, so that they could be compared on like variables and measurement units. The conception we offer provides the possibility of making accessible for

---

[31] Stanley E. Seashore and Ephraim Yuchtman, "The Elements of Organizational Performance." A paper prepared for a symposium on "People, Groups and Organizations: An Effective Integration of Knowledge," Rutgers University, November 1966. (This paper will appear in *Administrative Science Quarterly,* 1967.); Ephraim Yuchtman, *A Study of Organizational Effectiveness,* unpublished Ph.D. dissertation, the University of Michigan, 1966.

study the large middle range of comparisons involving organizations that they compete with respect to some but not all of their relevant and crucial resources.

3. Case studies of single organizations will be aided by the provision of a conceptual basis for treating a more inclusive and more realistic range of variables that bear on the effectiveness of the organization.

4. The meaning of some familiar variables will need to be reassessed and in some cases changed. For example, distributed profit, a favorite variable for the comparative assessment of business organizations, will be more widely recognized as a cost of organizational activity and not as an unequivocal sign of success or goal achievement. Some managers have already adopted this view. Similarly, growth in size, usually interpreted as a sign of organizational achievement, can now be better seen as a variable whose meaning is tied closely to environmental factors and to the position of the organization with respect to certain other variables; the conception we have presented highlights the idea that growth in size is not in itself an unmitigated good, even though it may mean greater effectiveness under some conditions. In a similar fashion, it will be seen as necessary that the judgment of the meaning of each criterion variable rests not upon an absolute value judgment or a universal conceptual meaning, but rather upon the joint consideration of an extensive integrated set of organizational performance and activity variables.

Warren G. Bennis

29. Towards a "Truly" Scientific Management: The Concept of Organization Health

> *Muggeridge:* Now, Charles, you, because you're a scientist . . . you have this idea, as I understand from your writings, that one of the failings of our sort of society, is that the people who exercise authority, we'll say Parliament and so on, are singularly unversed in scientific matters.
> *Snow:* Yes, I think this is a terrible weakness of the whole of Western society, and one that we're not going to get out of without immense trouble and pain.
> *Muggeridge:* Do you mean by that that, for instance, an M. P. would be a better M. P. if he knew a bit about science?
> *Snow:* I think some M. P.'s ought to know a bit about science. They'd be better M. P.'s in the area where scientific insight becomes important. And there are quite a number of such areas (52).

EXTOLLING SCIENCE has become something of a national and international past-time which typically stops short of the truly radical reforms in social organization the scientific revolution implies. Knowing "a bit about science" is a familiar and increasingly popular exemplar of this which C. P. Snow treats in his *Two Cultures* (69). But if culture is anything it is a way of life, the way real people live and grow, the way ideals and moral imperatives are transmitted and infused. Culture is more *value* than knowledge ("a bit of science"). Dr. Bronowski, who shares with Snow the view that "humanists" tend to be ignorant of and removed from science (they cannot discuss the Second Law of Thermodynamics) understands more than Snow seems to that a fundamental unification of cultural outlook is what is required (18). The connective tissue required, then, is cultural, social, institutional—not grafted-on evening courses on Science.

In this connection, and closer to some of the general aims of this paper, Nevitt Sanford has said:

> The ethical systems of other professions, such as business or the military, have become models for whole societies. Why should not the practice of science become such a model? After we have shown, as we can, that joy and beauty have their places in this system? At any rate, anyone who takes it

507

upon himself to be a scientist, and succeeds in living up to its requirements, may be willing for his behavior to become a universal norm (61).

This foreshadows the general theme of this essay: the recognition that the *institution* of science can and should provide a viable model for other institutions not solely concerned with developing knowledge. To demonstrate the proposition, this paper first discusses the criterion problem in relation to organizations.[1] An attempt is made to show that the usual criteria for evaluating organizational effectiveness, "enhancement of satisfaction on the part of industry's participants and improvement of effectiveness of performance" (36, p. 238), are inadequate, incorrect, or both, as valid indicators of organizational "health." (For the moment let us use the term "health" in the same vague way as "effectiveness." Organizational health is defined later in this paper.) Next it is suggested that an alternative set of criteria, extracted from the normative and value processes of science, provides a more realistic basis for evaluating organizational performance. These criteria are related to those of positive mental health, for it will be argued that there is a profound kinship between the mores of science and the criteria of health for an individual. From this confluence is fashioned a set of psychologically-based criteria for examining organizational health. Finally a discussion is presented of some of the consequences of these effectiveness criteria for organizational theory and practice.

THE SEARCH FOR EFFECTIVENESS CRITERIA

There is hardly a term in current psychological thought as vague, elusive, and ambiguous as the term 'mental health.' That it means many things to many people is bad enough. That many people use it without even attempting to specify the idiosyncratic meaning the term has for them makes the situation worse, . . . for those who wish to introduce concern with mental health into systematic psychological theory and research.—M. Jahoda (34, p. 3)

. . . no one can say with any degree of certainty by what standards an executive ought to appraise the performance of his organization. And it is questionable whether the time will ever arrive when there will be any pattern answers to such a question—so much does the setting of an organization and its own goal orientation affect the whole process of appraisal.—J. M. Pfiffner and F. P. Sherwood (56, p. 422).

Raising the problem of criteria, the standards for judging the "goodness" of an organization seldom fails to generate controversy and despair. Establishing criteria for an organization (or, for that matter, education, marriage, psychotherapy, etc.) accentuates questions of value, choice, and normality and all the hidden assumptions that are used to form judgments of operation. Often, as Jahoda has said in rela-

---

From *General Systems*, 1962, vol. 7, pp. 269–82. Reprinted with permission of the author and the publisher.

[1] For the purposes of this discussion, "organization" is defined as any institution from which one receives cash for services rendered. This paper deals with all such supra-individual entities, although reference is made mostly to industrial organizations.

tion to mental health criteria, the problem "seems so difficult that one is almost tempted to claim the privilege of ignorance" (34, p. 77).

However, as tempting as ignorance can be, research on organizations —particularly industrial organizations—has heroically struggled to identify and measure a number of dimensions associated with organizational effectiveness (74). Generally, these dimensions have been of two kinds: those dealing with some index of organizational performance, such as profit, cost, rates of productivity, individual output, etc., and those associated with the human resources, such as morals, motivation, mental health, job commitment, cohesiveness, attitudes toward employer or company, etc. In short, as Katzell pointed out in his 1957 review of industrial psychology, investigations in this area typically employ measures of *satisfaction* and *performance* (36). In fact, it is possible to construct a simple twofold table that adequately accounts for most of the research on organizations that has been undertaken to date, as shown in Table 1. On one axis are located the criteria variables: or-

TABLE 1

Major Variables Employed in the Study of Organizational Behavior

		Criteria Variables	
		Organizational Efficiency	*Satisfaction or Health*
Independent variables	Technology (rationalized procedures)	Management science: systems research, operations research, decision processes, etc.	Human engineering
	Human factors	Personnel psychology, training, and other personnel functions	Industrial social psychology and sociology

ganizational efficiency (the ethic of work performance) and member satisfaction (the ethic of "health"). On the other axis are located the two main independent variables employed, human and rationalized procedures. In other words, it is possible to summarize most of the research literature in the organizational area by locating the major independent variables (technological or human) on one axis and the dependent variables (efficiency or health) on the other.

This classification is necessarily crude and perhaps a little puzzling, principally for the reason that research on organizations lacks sufficient information concerning the empirical correlation between the two dependent variables, organizational efficiency and health factors. For a time it seemed (or was hoped) that personnel satisfaction and efficiency were positively related, that as satisfaction increased so did performance. This alleged correlation allowed the "human relations" school and the industrial engineers (Taylorism being one examplar[2]) to pro-

---

[2] For a recent historical review, see Aitken (1).

ceed coterminously without necessarily recognizing the tension between "happy workers" and "high performance." As Likert put it:

> It is not sufficient merely to measure morale and the attitudes of employees toward the organization, their supervision, and their work. Favorable attitudes and excellent morale do not necessarily assure high motivation, high performance, and an effective human organization. A good deal of research indicates that this relationship is much too simple (41, p. 49).

Indeed today we are not clear about the relation of performance to satisfaction, or even whether there is any interdependence between them. Likert and his associates have found organizations with all the logical possibilities—high morale with low productivity, low productivity with low morale, etc. Argyris' work (2, 4), with a popular assist from William H. Whyte, Jr. (76), clouds the picture even further by postulating the inevitability of conflict between human need-satisfaction and organizational performance (as formal organizations are presently conceived). This creates, as Mason Haire has recognized (31), a calculus of values: how much satisfaction or health is to be yielded for how many units of performance?

Generally speaking, then, this is the state of affairs: two criteria, crudely measured, ambiguous in meaning, questionable in utility, and fraught with value connotations (35). In view of these difficulties, a number of other, more promising, approaches have been suggested. The most notable of these are the criterion of multiple goals, the criterion of the situation, and the criterion of system characteristics.

The Criterion of Multiple Goals

This approach rests on the assumption that ". . . organizations have more than a single goal and that the interaction of goals will produce a different value framework in different organizations" (57, p. 42). Likert, who is a proponent of the multiple criterion approach, claims that very few organizations, if any, obtain measurements that clearly reflect the quality and capacity of the organization's human resources. This situation is due primarily to the shadow of traditional theory, which tends to overlook the human and motivational variables and the relatively new developments in social science that only now permit measurements of this type. Likert goes on to enumerate twelve criteria, covering such dimensions as loyalty and identification with the institution and its objectives, degree of confidence and trust, adequacy and efficiency of communication, amount and quality of teamwork, etc. (41). By and large, Likert's criteria are psychologically based and substantially enrich the impoverished state of effectiveness criteria.[3]

The Criterion of the Situation

This approach is based on the reasoning that organizations differ with respect to goals and that they can be analytically distinguished in

---

[3] See also Kahn, Mann, and Seashore (35), Introduction, for other suggestions for criteria.

terms of goal orientation. As Parsons pointed out: "As a formal analytical point of reference, *primacy of orientation to the attainment of a specific goal is used as the defining characteristic of an organization* which distinguishes it from other types of social systems" (55, p. 64).

In an earlier paper by Bennis (9), a framework was presented for characterizing four different types of organizations based on a specific criterion variable. These "pure" types are rarely observed empirically, but they serve to sharpen the difference among formally organized ac-

TABLE 2

Typology of Organization*

Type of Organization	Major Function	Examples	Effectiveness Criterion
Habit	Replicating standard and uniform products	Highly mechanized factories, etc.	No. of products
Problem-solving	Creating new ideas	Research organizations; design and engineering divisions; consulting organizations, etc.	No. of ideas
Indoctrination	Changing people's habits, attitudes, intellect, behavior (physical and mental)	Universities, prisons, hospitals, etc.	No. of "clients" leaving
Service	Distributing services either directly to consumer or to above types	Military, government, advertising, taxi companies, etc.	Extent of services services performed

* From W. G. Bennis, Leadership Theory and Administrative Behavior: The Problem of Authority, *Admin. Sci. Quart.*, vol. 4, no. 3, p. 299, December 1959.

tivities. Table 2 represents an example of developing effectiveness variables on the basis of organizational parameters.

The Criterion of System Characteristics

This approach, most cogently advanced by sociologists, is based on a "structural-functional" analysis. Selznick, one of its chief proponents, characterizes the approach in the following way:

Structural-functional analysis relates contemporary and variable behavior to a presumptively stable system of needs and mechanisms. This means that a given empirical system is deemed to have basic needs, essentially related to self-maintenance; the system develops repetitive means of self-defense; and day-to-day activity is interpreted in terms of the function served by that activity for the maintenance and defense of the system (62, p. 28).

Derivable from the system model are basic needs or institutional imperatives that have to be met if the organism is to survive and "grow." Selznick, for example, lists five:

(1) The security of the organization as a whole in relations to social forces in its environment. (2) The stability of the lines of authority and communi-

cation. (3) The stability of informal relations within the organization. (4) The continuity of policy and of the sources of its determination. (5) A homogeneity of outlook with respect to the meaning and role of the organization (62, pp. 29–30).

Caplow, starting from the fundamental postulate that organizations tend to maintain themselves in continuous operation, identifies three criteria of organizational success: the performance of objective functions, the minimization of spontaneous conflict, and the maximization of satisfaction for individuals (22). Obviously, with the exception of the second criterion, these resemble the old favorites, performance and satisfaction.

The preceding summaries do not do full justice to the nuances in these three approaches or the enormous creative effort that went into their development. Nor do they include the ideas of many thoughtful practitioners.[4] Despite these limitations, the discussion of multiple criteria, situational parameters, and system characteristics represents the main attempts to solve the criterion problem.[5]

INADEQUACY OF CRITERION VARIABLES FOR THE MODERN ORGANIZATION

One thing that is new is the prevalence of newness, the changing scale and scope of change itself, so that the world alters as we walk in it, so that the years of man's life measure not some small growth or rearrangement or moderation of what he learned in childhood, but a great upheaval. . . To assail the changes that have unmoored us from the past is futile, and in a deep sense, I think it is wicked. We need to recognize the change and learn what resources we have.—Robert Oppenheimer (53, pp. 10–11)

The history of other animal species shows that the most successful in the struggle for survival have been those which were most adaptable to changes in their world.—H. Bronowski (17, p. 137)

The present ways of thinking about and measuring organizational effectiveness are seriously inadequate and often misleading. These criteria are insensitive to the important needs of the organization and out of joint with the emerging view of contemporary organization that is held by many organizational theorists and practitioners. The present techniques of evaluation provide static indicators of certain output characteristics (i.e. performance and satisfaction) without illuminating the processes by which the organization searches for, adapts to, and solves its changing goals (56). However, it is these dynamic processes of problem-solving that provide the critical dimensions of organizational health, and without knowledge of them output measurements are woefully inadequate.[6]

---

[4] See, for example, Urwick (71).

[5] Another approach, advocated by A. L. Comrey, is the deliberate (and often wise) avoidance of a definition of effectiveness or health by obtaining judgments of knowledgeable observers. "This method of defining 'effectiveness' seems to be the only feasible course of action in view of the tremendous number of meanings involved in a conceptual definition of this term and the obvious impossibility of providing a criterion which would reflect all or most of these meanings" (23, p. 362).

[6] See Ridgway (59) for other criticisms of the use of performance measurements.

This rather severe charge is based upon the belief that the main challenge confronting the modern organization (and society) is that of coping with external stress and change. This point hardly needs elaboration or defense. Ecclesiastes glumly pointed out that men persist in disordering their settled ways and beliefs by seeking out many inventions. The recent work in the field of organizational behavior reflects this need and interest; it is virtually a catalogue of the problems in organizational change.[7] In a 1961 monograph on managing major change in organizations, Mann and Neff stated the issue this way: "Among the most conspicuous values in American culture of the twentieth century are progress, efficiency, science and rationality, achievement and success. These values have helped to produce a highly dynamic society—a society in which the predominant characteristic is *change*" (45, p. 1). Kahn, Mann, and Seashore, when discussing a criterion variable, "the ability of the organization to change appropriately in response to some objective requirement for change," remarked: "Although we are convinced of the theoretical importance of this criterion, which we have called organizational flexibility, we have thus far been unable to solve the operational problems involved in its use" (35, p. 4).

The basic flaw in the present effectiveness criteria is their inatten-

FIGURE 1

Two Types of Communication Networks for Problem-Solving by a Group of Five Persons

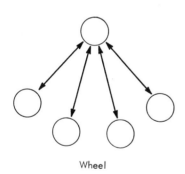

Wheel

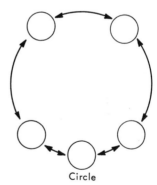

Circle

tion to the problem of adapting to change. To illuminate some of the consequences of this omission, let us turn to one rather simple example. The example is drawn from an area of research that started at the Massachusetts Institute of Technology about 1949 on the effects of certain organizational patterns (communication networks) on problem-solving by groups (38). Two of these networks, the Wheel and the Circle, are shown in Figure 1.

---

[7] See, for example: March and Simon (46), chap. 7; Argyris (3); Shepard (65); Gibb and Lippitt (29); Lippitt, Watson, and Wesley (42); Bennis, Benne, and Chin (10); Walker (73).

The results of these experiments showed that an organization with a structure like the Wheel can solve simple tasks (e.g. identification of the color of a marble that is common to all five group members) more rapidly, more clearly, and more efficiently than an organization like the Circle. Thus the Wheel arrangement is plainly superior in terms of the usual criteria employed to evaluate effectiveness. However, if we consider two other criteria of organizational effectiveness that are relevant to the concern with change-flexibility and creativity, we discover two interesting phenomena. First, the rapid acceptance of a new idea is more likely in the Circle than in the Wheel. The man in the middle of the Wheel is apt to discard an idea on the grounds that he is too busy or the idea is impractical. Second, when the task is changed, for example by going from "pure" color marbles to unusual color marbles (such as ginger-ale color or blue-green), the Circle organization is better able to adapt to this change by developing a new code (68). As Leavitt pointed out:

. . . by certain industrial engineering-type criteria (speed, clarity of organization and job descriptions, parsimonious use of paper, etc.), the highly structured, highly routinized, non-involving centralized net seems to work best. But if our criteria of effectiveness are more ephemeral, more general (like acceptance of creativity, flexibility in dealing with novel problems, generally high morale and loyalty), then the more egalitarian or decentralized type net seems to work better (39, p. 22).

If we view organizations as adaptive, problem-solving, organic structures, then inferences about effectiveness have to be made, not from static measures of output, though these may be helpful, but on the basis of the processes through which the organization approaches problems. In other words, no single measurement of organizational efficiency or satisfaction—no single time-slice of organizational performance—can provide valid indicators of organizational health. An organization may be essentially healthy despite measurements that reveal that its performance and satisfaction measurements are lower than last month's; it can be unhealthy even if its performance and efficiency figures are higher than last month's. Unhealthy and healthy, that is, in relation to the ability to cope with change, with the future. Discussing the neurotic processes, Kubie makes the same point:

There is not a single thing which a human being can do or feel, or think, whether it is eating or sleeping or drinking or fighting or killing or hating or loving or grieving or exulting or working or playing or painting or inventing, which cannot be either sick or well . . . . The measure of health is flexibility, the freedom to learn through experience, the freedom to change with changing internal and external circumstances, to be influenced by reasonable argument, admonitions, exhortations, and the appeal to emotions, the freedom to respond appropriately to the stimulus of reward and punishment, and especially the freedom to cease when sated. The essence of normality is flexibility in all of those vital ways (37, p. 20).

Any moment of behavior is unhealthy if the processes that set it in motion predetermine its automatic repetition, regardless of the environmental stimuli or consequences of the act. For example, it is plausible that lowering efficiency in order to adjust to some product change may be quite appropriate when market demands are considered. It is equally plausible that morale, or whatever measure is used to gauge the human factor, may similarly plummet during this period. In fact, maintaining the same level of efficiency and morale in new circumstances may be dysfunctional for the health of the organization.

Let us review the argument thus far. The main challenge confronting today's organization, whether it is a hospital or a business enterprise, is that of responding to changing conditions and adapting to external stress. The salience of change is forced on organizations because of the growing interdependence between their changing boundary conditions and society (a point that will be elaborated later) and the increasing reliance on scientific knowledge. The traditional ways that are employed to measure organizational effectiveness do not adequately reflect the true determinants of organizational health and success. Rather, these criteria yield static time-slices of performance and satisfaction, which may be irrelevant or misleading. These static, discrete measurements do not provide viable measures of health, for they tell us nothing about the processes by which the organization copes with its problems. Therefore, different effectiveness criteria have to be identified, criteria that reveal the processes of problem-solving. This point is corroborated by some recent works on organizational theory. Consider, for example, these remarks by Wilfred Brown, Chairman and Managing Director of the Glacier Metal Company:

Effective organization is a function of the work to be done and the resources and techniques available to do it. Thus changes in methods of production bring about changes in the number of work roles, in the distribution of work between roles and in their relationship to one another. Failure to make explicit acknowledgement of this relationship between work and organization gives rise to non-valid assumptions, e.g. that optimum organization is a function of the personalities involved, that it is a matter connected with the personal style and arbitrary decision of the chief executive, that there are choices between centralized and decentralized types of organization, etc. Our observations lead us to accept that optimum organization must be derived from an analysis of the work to be done and the techniques and resources available (19, pp. 18–19).

The work of Emery and Trist, which has influenced the thinking of Brown, stressed the "sociotechnical system," based on Bertalanffy's "open system" theorizing (13). They conclude that:

. . . the primary task of managing an enterprise as a whole is to relate the total system to its environment, and not internal regulation per se (24, p. 10).

If management is to control internal growth and development it must in the first instance control the "boundary conditions"—the forms of exchange between the enterprise and the environment. . . The strategic objective should be to place the enterprise in a position in its environment where it has some assured conditions for growth—unlike war the best position is not necessarily that of unchallenged monopoly. Achieving this position would be the primary task or overriding mission of the enterprise (24, p. 12).

In reference to management development, A. T. M. Wilson, Former Director of Tavistock Institute, pointed out:

> One general point of high relevance can be seen in these discussions of the·firm as an institution. The tasks of the higher level managers center on problems in which there is a continuously high level of uncertainty; complex value decisions are inevitably involved; and this has a direct bearing on the requirements of personality for top level management . . . (77, p. 13).

And H. J. Leavitt said on the same subject:

> Management development programs need, I submit, to be oriented much more toward the future, toward change, toward differences from current forms of practice and behavior . . . We ought to allocate more of the effort of our programs to making our student a more competent analyst. We ought, in other words, to try to teach them to think a little more like scientists, and indeed to know a good deal more about the culture and methods of scientists (39, pp. 32–33).[8]

What relevance have these quotations to the main theme of this essay? Note, first of all, that these theorists all view the organization (or institution) as an adaptive structure actively encountering many different environments, both internal and external, in their productive efforts. Note also the key terms: change, uncertainty, future, task, mission, work to be done, available resources, exchanges between the enterprise and environment. There is no dialogue here on the relation between "productivity" and "satisfaction," no fruitless arguments between the human relationists and scientific management advocates. Indeed, it seems that it is no longer adequate to perceive organization as an analogue to the machine as Max Weber indicated: ". . . (bureaucracy is like) a modern judge who is a vending machine into which the pleadings are inserted together with the fee and which then disgorges the judgment together with its reasons mechanically derived from the code" (8, p. 421). Nor is it reasonable to view the organization solely in terms of the sociopsychological characteristics of the persons involved at work, a viewpoint that has been so fashionable of late.[9] Rather, the approach that should be taken is that of these quoted writers: organizations are to be viewed as "open systems" defined by their primary task or mission and encountering boundary conditions that are rapidly changing their characteristics.[10] Given this rough definition, we must locate some effectiveness criteria and the institutional prerequisites that provide the conditions for the attainment of these criteria.

---

[8] Although not quoted here, a book by Selznick is also directly relevant. See (63).

[9] See Bennis (9) for elaboration of this point.

[10] Wilson lists six "areas of social activity, each of which contains a number of significant social institutions and social groups. These areas may be rather summarily labelled as: (i) Government, (ii) Consumers, (iii) Shareholders, (iv) Competitors, (v) Raw material and power suppliers, and (vi) Groups within the firm" (77, p. 3). These represent some of the boundary conditions for the manager.

THE SPIRIT OF INQUIRY AS A MODEL FOR ORGANIZATION

Findings are science's short-range benefits, but the method of inquiry is its long-range value. I have said that the invention of organization was Man's first most important achievement; I now add that the development of inquiry will be his second. Both of these inventions change the species and are necessary for its survival. But both must become a part of the nature of Man himself, not just given house room in certain groups. Organization is by now a part of every man, but inquiry is not. The significant product of science and education will be the incorporation within the human animal of the capability and habit of inquiry.—H. Thelen (70, p. 217)

Whether our work is art or science or the daily work of society, it is only the form in which we explore our experience which is different; the need to explore remains the same. This is why, at bottom, the society of scientists is more important than their discoveries. What science has to teach us here is not its techniques but its spirit: the irresistible need to explore.—J. Bronowski (18, p. 93)

It has been asserted throughout this paper that organizations must be viewed as adaptive, problem-solving structures operating and embedded in complicated and rapidly changing environments. If this view is valid, then it is fair to postulate that the methodological rules by which the organization approaches its task and "exchanges with its environments" are the critical determinants of organizational effectiveness. These methodological rules or operating procedures bear a close resemblance to the rules of inquiry, or scientific investigation. Therefore, the rules and norms of science may provide a valuable, possibly necessary model for organizational behavior.

First, it should be stated what is meant and what is not meant by "science" in this context. It is not the findings of science, the vast array of data that scientists produce. Nor is it a barren operationalism—what some people refer to as "scientism"—or the gadgetry utilized for routine laboratory work. Rather it is what may be called the scientific "temper" or "spirit." It is this "spirit of inquiry," which stems from the value position of science, that such authors as Dewey, Bronowski, Geiger, and Sanford have emphasized must be considered if our world is to survive. This position says essentially that the roles of scientist and citizen cannot be sharply separated. As Waddington put it:

> The true influence of science is an attitude of mind, a general method of thinking about and investigating problems. It can, and I think it will, spread gradually throughout the social consciousness without any very sharp break with the attitudes of the past. But the problems for which it is wanted face us already; and the sooner the scientific method of handling them becomes more generally understood and adopted, the better it will be. (72, p. xiii).

Now it is necessary to look a bit more closely at what is meant by this "scientific attitude." Relevant here are two important aspects of the scientific attitude, one having to do with the methodology of science and one related to the social organization of science. The former is a complex of human behavior and adjustment that has been summed up as the "spirit of inquiry." This complex includes many elements, only two of which are considered here. The first may be called the hypothetical spirit, the feeling for tentativeness and caution, the respect for probable error. As Geiger says: ". . . the hypothetical spirit is the unique contribution scientific method can offer to human culture; it certainly is the only prophylactic against the authoritarian mystique so symptomatic of modern nerve failure" (28, p. 11).

The second ingredient is experimentalism, the willingness to expose ideas to empirical testing, to procedures, to action. The hypothetical stance without experimentalism would soon develop into a rather arid scholasticism. Experimentalism without the corrective of the hypothetical imagination would bring about a radical, "dustbowl" empiricism lacking significant insight and underlying structures capable of generalization. These two features, plus the corrective of criticism, is what is meant by the methodological rules of science; it is the spirit of inquiry, a love of truth relentlessly pursued, that ultimately creates the objectivity and intelligent action associated with science.

The second important aspect of the scientific attitude is that concerning the social organization of science, the institutional imperatives of the scientific enterprise. A number of social scientists, inspired by the work of Parsons (54) and Merton (49), (50), have examined the society of scientific enterprise (5), (21), (47), (60). What they have said is important for the argument presented here. Only when the social conditions of science are realized can the scientific attitude exist. As Sanford pointed out:

> Science flourishes under that type of democracy that accents freedom of opinion and dissent, and respect for the individual. It is against all forms of totalitarianism, of mechanization and regimentation. . . In the historical development of the ends that are treasured in Western societies there is reason to believe that science has had a determining role. Bronowski again: Men have asked for freedom, justice and respect precisely as science has spread among them (61, p. 9).

Furthermore, Parsons states:

> Science is intimately integrated with the whole social structure and cultural tradition. They mutually support one another—only in certain types of society can science flourish and conversely without a continuous and healthy development and application of science such a society cannot function properly (5, p. 83).

What are the conditions that comprise the ethos of science? Barber identifies five that are appropriate to this discussion: Rationality, universalism, individualism, communality, and disinterestedness (6). A brief word about each of these is in order. The goal of science is understanding—understanding in as abstract and general a fashion as possible. Universalism, as used here, means that all men have morally equal claims to discover and to understand. Individualism, according to Barber, expresses itself in science as anti-authoritarianism; no authority but the authority of science need be accepted or trusted. Communality is close to the utopian communist slogan: "From each according to his abilities, to each according to his needs." This simply means that all scientific peers have the right to share in existing knowledge; withholding knowledge and secrecy are cardinal sins. The last element, disinterestedness, is to be contrasted with the self-interest usually associated with organizational and economic life. Disinterestedness in science requires that role incumbents serve others and gain gratification from the pursuit of truth itself. These five conditions comprise the moral imperatives of the social organization of science. They are, of course, derived from an "ideal type" of system, an empirically imaginable possibility but a rare phenomenon. Nevertheless, insofar as they are imperatives, they do in fact determine significantly the behavior of scientific organization.

There are two points to be made in connection with this model of organization. The first was made earlier but may require reiteration: the spirit of inquiry can flourish only in an environment where there is a commitment toward the five institutional imperatives. The second point is that what is now called the "human relations school"[11] has been preoccupied primarily with the study of those factors that this paper has identified as the prerequisites of the science organization. In fact, only if we look at the human-relations approach with this perspective do we obtain a valid view of their work. For example, a great deal of work in human relations has focused on "communication" (11), "participation" (43), and "decision-making." Overgeneralizing a bit, we can say that most of the studies have been (from a moral point of view) predicated on and lean toward the social organization of science as has been outlined here. Note, for instance, that many studies have shown that increased participation, better communication (keeping worker "informed"), more "self-control," and decreased authoritarianism are desirable ends. Because of their emphasis on these factors, the researchers and theoreticians associated with human-relations research have sometimes been perceived as "soft-headed," unrealistic, too academic, and even utopian. In some cases, the social scientists themselves have invited these criticisms by being mainly interested in demonstrating that these participative beliefs would lead to heightened morale and, on occasion, to increased efficiency. So they have been accused by many writers as advocates of "happiness" or a moo-cow psychology.[12]

These are invalid criticisms, mainly because the issue is being fought on the wrong grounds. The root of the trouble is that the social scientists have not been entirely aware or prescient enough to see the implications of their studies. Rather than debating the viability of sociopsychological variables in terms of the traditional effectiveness variables, which at this point is highly problematical, they should be saying that the only way in which organizations can develop a scientific attitude is by providing conditions where it can flourish. In short, the norms of science are both compatible and remarkably homogeneous with those of a liberal democracy. We argue, then, that the way in which

---

[11] See Bennis (9).

[12] See Baritz (6), bibliographies for chaps. 6, 9, and 10.

organizations can master their dilemmas and solve their problems is by developing a spirit of inquiry. This can flourish only under the social conditions associated with the scientific enterprise, i.e., democratic ideals. Thus it is necessary to emphasize the "human side of enterprise," that is, institutional conditions of science, if organizations are expected to maintain mastery over their environment.[13]

Now, assuming that the social conditions of science have been met, let us return to the designated task of identifying those organizational criteria that are associated with the scientific attitude.

THE CRITERIA OF SCIENCE AND MENTAL HEALTH APPLIED TO ORGANIZATIONS

Perhaps no other area of human functioning has more frequently been selected as a criterion for mental health than the individual's reality orientation and his efforts at mastering the environment.—M. Jahoda (34, p. 53)

I now propose that we gather the various kinds of behavior just mentioned, all of which have to do with effective interaction with the environment, under the general heading of competence.—Robert White (75, p. 317)

. . . all aspects of the enterprise must be subordinated to . . . its *primary task*. It is not only industrial enterprises, however, which must remain loyal to their primary tasks. This is so of all human groups, for these are all compelled, in order to maintain themselves in existence, to undertake some form of appropriate action in relation to their environment . . . An organism, whether individual or social, must do work in order to keep itself related to its external environment, that is, to meet reality.—Eric Trist (19, p. xvi)

These quotations provide the framework for the following analysis. They express what has been the major concern throughout this paper: that, when organizations are considered as "open systems," adaptive structures coping with various environments, the most significant characteristic for understanding effectiveness is competence, mastery, or as the term has been used in this essay, problem-solving. It has been shown that competence can be gained only through certain adaptations of science: its attitude and social conditions. It is now possible to go a step further by underlining what the above quotations reveal, that the criteria of science bear a close kinship to the characteristics of what mental-health specialists and psychiatrists call "health."

There is an interesting historical parallel between the development of criteria for the evaluation of mental health and the evolution of standards for evaluating organizational health. Mastery, competence, and adaptive, problem-solving abilities are words relatively new to both fields. In the area of organizational behavior these words are replacing the old terms "satisfaction" and "work competence." Similarly, an important change has taken place in the mental-health field, which has had some of the same problems in determining adequate criteria. Rather

---

[13] Shepard notes the irony that as research organizations expand their operations they become more like the classical, ideal-type bureaucracy. See (64) for another approach to the social conditions of science.

than viewing health exclusively in terms of some highly inferential intrapsychic reconstitutions, these specialists are stressing "adaptive mechanisms" and "conflict-free," relatively antonomous ego-functioning, independent of id energies. The studies of White (75), Rapaport (58), Erikson (25), Hartmann (33), and other so-called ego-psychologists all point in this direction.

The main reason for the confluence of organizational behavior and mental health is at bottom quite simple. Both the norms of science and the methodology of psychotherapeutic work have the same goal and methodology: to perceive reality, both internal and external; to examine unflinchingly the positions of these realities in order to act intelligently. It is the belief here that what a patient takes away and can employ *after* treatment is that methodology of science, the ability to look facts in the face, to use the hypothetical and experimental methods—the spirit of inquiry—in understanding experience. Sanford has said in this connection:

> . . . most notably in Freud's psychoanalytic method of investigation and treatment. (This method is in my view, Freud's greatest, and it will be his most lasting, contribution.) By the method I mean the whole contractual arrangement according to which both the therapist and patient become investigators, and both objects of careful observation and study; in which the therapist can ask the patient to face the truth because he, the therapist, is willing to try to face it in himself; in which investigation and treatment are inseparable aspects of the same humanistic enterprise (61, p. 12).

and in Freud's words:

> Finally, we must not forget that the relationship between analyst and patient is based on a love of truth, that is, on the acknowledgement of reality, and that it precludes any kind of sham or deception (26, pp. 351–52).

It is now possible to postulate the criteria for organizational health. These are based on a definition by Marie Jahoda, according to which a healthy personality ". . . actively masters his environment, shows a certain unit of personality, and is able to perceive the world and himself correctly." (25, p. 51). Let us take each of these elements and extrapolate organizational criteria from them:

1. "Actively masters his environment": *Adaptability*. In the terms of this paper, this characteristic coincides with problem-solving ability, which in turn depends upon the organization's flexibility. Earlier it was pointed out that flexibility is the freedom to learn through experience, to change with changing internal and external circumstances. Another way of putting it, in terms of organizational functioning, is to say that it is "learning now to learn." This is equivalent to Bateson's notion of "deutero-learning," the progressive change in rate of simple learning (7).

2. "Certain unit of personality": *The Problem of Identity*. In order for an organization to develop adaptability, it needs to know who it is and what it is to do; that is, it has to have some clearly defined identity.[14] The problem of identity, which is central to much of the

---

[14] See Selznick (63), chap. 3, for similar emphasis.

contemporary literature in the mental-health field, can in organizations be examined in at least two ways: (*a*) determining to what extent the organizational goals are understood and accepted by the personnel, and (*b*) ascertaining to what extent the organization is perceived veridically by the personnel.

As to the problem of goals, Selznick pointed out:

> The aims of large organizations are often very broad. A certain vagueness must be accepted because it is difficult to foresee whether more specific goals will be realistic or wise. This situation presents the leader with one of his most difficult but indispensable tasks. *He must specify and recast the general aims of his organization so as to adapt them, without serious corruption, to the requirements of institutional survival.* This is what we mean by the definition of institutional mission and role. (63, p. 66).

The same point is made by Simon, Smithburg, and Thompson: "No knowledge of administrative techniques, then, can relieve the administrator from the task of moral choice—choice as to organizational goals and methods and choice as to his treatment of the other human beings in his organization" (67. p. 24).

In addition to the clear definition of mission, which is the responsibility of the leader to communicate, there also has to be a working consensus on the organization of work. Wilfred Brown's work is extremely useful in this connection. He enumerates four concepts of organization: the *manifest* organization, the one that is seen on the "organization chart" and is formally displayed; the *assumed* organization, the one that individuals perceive as the organization (were they asked to draw their phenomenonological view of the way that things work); the *extant* organization, the situation as revealed through systematic investigation, say by a student of organizations; and the *requisite* organization, or the situation as it would have to be if it were "in accord with the real properties of the field in which it exists."

"The ideal situation," Brown goes on to say, "is that in which the manifest, the assumed, the extant, and the requisite are as closely as possible in line with each other" (19, p. 24). Wherever these four organizational concepts are in contradiction, we find a case of what Erikson calls "identity diffusion" (25). Certainly this phenomenon is a familiar one to students and executives of organizations. Indeed, the great attention paid to the "informal group" and its discrepancy with the formal (difference between the manifest and the assumed organizations or between the manifest and the extant) testifies to this.

Another useful analogy to the mental-health field shows up in this discussion. Many psychotherapeutic schools base their notions of health on the degree to which the individual brings into harmony the various "selves" that make up his personality. According to Fromm-Reichmann, ". . . the successfully treated mental patient, as he then knows himself, will be much the same person as he is known to others" (27, p. 188). Virtually the same criterion is used here for organizational health, i.e. the degree to which the organization maintains harmony—and knowledge—about and among the manifest, assumed, extant, and requisite situations. This point should be clarified. It is not necessary to organizational health that all four concepts of organization be identical. Rather, all four types should be recognized and allowance made for all the tensions attendant upon their imbalance. It is doubtful that there will always be total congruence in organizations. The important factor is recognition; the executive function is to strive toward congruence insofar as it is possible.

3. "Is able to perceive the world and himself correctly": *Reality-Testing.* If the conditions for requisite organizations are to be met, the organization must develop adequate techniques for determining the "real properties" of the field in which it exists." The field contains two main boundaries, the internal organization and the boundaries relevant to the organization. March and Simon, in their cognitive view of organization, place great emphasis on adequate "search behavior." Ineffective search behavior—cycling and stereotypy—are regarded as "neurotic" (46, p. 50).

However, it is preferable here to think about inadequate search behavior in terms of perception that is free from need-distortion (36). Abraham Maslow places this in perspective:

> Recently Money-Kyrle, an English psychoanalyst, has indicated that he believes it possible to call a neurotic person not only *relatively* inefficient, simply because he does not perceive the real world as accurately or as efficiently as does the healthy person. The neurotic is not only emotionally sick —he is cognitively *wrong!* (34, p. 50).

The requisite organization requires reality-testing, within the limits of rationality, for successful mastery over the relevant environment.[15]

In summary, then, I am saying that the basic features of organization rely on adequate methods for solving problems. These methods stem from the elements of what has been called the scientific attitude. From these ingredients have been fashioned three criteria or organizational mechanisms, which fulfill the prerequisites of health. These criteria are in accord with what mental-health specialists call health in the individual.

Undeniably, some qualifications have to be made. The mensuration problem has not been faced, nor have the concrete details for organizational practice been fully developed. Nonetheless, it has been asserted that the processes of problem-solving—of adaptability—stand out as the single most important determinant of organizational health and that this adaptability depends on a valid identity and valid reality-testing.[16]

---

[15] See March and Simon (46) for a formal model of search behavior (p. 50) and an excellent discussion of organizational reality-testing (chap. 6).

[16] Dr. M. B. Miles has suggested that an important omission in this approach is organization "memory" or storage of information. Organizations modelled along the lines suggested here require a "theory' based on an *accumulated* storage of information. This is implied, I believe, in the criterion of adaptability.

SOME IMPLICATIONS OF THE SCIENCE MODEL FOR ORGANIZATIONAL BEHAVIOR

There is one human characteristic which to-day can find a mode of expression in nationalism and war, and which, it may seem, would have to be completely denied in a scientific society. That is the tendency to find some dogma to which can be attached complete belief, forthright and unquestioning. That men do experience a need for certainty of such a kind can scarcely be doubted. . . Is science, for all its logical consistency, in a position to satisfy this primary need of man?—C. H. Waddington (72, pp. 163–164)

We are not yet emotionally an adaptive society, though we try systematically to develop forces that tend to make us one. We encourage the search for new inventions; we keep the mind stimulated, bright, and free to seek out fresh means of transport, communication, and energy; yet we remain, in part, appalled by the consequences of our ingenuity and, too frequently, try to find security through the shoring up of ancient and irrelevant conventions, the extension of purely physical safeguards, or the delivery of decisions we ourselves should make into the keeping of superior authority like the state. These solutions are not necessarily unnatural or wrong, but historically they have not been enough, and I suspect they will never be enough to give us the serenity and competence we seek . . . we may find at least part of our salvation in identifying ourselves with the adaptive process and thus share . . . some of the joy, exuberance, satisfaction and security . . . to meet the changing time.—E. Morison (51, p. 11).

The use of the model of science as a form for the modern organization implies some profound reforms in current practice, reforms that may appear to some as too adventurous or utopian. This criticism is difficult to deny, particularly since not all the consequences can be clearly seen at this time. However, before necessity diminishes the desirability of using the science model, let us examine a few consequences that stand out rather sharply.

1. The problem of commitment and loyalty. Although the viewpoint does have its critics, such as William H. Whyte, Jr., most administrators desire to develop high commitment and loyalty to the organization. Can the scientific attitude, with its ascetic simplicity and acceptance of risk and uncertainty, substitute for loyalty to the organizations and its purpose? Can science, as Waddington wonders, provide the belief in an illusion that organizational loyalty is thought to provide? The answer to this is a tentative "yes and no." Substituting the scientific attitude for loyalty would be difficult for those people to whom the commitment to truth, to the pursuit of knowledge, is both far too abstract and far too threatening. For some, the "escape from freedom" is a necessity, and the uncertain nature of the scientific attitude would be difficult to accept. However, it is likely that even these individuals would be influenced by the adoption of the science model by the organization. Loyalty to the organization per se would be transformed into loyalty and commitment directed to the spirit of inquiry. Hence, a higher rate of mobility is envisaged for organizations based on movement towards those environments in which the social conditions of science exist. Gouldner, in another context, has discussed this difference

between individuals in terms of the split of organizational roles into "locals and cosmopolitans" (30). The cosmopolitan derives his rewards from inward standards of excellence, internalized and reinforced through professional (usually scientific) identification. On the other hand, the local (what Marvick calls the "bureaucratic orientation" (48) derives his rewards from manipulating power within the hierarchy. The former are considered to be better organization men than the latter. Loyalty within the scientific organizational conditions specified here, would be directed not to particular ends or products or to work groups but to identification with the adaptive process of the organization.

2. Recruitment and training for the spirit of inquiry. There are some indications that the problems of recruitment and training for the social organization of science are not as difficult as has been expected. For one thing, as Bruner has shown (20), today's schoolchildren are getting more and better science teaching. It is to be hoped that they will learn as much about the attitudes of science as they will about its glamour and techniques. In addition, more and more research-trained individuals are entering organizations.[17] As McGregor points out: "Creative intellectual effort by a wide range of professional specialists will be as essential to tomorrow's manager as instruments and an elaborate air traffic control system are to today's jet pilot" (44, p. 27). Individuals trained in scientific methodology can easily adapt to, in fact will probably demand, more and more freedom for intellectual inquiry. If McGregor's and Leavitt's and Whisler's (40) prognostications are correct, as they presently seem to be, then there is practically no choice but to prepare a social milieu in which the adaptive, problem-solving processes can flourish.

As to training, only a brief word needs to be said. The training program of the National Training Laboratories (16) and the work of Blake (13), Blansfield (15), and Shepard (65) are based rather specifically on developing better diagnosticians of human behavior. It is apparent from such training studies that the organization of tomorrow, heavily influenced by the growth of science and technology and manned by an increasing number of professionals, appears to have the necessary requirements for constructing organizations based on inquiry.

3. Intergroup Competition. Blake and Mouton, guided partly by the work of the Sherifs (66), have disclosed for examination one of organization's most troublesome problems, intergroup conflict and collaboration. These perseverating conflicts, usually based on a corrupt practice of vested interests, probably dissipate more energy and money than any other single malady caused by humans. Intergroup conflict, with its "win-lose" orientation, its dysfunctional loyalty (to the group or product, not to the truth), its cognitive distortions of the outsider, and its inability to reach what has been called creative compromise, effectively disrupts the commitment to truth. By means of a laboratory approach Blake and Mouton have managed to break

---

[17] See Harbison (32) on this point.

. . . the mental assumptions underlying win-lose conflict. Factually based mutual problem identification, fluidity in initial stages of solution-proposing rather than fixed position taking, free and frequent interchange between representatives and their constituent groups and focusing on communalities as well as differences as the basis for achieving agreement and so on, are but a few of the ways which have been experimentally demonstrated to increase the likelihood of arriving at mutually acceptable solutions under conditions of collaboration between groups (14).

What the authors do not explicitly say but only imply is that the structure of their experimental laboratory approach is based on the methods of inquiry that have been advocated in this paper. Theirs is an action-research model, in which the subjects are the inquirers who learn to collect, use and generalize from data in order to understand organizational conflict. Rational problem-solving is the only prophylaxis presently known to rid organizations of perseverating intergroup conflict.

Loyalty, recruitment and training, and intergroup hostility are by no means all the organizational consequences that this paper suggests. The distribution of power, the problems of group cohesiveness,[18] the required organizational fluidity for arranging task groups on a rational basis, and the change in organizational roles and status all have to be considered. More time and energy than are now available are needed before these problems can be met squarely. However, one thing is certain: whatever energy, competence, and time are required, it will be necessary to think generally along the directions outlined here. Truth is a cruel master, and the reforms that have been mentioned or implied may not be altogether pleasant to behold. The light of truth has a corrosive effect on vested interests, outmoded technologies, and rigid, stereotypic patterns of behavior. Moreover, if this scientific ethos is ever realized, the remnants of what is now known as morale and efficiency may be buried. For the spirit of inquiry implies a confrontation of truth that may not be "satisfying" and a deferral of gratification that may not, in the short run, be "efficient." However, this is the challenge that must be met if organizations are to cope better within their increasingly complicated environments.

REFERENCES

1. Aitken, H. G. J. "Taylorism at Watertown Arsenal: Scientific Management in Action 1908–1915." Harvard University Press, Cambridge, Mass., 1960.

2. Argyris, C. The Integration of the Individual and the Organization. Paper presented at the University of Wisconsin, Madison, Wis., May, 1961

3. ———: Organizational Development—An Inquiry into the Esso Approach. Paper presented at Yale University, New Haven, Conn., July, 1960.

---

[18] It is suspected that group cohesiveness will decrease as the scientific attitude infuses organizational functioning. With the depersonalization of science, the rapid turnover, and some expected individualism, cohesiveness may not be functional or even possible.

4. ———: "Personality and Organization." Harper and Brothers, New York, 1957.

5. Barber, B. "Science and the Social Order." Free Press, Glencoe, Ill., 1952.

6. Bartiz, L. "The Servants of Power." Wesleyan University Press, Middletown, Conn., 1960.

7. Bateson, G. Social Planning and the Concept of "Deutero-Learning," in "Readings in Social Psychology," ed. T. M. Newcomb and E. L. Hartley, 1st. ed., pp. 121–28, Henry Holt and Co., Inc., New York, 1947.

8. Bendix, R. "Max Weber: An Intellectual Portrait." Doubleday & Company, Inc., Garden City, N.Y., 1960.

9. Bennis, W. G. Leadership Theory and Administrative Behavior: The Problem of Authority. *Adm. Sci. Quart.*, vol. 4, no. 3, pp. 259–301, December 1959.

10. ———, K. Benne, and R. Chin. "The Planning of Change." Henry Holt and Co., Inc., New York, 1961.

11. Berkowitz, N., and W. Bennis. Interaction in Formal Service-Oriented Organizations. *Adm. Sci. Quart.*, vol. 6, no. 1, pp. 25–50, June 1961.

12. Bertalanffy, L. V.: The Theory of Open Systems in Physics and Biology, *Science*, vol. 111, pp. 23–29, 1950.

13. Blake, R. R., and J. S. Mouton. Developing and Maintaining Corporate Health through Organic Management Training, unpublished paper, University of Texas, Austin, Tex., 1961.

14. ———, and ———: From Industrial Warfare to Collaboration: A Behavioral Science Approach, Korzybski Memorial Address, April 20, 1961.

15. Blansfield, M. G., and W. F. Robinson. Variations in Training Laboratory Design: A Case Study in Sensitivity Training. *Personn. Adm.*, Vol. 24, no. 2, pp. 17–22, 49, March–April, 1961.

16. Bradford, L. (ed.). "Theories of T-Group Training." National Training Laboratories, New York University Press, New York, in press for Winter, 1962.

17. Bronowski, J. "The Common Sense of Science." Modern Library, New York, no date.

18. ———. "Science and Human Values." Harper and Brothers, New York, 1959.

19. Brown, W. "Exploration in Management." John Wiley & Sons, Inc., New York, 1960.

20. Bruner, J. "The Process of Education." Harvard University Press, Cambridge, Mass., 1961.

21. Bush, G. P., and D. H. Hattery. "Teamwork in Research." American University Press, Washington, D.C., 1953.

22. Caplow, T. The Criteria of Organizational Success, in "Readings in Human Relations," ed. K. Davis and W. G. Scott, p. 96, McGraw-Hill Book Company, Inc., New York, 1959.

23. Comrey, A. L. A Research Plan for the Study of Organizational Effectiveness, in "Some Theories of Organization," ed. A. H. Rubenstein and C. J. Haberstroh, Dorsey-Irwin Press, Homewood, Ill., 1960.

24. Emery, F. E., and E. L. Trist. Socio-Technical Systems, paper presented at the 6th Annual International Meeting of the Institute of Management Sciences, Paris, France, September, 1959.

25. Erikson, E. Identity and the Life Cycle. *Psychol. Issues,* vol. I, no. 1, Monograph 1, 1959.

26. Freud, S. Analysis Terminable and Interminable, in "Collected Papers," ed. E. Jones, vol. 5, Basic Books, Inc., New York, 1959.

27. Fromm-Reichmann, F. "Principles of Intensive Psychotherapy." University of Chicago Press, Chicago, Ill., 1950.

28. Geiger, G. Values and Social Science. *J. Soc. Issues,* vol. VI, no. 4, pp. 8–16, 1950.

29. Gibb, J. R., and R. Lippitt (ed.). Consulting with Groups and Organizations. *J. Soc. Issues,* vol. XV, no. 2, pp. 1–74, 1959.

30. Gouldner, A. Locals and Cosmopolitans: Towards an Analysis of Latent Social Roles—I. *Adm. Sci. Quart.,* vol. 2, pp. 281–306, 1957.

31. Haire, M. What Price Value? *Contempt. Psychol.,* vol. 4, pp. 180–82, June 1959.

32. Harbison, F. H. Management and Scientific Manpower. Paper presented at the Centennial Symposium on Executive Development, School of Industrial Management, Massachusetts Institute of Technology, Cambridge, Mass., April 27, 1961.

33. Hartmann, H. "Ego Psychology and the Problem of Adaption." International Universities Press, Inc., New York, 1958.

34. Jahoda, M. "Current Concepts of Positive Mental Health." Basic Books, New York, 1958.

35. Kahn, R., F. C. Mann, and S. Seashore. Human Relations Research in Large Organizations—II. *J. Soc. Issues,* vol. XII, no. 2, p. 4, 1956.

36. Katzell, R. A. Industrial Psychology, in "Annual Review of Psychology," ed. P. R. Farnsworth, vol. 8, pp. 237–68, Palo Alto, Calif., 1957.

37. Kubie, L. S. "Neurotic Distortions of the Creative Process." Porter Lectures, Series 22, University of Kansas Press, Lawrence, Kan., 1958.

38. Leavitt, H. J. Effects of Certain Communication Patterns on Group Performance. *J. Abnorm. Soc. Psychol.,* vol. 46, pp. 48–50, 1951.

39. ———. Unhuman Organizations. Address presented at the Centennial Symposium on Executive Development, School of Industrial Management, Massachusetts Institute of Technology, Cambridge, Mass., April 27, 1961.

40. ———, and T. L. Whisler. Management in the 1980s. *Harv. Busin. Rev.,* vol. 36, pp. 41–48, November-December 1958.

41. Likert, R. Measuring Organizational Performance. *Harv. Busin. Rev.,* vol. 36, pp. 41–50, March–April 1958.

42. Lippitt, R., J. Watson, and B. Westley. "The Dynamics of Planned Change." Harcourt, Brace & Company, New York, 1958.

43. McGregor, D. "The Human Side of Enterprise." McGraw-Hill Book Company, Inc., New York, 1960.

44. ———. New Concepts of Management. *Technol. Rev.,* vol. 63, no. 4, pp. 25–27, February 1961.

45. Mann, F. C., and F. W. Neff: "Managing Major Change in Organiza-

tions," Foundation for Research on Human Behavior, Ann Arbor, Mich., 1961.

46. March, J., and H. Simon. "Organizations." John Wiley and Sons, Inc., New York, 1958.

47. Marcson, S. The Scientist in American Industry, Industrial Relations Section, Princeton University, Princeton, N.J., 1960.

48. Marvick, D. "Career Perspectives in a Bureaucratic Setting." University of Michigan Governmental Study 27, University of Michigan Press, Ann Arbor, Mich., 1954.

49. Merton, R. The Professions and Social Structure, in "Essays in Sociological Theory," chap. VIII, Free Press, Glencoe, Ill., 1949.

50. ———. The Sociology of Knowledge (chap. VIII) and Science and Democratic Social Structure (chap. XII). In "Social Theory and Social Structure," Free Press, Glencoe, Ill. 1949.

51. Morison, E. A Case Study of Innovation. *Engng. Sci. Mon.*, vol. 13, pp. 5–11, April 1950.

52. "Muggeridge and Snow," *Encounter*, vol. 27, p. 90, February 1962.

53. Oppenheimer, R. Prospects in the Arts and Sciences. *Perspectives USA*, vol. 11, pp. 5–14, Spring, 1955.

54. Parsons, T. "The Social System," chap. VIII, Free Press, Glencoe, Ill., 1951.

55. ———. Suggestions for a Sociological Approach to the Theory of Organizations—I. *Adm. Sci. Quart.*, vol. 1, pp. 63–85, 1956.

56. Paul, B. Social Science in Public Health. *Amer. J. Publ. Hlth.*, vol. 46, pp. 1390–93, November 1956.

57. Pfiffner, J. M., and F. P. Sherwood. "Administrative Organization." Prentice-Hall, Inc., Englewood Cliffs, N.J., 1960.

58. Rapaport, D. The Theory of Ego Autonomy: A Generalization. *Bull. Menninger Clin.*, vol. 22, no. 1, pp. 13–35, January 1958. (See also The Structure of Psychoanalytic Theory: A Systematizing Attempt, *Psychol. Issues*, vol. II, no. 2 Monograph 6, 1960.)

59. Ridgway, V. F. Dysfunctional Consequences of Performance Measurements. In "Some Theories of Organization," ed. A. H. Rubenstein and C. J. Haberstroh, pp. 371–77, Dorsey-Irwin Press, Homewood, Ill., 1960.

60. Rubenstein, A. H., and H. A. Shepard. Annotated Bibliography on Human Relations in Research Laboratories. School of Industrial Management, Massachusetts Institute of Technology, Cambridge, Mass., February, 1956.

61. Sanford, N. Social Science and Social Reform. Presidential address for SPSSI at the Annual Meeting of the American Psychological Association, Washington, D.C., August 28, 1958.

62. Selznick, P. Foundations of the Theory of Organization. *Amer. Sociol. Rev.*, vol. 13, pp. 25–35, 1948.

63. ———. "Leadership in Administration." Row, Peterson & Company, Evanston, Ill., 1957.

64. Shepard, A. H. Superiors and Subordinates in Research. *J. Busin.*, vol. 29, pp. 261–67, October, 1856.

65. ———. Three Management Programs and the Theories behind Them, in "An Action Research Program for Organization Improvement,"

Foundation for Research on Human Behavior, Ann Arbor, Mich., 1960.

66. Sherif, M., and C. Sherif. "Groups in Harmony and Tension." Harper and Brothers, New York, 1953.

67. Simon, H. A., D. W. Smithburg, and V. A. Thompson. "Public Administration." Alfred A. Knopf, Inc., New York, 1950.

68. Smith, S. Communication Pattern and the Adaptability of Task-Oriented Groups: An Experimental Study. Unpublished paper, Massachusetts Institute of Technology, Cambridge, Mass., 1950.

69. Snow, C. P. *The Two Cultures and the Scientific Revolution.* Mentor Books, New York,' 1962.

70. Thelen, H. "Education and the Human Quest." Harper and Brothers, New York, 1960.

71. Urwick, L. F. The Purpose of a Business, in "Readings in Human Relations." ed. K. Davis and W. G. Scott, pp. 85–91, McGraw-Hill Book Company, Inc., New York, 1959.

72. Waddington, C. H. "The Scientific Attitude." Penguin Books, Inc., Baltimore, Md., 1941.

73. Walker, C. (ed.). "Modern Technology and Civilization: An Introduction to Human Problems of a Machine Age." McGraw-Hill Book Company, Inc., New York, in press for December 1961.

74 Wasserman, P. Measurement and Evaluation of Organizational Performance. McKinsey Foundation Annotated Bibliography, Graduate School of Business and Public Administration, Cornell University, Ithaca, N.Y., 1959.

75. White, R. W. Motivation Reconsidered: The Concept of Competence. *Psychol. Rev.*, vol. 66, no. 5, pp. 297–33, September 1959.

76. Whyte, W. H., Jr. "The Organization Man." Simon & Schuster, Inc., New York, 1956.

77. Wilson, A. T. M. The Manager and His World, paper presented at the Centennial Symposium on Executive Development, School of Industrial Management, Massachusetts Institute of Technology, Cambridge, Mass., April 27, 1961.

INDEXES

Name Index

Bush, G. P., 527
Bush, R. R., 96

C

Cadwallader, M. L., 112
Campbell, D. T., 381, 392
Campbell, N. R., 97
Cannon, W. B., 71, 72, 165
Caplow, T., 471, 512, 527
Carlson, R. O., 216, 297, 319, 384, 385, 393
Cartwright, D., 72, 192
Caudill, W., 477, 487
Caws, P., 3, 23
Cervinka, V., 59
Chandler, A. D., 301, 319
Chandler, M. K., 19, 24
Chapple, E. D., 436
Chein, I., 170, 177
Cherry, C., 92
Chin, R., 318, 394, 395, 396, 397, 399, 402, 513, 527
Christie, L. S., 58
Churchman, C. W., 97, 174, 175, 177, 319
Clark, B., 385, 393
Clark, B. R., 183, 231
Clark, D. F., 91
Clemmer, D., 465
Coch, L., 106
Cohen, J. E., 63
Coleman, J. C., 62, 199
Coleman, J. S., 214, 467
Comrey, A. L., 512, 527
Comte, 116
Coombs, C. H., 97
Costello, T. W., 241
Cottrel, W. F., 105
Cottrell, L. S., Jr., 462
Cressey, D. R., 464, 465
Crozier, M., 10, 23
Crutchfield, R., 323, 340
Cumming, E. and J., 464, 465, 479, 487
Cyert, R. M., 158, 161, 184, 217, 493, 494

D

Dahl, R. A., 89
Dalton, M., 104, 134, 138, 300, 319
Daniel, D. R., 243
Daniels, L. R., 137, 138
Darwin, 178, 195, 408
Davis, K., 103, 106, 152, 470, 527
Davis, R. C., 79, 101, 162
Dearborn, D. C., 300, 319
de Chardin, P. T., 61
de Charms, R., 125, 138
Deemer, W. L., Jr., 91
Delany, W., 459
Dent, J. K., 491
Deutsch, K. W., 111, 112
Dewey, 517
Dickson, W. J., 89, 103, 107
Dill, W. R., 11, 12, 23, 182, 183, 192, 196, 214, 240, 381, 393
Dimock, M. E., 365
Dirlam, J. B., 493

Doutt, J. T., 107
Downs, A., 296, 302, 319
Driver, M. J., 213, 215
Drucker, P. F., 181, 260, 436, 491
Dublin, R., 467
Duncan, R. B., 11, 12, 23
Dunphy, D., 327
Durkheim, E., 186
Dutton, J. M., 198, 214, 215

E

Easton, D., 3, 23, 115
Eaton, J. W., 216, 228, 231
Eddington, A. S., 72, 116, 117
Eisenstadt, S. N., 217, 227
Elling, R. H., 183, 395, 402
Emery, F. E., 9, 10, 14, 163, 165–77, 178, 179, 180, 185, 186, 190, 196, 214, 247, 249–60, 261, 282, 293, 296, 297, 298, 299, 319, 401, 402, 480, 487, 515, 528
Erikson, E., 521, 528
Etzioni, A., 15, 21, 22, 24, 123, 127, 138, 181, 223, 365, 387, 388, 389, 393, 457, 459–75, 478, 481, 487, 490, 491, 492, 493
Evan, W. M., 16, 24, 187, 189, 194, 200, 214, 396, 400, 402

F

Fagen, R. E., 4, 59, 74
Fanshel, D., 227
Fayol, H., 430, 431, 432
Feld, S., 228
Fermi, E., 72
Festinger, L., 105
Fiedler, F., 327, 340
Fisher, L. H., 450, 467
Flood, M. M., 96
Form, W. H., 103, 365
Forrester, J. W., 7, 24, 436
Foster, W. Z., 460
Francis, R. G., 222
French, J. R. P., Jr., 20, 106, 425–37
Freud, S., 528
Friedson, E., 230, 476, 478, 487
Fromm-Reichmann, F., 522, 528

G

Gabarro, J. J., 12, 163, 196–215
Gagne, R. M., 428
Gallagher, A., 297, 319
Gamson, W. A., 379, 501
Garceau, O., 466
Gardner, B. B., 103, 105
Gardnere, J. W., 181
Garrison, J. S., 209, 214
Geiger, G., 517, 518, 526, 528
Georgopoulos, B. S., 83, 120–40, 469, 490
Gerard, R. W., 59
Gerth, H., 383, 393
Gibb, J. R., 513, 528
Gibbs, W., 74
Gilson, T. Q., 436
Gittell, M., 196, 214
Glazer, N., 128, 139

Subject Index

Turbulence—*Cont.*
  evidence for, 181–84
  and linkage, 315
Turbulent field, 10, 171, 173, 178–95, 297
Type, definition of, 39
Typology, elements of, 218

U

Units, of conceptual system, 34
Urban school systems, and organizational adaptation to changing environments, 196–215

V

Value pattern, 495
Value projection, 471–72
Values
  and open social systems, 121
  reformulation of for social change, 136–38
Violence and the definition of organizational membership, 386–87

W

Work group autonomy, 255
Work groups, 325
Work organization, 253, 254

This book has been set in 9 point and 8 point Primer, leaded 2 points. Section numbers are in 24 point Scotch Roman; section titles and reading numbers and titles are in 18 point Scotch Roman. The size of the type page is 27 x 46½ picas.